Verw.: RC
Af - B - III - 3 - 38

The Agriculture of Egypt

Centre for Agricultural Strategy Series

Editors: C. R. W. Spedding and L. J. Peel

Published in conjunction with the Centre for Agricultural Strategy, University of Reading

1. G. M. Craig (editor) *The Agriculture of the Sudan*
2. Xu Guohua and L. J. Peel (editors) *The Agriculture of China*
3. G. M. Craig (editor) *The Agriculture of Egypt*

THE AGRICULTURE OF
Egypt

Edited by
G. M. CRAIG
Centre for Agricultural Strategy
University of Reading

OXFORD UNIVERSITY PRESS
1993

Oxford University Press, Walton Street, Oxford OX2 6DP
Oxford   New York   Toronto
Delhi   Bombay   Calcutta   Madras   Karachi
Kuala Lumpur   Singapore   Hong Kong   Tokyo
Nairobi   Dar es Salaam   Cape Town
Melbourne   Auckland   Madrid
and associated companies in
Berlin   Ibadan

Oxford is a trade mark of Oxford University Press

Published in the United States
by Oxford University Press Inc., New York

© The contributors listed on pp. xiii–xiv, 1993

All rights reserved. No part of this publication may be
reproduced, stored in a retrieval system, or transmitted, in any
form or by any means, without the prior permission in writing of Oxford
University Press. Within the UK, exceptions are allowed in respect of any
fair dealing for the purpose of research or private study, or criticism or
review, as permitted under the Copyright, Designs and Patents Act, 1988, or
in the case of reprographic reproduction in accordance with the terms of
licences issued by the Copyright Licensing Agency. Enquiries concerning
reproduction outside those terms and in other countries should be sent to
the Rights Department, Oxford University Press, at the address above.

This book is sold subject to the condition that it shall not,
by way of trade or otherwise, be lent, re-sold, hired out, or otherwise
circulated without the publisher's prior consent in any form of binding
or cover other than that in which it is published and without a similar
condition including this condition being imposed
on the subsequent purchaser.

A catalogue record for this book is available from the British Library

Library of Congress Cataloging in Publication Data
The Agriculture of Egypt / edited by Gillian M. Craig
(Centre for Agricultural Strategy Series ; 3)
Includes bibliographical references and index.
1. Agriculture–Egypt   I. Craig, G. M.   II. Series
S473.E38A37   1993   338.1'0962–dc20   93–12780
ISBN 0–19–859203–5

Typeset by The Electronic Book Factory, Fife, Scotland
Printed in Great Britain by
Biddles Ltd, Guildford and King's Lynn

Preface

Egyptian agriculture, one of the oldest agricultural production systems in the world, is undergoing the most significant changes in decades. Traditionally, government intervention has been a distinctive feature of Egyptian agriculture and, for the past thirty years, a policy of tight centralized control over the whole agricultural production and food distribution process has been in place.

Since the late 1980s, with stagnating agricultural development, declining food self-sufficiency and a massive food import bill, worsening economic conditions, and continual pressure from international lending institutions, the Egyptian Government has begun to implement a major agricultural reform programme, phasing out rigid crop procurement and agricultural input policies, liberalizing the agricultural sector, and creating a more favourable environment for Egyptian farmers. However, Egypt's large and rapidly increasing population, in excess of fifty-seven million in 1992, the shortage of cultivable land, and agriculture's dependence on irrigation water, primarily from the Nile, are major constraints in the task of increasing agricultural production and food self-sufficiency.

As this book reaches completion, the agricultural reform programme is still in process and the outcome in its widest sense will remain unclear for some years. The aim of this book has been to provide a source of background information which will enable the changes that are occurring in Egyptian agriculture today, and which will shape the country's agriculture in the future, to be better understood. Many factors can interact to shape an agricultural change, and this book deliberately ranges widely, encompassing details about the natural environment, systems of agricultural production, and political and economic matters.

The book is aimed at all who visit Egypt to work or advise on agricultural projects or who wish to know more about the agriculture and rural economy of the country. It will provide a ready source of reference for those who need information on Egyptian agriculture but have no opportunity to visit the country and are reluctant to search the numerous reports and occasional publications in which much of the information on Egypt is published.

There are twenty-two contributors to the volume, each with different interests and experiences, but all with first-hand knowledge of Egypt. I have aimed at consistency and uniformity of style whenever possible, but I have assumed that the terminology used by authors in their various subject specialisms is, for the most part, that which would be encountered by any newcomer to Egypt. The Arabic and vernacular words used most frequently have been standardized and are included in the glossary.

Every effort has been made to provide reliable and up-to-date statistics, and Mr John Parker of the Agriculture and Trade Analysis Division, United States Department of Agriculture, Economic Research Service has been especially helpful in providing material. However, in the case of Egypt, the task of data collection and collation is a difficult one and sets of similar data from different sources may vary. Similarly, where data series are being continually revised inconsistencies can occur when the figures used in calculations change.

In 1987–1988 the Ministry of Land Reclamation merged with the Ministry of Agriculture to form the Ministry of Agriculture and Land Reclamation (MOALR). This new organization is still often referred to simply as the Ministry of Agriculture (MOA), and it should be noted that after 1987–1988 MOALR and MOA are the same organization.

I wish to record my thanks to all the authors who have contributed to the book, updated material, and have been patient while I have dealt with the editing procedures for all the chapters.

Acknowledgements are also due to Dr Lynnette J. Peel for her continuous support and frequent editorial advice, to Dr Adel Beshai, without whose liaison in Cairo the volume may not have reached completion, and to Peter Llewellyn of the British Council for his willing help when communications with Egypt were problematic.

Reading   G. M. C.
May 1993

Contents

List of contributors	xiii
1. Physiography, geology, and soils	**1**
A. C. Millington	
Introduction	
Physiography and geology	
Soils	
Conclusion	
2. Climate and hydrology	**16**
P. Beaumont	
Climate	
Hydrology	
Conclusion	
3. Natural vegetation	**39**
M. N. El Hadidi	
Introduction	
The Mediterranean	
The Deserts	
The Nile Basin	
4. Domesticated livestock and indigenous animals in Egyptian agriculture	**63**
M. B. Aboul-Ela	
Introduction	
The domesticated livestock population	
Fish production	
Insects and animal pests significant in Egyptian agriculture	
Conclusion	
5. Ethnic origins of the people and population trends	**80**
M. M. Wahba	
Introduction	
Historical origin and ethnic distribution of the population	
Population trends	
Conclusions	

6. Land tenure   94
A. Richards
Introduction: the importance of land tenure
The historical legacy
Nasser's land reforms and the creation of the modern system
Land tenure through the oil boom and bust

7. Food preferences and nutrition   114
H. Alderman
Introduction
Diet composition
The impact of government policies on food consumption
Nutritional concerns
Note on recent trends in health and nutrition

8. Government and the infrastructure in Egyptian agriculture   128
N. N. Ayubi
Introduction
Main elements of the governmental structure
The infrastructure
Agriculture in the development plans
The Egyptian Government and rural development

9. The role and impact of government intervention in Egyptian agriculture   146
T. A. Moursi
Introduction
Government control and measures in the 1940s
Government intervention in agriculture since the revolution
The impact of government intervention in agriculture
Liberalization and recent changes in government policy
Conclusion

10. Agriculture in the national economy   170
J. M. Antle
Introduction
Key events in Egyptian agricultural history
Recent developments in Egyptian agriculture
Agricultural contribution to economic growth
Agriculture and trade
Conclusion

11. The history of agricultural development 188
T. Ruf
    Introduction
    The significance of the changing use of the river Nile
    Population growth and the revolution in land ownership
    Developing systems of production
    The link between agriculture and livestock production:
      the key to understanding the development of peasant
      agriculture
    Recent history: changes and continuity under Nasser, Sadat,
      and Mubarak

12. Systems of agricultural production in the Northern Littoral Region 209
M. J. Martin
    Introduction
    The north-west coastal zone
    The north-east coastal zone
    Future

13. Systems of agricultural production in the Delta 229
P. N. Ward
    Introduction
    Cropping patterns
    Government policy
    Agricultural areas of the Delta
    Crop production
    Animal production
    Future

14. Systems of agricultural production in Middle and Upper Egypt 265
A. A. Beshai
    Introduction
    Ethnic identity of Upper Egyptians, the household, and
      division of labour
    Population, land area, and land quality
    The dynamics of agricultural production
    The importance of sugar cane in Upper and Middle Egypt
    Causes and effects of the low standard of living in Upper
      Egypt

15. The deserts of Egypt: desert development systems   278
A. Bishay
    Introduction
    Natural environment and resources
    Desert development
    Desert farming systems
    The future of desert development

16. Irrigation development programmes   345
M. D. Skold and E. V. Richardson
    Introduction
    Water delivery and removal system
    Reclamation of land
    Irrigation of the Old Lands
    Implications

17. Fertilizers   363
M. M. El-Fouly
    Introduction
    Production and importation of fertilizers
    Distribution of fertilizers
    Fertilizer requirements, distribution rates, and crop needs
    Foliar fertilizers and micronutrients
    Future outlook

18. Feedstuffs   383
T. J. Barker
    Introduction
    Fodder and the cropping pattern
    Concentrates
    Crop by-products
    Crop residues
    Minerals and supplements
    Trends

19. Agricultural mechanization and farm equipment supply   395
A. U. Khan
    Introduction
    Current level of mechanization
    Local manufacturing
    Imports of tractors and equipment

Credit availability
Repair, maintenance, and training
Research and testing
Extension
Mechanization policies
Conclusion

20. Marketing channels and price determination for agricultural commodities 420
J. Rowntree
Introduction
Development of contemporary agricultural marketing channels
The government's food subsidy and distribution system
Regulated agricultural markets
Unregulated agricultural markets
Problems and inadequacies of the agricultural markets
Conclusions

21. Agricultural education, research, and extension 445
M. E. I. Mansour and S. A. M. Ismail
Agricultural education
Agricultural research and extension
Bottlenecks between agricultural education, research, and extension
Conclusions

Glossary 469

Appendix 1 472
Price weights and tonnages of agricultural products, value of agricultural imports and exports, orange and potato exports by destination

Appendix 2 477
Cotton production and trade
J. B. Parker
Cotton area harvested, production, yield, imports and exports

Appendix 3 485
Wheat area harvested, production, yield, purchase prices, wheat and flour imports and distribution flow chart, maize production and imports, rice consumption

Appendix 4 490
Production, imports, exports, and consumption of fertilizers, and current situation and expectations until the year 2010

Index 497

Contributors

M. B. Aboul-Ela, Professor, Department of Animal Production, Faculty of Agriculture, Mansoura University, El-Mansoura

H. Alderman, Economist, Agricultural and Rural Development Department, The World Bank, Washington, DC

J. M. Antle, Professor, Department of Agricultural Economics, Montana State University

N. N. Ayubi, Reader, Department of Politics, University of Exeter

P. Beaumont, Professor, Department of Geography, St David's University College, University of Wales, Lampeter

T. J. Barker, The Old White Horse, Laxfield, Woodbridge, Suffolk, previously Animal Feed Improvement Project, Cairo

A. A. Beshai, Professor and Head of the Department of Economics, The American University in Cairo

A. Bishay, Professor and Director General, Desert Development Center, The American University in Cairo

G. M. Craig, Research Fellow, Centre for Agricultural Strategy, University of Reading

M. M. El-Fouly, Professor, National Research Centre, German Agency for Technical Cooperation, Cairo

M. N. El-Hadidi, Professor, Department of Botany, University of Cairo

S. A. M. Ismail, Senior Researcher, Agricultural Economics Research Institute, Ministry of Agriculture and Land Reclamation, Cairo

A. U. Khan, Agricultural Mechanization Adviser, National Agricultural Research Project, Agricultural Engineering Research Institute, Cairo

M. E. I. Mansour, Director, Agricultural Economics Research Institute, Ministry of Agriculture and Land Reclamation, Cairo

M. J. Martin, 31 Heysham Park, Heysham, Lancashire, previously Adviser, ODA, Alexandria University Milk and Meat Project

A. C. Millington, Reader, Department of Geography, University of Reading

T. A. Moursi, Lecturer, Faculty of Economics and Political Science, University of Cairo

J. B. Parker, Agriculture and Trade Analysis Division, USDA, Washington

A. Richards, Professor, Department of Economics, University of California, Santa Cruz

E. V. Richardson, Professor, Department of Civil Engineering, Colorado State University

J. Rowntree, Programme Director, Fisheries Stock Assessment CRSP, University of Maryland

T. Ruf, Chargé de Recherches, Laboratoire d'Etudes Agraires, ORSTOM, Montpellier

M. D. Skold, Professor, Department of Agricultural and Resource Economics, Colorado State University

M. M. Wahba, Department of Economics, The American University in Cairo

P. N. Ward, Tempellow, Liskeard, Cornwall, previously Adviser, ODA, Alexandria University Milk and Meat Project

1
Physiography, geology, and soils
A. C. MILLINGTON

Introduction

Egypt is a large, arid, and, for the most part, low-lying country on the north-east African peneplain. Exceptions to this general rule can be found along the northern coast, where the climate is semi-arid, and in the tectonically active Jordan and Red Sea Rift Valley zone, where earth movements have created rugged mountainous terrain bordering the Red Sea and in the Sinai Peninsula. The total land area is 1 001 449 km^2.

Physiography and geology

The physiography is closely related to the underlying geology; consequently these two aspects are dealt with together. Egypt can be divided into five main physiographic units (Fig. 1.1):

(1) Western Desert;

(2) Nile Valley;

(3) Nile Delta;

(4) Eastern or Arabian Desert;

(5) Sinai Peninsula.

The major units and their main subdivisions (Table 1.1) are discussed below. The geology of Egypt has recently been reviewed by Said (1989).

Western Desert

Like much of north-east Africa, the Western Desert is underlaid by level to gently dipping sedimentary rocks deposited on the Precambrian basement.

2　THE AGRICULTURE OF EGYPT

These rocks consist of continental sandstones with thin beds of marine limestones, sandstones, and marls. This sequence is known as the Nubian sandstone (Issawi 1973) and was deposited between the Lower Paleozoic and the Upper Cretaceous, the marine beds dating from the Carboniferous to Lower Cretaceous. The Nubian sandstone thickens northwards from about 50 m near Aswan to 600 m in Sinai (Furon 1963).

Between the Lower Miocene and Oligocene, the Nubian sandstone surface was covered in many places by a series of marine transgressions, each of

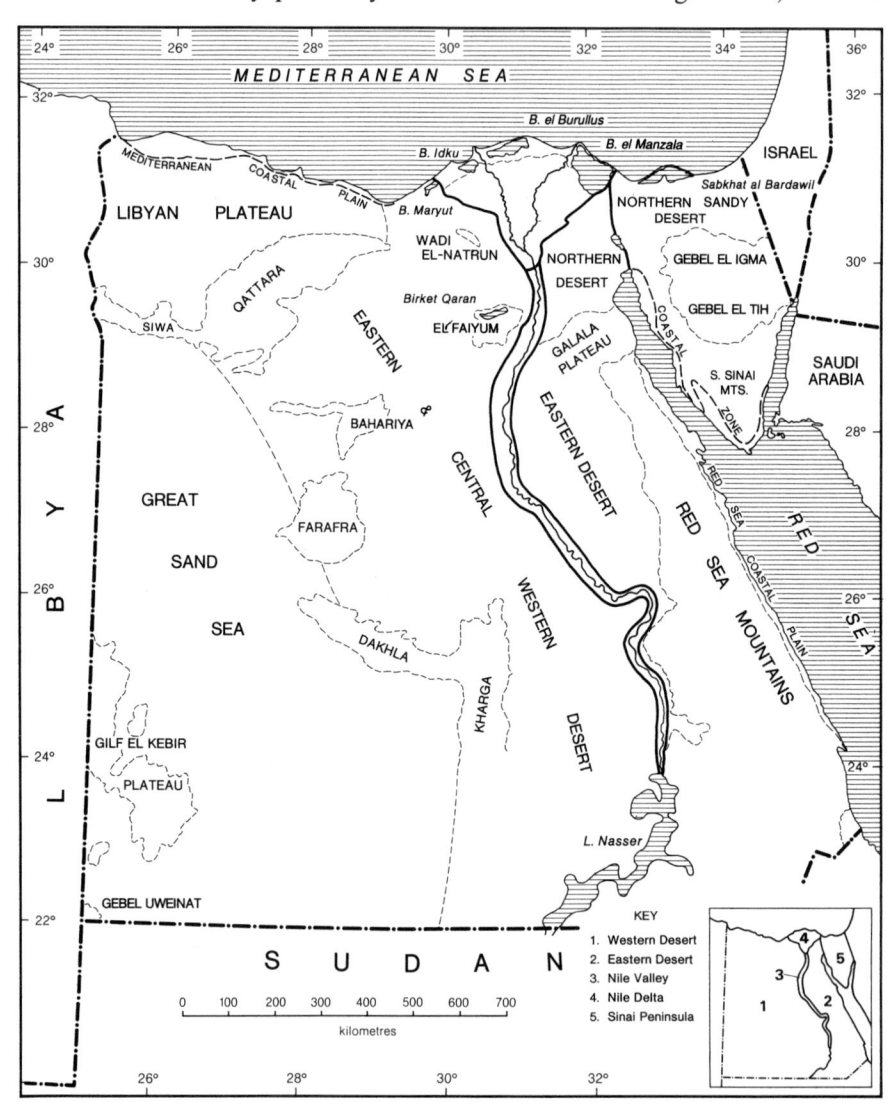

FIG 1.1.　Egypt: main physiographic units

TABLE 1.1. Egypt: physiographic units

Major physiographic units	Main subdivisions
Western Desert	Libyan Plateau
	Great Sand Sea
	Southern and Eastern Desert
	Large depressions
	El Faiyum
	Bahariya
	Dakhla
	Farafra
	Kharga
	Qattara
	Siwa
	Wadi el Natrun
	South-Western Mountains
	Gilf el Kebir Plateau
	Gebel Uweinat
Eastern or Arabian Desert	Red Sea Coastal Plain
	Red Sea Mountains
	Northern Desert and Galala Plateaux
	Eastern Desert
Nile Valley	
Nile Delta	Upper and Central Delta
	Coastal Zone
Sinai Peninsula	Southern Mountains
	Gebels el Igma and el Tih
	Northern Sandy Desert
	Coastal Zone

which deposited fluvio-marine and marine sediments. These are represented by sandstones, limestones, and marls. The marine transgressions extended southwards from the Mediterranean, each one covering a slightly different area. For instance, the Lower Eocene transgressions extended as far south as Aswan, the Middle Eocene reached the middle Nile Valley, while the Oligocene transgression penetrated only as far as Cairo.

There has been little tectonic activity in this area since deposition, and the rocks are either level or gently dipping. When these strata are eroded a distinctive tableland landscape with steep escarpments is formed; this is best developed in the Gilf el Kebir Plateau.

The Western Desert consists of five main physiographic units. Three of these (the Libyan Plateau, the Great Sand Sea, and the eastern and southern areas) are flat to undulating surfaces, dominated by wind action. They vary in altitude from 100 to 500 m above sea level (m a.s.l.).

The Libyan or El-Diffa Plateau in north-western Egypt extends from the Libyan border to the Qattara Depression, north of 29°N, reaching a height of 215 m a.s.l. It is a *hamada*, a level to gently sloping surface covered mainly by stones and gravel, formed by the deflation of fine-grained sediment over thousands of years leaving behind a lag deposit of coarse stones. Towards the Mediterranean coast the plateau ends in an escarpment and the coastal plain varies in width from a few kilometres to about 50 km. The groundwater conditions, proximity to moisture-bearing winds, and cooler temperatures lead to moderate winter vegetation cover on the coastal plain, making it an important grazing area.

The area to the south of the Libyan Plateau is dominated by the Great Sand Sea. It sweeps down the western half of the desert joining the Great Selima Sand Sea in Sudan (El-Baz and Issawi 1982). This part of the desert is dominated by mobile sand and has little agricultural potential.

The eastern and southern parts of the desert have areas of mobile sand and stony plateau surfaces, which characterize the arid desert environment in this area and are formed by aeolian geomorphological processes. The plateaux exceed 500 m a.s.l. and usually terminate in marked escarpments. The areas of mobile sand are less common and are mainly orientated north-west/south-east or north/south. The most significant of these is the Ghard Abu Muharik which exceeds 300 km in length.

The Western Desert is also characterized by a series of large depressions, the most famous being the Qattara which, at its lowest point is 134 m below sea level (m b.s.l.). It sweeps in an arc from approximately 50 km south of El Alamein on the Mediterranean coast to the Siwa Oasis, south of the Libyan Plateau; it extends further westwards into Libya. Other large depressions occur in the central and eastern parts of the Western Desert, and in areas adjacent to the Nile Valley and Delta (Table 1.1 and Fig. 1.1). All of these depressions have been formed by wind erosion and are termed deflation hollows; the areas of weaker rock in the Nubian sandstone sequence are preferentially eroded. The most important weakening factor is the geological structure (which is mainly faulted and jointed), but the surface expression of rocks such as chalk, shales, and marls, which are not resistant to wind erosion, is also important (El-Etr and Moustafa 1982). Depressions such as Moghara Oasis, Wadi el Natrun, and El Faiyum have thick sequences of Lower Miocene and Pliocene sandstones, limestones, and gypsiferous marls with occasional basalt flows; the latter depression contains a well documented sequence of gypsiferous clays, limestones, and clayey sand. One of the longest sequences, however, from the Cretaceous to the Oligocene, is found in Farafra Oasis, where it exceeds 225 m in thickness. Many of the depressions are geologically related along lines of extensive jointing or faulting, and smaller depressions can be found along

PHYSIOGRAPHY, GEOLOGY, AND SOILS 5

these alignments. For instance, the Qattara and Siwa Depressions are formed along the same structural trend; the Bahariya and Farafra Depressions, and the Farafra, Dakhla, and Kharga Depressions, form other structurally related groups. The large areal extents of these depressions leads to the presence of a number of geomorphological environments. First, there are stony *hamada* surfaces similar to those in the surrounding desert; second, areas of mobile sand dunes; and third, closed depressions with saline lakes or playas. Playas form when the wind erodes the rocks in the depressions, exposing aquifers. In some cases the groundwater from these aquifers can be a valuable, although slightly saline, water resource; in other depressions the salts are extracted for their mineral content. The combination of local groundwater resources for irrigation, and soils developed on sediments that have washed down or have been blown in, makes these closed basins suitable for oasis cultivation. A number of depressions (especially Dakhla and Kharga) are being developed by the Egyptian Government at the present time.

While there are isolated hills and low plateaux throughout the Western Desert formed by erosion of the Nubian sandstone, the only extensive mountainous area is found in the extreme south-west. This area can be divided into the more northerly sandstone Gilf el Kebir Plateau, and the granite Gebel Uweinat Mountains on Egypt's border with Libya and Sudan. The latter mountains are higher, Gebel Uweinat itself reaching 1893 m a.s.l. The area has been the site of many scientific investigations since its discovery in 1923, the most comprehensive work being that of El-Baz *et al.* (1980) and El-Baz and Maxwell (1982). Although this is a hyperarid region (with rainfall occuring only once in 10–20 years or more), the plateaux and mountains exhibit strong evidence of Quaternary fluvial erosion, particularly in the form of wadis and gorges (the latter being known as *kankurs*). The *kankurs* are so deep that groundwater-bearing rocks are often exposed and springs are present. The groundwater in the Gebel Uweinat *kankurs* supports a perennial savanna-like vegetation of grasses, sedges, small herbaceous shrubs, and small trees reaching 4–5 m in height (Boulos 1982). The occasional flash floods in the wadis give rise to blooms of ephemeral desert vegetation, which can last for up to three years, based on the moisture trapped in the wadi-floor flood sediments. These outposts of vegetation in the hyperarid desert are exploited by nomadic herdsmen for sheep, goat, and camel grazing.

Nile Valley

The underlying geology of the Nile Valley is similar to the Nubian sandstone although it has been strongly eroded by the Nile, creating a deep valley with escarpments of the Eastern and Western Desert on either side in many places.

The Nile Valley extends from the Sudanese border in the south to Cairo in the north. To the south of the Aswan High Dam the upper valley is

6   THE AGRICULTURE OF EGYPT

now flooded. The remaining valley follows the Nile swinging first east, and then west in an elongated flattened S-shape before reaching Cairo; the only major kink in this trend is a sharp eastward section around Qena in the upper valley. The valley length, from the High Dam to Cairo, is about 825 km and the valley is never more than 19 km wide. The valley floor has been extensively cultivated for thousands of years and little evidence of the backwaters, marshes, abandoned channels, and lakes typical of river floodplains can be seen; river terrace sequences, relicts of old valley floors have, however, been identified (Furon 1963). The valley sediments have been formed by the deposition of overbank deposits (sands, silts, and clays) when the Nile flooded. The sediments are the products of extensive erosion in the mountainous Nile headwaters in east Africa. River regulation has stopped this annual sediment influx, and the flood plain is now dissected by a network of irrigation canals.

Nile Delta

The Nile Delta has strong geological similarities with the desert to the west and the Nile Valley to the south. However, the Nile Delta is actively extending into the Mediterranean and much of the underlying sediment is modern and relates to this seaward extension.

The apex of the Nile Delta is the city of Cairo, and from here it extends north-westwards 160 km to Alexandria and north-eastwards 150 km to Port Said. There is a complex pattern of distributary channels in the delta but two distributaries are dominant, the Rosetta Branch in the west and the Damietta Branch in the east. Many of the minor channels have been channellized and additional canals have been built. Waterway engineering works have been carried out to provide both irrigation water for the fertile delta farms and efficient drainage. Much of the delta is dominated by similar alluvial sediments to those found in the Nile Valley and it is an extremely important agricultural area. However, at the coast there are a series of saline lagoons and salt flats trapped behind coastal sand bars with little agricultural potential. The biggest of these are Bahra el Burullus, between the mouth of the Rosetta and Baltrim, and Bahra el Manzala, between the mouth of the Damietta and the Suez Canal. Two smaller lakes, Bahra Maryut and Bahra el Idku, occur in the north-western delta (Fig. 1.1).

Eastern Desert

The northern and north-western parts of the Eastern Desert have a similar underlying geology to the Western Desert. In much of the desert, however, highly folded and faulted igneous and metamorphic Precambrian rocks are dominant. These form the Red Sea Mountains and are related to the tectonic activity associated with the Red Sea Rift Valley. The main volcanic rocks are gabbro, gneiss, serpentine, and dactitic and andesitic lava flows. The main metamorphic rocks are schist, quartzite, and marble. Both groups have been extensively intruded by granites.

The Eastern or Arabian Desert extends eastwards from the Nile Valley and Delta to the Red Sea, Gulf of Suez, and Suez Canal and can be divided into four regions (Table 1.1).

Between the Nile Delta and the Suez Canal the desert is quite low lying. Here it is underlain by sandstone, limestone, and marls and has been little affected by tectonic activity. The southern part of this area is dominated by two large plateaux, the Gebel el Galala el Bahariya and the Gebel el Galala el Qibliya developed on Middle Eocene rocks.

A second, geomorphologically similar, area is found in the western part of the desert from about 30°N to around Qena on the Nile. This area is developed on Lower and Middle Eocene sedimentary strata and consists of *hamada* surfaces ending in a steep escarpment overlooking the Nile. The escarpment zone is dissected by numerous small wadis.

South of 30°N, in the central and eastern parts of the desert, there are rugged and well dissected mountains developed on Precambrian rocks. These form a sharp contrast in relief with the gently sloping, lower desert to the west and the boundary is accentuated by the north-south-trending Wadi Qena. The Red Sea Mountains extend down the Red Sea coast into Sudan. They are highest in the south with peaks in excess of 2000 m a.s.l. The highest mountain is Gebel Shayib el Banat (2189 m a.s.l.). These mountains form an important drainage divide for the large ephemeral wadi systems which drain these mountains after storms. The longer wadis, such as Wadi el Tarfa, Wadi el Asyut, Wadi Qena, Wadi Abbad, Wadi el Kharit, and Wadi el Allaqi, flow westwards to the Nile. The shorter wadis, such as Wadi Arabah, Wadi Jimal, and Wadi Hudayn, flow eastwards to the sea.

Further to the east, the Red Sea coast is bordered by a narrow coastal plain developed on Tertiary and Quaternary sediments.

Sinai Peninsula

The Sinai Peninsula is a roughly triangular area at the head of the Red Sea, with both mountainous and lowland desert landscapes. It has a maximum north-south length of about 400 km, and the greatest width between the Suez Canal and Egyptian-Israeli border is approximately 220 km.

The Sinai Peninsula is the most geologically complex area in Egypt. Not only are most of the rock types found in Egypt represented there but tectonic activity, associated with the Red Sea Rift Valley, has resulted in a complex geological structure. The mountainous southern part of the peninsula is developed on Precambrian igneous and metamorphic rocks, similar to those found in the Red Sea Mountains. In central Sinai, the Nubian

8 THE AGRICULTURE OF EGYPT

sandstone reaches a thickness of 600 m (Furon 1963) and exhibits substantial post-Cretaceous folding. Post-folding erosion has led to the formation of many resistant limestone cuestas with low-lying areas between them representing exposures of less resistant marls and sandstones. Tertiary rocks are found along the western coast and northern Sinai and are mainly covered with Quaternary sediments.

The mountains are restricted to the southern half of the peninsula. The highest range, which has nine peaks over 2000 m a.s.l., is found between the southern tip of the peninsula and 28°40′N. The highest peak in the range is Gebel Katherina (2642 m a.s.l.), which is also the highest mountain in Egypt; the more famous Gebel Musa (Mount Sinai) is only 2285 m a.s.l. These mountains slope off steeply to the Gulfs of Aqaba and Suez with many short, steep ephemeral wadis flowing into the sea. To the north the decline in relief is more steady, leading into the gently sloping Gebel el Igma and, further north in central Sinai, the Gebel el Tih. The physiography here is one of gently northward-sloping hills decreasing in height, from about 1600 m a.s.l. in the south to about 750 m a.s.l. in the north, over a distance of about 110 km. The area is dominated by drainage into northern Sinai. The western coast of central Sinai is dominated by short wadis flowing into the Gulf of Suez and on the Egyptian–Israeli border a number of wadis flow into the Jordan Rift Valley.

In northern Sinai the land is much lower than in the central and southern peninsula; it lies mainly between 200 and 300 m a.s.l. in height, although there are isolated hills reaching heights of between 800 and 1100 m a.s.l. Adjacent to the coast throughout northern Sinai the desert is dominated by extensive areas of mobile sand. The Mediterranean coastline can be split into two parts. The western part is dominated by bars and saline coastal mudflats and lagoons known as *sabkhas*; the largest of these is the Sabkhat el Bardawil. The eastern part of the coastline sweeps up towards the Gaza Strip in a long curve.

Soils

The most comprehensive review of Egyptian soils is provided in the FAO *Soil map of the world* (FAO/UNESCO 1977). It is based on detailed soil maps of the Nile Valley (FAO 1966*a*, 1966*b*) and Nile Delta, generalized maps (Elgabaly 1969; Veenenbos 1963), and work by Ghaith (1959). It provides a comprehensive review of the distribution of Egyptian soils up to the mid-1970s. The soil terminology used in this chapter is based on the FAO System of Soil Classification, excellent summaries of which appear in FAO/UNESCO (1974) and Young (1976).

In the late 1980s the legend for the FAO/UNESCO *Soil map of the world* was revised (FAO/UNESCO 1988). This revised legend affects many of the

PHYSIOGRAPHY, GEOLOGY, AND SOILS

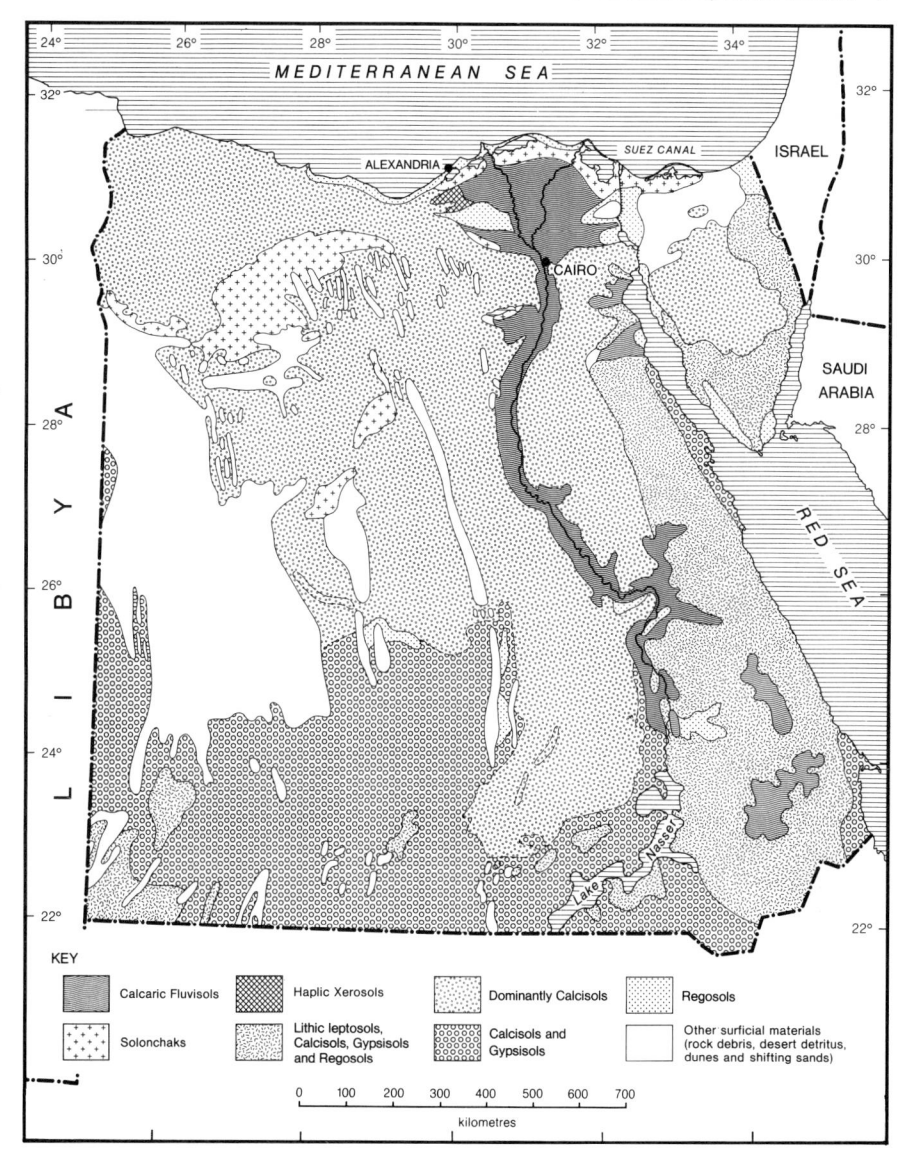

FIG 1.2. Egypt: soils

soils in Egypt. Therefore, although Fig 1.2 is based on FAO/UNESCO (1977) because a new map of the soils of Egypt has not yet been produced, the key to this figure has been revised, as has the table of different soil associations (Table 1.2).

Mappable soils, those with even the slightest degree of horizonation, cover 79 599 km², or 83.15 per cent of Egypt (FAO/UNESCO 1977).

The remainder of the country is covered by dunes and shifting sands (16 084 km^2) and rock debris and desert detritus (86 km^2) (Fig. 1.2 and Table 1.2), which exhibit no evidence of soil formation. Although the picture seems favourable, many of the mappable soils have considerable management problems and present severe limitations to agricultural activity.

The main Egyptian soil associations, in order of decreasing area, are: calcisols and gypsisols; mixed leptosols and regosols; calcaric fluvisols; solonchaks; regosols; lithic leptosols and mixed lithic leptosols; and calcisols and gypsisols (Table 1.2). In addition there are minor occurrences of gleysols, vertisols, and solonetz soils within the major associations.

TABLE 1.2. Egypt: soil association distribution

Major soil associations[a]	Area	
	km^2	% land area
Calcaric fluvisols	5771	6.03
Solonchaks	3283	3.43
Gleyic	797	0.83
Haplic	2307	2.41
(Takyric)[b]	179	0.19
(Haplic xerosols)[c]	107	0.11
(Yermosols)[c]	53 351	55.71
(Not subdivided)[c]	19 695	20.57
(Calcic)[c]	32 335	33.76
(Haplic)[c]	1321	1.38
Regosols	2084	2.18
Calcaric	628	0.66
Eutric	1456	1.52
Lithic leptosols	1214	1.27
Lithic leptosols and (calcic yermosols)[c]	720	0.75
Lithic leptosols and regosols	13 069	13.65
Rock debris and desert detritus	86	0.09
Dunes and shifting sands	16 084	16.78

[a] This table and the accompanying Map (Fig. 1.2) were based originally on the major soil types in the FAO *Soil map of the world* (FAO/UNESCO 1977). Since then the legend for the FAO *Soil map of the world* has changed but a new map covering Egypt has not yet been produced. The overall calculations are therefore based on the original map and legend, although new soil types have been used whenever possible.

[b] Takyric solonchaks have been deleted from the revised legend and these soils are now classified as takyric phases of other types of solonchaks.

[c] Xerosols and yermosols were entirely deleted in the revised legend. Most soils originally classified as xerosols and yermosols are now classified as calcisols and gypsisols. These are subdivided into haplic, luvic, and petric calcisols and haplic, calcic, luvic, and petric gypsisols.

Source: FAO/UNESCO (1977); FAO/UNESCO (1988).

Calcisols and gypsisols

Large areas of calcisols and gypsisols are found in Egypt, especially in the Western Desert, on the Libyan Plateau, in the north-western and south-western parts of the Eastern Desert, along the southern Red Sea coast, and in Central Sinai (Fig. 1.2).

Calcisols are dominated by a calcium carbonate-rich horizon within 1.25 m of the soil surface. Their subsoil properties are variable, ranging from those with subsoil clay accumulation to those with no B-horizon development at all. The organic matter contents vary but they are never saline and neither do they exhibit evidence of gleying in the upper metre.

Gypsisols are similar to calcisols in terms of most diagnostic properties with the very important exception of the type of calcium accumulation in the upper 1.25 m. In gypsisols this zone is dominated by gypsum (calcium sulphate). The lack of clay in many calcisols and gypsisols combined with the low amounts of organic matter means that most of these soils are characterized by inherently low fertility and poor water-holding capacity. Nevertheless, with irrigation and manuring or fertilizer application, they are being reclaimed on the desert land adjacent to the Nile Delta and in some of the Western Desert oases. The high calcium carbonate and calcium sulphate content of some calcisols and gypsisols can create further soil management problems. In many calcisols and gypsisols the calcium carbonate and sulphate has aggregated and hardened to form calcrete or gypcrete, respectively. These are rock-like materials which create severe problems for root penetration and ploughing, especially as they are usually found in the upper 1.25 m of the soil. A further physical problem, which is particularly prevalent when these soils have a high silt content and are irrigated, is surface crusting which dramatically reduces the infiltration rate. Other management problems are concerned with soil chemistry. The levels of available phosphorus are low due to the high pH, and the micronutrients such as copper, iron, manganese, and zinc also have low availabilities. There are often potassium and magnesium supply problems due to the calcium imbalance, and the soils have very low levels of microorganisms.

Lithic leptosols

Lithic leptosols have severe depth limitations. Continuous outcrops of consolidated rock can be found within 10 cm of the surface and consequently there is minimal profile development.

Lithic leptosols are restricted mainly to mountainous south-west Egypt. They are found in combination with calcisols and gypsisols in the most mountainous parts of the Eastern Desert and Red Sea Mountains, in a belt extending from the Sudanese border north to about 29°N, and on the southern tip of the Sinai Peninsula. Other isolated areas are found in the southern parts of the Western Desert, and a smaller area of lithic leptosols and regosols is restricted to the Egyptian–Israeli border in eastern Sinai (Fig. 1.2).

Cultivation is impossible on these soils due to their shallowness, limited nutrient supply, and very low water-holding capacity. Only localized land-use options, such as rough wet-season grazing, indigenous forestry, gathering plant products and hunting, or water catchment are available.

Calcaric fluvisols

Calcaric fluvisols are relatively young soils developed on recently deposited colluvial, fluviatile, lacustrine, or marine sediments in the Nile Valley and Delta, and on some of the coastal plains. In addition, four isolated but extensive areas of calcaric fluvisols are found in the Red Sea Mountains (Fig. 1.2).

Fluvisols still show some sedimentary stratification. Organic matter content decreases irregularly with depth (although it remains above 0.35 per cent in the upper 1.25 m) and the soils have sulphide-rich material within 125 cm of the surface. Generally fluvisols exhibit little horizonation, except for a weakly developed A-horizon and peaty horizons. Calcaric fluvisols, however, are strongly calcareous, having significant amounts of free calcium carbonate at depths of 20–50 cm and pH 7. These are the most intensively farmed soils in Egypt and have a high development potential due to the ease of irrigation, low water erosion potential, and their ability to be double-cropped. They do not, however, have very high nutrient levels, so the maintenance of fertility by traditional manuring practices or by high rates of fertilizer application is of particular importance in crop production. There are also potential wind erosion problems in silt-rich areas if the topsoil is allowed to dry out. The major management task is to control water supply and conserve soil moisture. The Nile fluvisols are extensively irrigated and the management of irrigation scheduling and drainage is time-consuming. In addition, in areas with a high clay content, poor irrigation practices often lead to subsoil compaction and pan formation, secondary salinization, and gleying.

Solonchaks

Solonchaks are the saline soils formed on recent alluvial and lacustrine material, often in closed basins, such as those found in the desert oases, or in the coastal *sabkhas*. They are commonly found in many of the major oases and depressions in the Western Desert, particularly in the Bahariya, Farafra, and Siwa Oases, the El Faiyum and Qattara Depressions, and in Wadi el Natrun. There is also a wide belt of solonchaks along the upper Nile Delta extending eastwards to the Sabkhat al Bardawil on the northern Sinai coast (Fig. 1.2).

A number of types of solonchaks are found in Egypt, but the main diagnostic feature of all solonchaks is their high soluble salt content. Gleyic solonchaks exhibit hydromorphic properties, such as waterlogging and reducing conditions, within 0.5 m of the surface. Some solonchaks lack hydromorphic properties but have extensive crystalline salt crusts with blisters, pressure ridging, and thrust polygons. When these features occur the solonchak is described as having a takyric phase. Orthic solonchaks lack salt crusts with either takyric or hydromorphic features in the top 50 cm of the soil, but do have a weakly developed organic A-horizon.

Solonchaks are extremely difficult soils to manage due to their severe salinity problems. The high soluble salt content affects crop growth in two ways. First the osmotic balance of the soil solution is altered, making it extremely difficult for roots to extract nutrients from the soil. This results in stunted growth and depressed yields unless salt-tolerant crop varieties are grown. Second, at soluble salt contents in excess of about 2 per cent, toxic effects become important. In particular, accumulations of chloride and boron are important, while sodium only really affects the ionic balance of the soil solution. Soil management is concerned with the amelioration of the high soluble salt contents in areas where there are no toxicity problems. Various strategies such as flushing out salts with irrigation water, the addition of less saline soil material, and the use of salt-tolerant crops are favourable measures for the less saline solonchaks, but where the salt contents are highest agriculture is impossible.

Regosols

Regosols are developed on sandy substrates and, apart from a rudimentary A-horizon with less than 1 per cent organic matter, show no horizonation. Two types of regosols are found in Egypt—calcaric regosols and eutric regosols. The former have accumulations of calcareous material and the latter are not calcareous at depths of 20–50 cm, but have a base saturation in excess of 50 per cent.

In Egypt, regosols are important only locally in a small number of restricted areas. Pockets of calcaric regosols are found in a narrow strip along the Mediterranean coast between Ras el Kanayis and Borj el Arab, in north-east Sinai, and in the El Kharga Oasis. Eutric regosols are more extensive; pockets are found along the Mediterranean coast from Ras el Kanayis to the Damietta Branch of the Nile, to the east and west of the Nile Delta, on the western and southern shores of Sinai, in the Dakhla Oasis, and on the Sudanese border (Fig. 1.2).

Regosols have low available water capacities and low inherent fertility. Their management involves irrigation together with the addition of clay and manure. Irrigation alone provides only part of the answer, because without the addition of clay and organic material most of the water quickly enters the groundwater leaving little held in the rooting zone for plants. The clay and organic matter also provide nutrients and, in particular, provide sites to hold nutrients in the rooting zone.

Other surficial materials

The non-soil units identified in Egypt, rock debris, desert detritus, dunes, and shifting sands, are found mainly in the inner Western Desert and the northern Sinai Peninsula. In these areas the soil management problems preclude any agriculture, apart from some desert nomadism.

Conclusion

Physiographic and pedological constraints to agriculture in Egypt are immense. These include salinization, alkalinization, and waterlogging of soils in valleys and depressions. Upland soils fare little better; in these areas the low water-holding capacities and inherent low fertility make soil management difficult. Dune encroachment on to agricultural land and shifting sand present problems all over the country. Consequently, agricultural activity is restricted mainly to the Nile Valley, Nile Delta, and some of the depressions in the Western Desert.

Much of Egypt remains uncultivated, although uncultivated land is often used for extensive grazing and, in some areas, is the focus of desert reclamation. The Egyptian Government has an ambitious target to reclaim 2.8 million feddans of uncultivated desert by the end of the twentieth century. Much of this will rely not only on supplies of suitable irrigation water but also on the appropriate management of reclaimed soils—something that has not been a strong point in reclamation schemes in Egypt since 1954.

The soils that are cultivated or reclaimed require significant management inputs to control the high rates of soil moisture evaporation, keep the levels of soil salinity low, and attain and sustain high levels of fertility.

References

Boulos, L. (1982). The flora of Gebel Uweinat and neighbouring regions, southwest Egypt. In *Desert landforms of southwest Egypt: a basis for comparison with Mars*, NASA CR-3611 (ed. F.A. El-Baz and T.A. Maxwell), pp. 341–8. National Aeronautics and Space Administration, Washington DC.

El-Baz, F.A. and Issawi, B. (1982). Wind patterns in the Western Desert. In *Desert landforms of southwest Egypt: a basis for comparison with Mars*, NASA CR-3611 (ed. F.A. El-Baz and T.A. Maxwell), pp. 119–40. National Aeronautics and Space Administration, Washington DC.

El-Baz, F.A. and Maxwell, T.A. (ed.) (1982). *Desert landforms of southwest Egypt: a basis for comparison with Mars*, NASA CR-3611. National Aeronautics and Space Administration, Washington DC.

El-Baz, F.A., Boulos, L., Breed, C., Dardir, A., Dowidar, H., El-Etr, H., Embabi, M., Grolier, M., Haynes, V., Ibrahim, M., Issawi, B., Maxwell, T., McCauley, J., McHugh, W., Moustafa, M. and Yousif, M. (1980). Journey to the Gilf Kebir and Uweinat, southwest Egypt, 1978. *Geographical Journal*, **146**, 51–93.

El-Etr, H.A. and Moustafa, A.R. (1982). Lineation patterns of the central Western Desert. In *Desert landforms of southwest Egypt: a basis for comparison with Mars*, NASA CR-3611, (ed. F.A. El-Baz and T.A. Maxwell), pp. 51–6. National Aeronautics and Space Administration, Washington, DC.

Elgabaly, M.M. (1969). *Soil map and land resources of U.A.R.* Alexandria University Research Bulletin No. 21. University of Alexandria.

Elgabaly, M.M. (1969). *Soil map and land resources of U.A.R.* Alexandria University Research Bulletin No. 22. University of Alexandria.

FAO (1966a). *High dam survey: U.A.R. Vol. 2. The reconnaisance soil survey* (Maps at a scale of 1:1 000 000). FAO, Rome.

FAO (1966b). *High dam survey: U.A.R. Vol. 3. The semi-detailed soil survey* (Maps at scales of 1:100 000 and 1:200 000). FAO, Rome.

FAO/UNESCO (1974). *Soil map of the world at a scale of 1:5 000 000*, Vol. 1. Legend. UNESCO, Paris.

FAO/UNESCO (1977). *Soil map of the world, at a scale of 1:5 000 000*, Vol. 6. Africa. UNESCO, Paris.

FAO/UNESCO (1988). *FAO-UNESCO soil map of the world. Revised legend.* World Soil Resources Report No. 60. FAO, Rome.

Furon, P. (1963). *Geology of Africa.* Oliver & Boyd, Edinburgh.

Ghaith, A.M. (1959). *Soil survey and land classification map of Kraf-el-Sheik.* Ministry of Agriculture, Cairo.

Issawi, B. (1973). Nubia sandstone, type section. *Bulletin of the American Association of Petroleum Geologists*, **57**, 741–5.

Said, R. (1989). *The geology of Egypt.* A.A. Balkema, Amsterdam.

Veenenbos, J.S. (1963). *Soil association map of U.A.R., 1:2 000 000.* FAO, Rome.

Young, A. (1976). *Tropical soils and soil survey*, Cambridge University Press, Cambridge.

2

Climate and hydrology

P. BEAUMONT

Climate

Introduction

The climate of Egypt is governed mainly by its location in the north-eastern part of Africa on the margin of the largest desert in the world. Its latitudinal position, between 22 and 32°N places it firmly in the sub-tropical dry belt, although conditions on its northern coast are ameliorated by the presence of the Mediterranean Sea. Egypt's climate can best be expressed as a contest between the hot, dry air masses over the Sahara and the cooler, damper maritime air masses from the north carried by eastward moving depressions. Throughout most of the year the hot, dry tropical continental air masses dominate, but during the winter period air masses of both tropical maritime and polar maritime origin make brief incursions into Egypt from the north, frequently bringing rain with them.

The hot, dry tropical continental air mass over the Sahara is caused by the general circulation of the atmosphere. Close to the equator the warm air rises and then flows northwards before subsiding. In the northern hemisphere the main area of subsidence is between latitudes 15 and 30°N. As the air descends it becomes warmer and drier to produce the stable airmass known as tropical continental. In winter, in response to the movement of the overhead sun, the inter-tropical convergence zone, which marks the boundary between tropical moist Atlantic air and hot dry air over the Sahara, moves southwards. This movement of the main climatic systems means that the Mediterranean area is covered by meridional circulation associated with the westerlies. In reality the westerlies are a series of eddies in the atmosphere moving in an easterly direction. They are identified as low pressure systems or depressions around which the wind blows in an anticlockwise direction. Individual depressions follow curving paths, so permitting air masses with very different characteristics to be drawn into their circulation patterns.

During winter the southernmost tracks of these depressions are found over the northern fringes of the African continent. This can lead to tropical continental air from the Sahara being drawn north before their arrival, while after their passage cooler and moister air from the Atlantic can be brought southwards over the northern African coastal fringe. The number of depressions moving across northern Africa varies greatly from year to year and this is reflected in the differing rainfall amounts experienced along the northern African coastal fringe.

In contrast, summer is a time of more settled conditions over Egypt when the influence of the subtropical subsidence zone predominates. At this time, temperatures are high and uniform and rainfall seldom if ever occurs. During late autumn the influence of the subtropical subsidence zone decreases in Egypt as it moves southwards, and from October onwards the first of the winter depressions begins to loop southwards over the Mediterranean.

Temperature

The climate of a country is most simply described in terms of average values for variables such as temperature, humidity, and precipitation. There is, however, a danger with this approach of overlooking significant fluctuations which are of vital importance to agricultural activity. Given its latitudinal position it is not suprising that over much of Egypt mean annual temperatures are high and register between 20 and 25 °C. Major variations occur between summer and winter temperatures, as well as between coastal and interior locations.

Along the coast mean maximum temperatures vary from 18–19 °C. in January to 30–31 °C in July and August (Fig. 2.1). For these same stations the mean minimum temperatures show variations from 9–11 °C in January, to 21–25 °C in July and August. The temperature range between the maximum and minimum figures is between 6 and 10 °C. At a distance from the moderating influence of the Mediterranean at inland stations such as Minya, Kharga, and Aswan, the temperature range attains values as high as 16–17 °C. Here the mean maximum temperatures are much higher in summer reaching figures of 37–42 °C in June, July, and August. In January, mean maximum temperatures are 20–24 °C. Mean minimum temperatures are lower in winter than at the coastal stations, with values in January of 4–9 °C. At this time the coldest places are the oases of the Western Desert. In summer, the mean minimum temperatures rise to between 20 and 26 °C, to give values close to those of their coastal equivalents. In effect the inland stations have higher maximum temperatures in summer and lower minimum temperatures in winter.

The mean monthly temperatures do, however, conceal the very large daily fluctuations in temperature which occur in Egypt at certain times of year. This is so particularly in spring and early summer when khamsin winds prevail. The khamsin is a very hot and dry wind which brings dusty unpleasant conditions to much of upper Egypt, and in the spring can cause considerable damage to newly emerging plants. Such winds are produced by eastward-moving depressions with centres tracking along the coastal fringe of northern Africa. In front of the depression anticlockwise winds draw in tropical continental

18   THE AGRICULTURE OF EGYPT

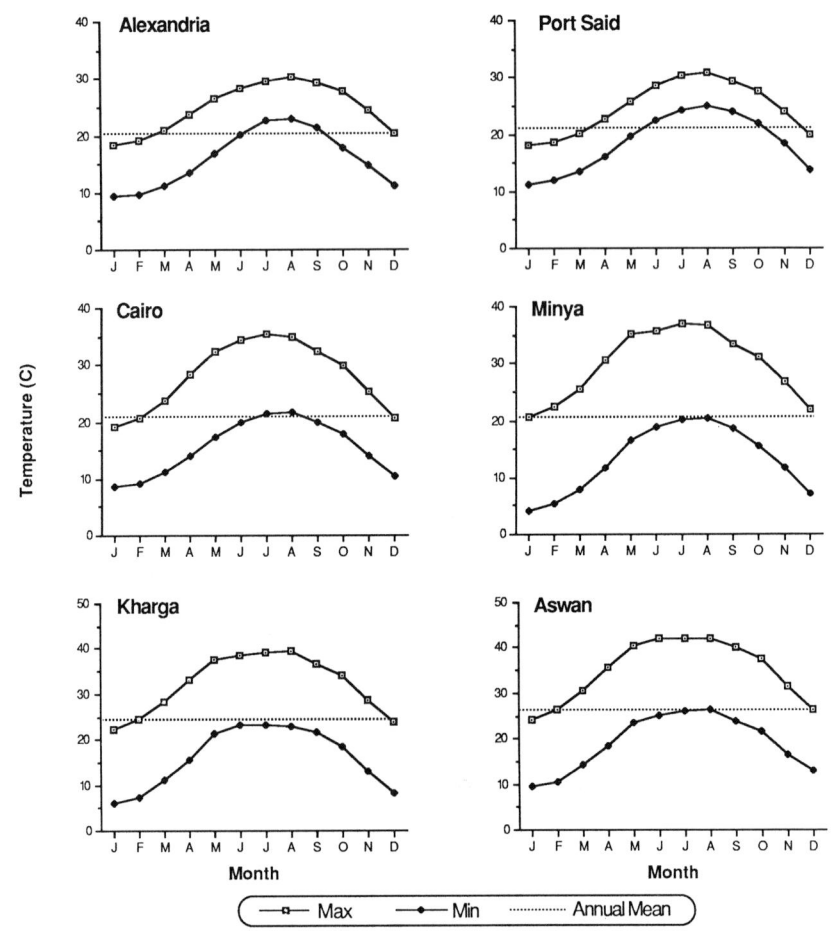

FIG 2.1.   Egypt: mean maximum and mean minimum temperatures at selected stations, 1931–60

air from over the Sahara. These winds blow from the south and south-east and can bring temperature rises of up to 10 °C in a few hours (Fig. 2.2). Such hot and dry conditions do not persist for long because cooler and moister air of Atlantic origin follows behind the depression causing temperatures to dip markedly. These cold fronts are often associated with squalls which produce ferocious, but short-lived dust storms.

Khamsin conditions occur most frequently between February and June when there is still a marked temperature gradient between the air over the Mediterranean Sea and that over the Sahara. After June, conditions throughout the Mediterranean are much warmer and the depression tracks are displaced further northwards. In autumn the depressions move south once again but, because warm temperatures still prevail over much of the region, the khamsin effect, although still present, is not as marked. The timing of the khamsin is seen clearly in Table 2.1 which shows the number of occasions each month when the maximum daily temperature on consecutive days registered an increase of more than 5 °C.

The range of temperature in the top layers of the soil is critical for crop growth. Details published by the Meteorological Department in Cairo (1968) reveal that with dry soils maximum temperatures can reach as high as 67 °C in July at depths of 0.3 cm (Fig. 2.3). Even in January, temperatures as high as 38.5 °C are recorded. In contrast, minimum temperatures at 0.3 cm varied between 20 °C in July and −4 °C in January. With increasing depth in the soil the maximum recorded temperatures decreased, while the minimum temperatures increased. At 3 m the maximum and minimum temperatures only varied from 22 to 27.3 °C. With wet soils the maximum temperatures near the surface were considerably reduced, while the minimum temperatures were higher. At depth, the soil temperatures in the wet field were a few degrees lower than in the equivalent dry one. In dry soils, minimum temperatures below 6 °C were noted in the surface horizons in five months of the year compared with four months in wet soils. Such low temperatures might check crop growth on a temporary basis.

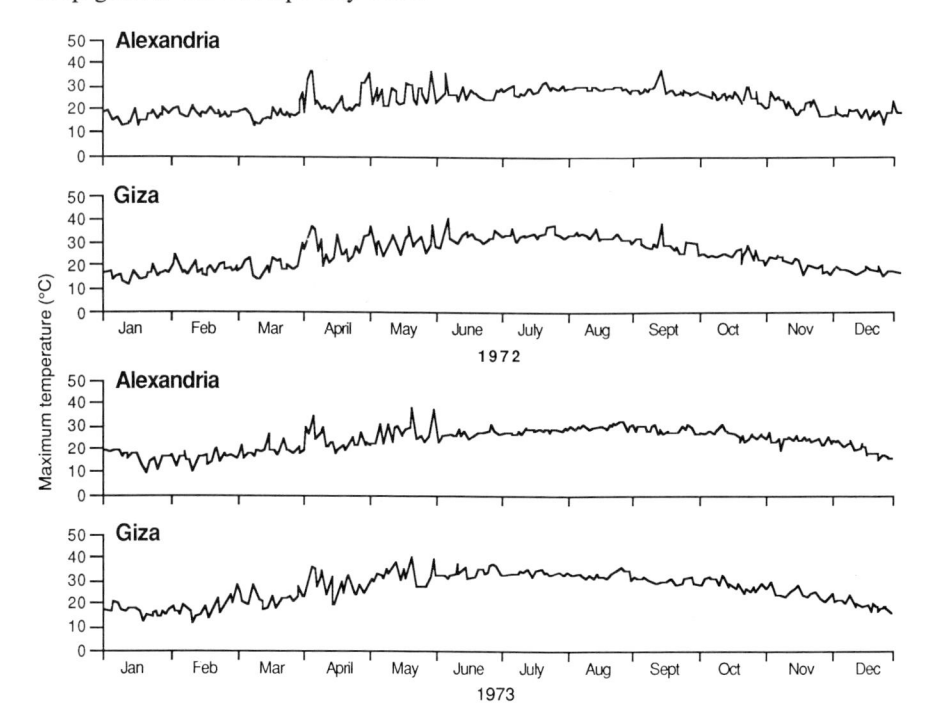

FIG 2.2. Egypt: maximum daily temperatures at selected stations to illustrate the effects of khamsin depressions in the period from February to June

TABLE 2.1. Number of occurrences when the maximum temperature on consecutive days registered an increase of >5 °C (1970–3)

Month	Alexandria	Giza
January	0	2
February	8	4
March	18	17
April	17	20
May	13	15
June	10	8
July	1	2
August	3	1
September	0	0
October	7	6
November	2	2
December	1	1

Precipitation and evapotranspiration

The key feature of the precipitation in Egypt is that there is very little and it is concentrated along the northern coastal zone. The rainy season is the winter period from October to May when the depressions follow their southern tracks over the Mediterranean region. Most of the precipitation is associated with the warm and cold fronts of these systems. Many of the fronts are weak by the time they reach Egypt and rainfall is light and showery. Rainy periods usually last for one to four days. Occasionally, however, an active system can produce substantial amounts of rainfall. Although such falls are exceptional, Alexandria has received 47.9 mm in a single day and Giza 53.2 mm.

The highest annual precipitation totals, reaching 180 mm, are recorded around Alexandria (Fig. 2.4). To the west the Mediterranean coastal strip receives between 120 and 150 mm. From Alexandria eastwards annual totals decline to about 80 mm at Port Said. Inland there is a very sharp precipitation gradient with only 50 mm falling in the middle of the Nile Delta. By the time Cairo is reached the annual total has dropped to 22 mm. Further inland it continues to decline until at Aswan a value of 1 mm is recorded. Over most of the interior of Egypt it is not unusual for a year to pass without any precipitation at all being recorded.

Throughout Egypt rainfall reveals considerable variability over time and space. For example, at Alexandria a rainfall of 168 mm has been recorded in a single month, which represents about 95 per cent of the long-term annual average. On the other hand, during ten months of the year at Alexandria no can be recorded at all and the remaining two months can have minimum values of 1 mm. At almost every other station in Egypt any month can be totally dry in a given year. Equally there are considerable variations over short distances in any one year. With the weak frontal systems it is possible for it to rain at one locality, while a few kilometres away no precipitation at all will be received. From the agricultural point of view precipitation is of little value as everywhere in Egypt it is too meagre to permit reliable crop production.

As crop growth is only possible with irrigation, a knowledge of water losses is essential for planning water application rates. Unfortunately relatively little accurate information is available on evaporation and transpiration rates in

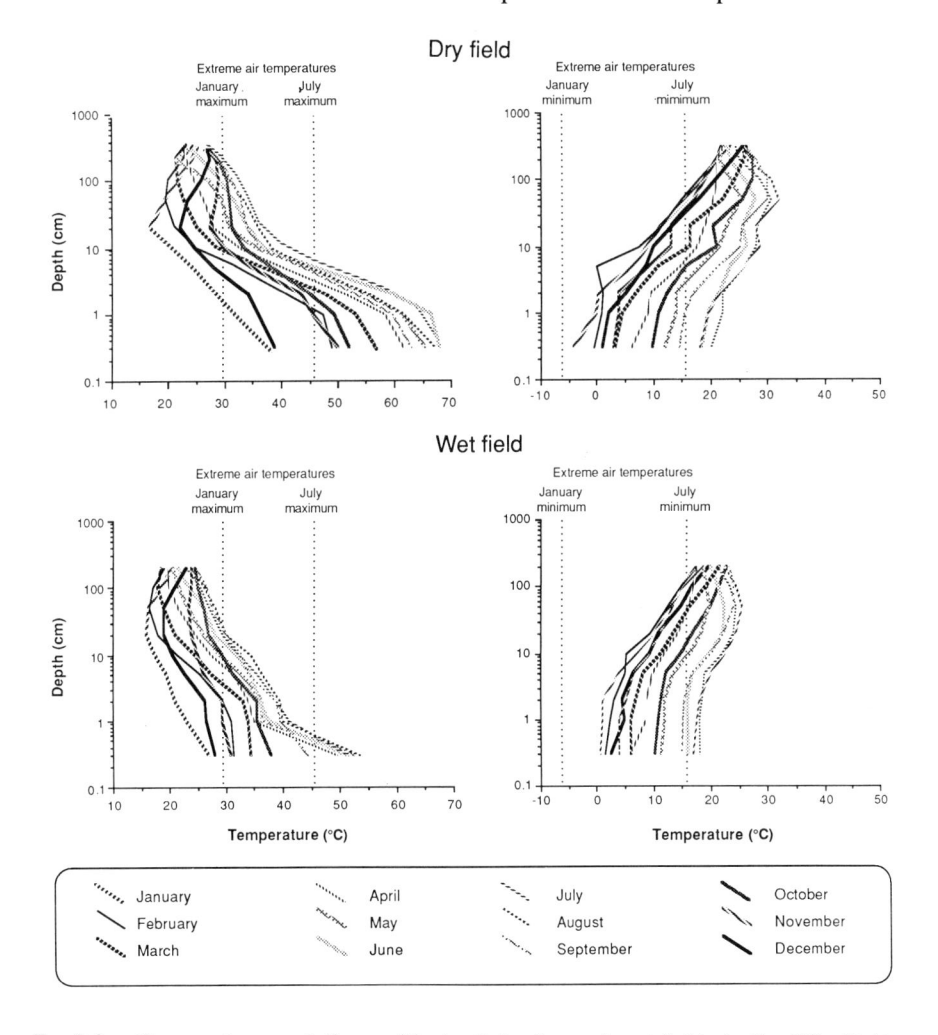

Fig 2.3. Temperature variations with depth in dry and wet fields in the Nile Delta region, 1956–60

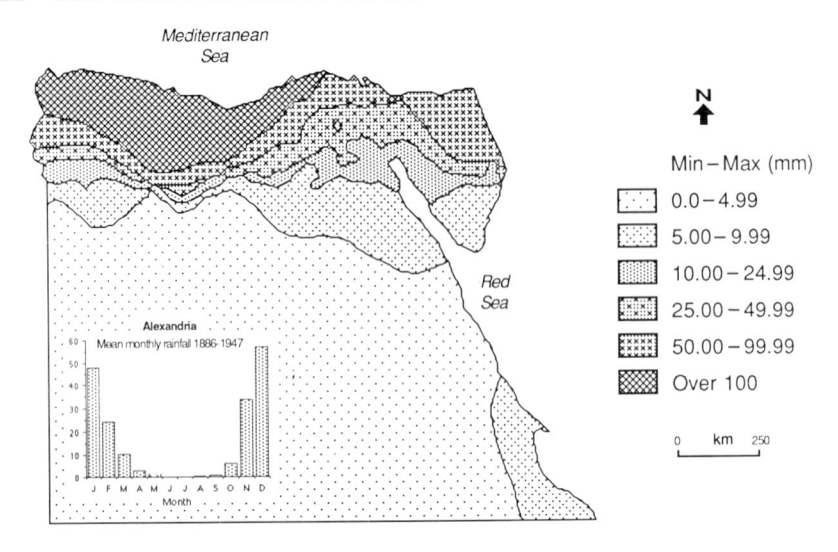

FIG 2.4. Egypt: mean annual precipitation, 1940–60

Egypt. Many meteorological stations in Egypt contain Piche evaporimeters, but the validity of the results obtained from these instruments as an indication of true environmental water losses has been questioned. Some data are available for class A evaporation pans at Giza and Aswan (Table 2.2). These reveal maximum daily values in June of 12.9 mm at Giza and 19.3 mm at Aswan. At Aswan the annual pan loss is close to 5 m, or about double that recorded at Giza.

Even class A pans produce values greater than measured evapotranspiration losses and so formulae have been designed to provide estimates of potential evapotranspiration using available climatic data. The two best known formulae are those of Penman (1948) and Thornthwaite (1948). Of the two, the Penman formula is the stronger scientifically as it includes a term to take account of the wind speed on the evapotranspiration rate. Calculations using the Penman formula reveal maximum potential evapotranspiration rates reaching 4.7 mm day^{-1} in July at Alexandria and 7.8 mm day^{-1} in June at Aswan (Table 2.3). In general these values seem to be less than one-half the values recorded by the class A pan results. At Giza the annual evapotranspiration rates, calculated from the daily values, are estimated to be 1.61 m and at Aswan to be 1.86 m. Such figures are believed to provide satisfactory monthly estimates of the volumes of water which are needed to ensure the maximum growth rates of crops.

Climate and crops

As far as agriculture is concerned the critical climatic factors are those which occur during the growing cycle of the crops. What happens at other times

TABLE 2.2. Class A pan evaporation rates for selected stations in Egypt

Month	Evaporation (mm day^{-1})	
	Giza	Aswan
January	3.0	6.3
February	4.2	8.5
March	6.4	12.3
April	8.6	15.2
May	11.4	17.5
June	12.9	19.3
July	11.1	19.0
August	10.3	18.8
September	9.1	16.5
October	6.9	12.8
November	4.3	19.1
December	3.1	7.3
Annual mean	7.6	13.6

Source: Ali (1978).

of year is relatively unimportant. Given the warm temperatures throughout the year and the perennial irrigation system, Egypt has been able to develop a cropping pattern which has permitted the cultivation of both winter and summer crops. Traditionally, before the advent of perennial irrigation in the nineteenth century, the Nile Valley was characterized by winter crops, of which the most important was wheat. With the development of summer cultivation, crops such as cotton, rice, sugar cane, and millet became important (Beaumont and McLachlan 1985).

The most widely grown winter crop is wheat, which is cultivated along the whole length of the Nile Valley. The seeds are usually sown in early November following fallow, maize, or cotton. Harvesting takes place in late April or early May. The first four weeks after sowing are a critical period for the wheat crop when the seeds germinate and the shoots emerge above ground level. This is followed by the main growth phase which includes maximum tillering (weeks 5–14), ear initiation (weeks 14–19), and flowering (weeks 19–22). Finally between weeks 22 and 26 the ripening of the wheat occurs.

Although sowing dates vary somewhat between different parts of the Nile Valley, it is interesting to compare the wide range of climatic conditions tolerated by the wheat crop during its growth. For example, around Alexandria the range of temperatures experienced is relatively small compared with elsewhere in Egypt, and in particular when compared with Aswan (Fig. 2.5). Similarly, the Alexandria and Nile Delta regions enjoy much more humid conditions than, for example, Aswan, where for most of the year relative humidity values are about one-half of those found in the north. With wind speed it

TABLE 2.3. Penman potential evapotranspiration rates for selected stations in Egypt

	Evapotranspiration (mm day^{-1})			
Month	Alexandria	Giza	Aswan	Kharga
January	1.2	2.0	2.2	2.2
February	1.7	2.7	3.2	3.1
March	2.5	3.5	4.1	4.2
April	3.1	4.9	5.5	5.8
May	4.1	6.2	6.8	7.0
June	4.6	6.7	7.8	7.9
July	4.7	6.8	7.6	7.5
August	4.5	6.3	7.5	7.6
September	3.5	5.3	6.4	6.5
October	2.4	4.0	4.4	4.6
November	1.4	2.3	3.0	3.1
December	1.1	1.6	2.2	2.2
Annual mean	2.5	4.4	5.1	5.1

Source: Ali (1978).

is the extremes of the country, at Alexandria and Aswan, where the highest values are recorded. Perhaps most interesting is the fact that during the first twelve weeks of the crop's growth, that is until approximately the end of the period of maximum tillering, the mean weekly temperature is falling. Then, for the ear initiation, flowering, and ripening stages, the temperature rises. During this period of increasing temperature nearly all the stations record a decline in relative humidity. Despite these wide climatic variations, wheat is cultivated successfully throughout Egypt.

Cotton is the most important cash crop grown in Egypt and, like wheat, is grown along the length of the Nile Valley. It is, however, a summer crop and is cultivated between early March and the end of August. For cotton, the first week after sowing, when germination and emergence of seedlings takes place is a critical time. This is the time of the khamsin winds which can do enormous damage to the crop by shrivelling up the tender shoots of cotton as they emerge. The vegetation stage covers the next six weeks, and this is followed by the main growth stage covering fourteen weeks, when maximum tillering, boll initiation, and flowering takes place. Finally there is the ripening phase over four or five weeks when the bolls attain their maximum size.

For the cotton crop, temperatures show a steady rise until the middle of the growth stage when boll initiation is taking place (Fig. 2.6). Thereafter, temperatures remain remarkably steady. In contrast, relative humidity decreases at all inland sites during the same period. As for wheat, the largest temperature ranges experienced by cotton occur at inland sites. These data suggest that despite some unusual circumstances, climatic conditions in Egypt do not seem to impose very serious restrictions on the growth of the two most important crops.

Hydrology

River flow

To a large extent the hydrology of Egypt is determined by factors outside the country. The very low annual rainfall means that no perennial rivers exist within the country with the exception of the Nile. Despite the huge size of the Nile basin in Africa, the Nile Valley in Egypt consists of a narrow north to south strip between the Red Sea Hills and the Western Desert. In all the region to the west of the Nile surface runoff takes place only following severe storms and then it is always short-lived. The waters of the Nile have always been the only water source available to Egyptian farmers for their irrigation

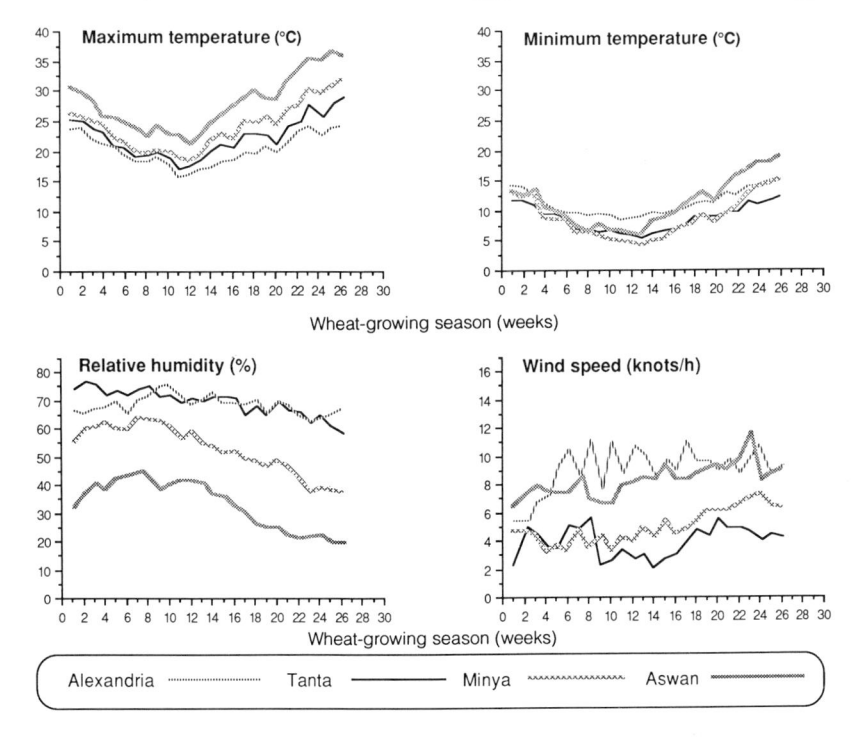

Fig 2.5. Variation in maximum and minimum temperature, relative humidity, and wind speed in different parts of Egypt throughout the wheat-growing season, 1960–74
Source: Ali (1978).

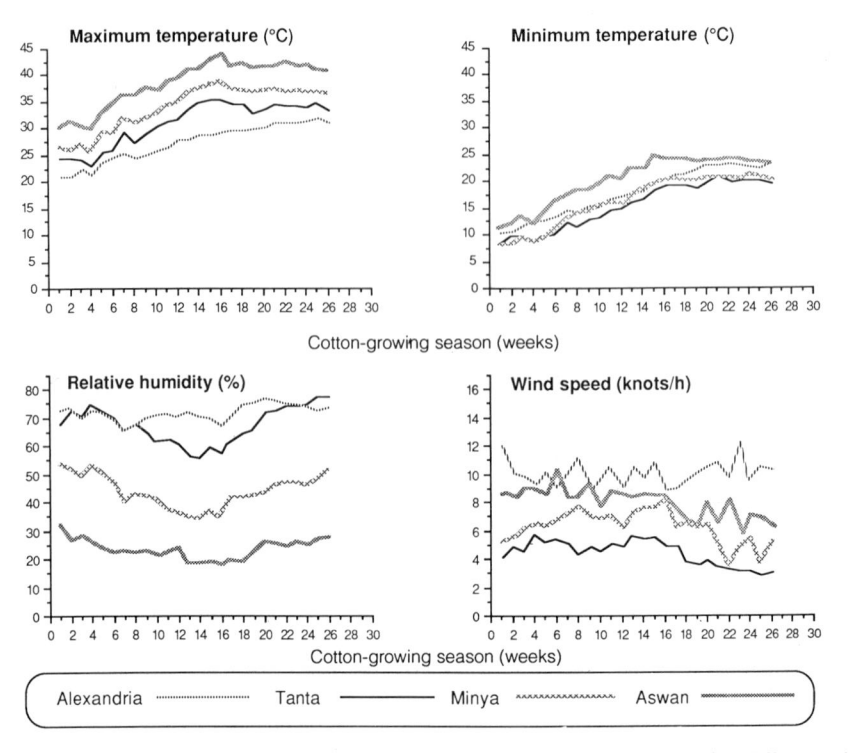

FIG 2.6. Variation in maximum and minimum temperature, relative humidity, and wind speed in different parts of Egypt throughout the cotton-growing season, 1960–74. Source: Ali (1978).

requirements. However, they have rarely been disappointed as the Nile has one of the largest discharges of any river in the world.

The basin of the Nile covers an area of almost 3 million km². The White Nile has its source in Lake Victoria on the East African Plateau, and then flows through a series of cataracts into Lake Albert. Downstream from Lake Albert the Nile flows into the huge swamp region of the Sudd. It then traverses the flat plains of the Sudan until it reaches Khartoum. Here the White Nile is joined by its major tributary, the Blue Nile, which rises in Lake Tana in Ethiopia. From Khartoum to the sea, a distance of 3000 km, the Nile does not receive any perennial tributaries. About 300 km north of Khartoum it is joined by the Atbara, but this only carries significant flows during the flood season.

The annual discharge from Lake Victoria into the White Nile is estimated at about 21 milliards (where 1 milliard = 1000 million m³) (Hammerton 1972). On leaving Lake Albert this has swollen to nearly 30 milliards. In crossing the Sudd the Nile loses enormous volumes of water as the result of evaporation, so that only 14 milliards enters the Sudan Plains section of the river (Sutcliffe

1974). The Sudd also acts as a large storage system and therefore reduces the size of the flood peaks. Maximum flows occur in October and November and minimum flows in April. The average annual discharge of the Blue Nile from Lake Tana is 3.8 milliards. Daily discharges vary considerably from 0.9 million m^3 in May and June to approximately 33 million m^3 in September (Talling and Rzoska 1967). From here it passes through a deep gorge section before emerging onto the Sudan Plains near Roseires. It is in this section of the river where most of the runoff is generated from the Ethiopian Plateau. Apart from Lake Tana there is little water storage capacity in the valley of the Blue Nile, but the altitude of the Ethiopian Highlands and the consequent lower temperatures limit evapotranspiration. When the summer monsoon arrives, torrents of water and eroded soil pour down into the valley of the Blue Nile to produce a huge flood peak at Khartoum, which is about forty times that of the period of minimum flow in April (Table 2.4). At this time the sediment concentrations of certain tributaries of the Blue Nile may be in excess of 10 000 parts per million (Worrall 1958).

Once the White and Blue Niles unite, their regimes are superimposed to produce a flow pattern at Aswan with a maximum in September and a minimum in April. Here maximum flow is reduced to only 10–12 times the lowest monthly discharge. The total annual discharge at Aswan averages 84 milliards. Over the years there have been significant variations in the discharges of the individual months as well as in the total water volumes from one year to another (Fig. 2.7). In the last century and a half the annual

TABLE 2.4. Monthly average discharge for the major rivers in the Nile catchment

	Monthly discharge (m^3 s^{-1})			
	Whie Nile	Blue Nile	Atbara	Nile–Aswan
October	1200	3040	340	5200
November	1200	1030	79	2270
December	1100	499	25	1400
January	829	282	8	1100
February	634	188	2	1020
March	553	156	0	834
April	525	138	0	819
May	574	182	1	698
June	742	461	35	1340
July	897	2080	640	1910
August	1030	5950	2100	6570
September	1130	5650	1420	8180
Annual average	872	1640	389	2650

Source: UNESCO (1969).

28 THE AGRICULTURE OF EGYPT

Nile discharge has varied from a low of 45.5 milliards in 1913 to a high of 137 milliards in 1879 (Hurst *et al.* 1966) (Fig. 2.8). A number of workers have attempted to discover if trends in annual flood levels could be distinguished. Much of this work is still inconclusive, though some evidence seems to point to a progressive decline in annual flood levels during the period of the First and Second Dynasties (*c.* 3100 to 2800 BC) (Bell 1970; Butzer and Hansen 1968).

Water use for irrigation

The traditional form of irrigation along the Nile was the use of basins, varying in size from 1000 to 4000 feddans. In turn these were divided into smaller units of 4 to 5 feddans running in parallel to the main irrigation canals. The flood wave along the Nile used to begin to affect upper Egypt by late July. Water levels rose rapidly in August and peaked in September. When the river reached the requisite height the irrigation canals were opened and the basins flooded to a depth of 1.5 m. This water was kept in the basins for 40–60 days and then drained back to the river, or, in times of water shortage, into the next basin downstream (Hurst 1952). Once the water had left the fields, no more water was added until the next flood. All growth of the crops was, therefore, dependent on the soil moisture reserves which had been replenished. With the

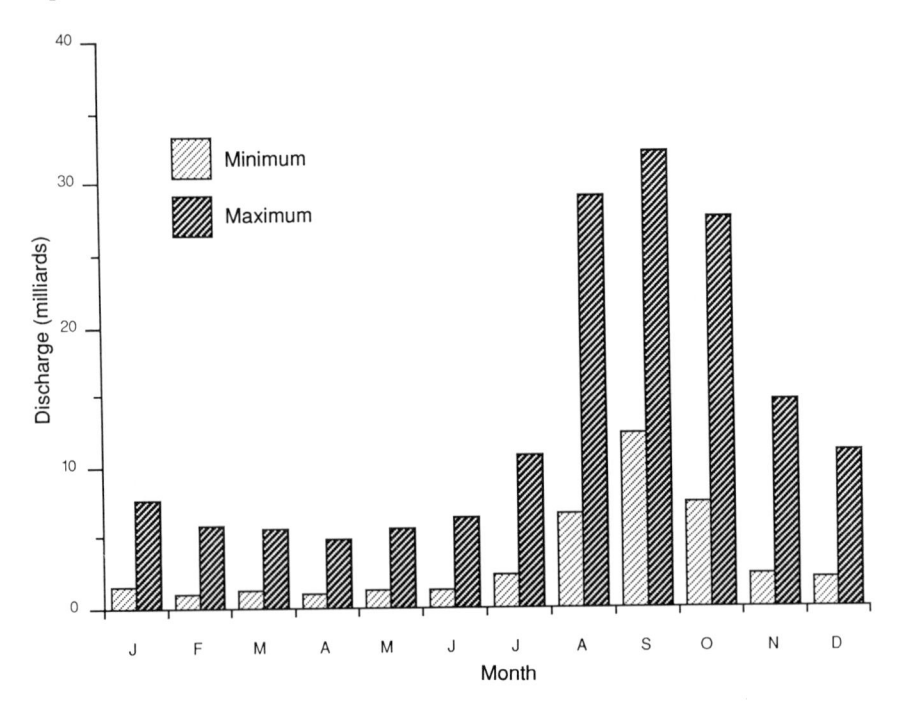

FIG 2.7. Maximum and minimum monthly discharge values for the River Nile at Aswan, 1871–1965

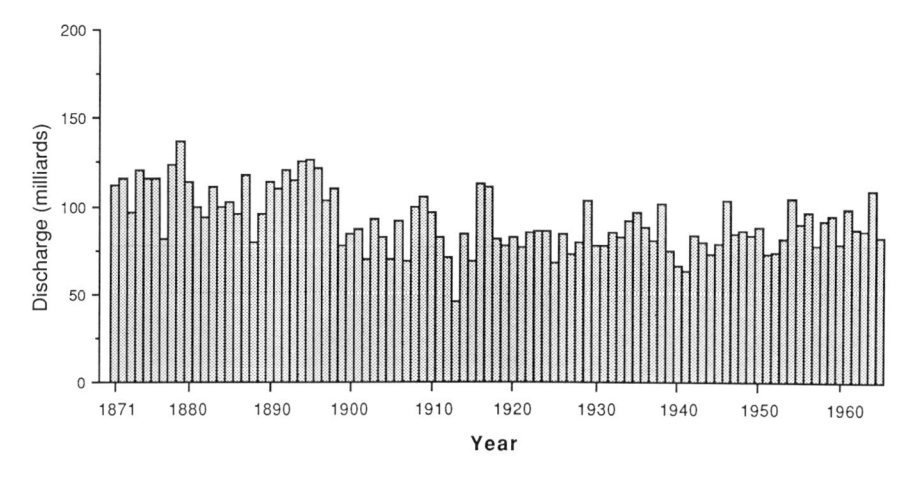

FIG 2.8. Annual discharge of the River Nile at Aswan, 1871–1965

long dry season the soil was broken up by dessication and little ploughing was necessary before the seeds were sown on the newly deposited silts and clay once the waters had receded. Plant growth took place in the winter period and the crop was harvested in late spring. The ground then remained fallow until the next crop was sown following the succeeding annual flood.

The traditional basin irrigation system which developed in Egypt only permitted cultivation in winter and so the range of crops which could be grown was limited. Cereals were by far the most important food crops. In order to produce a wider range of crops summer cultivation was also practised using waters from flood plain wells or from the Nile itself. The main difficulty with this type of operation was that the water often had to be lifted considerable distances and so the area which could be cultivated was severely restricted. It did, however, enable valuable crops such as sugar cane, rice, cotton, and tobacco to be grown. The basin irrigation system was designed to make use only of the flood wave. No control of the flow of the river was ever attempted and as a result most of the flood waters were discharged into the Mediterranean unused. During the flood the high stage levels of the river permitted water to be directed into the irrigation canals and then into the fields. It was, however, a robust system which was well adapted to the environment and so was able to last almost unchanged for 7000 years (Hamdan 1961).

The changeover from basin to perennial irrigation began in the nineteenth century and was completed with the closing of the Aswan High Dam in 1964. The great advantage with perennial irrigation was that it permitted the growth of valuable summer crops on a large scale. Initially in the 1820s the flood canals in the Nile Delta were deepened to allow water to enter them during the low summer water levels. Then in the 1840s barrages were built in the Delta to raise water flows and so permit a more assured summer flow. However, these did not work entirely satisfactorily until modified by British engineers in the 1890s. Throughout the nineteenth century the area of perennial irrigation increased mainly by making canals deeper or by raising water levels with barrages.

The first storage of the flood waters of the Nile came with the construction of the Aswan Dam in 1902. The storage capacity of the lake behind the dam was only 1 milliard but this allowed an extension of the area of perennial irrigation and also provided a more reliable water flow during the summer months. Downstream from Aswan a number of barrages were constructed to ensure higher water levels for the perennial irrigation systems. The Aswan Dam proved successful as a reservoir facility and its height was subsequently raised twice to improve its storage capacity. In 1912 the storage was increased to 2.5 milliards, and to 5.7 milliards in 1933. However, even with these extensions the reservoir was only able to hold about one-fifteenth of the annual flow of the Nile, although it was able to eliminate the worst of the flood dangers. Most of the discharge of the river still flowed unused into the Mediterranean.

The final stage in the control of the flow of the River Nile in Egypt was marked by the construction of the Aswan High Dam in the mid-1960s, 7 km upstream from the existing Aswan Dam (Little 1965). The decision to build the dam was made in 1954, but before work on the project could begin it was essential that a new agreement on the distribution of the waters of the Nile was made between Egypt and the Sudan in accordance with the Anglo-Egyptian Agreement of 1929. This agreement gave the Sudan rights to 4 milliards annually of the Nile's waters and Egypt 48 milliards, leaving an unused surplus of 32 milliards annually at Aswan. In 1959 a new treaty was signed between Egypt and the Sudan. With the completion of the dam it was predicted that the surplus at Aswan would be reduced to 22 milliards owing to evaporation and seepage losses from the reservoir. Of this surplus, 14.5 milliards was allocated to the Sudan and the remaining 7.5 milliards to Egypt. Bearing in mind the original divisions made under the 1929 agreement, this increased the total annual water rights of the Sudan to 18.5 milliards and of Egypt to 55.5 milliards (Waterbury 1979).

The aim of this new dam was to provide sufficient storage capacity for full regulation of the river's flow. Lake Nasser, the lake behind the Aswan High Dam, has a storage capacity of 164 milliards which represents approximately two year's annual flow of the river (Ministry of Culture and Information 1972). Evaporation and seepage losses from the lake are high and estimated to be about 15 milliards each year. Sediment deposition in the lake is believed to be around 60 million m^3 annually, so it is estimated that it will take hundreds of years to fill even the dead storage capacity of the reservoir. With the construction of the dam, the River Nile below it has been effectively changed into an irrigation canal (Beaumont 1981, 1989) (Fig. 2.9). Flood flows have been completely eliminated and since 1964 the pattern of flow has reflected actual demands for irrigation water. Before the construction of the Aswan High Dam, the River Nile in upper Egypt registered an annual water level fluctuation of about 6 m. Associated variations were also noted in the height of the water table of the flood plain, except that here the maximum levels were recorded between 15 and 60 days after the highest water level of the river (Hammad 1970). With the completion of the High Dam the fluctuation in river level has been markedly reduced.

The irrigation water demands of individual crops grown along the Nile Valley vary widely. Ministry of Agriculture estimates for winter crops show that wheat requires between 1150 and 1500 m^3 feddan^{-1}, compared with 2500 and 3400 m^3 feddan^{-1} for *bersim* (Clawson *et al.* 1971, p.29). The demands of summer crops are even higher with cotton requiring 3750–4650 m^3 feddan^{-1}; sugar cane requiring 6000 and 7650 m^3 feddan^{-1}, and rice about 16 000 m^3 feddan^{-1}.

The Aswan High Dam has enabled an expansion of the cultivated area of Egypt by 1.3 million feddans and the conversion of 700 000 feddans from basin to perennial irrigation. This has helped to alleviate, but has not solved, the problem of pressure on available land resources in Egypt. Since 1820 the cropped area of the country has risen more than fourfold, but there has been an 18-fold increase in population and the cropped area per caput has shown a continued decline (Fig. 2.10). In the future, the situation seems bound to deteriorate. By the year 2000, Egypt's population is expected to reach 67 million, yet the opportunities for significant expansion of the cropped area are limited.

The Aswan High Dam has not been without its problems. Unlike the Aswan Dam, the High Dam has no provision for the passage of sediment. This has meant that the sediment accumulates within Lake Nasser and the river water downstream from the Dam has more energy available for erosion because it is carrying less sediment. Many structures, such as bridges and barrages have begun to be undermined, and costly repairs have been incurred. The lack of sediment downstream from Aswan has also meant the loss of the fertilizing qualities of the Nile silt causing farmers to turn to inorganic fertilizers to ensure crop production. In the Nile Delta the decline in sediment has resulted in rapid coastal erosion and, offshore, the lower nutrient value of the Nile waters appears to have reduced fish populations significantly. A similar decline in fish catches has been observed along the lower Nile, but this has been counteracted by the development of a successful fisheries industry on Lake Nasser. Throughout much of the Nile Valley the spread of perennial irrigation has resulted in rising water tables. This has caused a rise in capillary water and severe soil salinity problems which are likely to get worse unless more efficient drainage systems are installed.

Since the completion of the Aswan High Dam, Egypt has been using the waters of the Nile within her boundaries to the maximum extent possible. Any further development of water resources will need basin-wide development

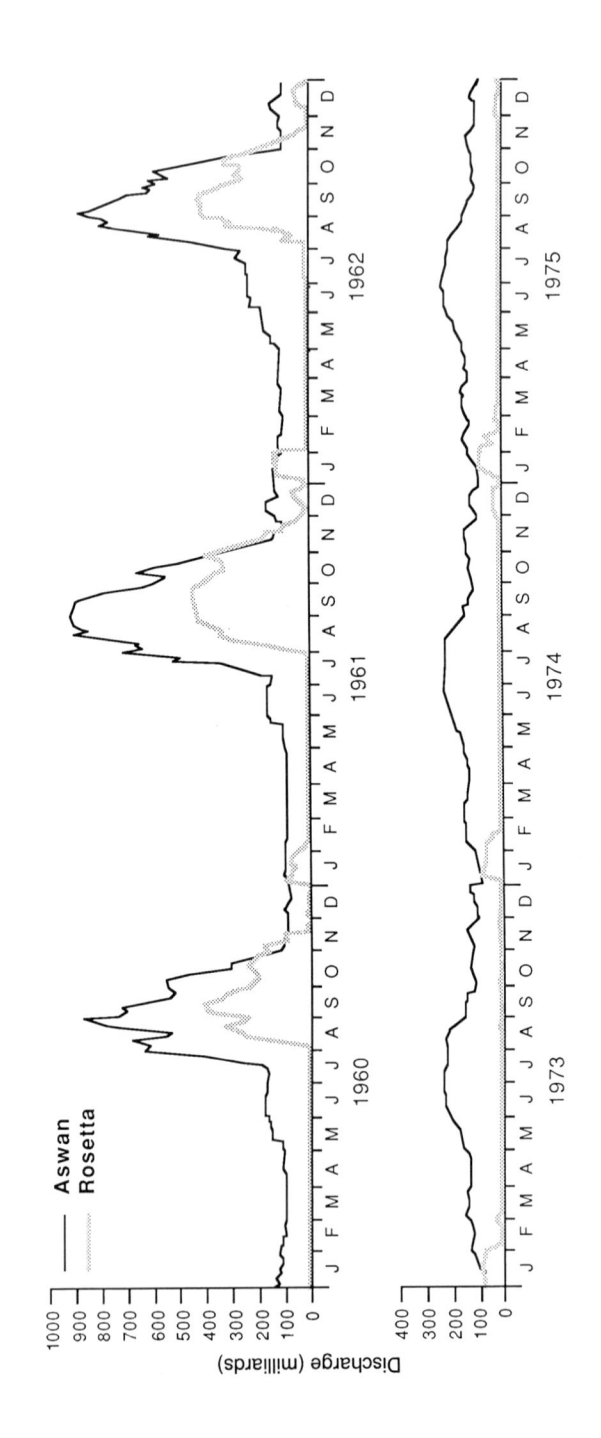

FIG 2.9. Daily discharge values for the River Nile at Aswan and Rosetta, 1960–2 and 1973–5

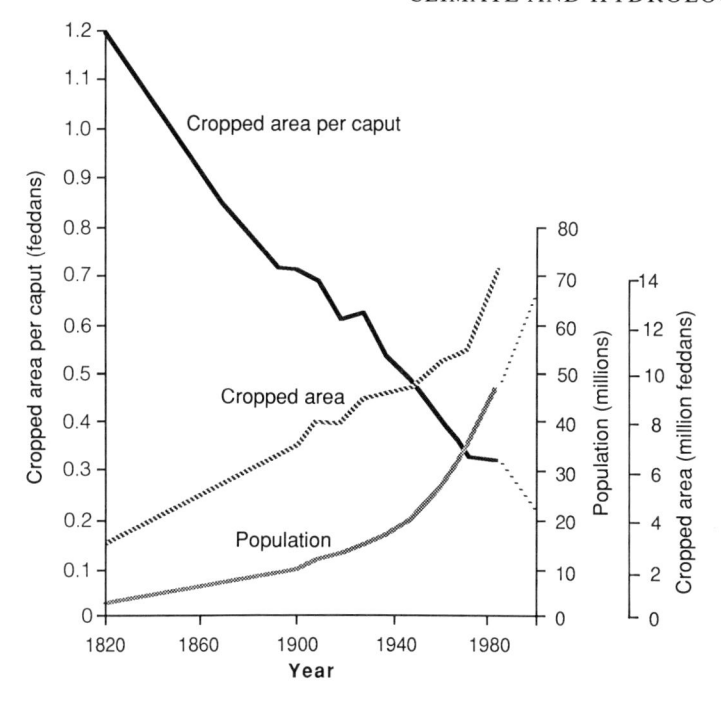

FIG 2.10. Egypt: relationship between cropped area and population growth, 1820–1980

involving all the countries which lie within the Nile watershed (Whittington and Haynes 1985). Over the years, many major projects aimed at more efficient use of water resources have been proposed; the relative merits of many of these schemes have been evaluated (Morrice and Allen 1959; Simaika 1967). One of the main projects aims to reduce the annual evaporation loss of 31 milliards which occurs in the upper part of the Nile Basin in the Sudd and Mashar swamps (Field 1973). Work has already begun on a diversion waterway known as the Jonglei Canal (Jonglei Investigation Team 1954) (Fig. 2.11), but this has had to cease because of civil war. A regulator to be built at the head of the Nile would direct about one-half of the discharge of the river into the canal, leaving the rest to flow along the main channel at below the level of bankfull discharge. Three other canal schemes with the similar aim of reducing evaporation losses have been suggested. If all the above schemes were implemented it is thought that the annual discharge of the Nile at Aswan would increase by 9 milliards (Field 1973). There has been considerable discussion about constructing regulation dams to control water flow throughout the Nile Basin. One such scheme, the Owens Falls Dam at the outlet of Lake Victoria, was built in the mid-1950s to supply hydroelectric power to Uganda and to control the flow of the Nile. Similar projects have

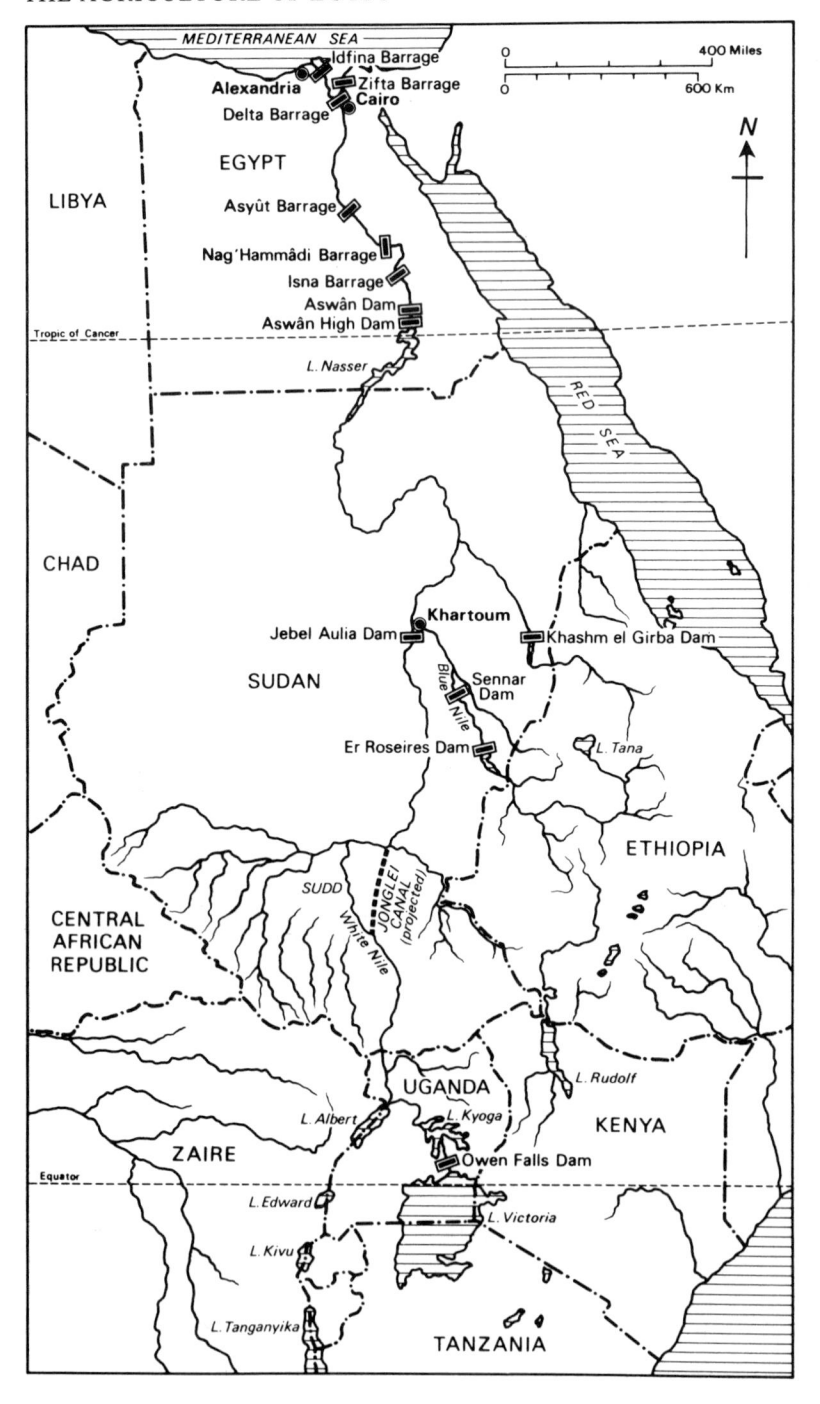

FIG 2.11. Major hydraulic works along the River Nile

been suggested for Lake Albert and Lake Kioga on the White Nile and Lake Tana on the Blue Nile, but none of them are likely to be implemented in the near future. A scheme for supplying more water from Lake Victoria to the White Nile has also been discussed.

Groundwater development

Groundwater is of local importance to agriculture in Egypt Throughout the Nile Valley alluvial aquifers are recharged by percolation from the river Nile, but the largest groundwater reserves, which consist of thick deposits of sand and gravel with a thin layer of clay and silt at the top of the succession, are found in the Nile Delta. In the central part of the Delta where the clay cap is thickest, a semi-confined aquifer system is found. On the desert fringes the clay cap wedges out, and an unconfined aquifer prevails. The main sources of recharge are the Rosetta and Damietta branches of the Nile, the irrigation canals, and seepage from irrigated fields. The major outlet of water from these delta aquifer systems is seepage into the Mediterranean Sea, although during low flow conditions the aquifers feed water into the Nile branches. Numerous pumped wells scattered throughout the Nile Delta take water from these aquifers and it seems likely that much larger abstractions can be made. Estimates suggest that each year about 740 million m^3 flow unused from the aquifers into the Mediterranean (Hammad 1970). If too much water is abstracted, sea water intrusion into these aquifers is a potential danger.

The other major groundwater source in Egypt is the Nubian sandstone aquifer which underlies most of the Western Desert. The Nubian series consist of sands, sandstones, clays, and shales, and attain thicknesses of up to 3500 m. Although a number of different aquifer units have been recognized, the whole seems to function as a single multilayered artesian system covering about 2.5 million km^2 (Hammad 1970; Himida 1970). Natural discharge from the aquifer system is towards the north-west, with water reaching the surface in the Qattara Depression and the Siwa, Farafra, Bahariya, and Dakhla Oases (Fig. 2.12). A fresh/salt water groundwater interface is found about 200 km inland from the coast of Libya and Egypt. The water in the aquifer system is believed to be the result of recharge generated from runoff occurring on the highlands of north-east Chad, the western uplands of the Sudan and the Ennedi and Tibesti Plateaux, and perhaps also from the Nile itself (Wafer and Labib 1973). Most of this water is regarded as fossil, with the recharge occurring during wetter phases of the Quaternary probably between 30 000 and 15 000 years ago. During the 1950s and 1960s, the Egyptian Government hoped to develop widespread irrigated agriculture in the oases of the Western Desert using groundwater from the Nubian aquifer. A number of schemes were implemented, but these ran into difficulties. Well outputs fell off considerably as artesian pressures fell and in some cases saline waters were met. Costs were much higher than expected and the managers did not like

36   THE AGRICULTURE OF EGYPT

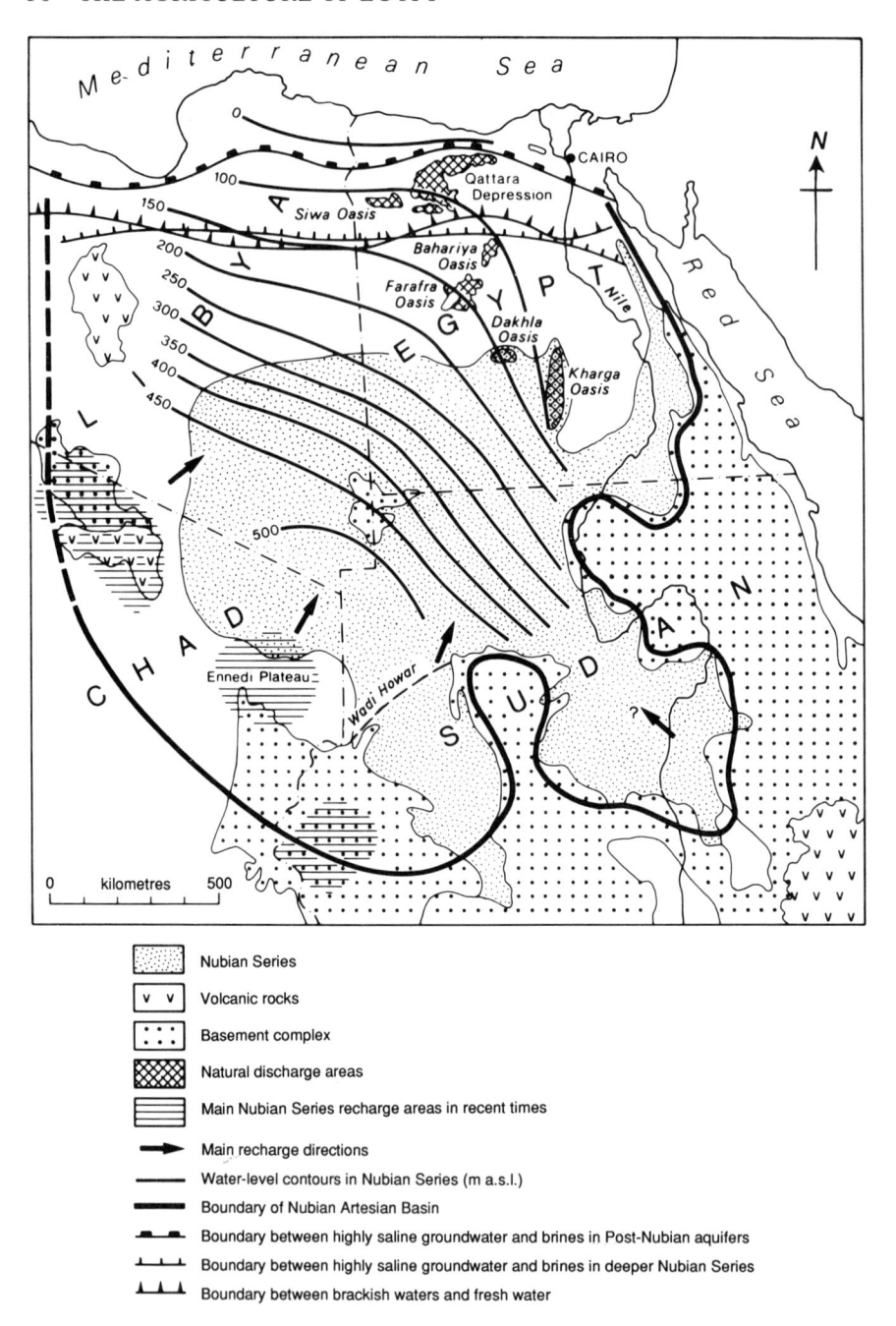

FIG 2.12. Water movement in the Nubian sandstone aquifer
Source: Redrawn from Himida (1970).

living under such harsh conditions. Overall, the project was not the success that had been hoped.

Conclusion

The thermal regime of the climate of Egypt is such that a range of crops with different temperature requirements can be grown at some time during the year. With careful planning it is even possible to cultivate three crops each year. Although Egypt only receives low annual precipitation totals, it is fortunate in having one of the largest rivers in the world, the Nile, flowing through its territory. Over the years this has permitted irrigated agriculture to flourish and has led to the creation of some of the earliest major civilizations. Since the late nineteenth century, perennial irrigation has become increasingly important gradually displacing the older type of basin irrigation. With the completion of the Aswan High Dam in the 1960s the storage of large volumes of water became possible, so permitting the spread of perennial irrigation to all parts of the Nile Valley.

References

Ali, A.A.A.H. (1978). A study of the climate of Egypt with special reference to agriculture. Unpublished Ph.D. thesis. University of Durham.
Beaumont, P. (1981). Water resources and their management. In *Change and development in the Middle East* (ed. J.I. Clarke, and H. Bowen-Jones), pp.40–72. Methuen, London.
Beaumont P. and McLachlan, K.S. (ed.) (1985). *Agricultural development in the Middle East*, John Wiley, Chichester.
Bell, B. (1970). The oldest records of the Nile floods. *Geographical Journal*, **136**, 569–73.
Butzer, K.W. and Hansen, C.L. (1968). *Desert and river in Nubia*. University of Wisconsin Press, Madison.
Clawson, M., Landsberg, H.H., and Alexander, L.T. (1971). *The agricultural potential of the Middle East*. American Elsevier, New York.
Field, M. (1973). Developing the Nile. *World Crops*, **25**, 11–15.
Hamdan, G. (1961). Evolution of irrigated agriculture in Egypt. In *A history of land use in arid regions* (ed. L. Dudley Stamp) Arid Zone Research, Vol.17, (ed. L. Dudley Stamp), pp.119–42. UNESCO, Paris.
Hammad, H.Y. (1970). *Ground water potentialities in the African Sahara and the Nile Valley*. Beirut Arab University, Beirut.
Hammerton, D. (1972). The Nile River—a case history. In *River ecology and man* (ed. R.T. Oglesby, C.A. Carlson, and J.A. McCann), pp.171–214. Academic Press, New York.
Himida I.H. (1970). The Nubian Artesian Basin, its regional hydrogeological aspects and palaeohydrological reconstruction. *Journal of Hydrology, New Zealand*, **9**, 89–116.

Hurst, H.E. (1952). *The Nile: a general account of the river and the utilization of its waters.* Constable, London.
Hurst, H.E., Black, R.P., and Simaika, Y.M. (1966). *The Nile Basin.* Ministry of Public Works, Cairo.
Jonglei Investigation Team (1954). *The Equatorial Nile Project and its effects on the Anglo-Egyptian Sudan.* Sudan Government, London.
Little, T. (1965). *High Dam at Aswan.* Methuen, London.
Meteorology Department, Cairo (1968). *Climatological normals for the UAR up to 1960.* Meteorological Department, Cairo.
Ministry of Culture and Information (1972). *The High Dam,* State Information Office Report, Ministry of Culture and Information, Cairo.
Morrice, H.A.W. and Allan, W.N. (1959). Planning for the ultimate hydraulic development of the Nile. *Proceedings of the Institution of Civil Engineers,* **14**, 101–56.
Penman, H.L. (1948). Natural evaporation from open water, bare soil and grass. *Proceedings of the Royal Society, Sect. A.*, **193**, 120–45.
Simaika, Y.M. (1967). Multi-purpose development of the Nile basin. In *Water for peace,* Vol.8, *Planning and developing water programs,* pp.214–22. US Government Printing Office, Washington, DC.
Sutcliffe, J.V. (1974). A hydrological study of the southern Sudd region of the Upper Nile. *Bulletin of the International Association of Scientific Hydrology,* **19**, 237–55.
Talling, J.F. and Rzoska, J. (1967). The development of plankton in relation to the hydrological regime of the Nile. *Journal of Ecology,* **55**, 637–62.
Thornthwaite, C.W. (1948). An approach towards a rational classification of climate. *Geographical Review,* **38**, 55–94.
UNESCO (1969). *Discharge of selected rivers of the world.* Vol.1, *General and regime characteristics of stations selected.* UNESCO, Paris.
Wafer, T.A. and Labib, A.H. (1973). Seepage from Lake Nasser. In *Man-made lakes: their problems and environmental effects,* American Geophysical Union, Geophysical Monograph, No. 17, (ed. W.C. Ackerman, G.F. White, E.B. Worthington, and J.L. Young), pp.287–91. American Geophysical Union, Washington, DC.
Waterbury J. (1979). *Hydropolitics of the Nile Valley.* Syracuse University Press, Syracuse.
Whittington, D. and Haynes, K.E. (1985). Nile water for whom? Emerging conflicts in water allocation for agricultural expansion in Egypt and Sudan. In *Agricultural development in the Middle East* (ed. P. Beaumont and K.S. McLachlan), pp.125–49. John Wiley, Chichester.
Worrall, G.A. (1958). Deposition of silt by the irrigation waters of the Nile at Khartoum. *Geographical Journal,* **124**, 219–22.

Further reading

Beaumont, P. (1989). *Environmental management and development in drylands.* Routledge, London.
Beaumont, P., Blake, G.H., and Wagstaff, J.M. (1988). *The Middle East—a geographical study.* David Fulton Publishers, London.
Howell, P.P. and Allen, J.A. (ed.) (1990). *The Nile—resource evaluation, resource management, hydropolitics, and legal issues.* Centre of Near and Middle Eastern Studies, School of Oriental and African Studies, University of London.

3

Natural vegetation

M. N. EL HADIDI

Introduction

Egypt is perhaps the most arid country in North Africa. According to Ayyad and Ghabbour (1986), hot desert ecosystems cover all of Egypt and extend south to latitude 12°N in Sudan. Such hot ecosystems are either arid or hyperarid and it is possible to distinguish between three main hyperarid and two arid provinces (see Fig. 3.1). Hyperarid provinces are characterized by a hot or very hot summer (mean temperature of 30 °C in the hottest month) and a winter rainfall averaging 30 mm per year. Arid provinces are characterized by a hot summer (mean temperature of 20–30 °C in the hottest month) and a winter rainfall ranging between 20 and 100 mm per year.

From a phytogeographical point of view, Egypt is the meeting point of floristic elements belonging to at least three phytogeographical regions: the African Sudano-Zambesian; the Asiatic Irano-Turanian; the Afro-Asiatic Saharo-Sindian and the Euro-Afro-Asiatic Mediterranean. Considering climatic, geomorphological, and phytogeographical variations, El Hadidi (1980a) distinguished between eight main phytogeographical (ecological) territories, two of which have subdivisions. These are shown in Fig. 3.2.

The main features of the vegetation (plant names after Täckholm 1974) can be outlined as follows.

The Mediterranean

The Mediterranean coastal land of Egypt extends for about 970 km between Sallum eastwards to Rafah with an average width ranging between 15 and 20 km in a north-south direction. If the plateau of Sallum near the Libyan frontier is excluded, the remainder of the Mediterranean coastal strip has virtually no conspicuous highlands. According to Zahran et al. (1985), the Mediterranean coastal land can be divided into three sectors: western (the Mareotis, extending for 550 km between Sallum and Alexandria), middle (Deltaic, extending for 180 km between Alexandria and Port Said), and eastern (Sinaitic, extending for 220 km between Port Said and Rafah).

Boulos (1975) estimated a total of 1095 species in the Mediterranean coastal land of Egypt, which constitute about 53 per cent of the native species in Egypt (2085 species: 130 families and 722 genera of angiosperms). Mashaly (1987) recorded 225 species from the Deltaic sector, while Gibaly (1988) recorded 382 species from the Sinaitic sector. At present, no exact estimate can be given for the number of species in the Mareotis sector, which can raise up to 600 species.

According to Mashaly (1987, p.176), 50 per cent of species in the Deltaic sector are from Mediterranean taxa, 32 per cent are from pluri-regional taxa, and 8 per cent are from Saharo-Sindian. The pluri-regional taxa of this sector are mainly common weeds of cultivation which exhibit a very similar spectrum to that known from the weed populations of the farmlands of the Nile Basin (Cosmopolitan taxa 12 per cent, Pantropical taxa 12 per cent, and Palaeotropical taxa 8 per cent).

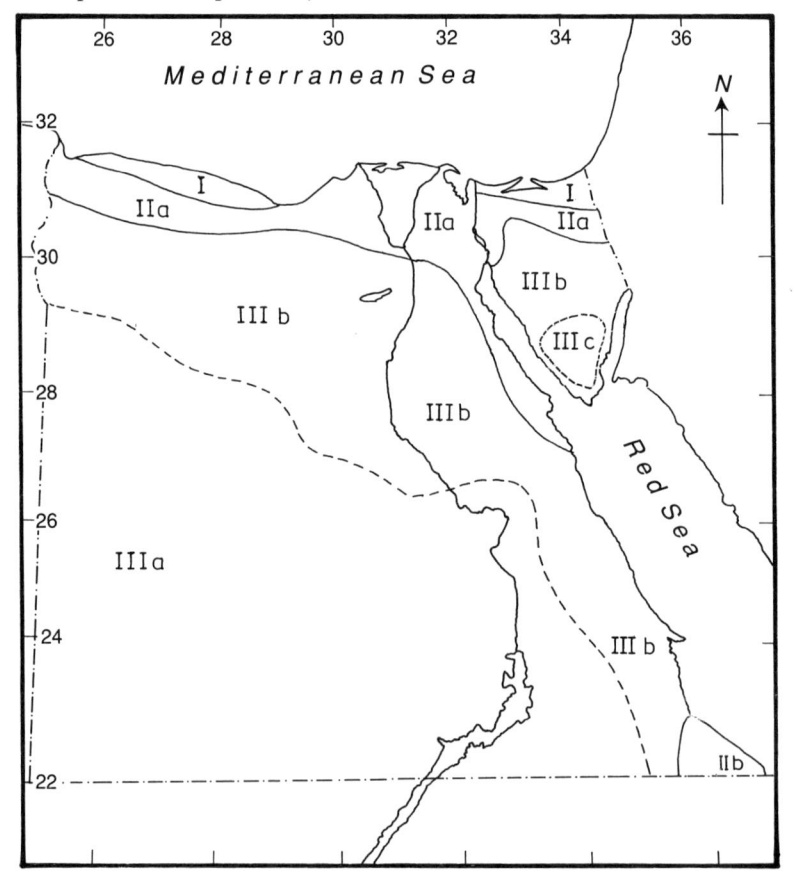

FIG 3.1. Egypt: hot desert ecosystems. I, semi-arid province; II, arid provinces; III, hyperarid provinces

NATURAL VEGETATION 41

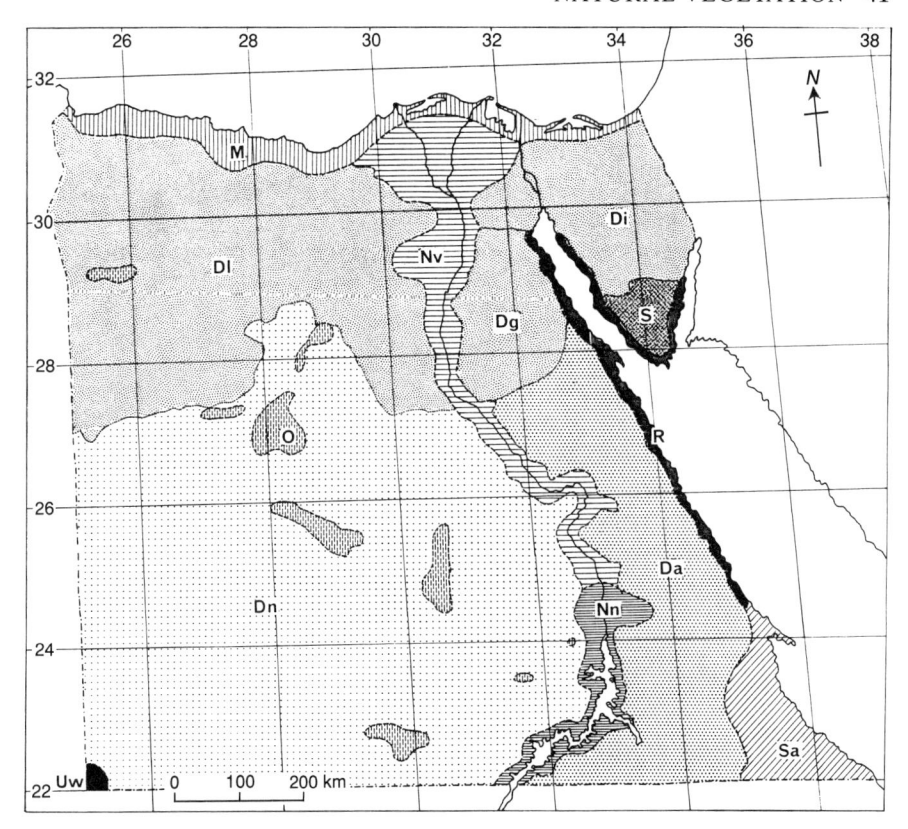

FIG 3.2. Egypt: phytogeographical territories. M, Mediterranean coastal belt. D, Deserts including: Dl, Libyan Desert; Dn, Nubian Desert; Di, Isthmic Desert; Dg, Galala Desert; Da, Arabian Desert. N, Nile land including: Nv, Nile Valley; Nn, Nubian Nile. O, Oases of Dl and Dn, including the Kisseiba-Shabb area of the latter. S, Sinai mountainous region between the Gulfs of Suez and Aqaba. R, Red Sea coastal plains including those of Dg, Da as well as those along the Gulfs of Suez and Aqaba. Sa, Sahelian scrub in Gebel Elba mountainous block, its coastal plains along the Red Sea, and their extension westwards through Da. Uw, Massif of Gebel Uweinat and the intersecting wadis

For the Sinaitic sector, Gibaly (1988) estimated that 45 per cent of species are from Mediterranean taxa, 45 per cent are from Saharo-Sindian taxa, and a lower fraction of 7 per cent are from the pluri-regional taxa including Cosmopolitan, Pantropical, and Palaeotropical species. The similar proportions of 45 per cent for both of the Mediterranean and Saharo-Sindian taxa in the Sinaitic sector can be attributed to its 'bimodal nature' (Gibaly 1988, pp.335–42). A western plain running for 160 km from El Qantara to El Arish is distinct from an eastern plain running for 60 km from El Arish to Rafah. Anthropogenic activities, demographic patterns, human settlements, and agriculture are all better developed in the humid eastern plain than the arid western one. The western plain is characterized by the prevalence of internal desert flora (Isthmic element) while in the eastern plain Mediterranean elements predominate. The cultivated areas of the eastern plain are characterized by a weed assemblage which is similar in its structure to that of the barley fields of the Mareotis sector (Tadros and Atta 1958).

According to Mashaly (1987, pp.180–93), seven habitat types are recognized in the Mediterranean coastal land. A synopsis of his studies are presented in Table 3.1.

In their recent account of the Western Desert, Bornkamm and Kehl (1990,

TABLE 3.1. Habitat types and the leading dominant species in the different sectors of the Mediterranean coastal land of Egypt

Habitat type	Sector			Leading dominant species
	W	D	S	
Coastal dunes	+	+	+	*Pancratium maritimum* *Thymelaea hirsuta* *Moltkiopsis ciliata*
Salt marshes	+	+	+	*Arthrocnemum macrostachyum* *Halocnemon strobilaceum* *Phragmites australis* *Juncus rigidus*
Non-saline depressions (fertile non-cultivated land)	+	+	+	*Anabasis articulata* *Zygophyllum album* *Plantago albicans* *Thymelaea hirsuta* *Atriplex halimus* *Alhagi graecorum*
Rocky ridges	+	–	–	*Gymnocarpos decandrum* *Thymus capitatus* *Globularia arabica* *Asphodelus microcarpus*
Wadis	+	–	+	*Ephedra foeminea* *Lycium europaeum*
Inland siliceous deposits	+	–	+	*Moltkiopsis ciliata* *Artemisia monosperma* *Convolvulus lanatus* *Astenatherum forskalii*
Reed swamps	+	+	–	*Typha domingensis* *Phragmites australis*

W, western, Mareotis sector; D, middle, Deltaic sector; S, eastern, Sinaitic sector pp.157–73) pointed out that, besides the plant communities of the littoral dunes, the other vegetation units recognized in the Mareotis sector (eventually Zone I of their account) belong to the *Atriplex halimus – Lycium europaeum* and *Thymelaea hirsuta–Plantago albicans* associations.

Littoral dunes are dominated by *Ammophila arenaria* and *Euphorbia paralias* on younger dunes, or by *Crucianella maritima* and *Ononis vaginalis* on older dunes.

The *Atriplex–Lycium* association is perhaps the most apparent association along the Mediterranean coastal land. It consists of perennial shrubby species at an average height of 100 cm or more. For comparison, the *Thymelaea – Plantago* association includes vegetation units with an average height of less than 100 cm. The *Atriplex – Lycium* association is found near the Mediterranean coast, while the *Thymelaea – Plantago* association is more widespread southwards.

Zahran et al. (1989) gave an account of the islands in Lake Manzala, one of a chain of five lakes extending along the Mediteranean coast of Egypt. Lake Manzala (31°00′–31°30′N, 31° 50′–32°20′E) is the largest lake in Egypt with an area of about 1400 km^2. It includes about 1000 islands which consist of clay, sand, or mollusc shells. The larger islands range in area between 2.5 and 5.0 km^2. The vegetation of these islands is essentially halophytic and seven community types were recognized. The dominant species were: *Phragmites australis*, *Juncus acutus*, *J. rigidus*, *Arthrocnemum macrostachyum*, *Halimione* (=*Atriplex*) *portulacoide*, *Halocnemon strobilaceum*, and *Zygophyllum aegyptium*.

The Deserts

The deserts represent the major part of Egypt (about 90 per cent of its area or ± 900 000 km^2), with a climatic range from arid close to the coasts of the Mediterranean and the Gulf of Suez (Red Sea), to extremely arid inland—there the rainfall limits plant growth and high temperatures prevail.

The Nile divides the deserts of Egypt into two distinct geomorphological units:

1. An eastern dissected plateau, which can be divided geographically into the Eastern Desert and the Peninsula of Sinai, separated by the Gulf of Suez.

2. A western flat expanse which forms an extension of the Libyan Desert.

The Eastern Desert

The Eastern Desert includes a northern and southern limestone plateau as well as a basement complex (mainly Nubian sandstone) which extends to the Sudano-Egyptian border (22°N). The northern plateau (Galala Desert **Dg** *sensu* El Hadidi 1980*a*) extends along the Gulf of Suez and comprises the heights of Gebel Ataqa (871 m), North Galala (1237 m), and South Galala (1464 m), which are separated by broad wadis. The southern plateau (Arabian Desert partially **Da** *sensu* El Hadidi 1980*a*) extends from Asyut to Qena and includes the peaks of Gebel Abu Dukhan (1705 m), Gebel Qattar (1963 m), and Gebel Shayeb El Banat (2187 m). The basement complex formations (Arabian Desert partially **Da** and Sahelian territory **Sa** *sensu* El Hadidi 1980*a*) to the south of Qena are represented by the Gebel Nugrus group (1200–1500 m), the Gebel Samiuki group (1280–1977 m), and the Gebel Elba group on the Sudano-Egyptian border. The latter is an extensive granite block comprising Gebel Elba (1428 m).

It is possible to distinguish between the following habitat types in the Eastern Desert:

(1) Gravel desert;

(2) Red Sea coastal plains;

(3) Limestone plateaux;

(4) Basement complex;

(5) Wadis.

Gravel desert

The desert country between the Moqattam Hills at Cairo and the Ataqa Mountains near Suez is a good example of this habitat type. Surface deposits consist of transported material which are usually globose gravels or pebbles, essentially siliceous, and are usually poor in salt content. The gravel desert is usually sterile except for lichen growth in certain localities. Its undulated surface allows the accumulation of sand sheets of variable thickness which allow growth of plants particularly after rainy seasons. Ephemeral species include *Centaurea aegyptiaca*, *Stipagrostis plumosa*, and *Fagonia glutinosa*. Deeper soils support the growth of perennials such as *Hamada elegans*, *Zilla spinosa*, and *Panicum turgidum*.

Red Sea coastal plains (Red Sea territory **R** *sensu El Hadidi 1980a).*

The shoreline morphology and climate of the Egyptian Red Sea coast, especially south of Hurghada, favour the growth of mangal vegetation which is fully described by Zahran (1967). *Avicennia marina* usually grows in pure stands, but may be found mixed with *Rhizophora mucronata* as a co-dominant (Kassas and Zahran 1967). Associate marine phanerogams include *Cymodocea ciliata*, *Diplanthera uninervis*, and *Halophila avalis*.

The vegetation of the littoral salt marshes is arranged into zones running parallel to the shoreline. Twelve communities are reported by Ayyad and

Ghabbour (1986, p.162), the most common of which are those dominated by *Halocnemon strobilaceum*, *Limonium pruinosum*, *Zygophyllum album*, *Nitraria retusa*, *Aeluropus lagopoides*, and *Suaeda vermiculata* which is replaced to the south by *S. monoica*.

Desert plains occupy the midland belt between the littoral salt marshes and the ranges of hills and mountains. Kassas and Zahran (1965) described several community types including desert grassland, communities dominated by *Hyparrhenia hirta*, *Lasiurus hirsutus*, *Pennisetum dichotomum* or *Panicum turgidum*, and woody perennial communities dominated by *Artemisia judaica* and *Iphiona mucronata*.

Reed swamps are found at the mouths of big wadis draining to the Red Sea and in the areas where the water from brackish springs combine. The reed swamps are dominated by *Phragmites australis* and *Typha domingensis*. The latter usually inhabits areas where the soil is relatively less saline. *P. australis* grows in soils with higher salt content (Zahran 1966). The common associate species include *Berula erecta*, *Samolus valerandi*, and *Scirpus tuberosus*.

Limestone plateaux

The slopes of the hills and mountain ranges are dissected by shallow runnels which are usually covered by sheets of sand, massive blocks, or boulders. They provide interesting enclaves for rare species of Irano-Turanian elements and which are confined in Egypt to the Galalas and mountainous southern Sinai (Hassan 1987, pp.410–19). Such species include *Tanacetum santolinoides*, *Poa sinaica*, *Silene oliveriana*, *Tulipa biflora*, and *Pistacia khinjuk* var. *glabra*. Hassan (1987, p.434) reports five taxa endemic to these highlands: *Dianthus guessfeldtianus*, *Colchicum guessfeldtianum*, *C. cornigerum*, *Origanum syriacum* var. *sinaicum*, and *Echinops galalensis*.

Basement complex

Communities of *Moringa peregrina* characterize the mountain slopes of the Gebel Nugrus group. In the Gebel Samiuki area in general, the flora is much richer in species composition and in plant cover, and *Moringa peregrina* reaches higher altitudes. The flora of the Gebel Elba group is much richer than that of the Gebel Samiuki group. Three altitudinal zones of the vegetation are recognized on the north and north-eastern slopes of Gebel Elba: a lower zone of *Euphorbia cuneata*; a middle zone of *E. nubica*; and a higher zone of *Acacia etbaica*, *Dodonaea viscosa*, *Dracena ombet*, *Euclea schimperi*, *Ficus salicifolia*, *Pistacia khinjuk*, and *Rhus abyssinica*. Within these higher altitudes ferns, mosses, and liverworts abound. The southern limits are notably drier and plant growth is confined mostly to the runnels of the drainage system where *Commiphora apobalsamum* dominates. At higher altitudes, individuals of *Acacia etbaica* and *Moringa peregrina* may be found (Kassas and Zahran 1971). The vegetation of the north-eastern slopes of Gebel Shindodai (Gebel Elba group) comprises four main zones from base to top: a zone dominated by *Caralluma retrospiciens*; a zone dominated by *Delonix elata*; a zone of *Moringa peregrina*; and a zone with bushes of *Dodonaea viscosa*, *Euclea schimperi*, and *Pistacia khinjuk*.

Wadis

The most pronounced features of the entire Eastern Desert are the wadis which dissect it and drain eastward to the Red Sea or westward to the Nile Valley. Among the notable wadis of this desert are the following.

Wadi Araba, separates the mountainous blocks of North and South Galalas and drains to the Red Sea. The main communities are dominated by *Farsetia aegyptiaca*, *Zilla spinosa*, *Hamada elegans*, *Zygophyllum coccineum*, and *Retama raetem* (Sharaf El Din and Shaltout 1985).

Wadi El Asyuti pours into a deltaic fan that joins the Nile Valley at Asyut. Communities of *Calligonum comosum*, *Leptadenia pyrotechnica*, *Zilla spinosa*, and *Zygophyllum coccineum* are reported from this wadi and its affluent, Wadi Habib (Girgis 1965).

Wadi Qena is unique on account of its north–south direction with several affluents in an east–west direction. Kassas and Girgis (1972) described the main communities dominated by *Crotalaria aegyptiaca*, *Aerva javanica*, *Acacia ehrenbergiana*, *Leptadenia pyrotechnica*, *Tamarix mannifera*, and *T. aphylla*.

Wadi Allaqi, the most extensive drainage system in the Egyptian deserts, flows westward into a deltaic fan composed mainly of Nile alluvium. Five communities were recognized by Sheded (1987): *Acacia braddiana–Aerva javanica*, *Acacia ehrenbergiana–Aerva javanica*, *Tamarix nilotica–Pulicaria crispa*, *Salsola imbricata*, and *Fagonia indica*.

According to Hassan (1987, p. 385), 433 species are recorded from the northern sector of the Eastern Desert extending from the Cairo–Suez road to the Idfu–Mersa Alam road. A total of 103 species were recorded by Sheded (1987) from the southern sector extending from the Idfu–Mersa Alam road to the Allaqi area. Fahmy (1936) reported about 340 species from the Gebel Elba district, and the number was increased to 402 species by Täckholm (1974).

Chorological analysis of the flora of the Eastern Desert is presented in Table 3.2. It is obvious that Saharo-Sindian taxa predominate in the Eastern Desert. Sudano-Zambesian and Palaeotropical elements are apparently well represented while Mediterranean and Irano-Turanian taxa decrease southwards.

The Sinai Peninsula

Apart from a coastal plain dealt with earlier (Sinaitic sector of the Mediterranean region), the Sinai Peninsula has a core near its southern end (Sinai proper) which is an intricate complex of very rugged igneous and metamorphic mountains (Said 1962). The northern two-thirds of the peninsula (the Isthmic Desert) is occupied by a great northward-draining limestone plateau which

TABLE 3.2. Chorological analysis of the flora of the Eastern Desert

Phytogeographical element	Northern (%)	Southern (%)	Elba (%)
Saharo-Sindian	54	48	42
Sudano-Zambesian	13	30	45
Mediterranean	16	8	2
Irano-Turanian	8	1	1
Cosmopolitan	2	3	2
Palaeotropical	1	8	8
Other elements	6	2	–

stretches southwards from the Mediterranean coast to a high escarpment on the northern flanks of the igneous core.

Considering the climatic and geomorphological variations, the following sectors may be distinguished:

The Isthmic Desert (or **Di** *sensu* El Hadidi 1980*a)*

The northern part of the plateau surface is broken by hill masses of Gebel Yillag (1090 m), Gebel Halal (890 m), and Gebel Maghara (735 m). Remnants of coniferous forests made up of *Juniperus phoenicia* are confined to these mountains. The slopes of the hills support the growth of rare species such as *Tulipa polychroma* or endemic species such as *Caralluma sinaica* and *Origanum isthmicum*. Sandy ground is dominated by *Anabasis articulata*, while mobile dunes are dominated by *Stipagrostis scoparia*. *Acacia raddiana* and *Tamarix nilotica* grow in large wadis.

The central areas of the plateau drain to the Mediterranean by numerous affluents of Wadi El Arish. The eastern and western edges are dissected by numerous narrow wadis draining into the Gulfs of Aqaba and Suez. These wadis are dominated by communities of *Tamarix nilotica, Acacia raddiana* or *A. gerrardii*.

The highest positions of the plateau include Gebel El Tih (1400 m) and Gebel El Igma (1600 m). The steep escarpments of the Tih Plateau support associations dominated by *Halogeton alopecroides* or *Anabasis setifera*. Xerohalophytic vegetation dominates the plateau with *Zygophyllum dumosum* and *Salsola tetrandra* as common species. *Pistacia atlantica* trees occur in wadis of Gebel Sahaba where *Loranthus acaciae* is recorded as a parasite on some of the *Pistacia* trees.

The Sinai Massif (Sinai proper **S** *sensu* El Hadidi 1980*a)*

A northern sandstone belt runs almost horizontal between the Gulfs of Suez and Aqaba. At elevations of 900–1200 m, *Varthemia montana* predominates.

Rare endemic species include *Polygala sinaica*, and *Ferula sinaica*. At lower elevations of 600–800 m, communities of *Launaea spinosa* and *Pituranthos triradiatus* grow.

The lower Sinai Massif is a transition area between the hot coastal plains of the Gulfs of Suez and Aqaba and the relatively cool central core (Upper Massif). Thermophilous species including *Acacia raddiana* and *Hamada elegans* do not grow above 1300 m. Higher elevations support the growth of communities dominated by *Anabasis articulata*, *Fagonia boveana*, and *Artemisia herba alba*.

Several oases populated by Bedouins are a prominent feature of this district. The Feiran Oasis is the largest receiving considerable amounts of runoff water from the drainage system of Wadi Feiran. Agriculture has been practised in this area for millenia.

The igneous core (Upper Massif) is formed of mountains which represent the highest peaks in Egypt. Gebel Katherina (2641 m), Gebel Um Shomar (2586 m), and Gebel Serbal (2070 m). The flora of this district is dominated by Irano-Turanian species of which *Artemisia herba alba* and *A. judaica* are the most common, accompanied by *Tanacetum santolinoides* and *Gomphocarpus sinaicus*. Rock vegetation is a prominent feature consisting of trees and shrubs. Important species are: *Crategus sinaicus*, *Ficus pseudosycamorus*, *Cotoneaster orbicularis*, *Rhus tripartita*, *Rhamnus dispermus*, and *Rosa arabica*.

Flowing water, waterfalls, springs, and pools are common throughout the area especially on red granite. Characteristic species growing near the springs and the pools include *Holoschoenus vulgaris*, *Mentha longifolia*, and *Equisetum campestre*. Small dripping springs on cliffs support the growth of dense thickets of mosses, *Adiantum capillus veneris*, *Hypericum sinaicum*, and *Primula boveana*. The latter two species are among the rare and endemic species of this district.

Characteristic of this district are the rock gardens raised and cared for by the monks of St Catherine Monastry and the Bedouins of the Jabaliya tribe. These are fenced areas where alluvium and runoff water accumulate, and many fruit trees including olives, figs, carob, pear, apple, almond, pomegranate, and grapes can be grown. In most of these gardens, vegetables are grown on the lower levels in the vicinity of wells to assist irrigation. Italian cypress and Italian poplar are grown for timber. These rock gardens are characterized by a special weed flora which was described by El Hadidi *et al.* (1970).

Coastal plains of the Gulf of Suez

Most of this area is composed of sandy alluvial fans of the wadis draining from the limestone plateau (for example, Wadi Sudr and Wadi Gharandal) or the igneous massif (for example Wadi Isla). The vegetation of the wadi beds is dominated by *Hamada elegans*, *Anabasis articulata*, *Zygophyllum coccineum*, or *Artemisia judaica*. The arboral components include *Acacia raddiana*, *Tamarix aphylla*, or *T. nilotica*. The salt marshes along the shore are dominated by communities of *Zygophyllum album*, *Nitraria retusa*, *Halocnemon srobilaceum*, and *Juncus rigidus*.

Coastal plains of the Gulf of Aqaba

Compared with the coastal plains of the Gulf of Suez, this district is a narrow, hilly strip of land with a great variety of rock and soil types. Dominant species of the coastal salt marshes include *Zygophyllum album*, *Suaeda vermiculata*, *Nitraria retusa*, and *Limonium axillare*.

The most interesting phenomenon of this district is the presence of mangal vegetation dominated by *Avicennia marina* on the southern shores from Ras Mohamad, northwards to Nabq.

Characteristic for this district are species of Sudanian elements including the annual *Schouwia thebaica*, the shrubby *Leptadenia pyrotechnica* and *Capparis decidua*, as well as trees of *Acacia tortilis* and the forked palm *Hyphaene thebaica*.

Among the notable works on the flora of Sinai are those of Boulos (1960), Danin (1972, 1978), and El Hadidi (1969).

Most outstanding is the work by Danin (1983) on the flora and vegetation of Sinai. He distinguished between eleven geomorphological districts in the peninsula, described its vegetation briefly, and gave a chorophytic analysis of its flora. Three, and possibly four, main phytochoria are represented in Sinai, of which the Saharo-Sindian elements are represented by about 30 per cent of the total number of species. Irano-Turanian and Sudanian elements are represented by about 25 per cent each, while Mediterranean elements contribute about 20 per cent of the total number of the species of the peninsula.

The Western Desert

The northern and major part of the Western Desert (the Libyan Desert **Dl** *sensu* El Hadidi 1980*a*) is a limestone plateau which slopes gradually to the northwest towards the Siwa Oasis and the great Qattara Depression, where the ground descends below sea level. Its southern boundry is a high escarpment which leads to the depressions of the Kharga and Dakhla Oases. In this limestone plateau, the great hollows containing the Farafra and Bahariya Oases are situated. To the northeast of Bahariya Oasis, the plateau rises to form Gebel Qatarain (565 m) overlooking the Nile-fed depression of Faiyum.

The southern part of the Western Desert (the Nubian Desert **Dn** sensu El Hadidi 1980*a*) is a sandstone ground which is delimited northwards by the Kharga-Dakhla depression. The mountainous mass of Gebel Uweinat (1907 m) lies at its south-west boundries.

A prominent feature of the Western Desert is the parallel belts of sand dunes that extend in a north-south direction for hundreds of kilometres. Extensive flat expanses of drifted sand, especially in the south and the west, have gained for the Western Desert the fame of being a sand sea.

There are a few gullies draining from the northern edge of the limestone plateau to the Mediterranean and to the western edge of the Nile, but well-marked drainage systems (wadis) comparable to those of the Eastern Desert are not found.

The climate ranges from arid, near the Mediterranean, to extremely arid further south. The aridity and the extremely high temperatures are the main reasons for the scarcity of vegetation in the Western Desert.

The Libyan Desert

According to Ayyad and Ghabbour (1986, p. 166), the transition between the arid attenuated and the arid accentuated provinces of the Mediterranean Western Desert (Libyan Desert) is characterized by communities dominated by *Anabasis articulata*, *Salsola inermis*, and *Thymelaea hirsuta* near the coastal region. Further south, the communities are dominated by *Artemisia monosperma*, *Convolvulus lanatus*, and *Helianthemum lippii*. Within the northern limit of the arid accentuated provinces, the communities become dominated by *Moltkiopsis ciliata*.

Abd El Ghani (1981) described the desert vegetation along the Giza-Bahariya road which dams one of the runoff collecting channels. Perennial plant growth is exemplified by communities of *Pulicaria crispa* and *Calligonum comosum*. The accidental type of plant growth (Kassas 1966), that appears after a short rainy season includes the therophytes *Monsonia nivea* and *Stipagrostis hirtigluma*.

The vegetation of the non-saline deserts of the Bahariya depression is fairly rich denoting a subsurface moisture which is not remote. Sand mounds and hillocks are dominated by communities of *Calligonum comosum*, *Stipagrostis vulnerans*, and *Tamarix nilotica*.

In a recent study of the vegetation of the Western Desert, Bornkamm and Kehl (1990) distinguished between five desert zones, most of them characteristic of extreme (hyperarid) deserts. These zones show a change along the precipitation gradient, and also exhibit some features caused by geomorphology. Zone I can be classified as semi-desert with an annual precipitation ranging between 100 and 150 mm. It is identical to the Mareotis sector of the Mediterranean coastal land, dealt with earlier in this chapter. Zone II is classified as full desert, while Zones III–V are classified as extreme deserts and numbered 1, 2, and 3, respectively.

The plant associations recognized by Bornkamm and Kehl (1990, p. 218) can be included in the order *Pituranthetalia tortuosi*, which at the present state of knowledge includes the following characteristic species: *Pituranthos tortuosus*, *Helianthemum lippii*, *Astragalus trigonus*, *Salvia aegyptiaca*, *Farsetia aegyptiaca*, and *Stipagrostis plumosa*.

Zone II (full desert) and Zone III (extreme desert 1) *sensu* Bornkamm and Kehl (1990) are identical in their area to that of the Libyan Desert *sensu* El Hadidi (1980*a*). The vegetation in Zone II (29–30°N) is permanently changing from diffuse in the northern half to contracted southwards. The *Pituranthos tortuosus – Gymnocarpos decandrum* association is the main vegetation unit in this zone, and a total of 90 species is recorded. The vegetation in Zone III (28–30°N) is accidental and contracted. It covers about 1 per cent of the total area. The *Stipagrostis plumosa* association, the *Suaeda fructicosa* association, and the *Cornulaca monacantha – Fagonia arabica* association grow in this zone, with a total of fifty-three species recorded.

Bornkamm and Kehl (1990, p.188) described a few stands of 'woodland' in the Western Desert as precipitation-dependent vegetation. In the northern part of the Qattara Depression, the *Zygophyllum coccineum – Acacia raddiana* association was reported as small stands of contracted vegetation. *Acacia raddiana* becomes very rare south of the Qattara Depression.

The Nubian Desert

A preliminary survey of the vegetation of the Nubian part of the Western Desert was carried out by El Hadidi (1980*b*). A total of 21 species of angiosperms are recorded from this area and these are confined mainly to the sand or clay accumulations near the widely scattered wells (*birs*), in the lowest areas of the desert plains and, particularly, after the very rare rain showers.

The vegetation around wells constitutes what may be regarded as the initial stage of an oasis (a small oasis) and this is discussed below.

The vegetation of the desert plains is very scarce and is restricted to the lower areas of large internally drained basins, particularly after short and sudden rain showers which may happen only once in 20–50 years. Among the annual species confined to moistened sand sheets are *Stipagrostis ciliata* and *Astragalus vogelii*. Several desert perennial shrublets are widespread on shallow sand accumulations mixed with pebbles. These include *Fagonia indica* and *Tribulus pentandrus*. In the lowest areas of the internally drained basins, quite thick layers of sand and silt provide a habitat for other desert perennials including *Panicum turgidum*, *Citrullus colocynthis*, *Aerva persica*, and *Crotalaria thebaica*. The highly tolerant desert shrub *Salsola baryosma* is common in many diverse arid habitats.

Zones IV and V (extreme deserts 2 and 3) *sensu* Bornkamm and Kehl (1990) are practically identical to the Nubian Desert *sensu* El Hadidi (1980*a*). The vegetation of Zone IV is strictly contracted and covers much less than 0.1 per cent of the total area of this zone. The majority of the vegetation is purely accidental. Typical vegetation types of this zone are the *Zygophyllum coccineum – Salsola imbricata* association and the *Stipagrostis vulnerans – Zilla spinosa* association. A total of thirty-nine species were recorded from this zone. Zone V is practically void of vegetation. Here two factors act together:

the rainless climate with an annual precipitation of 1 mm and the uniformity of the Selima sand shield.

Monotypic stands of *Acacia ehrenbergiana* were reported by Bornkamm and Kehl (1990, p. 190) to form large hillocks in the area between Bir Safsaf and Bir Kisseiba. El-Hadidi (1980b, p. 350) referred to these stands as *tarabeles*, each of which consists of a huge *Acacia ehrenbergiana* plantation, about 700 years old. These *tarabeles* are probably relics of the savanna vegetation prevailing in the area during the wet episodes of the Nubian Desert, several millennia ago.

Stands of *Capparis decidua* around Bir Kurayim and *Cocculus pendulus* in the northern part of the Farafra Oasis were reported by Bornkamm and Kehl (1990, p. 192). The latter species is more frequent in the Eastern Desert and in Farafra seems to reach the western limit of its distribution in Egypt.

*The Gebel Uweinat area (*Uw *sensu* El Hadidi 1980a)*

Boulos (1980) recorded some eighty species of angiosperms from this area.

The open plains are characterized by a dessicated vegetation of perennials including *Panicum turgidum*, *Zilla spinosa*, *Citrullus colocynthis*, and *Trichodesma africanum* var. *abyssinicum*. Ephemerals, which appear after occasional rains, include *Stipagrostis plumosa*, *Farsetia ramosissima*, and *Anastatica hierochuntica*.

The vegetation in the winding gorges (*kankurs*) depends on groundwater resulting from seepage. The reeds *Typha domingensis* and *Phragmites australis*, and the rush *Juncus rigidus* grow near springs and some annuals including *Eragrostis aegyptiaca*, *Polypogon monspeliensis*, and *Portulaca oleracea* grow on the muddy borders of the ponds.

A perennial layer of vegetation almost without trees is common in the gorges. Most characteristic are *Aerva javanica*, *Cassia italica*, *Pulicaria crispa*, *Crotalaria thebaica*, *Pergularia tomentosa*, *Cleome chrysantha*, and *Fagonia indica*.

Open thorny forests are confined to the *karkur* beds with luxuriant growth of *Acacia raddiana* and *A. ehrenbergiana*, and dense tufts of *Panicum turgidum*. *Maerua crassifolia* is also abundant while *Ficus salicifolia* appears at higher altitudes around 850 m a.s.l.

*The Oases (*O *sensu* El Hadidi (1980a)*

Oases are depressions which receive very little or almost no rain but are green patches within the surrounding sterile desert. Their size varies from a fraction of a square kilometre to hundreds or thousands of square kilometres. Oases owe their greeness to their perennial underground water supply, which appears on the soil surface in the form of springs or is pumped from wells. Some oases are inhabited, their population living in villages that depend mainly on agriculture and agricultural commercial products. Other oases have been abandoned as their supply of fresh underground water has run out. Some oases with a supply of natural resources may support seasonal inhabitants who collect dates and firewood and graze their herds.

Small oases constitute the initial stage of oasis vegetation which follow a characteristic general pattern. One or more palm groves grow near a water source (well or spring) which is clearly marked by the growth of the tall reed *Phragmites australis*. A grass community of *Stipagrostis vulnerans* or *Desmostachya bipinnata* extends over vast areas and is delimited in various directions by low phytogenic sand dunes that support acacias or tamarisks.

Bornkamm (1986) studied the flora of some small oases in the south-east part of the Nubian Desert. Only 14 species were recorded in an area of about 20 000 km^2, and most of these species built up monospecific stands.

At Bir Safsaf (El Hadidi 1980*b*, p. 350), remains of two palm groves (a date palm and a dom palm) grow near a water source. These are surrounded by a pure community of the Safsaf grass (*Phragmites australis*). The *bir* area is encircled by several *Acacia ehrenbergiana* mounds (*tarabeles*), some of which must be 500–700 years old. Other mounds have artefacts and traces of human settlements dating back to the Neolithic period at their bases.

The vegetation around Bir Kisseiba, a stopping point on the Darb el-Arba'in caravan route, clearly shows human influence. Tall palm groves cultivated by man and yielding edible dates grow with dwarf spontaneus groves producing inedible dates. Grasses including *Sorghum sudanense*, *Cynodon dactylon*, and *Imperata cylindrica* denote other human activities.

According to Bornkamm and Kehl (1990, p. 209) the ground-water-dependent plant communities in the smaller, eventually wild, oases are arranged in series according to the depth of seepage. The following sequence was reported: *Typha domingensis*, *Phragmites australis*, *Juncus rigidus*, *Cyperus laevigatus*, *Imperata cylindrica*, *Phoenix dactylifera*, *Alhagi graecorum*, *Sporobolus spicatus*, *Stipagrostis vulnerans*, *Tamarix* sp., *Nitraria retusa*, *Zygophyllum album*, and *Pulicaria* (=*Francoeuria*) *crispa*.

The ancient wells of the Kharga Oasis provide an example of smaller oases with ancient patterns of agriculture. According to Shamloul (1986), hundreds of these ancient wells dating back to Pharonic or Roman times occur in the Kharga town area. Most of these wells have been neglected, either because they have been overwhelmed by drifted sand, or because the artesian pressure could no longer raise water to the surface. About twenty wells still exist around Kharga town, each supporting the cultivation of a limited area of 0.5–5 acres. The method of irrigation follows a common pattern. The well opening is usually located in the middle of an elevated mound, and water flows from the opening through a principal canal of variable length to a storage basin (*mahbas*).

This distributes the water through a system of side canals to the different parts of the farmland. The excess irrigation water is usually drained to a shallow depression (*sabkha*).

The main opening of the well is encircled by thickets of *Phragmites australis* and a few *Acacia* trees which provide shade for the water source. The opening is also marked by a date-palm or a dom-palm grove. The higher levels of farmland are occupied by orchards (olives or oranges) or date-palm groves. The dense shade of these trees provides a habitat for a simple weed assemblage comprising *Trifolium resupinatum*, *Oxalis corniculata*, and *Solanum nigrum*. The lower levels of the farmland are cultivated with winter crops such as wheat or broad beans. During summer, these levels are left fallow or cultivated with melons. The cultivated plots are associated with weed assemblages which are simpler than those known from other farmlands of Egypt.

The Bahariya Oasis was the subject of a study by Abd El Ghani (1981), which provides a model for the larger oases with traditional land use. The oasis is 1800 km^2 in area, with a cultivated area of 2000 feddans, irrigated by about 150 ancient springs and 35 newly dug wells.

Plant life may be distinguished into three principal ecosystems: farmlands, *sabkhas*, and desert, with the habitat conditions ranging from hydrophytic to arid.

Farmland On farmland, agriculture follows the general pattern practised in the Nile Basin. There is a great variety of summer and winter crops, and 12 tree crops. Lucerne (*Medicago sativa*) is the principal fodder crop and is grown as a perennial for 2–3 years. Broad beans and wheat are the most important winter crops, while rice is a summer crop. Dates and olives are the principal tree cash crops.

The weed assemblages associated with field crops are principally the same as those known from the farmlands of the Nile Basin. The floristic composition of the weed communities varies between the oases and with the crop (Abd El Ghani 1985). According to the seasonability of crop production, winter and summer can be discerned (El Hadidi and Kosinová 1971). Kosinová (1975) described the *Astragalo corrugati – Plantaginetum lagopodis* as a weed community from the Dakhla and Kharga Oases. Ruderal habitats are common in waste places and are dominated by species including *Cynodon dactylon* and *Alhagi graecorum*.

The weed assemblages of the palm groves and orchards are simpler in composition with a limited number of species. The ground under the groves is usually covered with a carpet of *Stellaria pallida* or with preponderant growth of *Euphorbia peplus*, *Anagallis arvensis*, or *Oxalis corniculata*.

Reed swamp communities are abundant in shallow water around springs or drainage ditches. These are dominated by one of the following, or a combination of, *Phragmites australis*, *Typha domingensis*, and *Cyperus mundtii*. Sides of irrigation canals are inhabited by species including *Adiantum capillus-veneris*, *Samolus valerandi*, *Apium graveolens*, *Juncus hybridus*, and *Veronica anagallis-aquatica*. Submerged or floating water plants including *Marsilea minuta*, *Ottelia alisimoides*, *Lemna gibba*, or *Zannichellia palustris* are found in deeper water at springs or wells.

Sabkhas Sabkhas are areas which were originally cultivated but which became deserted due to salinization. The principal vegetation of the dry *sabkhas* includes communities dominated by *Tamarix nilotica* or *T. aphylla* (mound or hillock-forming plants), the Halfa grasses *Desmostachya bipinnata* or *Imperata cylindrica*, or halophytic succulents including *Arthrocnemum macrostachyum*, *Suaeda aegyptiaca*, or the rush *Juncus rigidus*. The wet *sabkhas* are characterized by lawn communities dominated by *Juncellus laevigatus* associated with *Eleocharis palustris*, *Carex divisa*, *Juncus subulatus*, *Lythrum hyssopifolia*, and *Ranunculus scleratus*.

Desert The deserts constitute the outskirts of inhabited villages and farmland. Communities of mound and hillock-forming species are most common, dominated by one or a combination of *Tamarix nilotica*, *Stipagrostis vulnerans*, *Calligonum comosum*, and *Sporobolus spicatus*. Associate species include *Alhagi graecorum*, *Prosopis farcta*, *Imperata cylindrica*, and *Calotropis procera*.

Bornkamm and Kehl (1990, p. 212) described the reclaimed desert plains as oases with a new land-use system. They pointed out that, unlike the traditional oases, this system is not bound to any of the geomorphological features. Agriculture is practised anywhere, as long as a water source and adequate soil are available. The floristic composition of the new farmlands depends on the age of cultivation (El Hadidi and Kosinová 1971) and also on the source of the cultivated crops. This can be shown by observations made by Bornkamm (1985) at the experimental farm at East Uweinat (22°27'N, 28°42'E), about 400 km south of Mut in the Dakhla Oasis. Most of the weeds which appeared on the Uweinat experimental farm were presumably introduced from the Dakhla Oasis, the source of most of the crops.

A special feature of this new land-use system is the *Casuarina* plantation which can be seen from a distance; it acts both as a sand binder and a wind-break. Bornkamm and Kehl (1990, p.212) argue that such plantations provide a new type of habitat which was not known earlier. The shady soil, densely covered with *Casuarina* twigs provides optimal conditions where *Imperata cylindrica* copes better.

The Nile Basin

The Nile Basin includes the lands bordering the main stream of the Nile and its Rosetta and Damietta branches. These lands can be divided into a southern Nile Valley (Upper Egypt) and a northern Nile Delta (Lower Egypt). The Nile Valley varies in width from a few hundred metres at Aswan to over 25 km in some parts of Minya or Beni Suef Provinces. The total area of the Nile Valley is about 13 000 km^2. The Nile Delta is about 170 km long and 220 km broad with a total area of 22 000 km^2.

El Hadidi (1980a) divides the lands of the Nile Basin into a southern Nile-Nubian phytogeographical territory (**Nn**) which extends from the Sudanese borders at Wadi Halfa, northwards to Kom Ombo, some 75 km north of Aswan town. The northern Nile Valley territory (**Nv**) extends from Kom Ombo northwards and includes the Nile-fed Faiyum area, and the Nile Delta with its extensions of reclaimed lands in Tahrir Province (west) and Salhiya Province (east).

The Nile Basin has witnessed the development of agriculture over the last 5–6 millennia. As a result, the natural vegetation (riverain) has been subjected to continuous change and it becomes difficult to recognize any of its original features. In Upper Egypt, however, it is possible to trace smaller remnants of what is believed to be the natural vegetation.

The riverain flora of the Nile-Nubian territory was described briefly by El Hadidi (1976). Before the High Dam in 1965, the narrow silty terraces of the Nile were inhabited by a scrub-palm vegetation which gave the flora of this area its conspicuous features. The Nile banks were inhabited by a rich weed flora with Sahelian elements predominating. Lower levels of the terraces were characterized by a scrub vegetation of *Acacia nilotica* or *A. seyal* and *Tamarix nilotica*. Gaps between trees are filled by populations of grasses including *Saccharum spontaneum* or *Desmostachya bipinnata*. Date-palm groves dominated the higher levels forming distinct zones which were occasionally interrupted by clusters of dom palms or *Acacia albida*. At the highest levels were xerophytic species including *Tamarix aphylla*, *Calotropis procera*, and *Leptadenia pyrotechnica*.

Within the farmland of Idfu (130 km north of Aswan), it is possible to recognize places with scrub vegetation, where acacias, tamarisks, or palms grow among a dense cover of Halfa grasses (*Desmostachya* or *Imperata*). Closer to the water courses, sycamore or sidder trees become more important, while silty banks are encircled by thickets of reeds such as *Cyperus alopecroides* or *Phragmites australis*. Among the characteristic weeds of embankments are *Ceruana pratensis* and *Senecio aegyptius* (El Hadidi 1982).

The farmland and the associated irrigation-drainage system are the two main ecosystems of the vegetation in the Nile Basin. Desert systems (including reclaimed land) are important in the desert outskirts of the farmland.

Farmlands

The farmlands support a wide range of habitats which are rather distinct in their floristic, edaphic, and structural criteria. The most important are: cultivated land, waste land, and roadsides.

Cultivated land

These areas have been farmed since ancient times and they are the main productive areas of the country. Field plots usually have weed assemblages

NATURAL VEGETATION 57

which are fairly specific for each crop. Winter weeds, abound in the cooler months of the year and are associated with winter crops. Summer weeds abound in the warmer months of the year and are associated with summer crops. Some summer and winter weeds are biologically active throughout the year, and are usually associated with perennial crops such as fruit crops and palms.

A preliminary survey of the weed flora of the cultivated lands of Egypt was published by El Hadidi and Kosinová (1971). The distributional patterns of the most common weeds (about 100 species) were outlined by Kosinová (1974), who pointed out that the omni-territorial and the riverain patterns were well represented but the extra-riverain species were less common.

The weed alliance of the winter crops of the Nile Basin is dominated by *Melilotus indicus* (Kosinová 1975). Associate species include *Convolvulus arvensis*, *Rumex dentatus*, *Cichorium pumilum*, *Spergularia marina*, *Chenopodium album*, *Eurphorbia peplus*, and *Vicia sativa*.

Common weeds of summer crops (El Hadidi and Kosinová 1971) include *Echinochloa colonum*, *Portulaca oleracea*, *Amaranthus viridis*, *Corchorus olitorius*, *Conyza bonariensis*, and *Sorghum virgatum*.

Weed assemblages of winter and summer crops were studied by El Amry (1981) in Minya Province, El Bakry (1982) in Suez Canal and Qalyubiya Provinces, Mahgoub (1985) in Sharqiya Province, Abd El Ghani (1985) in Faiyum Province, and Shaheen (1987) in Aswan Province. Their detailed studies showed that assemblages are essentially the same for each crop and are related to the season (summer or winter), type of soil (sand or clay), or geographical range (Upper or Lower Egypt).

Héjny and Kosinová (1977) described the main types of habitats and weed communities in irrigated lawns, gardens, parks, and flower beds in the Greater Cairo area. Leading community types include those of *Cynodon dactylon*, *Dactyloctenium aegyptium*, *Paspalum diltatum*, and *Oxalis corniculata*. Precise information on the earliest records of contemporary introductions of some weeds was given.

According to Kosinová (1974), the current weed flora of Egypt has a Mediterranean origin or distribution. Several Mediterranean species occurring in Egypt since prehistoric times are winter weeds playing a significant part in recent weed communities. The tropical species probably of recent introduction and naturalization are confined to the Nile Basin where they are characteristic weeds in the fields of summer crops. Kosinová (1974) estimated a total of 400 species as weeds of cultivated lands. Boulos and El Hadidi (1984) enumerated 163 species as the most abundant weeds in the Nile Basin.

Chorological analysis of the available data of the weed flora in the different provinces of the Nile Basin are presented in Table 3.3. The percentage of Mediterranean taxa is highest (25–33 per cent) in the provinces of the Nile Delta, and this decreases gradually southwards (to a minimum of 13.5 per

TABLE 3.3 Chorological analysis of the weed flora in some provinces of the Nile Basin

Element	Province				
	Aswan (%)	Minya (%)	Giza (%)	Qalyubiya (%)	Sharqiya (%)
Cosmopolitan	14.5	16.0	16.0	17.0	22.0
Palaeotropical	18.5	15.5	17.0	13.0	17.0
Pantropical	14.0	9.0	7.5	6.0	9.5
Mediterranean	13.5	18.0	19.5	25.0	33.0
Saharo-Sindian	12.0	8.5	5.0	4.0	2.5
Sudano-Zambesian	16.0	14.0	8.0	2.5	3.5
Endemic	–	1.2	–	0.8	0.6
Other elements	11.5	17.8	32.0	31.7	11.9
Total number of species	144	158	205	254	248

cent) in Aswan Province. The percentage of tropical taxa (Sahelian and Sudanian *sensu* Wickens 1977) is highest (28 per cent) in Aswan Province, decreasing gradually northwards (to a minimum of 6 per cent) in the Nile Delta. A major percentage (35–50 per cent) of the weed flora is represented by the widely spread Cosmopolitan, Palaeotropical, and Pantropical taxa.

Waste Land

Héjny and Kosinová (1977) distinguished between countryside and urban wastelands. The first are cultivated plots which have been abandoned due to salizination or inadequate drainage. Plant communities are dominated by *Alhagi graecorum* and associate species include *Cynodon dactylon, Convolvulus arvensis*, and *Conyza bonariensis*.

Urban wasteland includes sites of neglected gardens, demolished houses, and yards of factories, where soils are covered with organic and inorganic refuse. Communities dominated by *Desmostachya bipinnata* are characteristic of the gravelly – sandy substrate of these wastelands, and associate species include *Phragmites australis* subsp. *stenophyllus* and *Chrozophora plicata*. Soils with an organic substrate favour the growth of communities dominated by *Withania somnifera*, commonly associated with *Datura innoxia, Solanum nigrum*, and *Cynodon dactylon*.

Roadsides

The roadside habitat favours the growth of rhizomatous species which help in binding soils and protecting them against wind erosion. Leading species include

Imperata cylindrica, *Desmostachya bipinnata*, and *Phragmites australis*. Associated species include *Cynodon dactylon*, *Polygonum equisetiforme*, *Chenopodium murale*, and *Kochia indica*.

The irrigation – drainage system

The farmland of the Nile Basin is dissected by a complicated network of irrigation canals and drains. Major irrigation canals originate from the Nile or its Damietta and Rosetta branches. They divide repeatedly to bring fresh water to every plot of farmland. Smaller drains collect excess irrigation water from farm plots and bring it to larger drains which release their water into depressions such as Wadi Rayan in the Western Desert, or into the brackish lakes of the northern Delta.

In areas of fresh (irrigation system), brackish, or saline (drainage system) water, communities of phanerogamic water plants can cause serious blockages.

Most of the water plants of the Nile basin are from Cosmopolitan or Palaeotropical taxa which can grow in fresh or brackish water. Some species such as *Nymphaea lotus* and *Nymphaea coerulea* are confined to drains, where slow water currents provide the favourable habitat for their growth. Others with local distribution include *Vallisneria spiralis* confined to Aswan Province, *Potamogeton trichoides* in Asyut and Minya Provinces and *P. perfoliatus* at Sadd El Rawafaa in Sinai as well as the Aswan area. The introduction of these species in such places can be attributed to migratory birds (El Hadidi 1965, 1968). *Potamogeton schweinfurthii* and *Najas armata* are tropical species which are associated with the waters south of the Aswan Dams; these act as natural barriers for its spread northwards. The establishment of the Aswan High Dam in 1960–5 led to changes in environmental conditions which favoured the vigorous growth of some rare species such as *Myriophyllum spicatum* and *Eichhornia crassipes* which are becoming serious problems for irrigation and navigation.

The muddy banks of the Nile, irrigation canals and drains support the growth of populations of hygrophyllic species including the tall reeds *Cyperus alopecroides*, *Typha domingensis*, *Saccharum spontaneum*, and *Phragmites australis*. Shrubby species include *Pluchea dioscorides*, *Ambrosia maritima*, and *Sesbania sesban*. Herbaceous species include, *Mentha longifolia* subsp. *typhoides*, *Cotula anthemoides*, *Ageratum conyzoides*, *Coronopus niloticus*, and *Gnaphalium luteo-album*. *Ceruana pratensis* used to be rather abundant especially in Upper Egypt. After the construction of the Aswan High Dam, less silt was deposited in the Nile Basin north of the dam. This created unfavourable plant habitat and led to the disappearance of most of the populations. *Aster squamatus* was recently introduced to Egypt, is now completely naturalized, and is one of the most widespread species in the country.

Reclaimed deserts

Alluvial terraces of the Nile and the desert outskirts of farmland are the subjects of active land reclamation. In these newly reclaimed farmlands, alfalfa is one of the early crops, adding nitrogen to the otherwise very poor soil, and providing cover and green manure. Other conventional crops include tomatoes, potatoes, sweet potato, and groundnuts.

The weed populations of the newly farmed land include a number of desert plants that seem to flourish. Older fields seem to contain fewer desert populations, and the weeds characteristic of old farmland are established in place of desert species.

The system of irrigation influences the weed population. Drip irrigation provides limited scope for weed growth because the moisture is confined to the surroundings of the saplings. Sprinkers and flood irrigation which provide moisture to the large areas of the farmland, allow profuse weed growth.

References

Abd El Ghani, M.M. (1981). Preliminary studies on the vegetation of Bahariya Oasis, Egypt. Unpublished M.Sc. thesis. Faculty of Science, Cairo University.

Abd El Ghani, M.M. (1985). Comparative study on the vegetation of the Bahariya and Farafra Oases and the Faiyum region. Unpublished Ph.D. thesis. Faculty of Science, Cairo University.

Ayyad, A.A. and Ghabbour, S.I. (1986). Hot deserts of Egypt and the Sudan. In *Ecosystems of the world,* Volume 12B, *Hot deserts and arid shrublands, B.* (ed. Evenari *et al.*), pp. 149–205. Elsevier, Amsterdam.

Bornkamm, R. (1985). Beobachtungen uber die Vegetation einer Versuchsfarm in Sud-aegypten. *Tuexenia*, **5**, 81–7.

Bornkamm, R. (1986). Flora and vegetation of some small oases in South Egypt. *Phytocoenologia*, **14**, 275–84.

Bornkamm, R. and Kehl, H. (1990). The plant communities of the Western Desert of Egypt. *Phytocoenologia*, **19**, 149–231.

Boulos, L. (1960). *Flora of Gebel El Maghara, North Sinai.* Special Publication, Herbarium, Ministry of Agriculture, Cairo.

Boulos, L. (1975). The Mediterranean element in the flora of Egypt and Libya. In *La flore du Basin Mediterraneen, Colleques Internationaux du CNRS*, No. 235, pp. 119–24. Centre National de la Recherche Scientifique, Paris.

Boulos, L. (1980). Journey to the Gilf Kebir and Oweinat, south-west Egypt, 1978. IV. Botanical results of the expedition. *Geographical Journal*, **146**, 68–71.

Boulos, L. and El Hadidi, M.N. (1984). *The weed flora of Egypt.* American University in Cairo Press, Cairo.

Danin, A. (1972). Mediterranean elements in rocks of the Negev and Sinai deserts. *Notes from the Royal Botanical Garden, Edinburgh*, **31**, 437–40.

Danin, A. (1978). Plant species diversity and ecological districts of the Sinai desert. *Vegetatio*, **36**, 83–93.

Danin, A. (1983). *Desert vegetation of Israel and Sinai.* Cana Publishing House, Jerusalem.

El Amry, M.I.A. (1981). Plant life in Minya Province, Egypt. Unpublished M.Sc. thesis. Faculty of Science, Cairo University.
El Bakry, A.A. (1982). Studies on plant life in the Cairo—Ismailia region. Unpublished M.Sc. thesis. Faculty of Science, Cairo University.
El Hadidi, M.N. (1965). *Potamogeton trichoides* Cham. & Schlecht. in Egypt. *Candollea*, **20**, 159–65.
El Hadidi, M.N. (1968). *Vallisneria spiralis* L. in Egypt. *Candollea*, **23**, 51–8.
El Hadidi, M.N. (1969). Observations on the flora of Sinai mountain region. *Bulletin de la Société Géographie d'Egypte*, **40**, 123–55.
El Hadidi, M.N. (1976). The riverain flora in Nubia. In *The Nile, biology of an ancient river*, (ed. J. Rsoska), pp. 87–91. Junk, The Hague.
El Hadidi, M.N. (1980*a*). An outline of the planned flora of Egypt. *Taeckholmia additional series*, **1**, 1–12.
El Hadidi, M.N. (1980*b*). Vegetation of the Nubian Desert. In *Prehistory of the eastern Sahara*, (ed. F. Wendorf and R. Schild), pp. 345–51. Academic Press, New York.
El Hadidi, M.N. (1982). The predynastic flora of Hierakonpolis region. In *The predynastic of Hierakonpolis—an interim report*, Egyptian Studies Association 1, (ed. M.A. Hoffman), pp. 102–15, Cairo University Herbarium, Giza and Western Illinois University, Macomb.
El Hadidi, M.N. and Kosinová, J. (1971). Studies on the weed flora of cultivated land in Egypt. 1. Preliminary survey. *Mitteilungen Botanische Staatssammlung, Muenchen*, **10**, 354–67.
El Hadidi, M.N. Kosinová J., and Chrtek, J. (1970). Weed flora of southern Sinai. *Acta Universitatis Carolinae: Biologica*, **1969**, 367–81.
Fahmy, I.R. (1936). *Report on Gebel Elba*. Egyptian University, Cairo.
Gibaly, M.A.A. (1988). Studies on the flora of the Northern Sinai. Unpublished M.Sc. thesis. Faculty of Science, Cairo University.
Girgis, W.A. (1965). Studies on the plant ecology of the Eastern Desert (Egypt). Unpublished Ph.D. thesis. Faculty of Science, Cairo University.
Hassan, L.M. (1987). Studies on the flora of the Eastern Desert, Egypt. Unpublished Ph.D. thesis. Faculty of Science, Cairo University.
Héjny, S. and Kosinová, J. (1977). Contributions to synanthropic vegetation of Cairo. *Publication of the Cairo University Herbarium*, **7–8**, pp. 273–86.
Kassas, M. (1966). Plant life in deserts. In *Arid lands*, (ed. E.S. Hills), pp. 145–80. Methuen, London and UNESCO, Paris.
Kassas, M. and Girgis, W. A. (1972). Studies on the ecology of the Eastern Desert of Egypt. I. The region between latitude 27°30' and latitude 25°30'. *Bulletin de la Société Géographie d'Égypte*, **42**, 42–72.
Kassas, M. and Zahran, M.A. (1965). Studies on the ecology of the Red Sea coastal land. II. The district from El Galala El Qibliya to Hurghada. *Bulletin de la Société Géographie d'Égypte*, **38**, 155–73.
Kassas, M. and Zahran, M.A. (1967). On the ecology of the Red Sea littoral salt marshes, Egypt. *Ecological Monographs*, **37**, 297–316.
Kassas, M. and Zahran, M.A. (1971). Plant life on the coastal mountains of the Red Sea, Egypt. *Journal of the Indian Botanical Society*, **50A**, 571–89.
Kosinová, J. (1974). Studies on the weed flora of cultivated land in Egypt. 3. Distributional types. *Botanische Jahrbuecher fuer Systematik, Pflanzengeschichte and Pflanzengeographie*, **94**, 449–58.
Kosinová, J. (1975). Weed communities of winter crops in Egypt. *Preslia*, **47**, 58–74.

Mahgoub, A.M.A. (1985). Study on plant life in farmlands of the Isthmic region. Unpublished M.Sc. thesis. Faculty of Science, Cairo University.
Mashaly, I.A.I. (1987). Ecological and floristic studies of the Dakahlia-Damietta region. Unpublished Ph.D. thesis. Faculty of Science, Mansoura University.
Said, R. (1962). The geology of Egypt. Elsevier, Amsterdam.
Shaheen, A.M. (1987). Studies on the weed flora of the Aswan area. Unpublished M.Sc. thesis. Faculty of Science (Aswan), Assiut University.
Shamloul, A.M. (1986). Plant life around the ancient wells in the Kharga Oasis. Unpublished M.Sc. thesis. Faculty of Science (Sohag), Assiut University.
Sharaf El Din, A. and Shaltout, K.H. (1985). On the phytosociology of Wadi Araba in the Eastern Desert of Egypt. *Proceedings of the Egyptian Botanical Society*, **4**, 1311–25.
Sheded, M.G. (1987). Studies on the vegetation of Eastern Desert. Unpublished M.Sc. thesis. Faculty of Science (Aswan), Assiut University.
Täckholm, V. (1974). Students' flora of Egypt (2nd Edn), Cairo University, Cairo.
Tadros, T.M. and Atta, B.A.M. (1958). The plant communities of barley fields and uncultivated desert areas of Mareotis (Egypt). *Vegetatio*, **8**, 161–75.
Wickens, G.E. (1977). *The flora of Jebel Marra (Sudan Republic) and its geographical affinities. Kew Bulletin Additional Series V*. HMSO, London.
Zahran, M.A. (1966). Ecological study of Wadi Dunkul. *Bulletin de l'Institut du Désert d'Égypte*, **16**, 127–43.
Zahran, M.A., (1967). Distribution of the mangrove vegetation in U.A.R. (Egypt). *Bulletin de l'Institut du Désert d'Égypte*, **15**, 7–9.
Zahran, M.A., El Demerdash, M.A. and Mashaly, I. (1985). On the ecology of the Deltaic coast of the Mediterranean Sea, Egypt. I. General survey. *Proceedings of the Botanical Society of Egypt*, **4**, 1392–1402.
Zahran, M. A., Abu Ziada, M. E., El Demerdash, M. A., and Khedr, A. A. (1989). A note on the vegetation on islands in Lake Manzala, Egypt. *Vegetatio*, **85**, 83–8.

Further reading

El Hadidi, M.N., Abd El Ghani, M.M., and Fahmy, A.G. (1992). *The plant red data book of Egypt. Volume I. Woody perennials.* Cairo University Herbarium, Cairo.
Zahran, M.A. and Willis, A.J. (1992). *The vegetation of Egypt.* Chapman and Hall, London.

4

Domesticated livestock and indigenous animals in Egyptian agriculture

M. B. ABOUL-ELA

Introduction

With the intensive agricultural system applied in Egypt, cultivable land is utilized all the year round. With a rapidly growing human population, which had increased to about 57 million in 1992, domesticated livestock species have special importance as the main source of animal protein. A wide range of indigenous animal species is also present in Egypt. Representatives of most of the known classes of insect pests have significant relevance to crop productivity, while vertebrate pests such as birds and rodents have always been regarded as a serious threat to crop production. The sea areas bordering Egypt, the River Nile and its tributaries, Lake Nasser, and the Northern littoral lakes all provide a variety of saline and fresh water habitats for a range of fish species.

In this chapter the emphasis is on the characteristics of the different types of domesticated livestock, but some information on fish and on insect and other animal pests is included.

The domesticated livestock population

Table 4.1 illustrates the numbers of different species of farm animals in 1970 and 1988. Over that period there was an upward trend in the population of most species, with the exception of horses which showed a decline of 65 per cent. The rate of increase was highest for rabbits, goats, and poultry species. The density of livestock units per 1000 head of population in 1988 (calculations based on the local cow as the standard animal unit) was 121 units for large ruminants (buffalo and cattle), 22 units for small ruminants (sheep and goats), and 7.8 units for the small-size species (poultry and rabbits). On this basis, the overall density was 150.8 units of which buffalo and cattle accounted for 80.2 per cent. This illustrates the importance of large ruminants which have always played an important part in Egyptian rural life, within the irrigated

TABLE 4.1. Egypt: animal population, 1970 and 1988

Animal	1969/70 ('000 head)	1988 ('000 head)
Cattle	2115	2717
Buffaloes	2009	2460
Sheep	2006	2908
Goats	1155	2697
Camels	127	148
Pigs	15	28
Horses	35	13
Donkeys	1362	1813
Poultry		
Chicken	24541	33515
Ducks	3002	5205
Geese	2493	4895
Turkey	652	1307
Pigeons	2048	2505
Rabbits	2095	6231

Source: CAPMAS (1990).

lands of the Nile Valley and the Nile Delta. More than 90 per cent of the buffalo and cattle population is scattered over about one million farms, each farmer owning between one and three animals.

The animal production system in the irrigated lands of Egypt is mainly a smallholder production system. The importance of this characteristic of smallholder ownership will continue because each small farmer, regardless of the size of his holding, wishes to own a cow and/or buffalo for reasons of family nutrition, extra cash return, and prestige, as well as for draft use. In the desert and rain-fed areas of Egypt (the north-west coastal zone, Sinai, and the New Valley) sheep and goats are dominant and their population has increased markedly in the last decade due to expansion of lamb and kid production systems for exportation to the Gulf countries.

Buffalo

Buffalo are the most important livestock species in Egypt. They provide about 70 per cent of the total milk production in the country and contribute considerably to the red meat production. The Egyptian buffalo are river-type water buffalo (*Buffalo bubalus*). It is claimed that buffalo were introduced into Egypt after the Islamic ruling. About two-thirds of the buffalo population are in the Delta and the rest are in middle and upper Egypt. Wahby (1937) divided Egyptian buffalo into the Beheri of lower Egypt and Saidi of upper Egypt. The Beheri is a large, slate grey, animal with a smooth skin and it is claimed to be a better milker. The Saidi is smaller, almost black, more hairy, and a poor milker. Khishin (1951) and Zaki (1951) have attempted a further classification of the buffalo in lower Egypt into Beheri and Menoufi, but such distinctions are small and are difficult to maintain. In general, Egyptian buffalo are blackish grey in colour. Their horns vary from lyre-to sword-shaped, they are relatively short compared with other buffalo, curve backwards alongside the head, and bend upwards at the tip (Fig. 4.1).

Productive characteristics

The birth weight of a buffalo calf is about 40 kg (Ahmed and Tantawy 1954; Ragab and Abdel-Salam 1963). The weight of adult animals varies from 500 to 600 kg (Ahmed and Tantawy 1954). In research studies, the average milk yield was 1300–1800 kg over a lactation period of about 380 days (Mostageer *et al.* 1981; Mourad 1978), although higher levels were recorded in a small-scale study of small farmers' holding (Nigm *et al.* 1986). Buffalo milk, with its high fat percentage (7.5 per cent) is generally preferred by consumers throughout the country. The milk yield from the first lactation is usually much lower than that from subsequent lactations, and the lactation periods are shorter than for cattle. It is claimed that Egyptian buffalo have poor reproductive performance and this hinders their productivity throughout their productive life. Most of the data is obtained from herds kept at research stations where management is quite different from that of small farmers. However, wide variations are found in the published reports on the reproductive performance of Egyptian buffalo. Age at puberty varies from 12 to 22 months and age at first calving varies from 27 to 40 months. In well managed herds Mohamed *et al.* (1980) obtained a figure of 10 months for age at puberty and 27 months for age at first calving. Gestation length averages 318 days and calving interval has often been reported to be over 500 days. In the few studies made on animals belonging to small farmers, the calving interval was reported to be as low as 416 days (Aboul-Ela 1988; Nigm *et al.* 1986). Several authors have reported on the seasonality of reproduction in the buffalo, with lower reproductive activity during the summer season (Aboul-Ela 1988; Mostageer *et al.* 1981). Recent reports from the Ministry of Agriculture indicate that the small number of buffalo bulls in the country may contribute to the problem of poor fertility in the buffalo. About 40 per cent of the villages screened in a recent survey (Aboul-Ela 1988) had no buffalo bulls and females had to be taken to nearby villages to be served.

A large proportion of the male buffalo calves used to be slaughtered as veal calves at an early age. Recently, the Egyptian Government has encouraged farmers to fatten their buffalo calves before slaughter, keeping them until they reach a weight of about 450 kg at 18–20 months. The dressing percentage of slaughtered calves averaged 56 per cent and the growth rate of buffalo calves averaged 700–800 g day^{-1} during such fattening periods (El-Ashry et al. 1972; El-Naggar et al. 1972).

Cattle

Native cattle show large variations in their phenotypic characteristics. Their colour ranges from sandy yellow to dark brown or reddish brown, sometimes spotted with white or other coloured spots. Some have small horns of different shapes. The body is relatively high for its weight (Fig. 4.2). The Egyptian cattle have been arbitrarily classified into Damietta, Baladi (Menoufi), and Saidi. The Damietta cow is a relatively high yielder while the Saidi is the smallest in size and has a low milk yield. It is difficult, however, to distinguish between these different types and therefore native cattle are referred to, in general, as Baladi cattle.

For thousands of years, the Egyptian farmer has used native cattle in draft work and he will continue to use them until they are replaced by mechanization, a process which is taking place slowly but is being encouraged by the Egyptian Government.

Productive characteristics

The native Baladi cow is known to be a poor milker and the milk produced is often only enough to feed the calves. The average milk yield is 400–700 kg over a lactation period of about 230 days. Butter fat content is on average 4.3 per cent. Age at puberty is about 34 months and the calving interval is on average 410 days with a gestation period of 289 days (Ahmed and Tantawy 1959). One of the advantages of the native cattle is their good fecundity, compared with that of buffalo. The farmer relies on his cow to produce one calf annually. The birth weight of native calves averages 23–28 kg, while body weights at 6 and 12 months of age are about 100 and 200 kg, respectively. Higher weights have been reported in other studies (Mostageer 1982). Mature body weights are 450 and 550 kg for cows and bulls, respectively. Male calves are a main source of red meat and they are often fattened up to the age of 18–24 months when they are slaughtered at a weight of about 350–400 kg (El-Ashry et al. 1972). Dressing percentage is on average 55 per cent. In recent years, there have been changes in the cattle industry in Egypt. The draft contribution by the Baladi cow in Egyptian agriculture has begun to decline with the national mechanization programme, but cattle are still very important to the smallholder for family nutrition and prestige (Table 4.1). Because of their low level of milk production, mature local cattle are being crossed with other exotic breeds, mainly Friesian, of which large numbers of live animals have been imported over the last 15 years. Although these were raised mainly at specialized dairy units, the males produced were used for crossing with Baladi cattle. Recent Ministry of Agriculture cattle development programmes rely mainly on artificial insemination using Friesian semen. In some central areas

DOMESTICATED LIVESTOCK AND INDIGENOUS ANIMALS

Fig 4.1. Buffalo

Fig 4.2. Baladi cow

Fig 4.3. Ossimi sheep

of the Delta, Gharbiya, and Kafr el-Sheik governorates large numbers of crossbred cattle owned by small farmers are already replacing Baladi cows.

Sheep

All Egyptian sheep are fat-tailed coarse wool breeds. Illustrations of livestock from ancient Egypt suggest that the only sheep in the country until about 2000 BC were the thin-tailed, 'hairy' type. Descriptions from the twelfth Dynasty show fat-tailed sheep, which seemed to have entered the country at that time.

About 43 per cent of the sheep population are in middle and upper Egypt, 27 per cent are in the Nile Delta, and the rest are in the desert area mainly in the north-west coastal zone. Rahmani and Ossimi breeds are dominant in the Delta and middle Egypt while Barki is the only breed raised in the desert area. The Saidi and Ibeidi are raised in relatively smaller numbers in upper Egypt. A brief description of these breeds is given here, but detailed descriptions of the breeds are given by Ghanem (1980) and Mason (1967).

The Ossimi is the most numerous widespread breed in Egypt. It has a narrow shallow body on long spindly legs (Fig. 4.3). It has a white body with a reddish-brown head and semi-pendant ears. Rams have medium-sized horns and ewes are polled. The tail is heavy, fat, and twisted, ending abruptly in a thin end piece.

Rahmani is the dominant breed in the northern part of the Delta but is also found in the mid Delta area. It is brown in colour, occasionally with white spots on the head. The external ears are often absent. Rams have long spiral horns curved downwards and ewes are usually polled. The tail is the largest among Egyptian breeds, with an S-shaped end (Fig. 4.4).

Barki is the smallest Egyptian breed, well adapted to the desert environment. It has a shallow body, narrow back, and long legs. It is white in colour and the head is usually brown or black. Ears are semi-pendulous and rams have big horns twisted backwards. The fat-tail has a triangular shape with a twisted end (Fig. 4.5).

Saidi sheep are found only in upper Egypt, in relatively small numbers. They are either black or brown in colour, with long and cylindrical fat tails which often reach to the ground. Both rams and ewes have Roman noses and lack horns.

Productive characteristics

A summary of the productive characteristics of the three main Egyptian sheep breeds is given in Table 4.2.

One of the important characteristics of Egyptian sheep breeds is that they possess the ability to breed throughout the year (Aboul-Ela and Aboul-Naga 1987). When an accelerated lambing system is applied with three mating seasons every two years (September, May, and January) the highest reproductive performance is from the September mating (Aboul-Naga et al. 1987).

Egyptian sheep breeds are generally of good fertility but their prolificacy is relatively low and this deserves attention in any development programme for these sheep. Growth performance of lambs is generally low (growth rate from 4 to 12 months averages 70–100 g day^{-1}) although in some fattening trials much higher rates (170–200 g day^{-1}) were achieved.

Wool production from local sheep is of secondary importance while milk production is insignificant in the sheep industry. The coarse wool is used mainly for carpets, and sheep are usually sheared twice a year, in March and September. Annual greasy fleece weights range from 1.4 to 2.8 kg for different local breeds. Barki sheep have higher greasy fleece weights due to low shrinkage. Fibre length is 10–16 cm, fibre diameter 30–37 μm, and kemp content 2–8 per cent (Guirgis 1973). Milk produced from local sheep is used mainly for lamb suckling, and milk yield ranges from 30 to 65 kg over a lactation period of 12–16 weeks (Aboul-Naga et al. 1981).

Goats

Egyptian goats are hairy and medium-sized. There are three breeds: the desert type, Barki; the valley breed, Baladi; and the north Delta breed, the Egyptian Nubian.

The Barki goats are relatively small, have long black hair, medium length horns of sickle or scimitar shape, and long slender ears carried laterally. The Baladi goats are dominant within the Nile Valley and the Delta. They do not have distinct characteristics as they vary a lot in their colour and phenotypic appearance which makes it hard to describe the breed precisely. The Egyptian Nubian goat, often called the Zaraibi, is believed to be one of the progenitors

TABLE 4.2. Productive characteristics of the main Egyptian sheep breeds

Characteristic	Rahmani	Ossimi	Barki
Birth weight (kg)	2.7–3.1	2.95–3.6	2.6–3.4
Weaning weight (kg)	18–21	18–20	17–21
Yearling weight (kg)	39–42	32–37	30–35
Mature weight (kg)			
Rams	55–70	55–65	50–65
Ewes	45–55	45–50	35–50
Age at first lambing (month)	15–22	15–22	18–25
Conception rate (%)	86	83	88
Number of lambs	1.23	1.14	1.05

Source: Aboul-Naga (1976); Aboul-Naga and Aboul-Ela (1987); Aboul-Naga et al. (1987); Hafez (1953); Younis et al. (1984).

FIG 4.4. Rahmani sheep

FIG 4.5. Barki sheep

FIG 4.6. Zaraibi goat

TABLE 4.3. Productive characteristics of the main Egyptian goat breeds

Characteristic	Baladi	Barki	Egyptian-Nubian (Zaraibi)
Birth weight (kg)	2.1	1.9	2.0–2.4
Mature weight (kg)			
Males	40	35	42
Females	32	28	35
Conception rate (%)	86	88	90
Number of kids	1.8–1.9	1.5	1.9–2.1
Milk yield (kg)	50–75	30–50	80–150

Source: Aboul-Ela *et al.* (1988); Aboul-Naga *et al.* (1981, 1987); Heider (1982); Kandil *et al.* (1984).

of the Anglo-Nubian. It is the distinct local breed with its convex facial profile, extremely Roman nose, and long head with lapped ears (Fig. 4.6). Although its name indicates that it has originated from El-Nubah south of Aswan, it could only be found in small numbers in the north-eastern part of the Delta.

Productive characteristics

The results included in Table 4.3 indicate that Egyptian goat breeds are generally highly fertile. They have the ability to produce more than one kid crop per year (Aboul-Naga *et al.* 1987). They generally have good prolificacy, particularly the Zaraibi (Aboul-Ela *et al.* 1988).

In the Delta and Nile Valley, goats are raised as dual purpose animals for both kid production and as a household dairy animal. Kid growth performance is rather low for all three local breeds. Milk production is very low for both the Baladi and Barki goats but the Egyptian Nubian goat has potential for a high milk yield.

In the desert areas, Barki goats are considered the main source of milk for the human population. With their low level of milk production, the Ministry of Agriculture has begun a development programme in the north-west coastal zone to improve goat productivity by cross-breeding with the Damascus goat.

It is important to note that in recent years in Egypt more attention has been given to goats which were looked upon as scavengers in the past. This attention is reflected in the increase in goat population in Egypt.

Camels

Egypt has a low density of camels. Available information (Table 4.1) indicates a stability in the camel population over the last two decades. It is worth noting, however, that most of the camel meat consumed in Egypt is imported, primarily from the Sudan. Camels are nevertheless of some importance in two different areas: the Nile Delta and Valley where they are used mainly as draft animals, and in the desert. The so-called delta camel has various origins. Epstein (1971) mentioned three breeds of camels within the irrigated lands of Egypt; Sudani, Maghrabi, and Fellahi. Wilson (1984) recognized these as several types of one breed which originated from a mixture of different breeds. The delta camel is a general purpose baggage-type animal which has become used to green fodder and plenty of water. Most of the camels raised in the desert areas of Egypt are the same as the Maghrabi of Libya which is of medium size, but larger than the Sudanese riding camel. A large number of camels (estimated as 40 000 annually) are imported from the Sudan for slaughter. They usually make the 40-day trip from Darfur in Sudan to Asyut in upper Egypt via the notorious Darb el-Arba'in during which high losses in animal weight and deaths occur.

Equine

The number of horses raised in Egypt is very small, and the number has declined significantly during the last decade (Table 4.1). However, large numbers of donkeys are kept. Donkeys have always played an important role in the Egyptian farmer's life. They provide his main means of transport and are often used for carrying loads along the narrow village roads to the small cultivated area. Some politicians have called for a cut in the number of donkeys raised in Egypt and consequently a saving in the amount of feed consumed in a country where the shortage of feed resources is getting worse. However, the role of donkeys in the small farmer's life will remain in the forseeable future.

Poultry

Poultry products contribute significantly to the animal protein supplies in Egypt, mostly in the form of eggs and broiler meat from chickens but with a smaller contribution made by geese, ducks, and turkey.

There has been a tremendous increase in both broiler and egg commercial production units in Egypt since the late 1970s. However, rural production still plays a major part in the national poultry production. Most of the poultry breeds raised in rural areas are native breeds while those raised in commercial units are specialized lines whose parent stocks have been imported into the country.

Egyptian chicken breeds are the same as the Mediterranean breeds. Fayoumi is the most important breed. It is characterized by its good adaptation to the prevailing harsh conditions in rural areas. It has irregular white and black colour, the feathers are usually white at the neck in both males and females, and the tail is white-grey in males. Body weight is on average 1.4 and 1.6 kg in adult females and males, respectively. Egg weight (42 g) is lower and egg production (150 eggs year^{-1}) is less than in specialized egg-producing breeds. Fayoumi is considered a dual-purpose breed which has been included in the formation of other breeds to utilize its adaptability to the local conditions. Baladi is also a dominant breed. It is similar in size to Fayoumi but is very variable in its phenotypic characteristics which makes it hard to define. There are other breeds which have been formed by cross-breeding local (mainly Fayoumi) and imported breeds. Dokki-4 is the most common cross-breed, produced at Cairo University by crossing Fayoumi with Plymouth Rock. The cross-bred has been introduced widely in several areas of the country. It is a dual-purpose breed, but is superior to Fayoumi in both growth performance and egg production. Other breeds which are less widespread in the country are Alexandria, Golden Montazah, Silver Montazah, and Matrouh, bred by the Ministry of Agriculture. Chickens are raised in small numbers by most small farmers, and their products are mainly consumed by the family, although occasionally these products are sold at local village markets. The main supplies of chicks are those hatched at local old-style hatcheries. There are about 800 of these distributed throughout the Delta and Nile Valley with a total capacity of about 250 million eggs year^{-1}. They usually apply primitive methods which have the disadvantages of low hatchability (about 65 per cent) and inadequate sanitary conditions, with the consequence of disease. These local hatcheries deserve more attention in any development programme for rural poultry production.

Commercial broiler and layer units have increased by more than 300 per cent over the last ten years. In 1985 there were about 18 000 broiler units and 4550 layer units. The Egyptian Government has invested about £E2500million in commercial broiler and layer units as part of a policy to reduce the dependence on red meat and consequently reduce the area of fodder crops, replacing it by wheat. With such rapid development in the poultry industry, Egypt is now, in 1992, almost self-sufficient in poultry products.

Other poultry species are raised mainly by small farmers in rural areas. Ducks and geese are native breeds and are raised mainly for meat. Both ducks and geese are usually raised with little feed input as they are left foraging throughout most of the day in the small canals and streams around the villages. They contribute to the supply of animal protein for the farmers' families, particularly in the northern part of the Delta. Turkey breeding is not common in Egypt, although some farmers keep a small number of birds. The turkey breeds raised are mainly small, body weights averaging 4–6 kg for males and 3–4 kg for females.

Rabbits

In Egypt rabbit production had not received much attention from formal organizations or from individual producers until the mid-1980s. It is believed, however, that rabbits could help considerably in meeting the increasing demand for meat in the country. The number of rabbit producers and their unit sizes are unknown, but the figures available for the rabbit population are shown in Table 4.1. There are three main native breeds of rabbit in Egypt. Baladi rabbits are small but very variable in phenotypic characteristics. They are bred in small numbers by farmers as household animals. Mature body weights are about 2 kg and females produce 5–6 young per litter. The second breed is the Gabali, which is medium in size and at maturity has a liveweight of about 3.5–4.5 kg. It is grey in colour and is known to tolerate harsh conditions. Gabali rabbits have a litter size of about eight and are known to be more prolific than the Baladi. The third breed is Giza White, which was bred by Cairo University in the 1930s but is found in relatively smaller numbers than the other two breeds. The Giza White has soft dense white silky fur, and a body weight of 2.5–3.5 kg. Litter size averages 6–7 (Khalil 1986).

In general, the rabbits kept in rural areas are raised in inappropriate housing, and the inadequate sanitary and feeding conditions hinder their productive potential. Mortality between birth and weaning is high (30–50 per cent) among local breeds. The Egyptian Government has recently encouraged national plans for developing rabbit production by small farmers. In addition, modern rabbit production systems, which have been introduced as large commercial units in Egypt in the 1980s, are inceasing in different provinces.

Fish production

As indicated by the pictures engraved on the walls of ancient Egyptian tombs and temples, fishing has been a traditional activity for a very long period. The main fishing areas are the Red Sea and the Mediterranean Sea, the River Nile and its branches, and the several lakes (Manzala, Idku, Maryut, Karoun, and Bardawil). In the last two decades, Lake Nasser, south of Aswan, has been considered as an important fishing area.

Total fish production in Egypt amounts to 300 000 tonnes annually, but 25 per cent of this is produced from the Red Sea and the Mediterranean Sea, indicating that these sea areas are possibly underutilized. Almost 50 per cent of the total production is produced from the main six lakes, 10 per cent is from the Nile and its branches, and the remaining 5 per cent is produced from fish farms. Since the late 1970s the Egyptian Government has encouraged fish farming in areas which are unsuitable for cultivation, in rice fields, and in small lakes and ponds.

The main fresh water fish are the *Tilapia* sp. (*Tilapia nilotica* and *Tilapia zillii*), *Bagrus bayad*, *Mugil cephalus*, *Mugil capito*, *Clarias lazera*, *Clarias anguillaris*, and sole. Salt-water types include *Mugil* sp., *M. capito*, carp, and shrimps. Several carp species have been introduced in Egypt and these are used in many fish farms. The Egyptian Government has adopted a plan to develop fish production, particularly in underutilized waters, so that a significant part of the demand for animal protein from fish may be met.

Insects and animal pests significant in Egyptian agriculture

Insects

Of the many insect species found in Egypt in the order Orthoptera, the mole crickets (Gryllotalpae) and the short-horned grasshoppers and locusts (Acrididae) are the most important. Locusts are considered a very harmful pest in Egypt. *Anacridium aegypticum* and *Locusta migratoria danica* are less harmful than the desert locust (*Schistocerca gregaria*) which migrates from other parts of Africa. The last attack by desert locusts in Egypt was in 1961, but continuous efforts are made in co-operation with other African countries to control the desert locust usually before it approaches Egypt.

The order Hemiptera includes a wide variety of insects such as bugs, scale insects, mealbugs, white flies, aphids, and leaf hoppers. *Oxycarenus hyalinipennis* (Lygaeidae) attacks cotton seeds and okra fruits while *Nezara viridula* attacks cotton, maize, soyabeans, and some vegetable crops.

Empoasca decipiens is the most important of the Cicadellidae leaf hoppers as it attacks a variety of vegetable crops. *Icerya purchasi* (Margarodidae), in particular, attacks citrus trees and some vegetable crops while the Egyptian mealbug (*Icerya aegyptiaca*) affects citrus and other fruit trees. The family Aleyrodidae includes cotton and tomato white fly (*Bemisia tabaci*) which affects cotton and many late summer vegetable crops, and also citrus white fly (*Aleurotrachelus citri*). Plant lice (*Aphididae*) include the cotton aphid (*Aphis gossypii*) which attacks cotton and more than 50 other crops, and *Aphis craccivora* which affects legume food crops. Cotton thrips (*Thrips tabaci*) is the most common species in the order Thysanoptera and it has many plant species hosts.

In the order Lepidoptera, the family Gelechiidae includes the pink bollworm (*Pectinophora gossypiella*) which is a serious threat to the cotton crop and the potato tuberworm (*Phthorimaea operculella*) which attacks the potato crop. *Earias insulana* (Sphingidae) attacks the cotton crop at various stages of growth.

The Egyptian cotton leafworm (*Spodoptera littoralis* (Noctuidae)) is the most harmful pest in Egyptian agriculture. Its effect is not limited to the cotton crop as it also attacks other green crops such as *bersim*, maize, and vegetable crops. Control of *S. littoralis* is organized at the central national level, and is governed by specific by-laws. The sugar cane stemborer (*Sesamia cretica*) is another important species in this family. It attacks both maize and sugar cane.

In the order Coleoptera, the family Dermestidae includes *Trogoderma irroratum* which is an important pest of stored grains, particularly in upper Egypt. Ladybirds (Coccinellidae) include several species such as *Coccinella undecimpunctata* which have an economic value for their biological control of other insects. The family Bruchidae includes several species that attack legume seeds while the family Curculionidae includes species that affect stored grains particularly wheat and rice.

The order Diptera is the largest of the insect orders. The Mediterranian fruit fly (*Ceratitis capitata*) is of particular importance in Egypt on account of the damage it causes in fruit production. In recent years a national programme for its control has been launched.

Non-insect pests

Many mite species (*Acaridae*) are found in Egypt. The family Tetranychidae is the most common and *Tetranychus arabicus* and *T. neocaledonicus* are important because they attack a variety of plant species. Other species include *Oligonychus mangiferus,* which attacks mango and vine, and *Eutetranychus orientalis* which attacks citrus trees. Several species of mite constitute a threat to stored grains. Predaceous mites include *Euseius gossipi* (Phytoseiidae) *Agistemus exsertus* (Stigmaeidae), and *Cheletogenes ornatus* (Cheyletidae).

Plant nematodes are serious pests which affect the productivity of many different crops. Several types are present in Egypt including *Meloidogyne spp.* (root knot), *Tylenchulus semipenetrans* (citrus nematodes), and *Anguina tritici* (wheat nematodes).

Most of the rats and mice in Egypt are of the family Muridae. The most common species are *Rattus norvegicus, Arvicanthis niloticus, Rattus rattus, Acomys cahirinus,* and *Mus musculus.* The first two species have greater agricultural significance since they are common in fields and stores. Since the late 1970s, the rat population has inceased dramatically, and a national campaign has been in operation (Tantawy Omar 1984).

Birds have been documented in the drawings found in ancient Egyptian tombs and they have always been considered a major threat to productivity, particularly that of grain crops. Species of the family Fringilidae are the most significant and *Passer domesticus niloticus* is the most common. Its population has been estimated at about 30 million consuming in excess of 120 tonnes grain day^{-1} (Hosny *et al.* 1976). Modern methods of bird control in grain fields are being practised.

Conclusion

In Egypt, the quantity of food produced from animal products and crops falls far short of satisfying the needs of the rapidly increasing population. Since 1960, the gap between local supply and demand has increased progressively and the situation has resulted in the importation of greater quantities of food products.

There is a continuous shortage of animal protein available for the human population but recent government policy has been to increase the area of wheat grown for human consumption at the expense of *bersim*, which is the main fodder supply for the ruminant. Furthermore, the productivity of native animal breeds is generally low due to their low genetic potential, lack of adequate food resources, and improper managerial practices. Alternative food resources have to be provided for the livestock population, and use of crop residues and by-products seems to be the optimal solution for this ever growing problem. Genetic improvement of native breeds may be achieved, but such programmes face major obstacles at the farmer level and need to be carried out in conjunction with improved veterinary and extension services.

References

Aboul-Ela, M. B. (1988). Reproductive patterns and management in the buffalo. In *Ruminant production in the dry subtropics: constraints and potentials*, (ed. E.S.E. Galal, M.B. Aboul-Ela, and M.M. Shafie), pp. 174–9. PUDOC, Wageningen, Netherlands.

Aboul-Ela, M. B. and Aboul-Naga, A. M. (1987). Improving fecundity in subtropical fat-tailed sheep. In *New techniques in sheep production*, (ed. I. Marai and J. B. Owen), pp.163–71. Butterworths, London.

Aboul-Ela, M. B., Aboul-Naga, A. M., El-Nakhla, S. M., and Mousa, M. R. (1988). Cyclic activity, ovulation rate and breeding performance of the prolific Egyptian-Nubian goats. In *Proceedings of the 11th International Congress on Animal Reproduction and Artificial Insemination, Dublin, 26 June–1 July, 1988.*

Aboul-Naga, A. M. (1976). Location effect on the reproductive performance of three indigenous breeds of sheep under subtropical conditons of Egypt. *Indian Journal of Animal Science*, **46**, 630–6.

Aboul-Naga, A. M. and Aboul-Ela, M. B. (1987). The performance of Egyptian breeds of sheep, European breeds and their crosses. I. Egyptian sheep breeds. *World Review of Animal Production*, **23**, 75–82.

Aboul-Naga, A. M., El-Shobokshy, A. S., Marie, I. F., and Moustafa, M. A. (1981). Milk production from subtropical non-dairy sheep. 1. Ewe performance. *Journal of Agricultural Science, Cambridge* **97**, 297–301.

Aboul-Naga, A. M., Hassan, F., and Aboul-Ela, M. A. (1987). Reproductive performance of local Egyptian sheep and goat breeds and their crosses with imported temperate breeds. In *38th Annual Meeting of the European Association for Animal Production, 28 September – 1 October, Lisbon.*

Ahmed, J. A. and Tantawy, A. O. (1954). Growth in Egyptian cattle from birth to two years of age. *Alexandria Journal of Research*, **2**, 1–8.

Ahmed, I. A. and Tantawy, A. O. (1959). Breeding efficiency of Egyptian cows and buffaloes. *Empire Journal of Experimental Agriculture*, **27**, 17–23.

CAPMAS (Central Agency for Public Mobilization and Statistics) (1990). Livestock statistics, *In Statistical yearbook*, pp.67–71. CAPMAS, Cairo.

El-Ashry, M. A., Mogawer, H. H., and Khishin, S. S. (1972). Comparative study of meat production from cattle and buffalo male calves. *Egyptian Journal of Animal Production*, **12**, 99–107.

El-Naggar, A. A., El-Shazly, K., and Ahmed, I. A. (1972). Effect of early weaning on the performance of male buffalo and cattle calves. *Animal Production*, **14**, 171–6.

Epstein, H. (1971). *The origin of domestic animals of Africa*. Africana Publishing Corporation, New York.

Ghanem, Y. S. (1980). *Encyclopedia of animal health*, Part I, *Arab sheep breeds*. Arab Organization for Education, Culture, and Sciences. Arab Centre for the Studies of Arid Zones and Dry Lands, Damascus, Syria (in Arabic).

Guirgis, R. A. (1973). The study of variability in some wool traits in a coarse wool breed of sheep. *Journal of Agricultural Science, Cambridge*, **80**, 233–8.

Hafez, E. S. E., (1953). Ovarian activity in Egyptian fat-tailed sheep. *Bulletin of the Faculty of Agriculture, Cairo University* **34**, 27.

Heider, A. (1982). Studies of the performance of some breeds of goats and their crosses under desert conditions in Egypt. Unpublished Ph.D. thesis. University of Alexandria, Egypt.

Hosny, M. M., Asem, M. A., and Nasr, E. A. (1976). *Insects and agricultural animal pests*. Dar El-Maaref Publishers, Cairo (in Arabic).

Kandil, A. A., Eldanasory, M. S., and Salama, M. A. (1984). Reproductive performance of Baladi goats of Egypt. In *Proceedings of the 1st Egyptian-British Conference on animal and poultry production, Zagazig, 11–13 September 1984*, Vol. 2, pp.368–75.

Khalil, M. H. (1986). Estimation of genetic and phenotypic parameters for some productive traits in rabbits. Unpublished PhD thesis. Zagazig University, Moshtohor, Egypt.

Khishin, S. S. (1951). Studies on the Egyptian buffalo. I. Average age and calving interval. *Empire Journal of Experimental Agriculture*, **19**, 185–90.

Mason, I. L. (1967). *Sheep breeds of the Mediterranean*. Commonwealth Agricultural Bureau Publications, Farnham Royal, UK.

Mohamed, A. A., El-Ashry, M. A., and El-Serafy, A. M. (1980). Reproductive performance of buffalo heifers bred at young age. *Indian Journal of Animal Science*, **50**, 8–12.

Mostageer, A. (1982). On the introduction of the Pinzgauer cattle to Egypt. In *Proceedings Vth International Pinzgauer Cattle Breeders' Congress, Alzbury, Austria, 2nd July, 1982*. Ministry of Agriculture, Cairo.

Mostageer, A., Morsy, M. A. and Sadek, R. R. (1981). The productive characteristics of a herd of Egyptian buffaloes. *Zeitschrift fur Tier Zuchtung und Zuchtungsbiologie*, **98**, 220–6.

Mourad, K. A. (1978). Some productive and reproductive characteristics of the Egyptian buffaloes. Unpublished MSc thesis. Cairo University, Cairo.

Nigm, A. A., Soliman, I., Hamed, M. K., and Abdel-Aziz, A. S. (1986). Milk production and reproductive performance of Egyptian cows and buffaloes in small livestock holdings. In *Proceedings 7th Conference of the Egyptian Society of Animal Production, Cairo, September 16–18*, pp.290–7. Egyptian Society of Animal Production, Cairo.

Ragab, M. T. and Abdel-Salam, M. F. (1963). Relation and interrelation between body weights and growth rates of cattle and buffaloes at some different ages of life. *Journal of Animal Production UAR*, **3**, 27–31.

Tantawy Omar, M. (1984). The National Rat Control Campaign in Egypt. In *Proceedings of a conference on The Organisation and Practice of Vertebrate Pest*

Control, 30 August – 3 September 1982 Elvetham Hall, Hampshire, UK, (ed. A. C. Dubock), pp.443–58. ICI Plant Protection Division, Fernhurst, Haslemere,

Wahby, A. M. (1937). Some facts on dairying in Egypt. *Wissenschaffliche Berichte des XI Milchwirtschaftlichen Weltkongresses, 22. bis 28. August 1937, Berlin,* I, pp.364–70. Reichsminister fur Ernahrung and Landwirtschaft, Berlin.

Wilson, R. T. (1984). *The Camel.* Longman, London.

Younis, A. A., Abdel-Aziz, M. M., Afifi, E. A., and Khaiery, M. (1984). Biological efficiency of meat production in Barki sheep. *World Review of Animal Production,* **20**, 31–7.

Zaki, M. H. (1951). The Menoufi type of buffalo: a suggested score-card. *Empire Journal of Experimental Agriculture,* **19**, 131–3.

5

Ethnic origins of the people and population trends

M. M. WAHBA

Introduction

Egypt is, after Nigeria, the most heavily populated state in Africa. Within the group known as the Arab World, Egypt is by far the most populated of the states, with double the population of the second most populated Arab nation, Morocco. In addition, Egypt has the dubious distinction of having the largest city in Africa and the Arab World. The importance of the population of Egypt is reinforced by a paradoxical land situation, where of the one million km^2 forming Egypt only some 55 039 km^2 are inhabited, i.e. around 5.5 per cent of the total area (CAPMAS 1985a, p. 5). Some 99 per cent of the population live in the inhabited area, while the remainder are scattered among the oases of the Western Desert, the Mediterranean coastal regions where rainfall allows for agriculture, the Sinai, and the Red Sea littoral. As a consequence of the crowding of the quasi-totality of the population in the Nile Valley and Delta, population densities in Egypt are among the highest in the world, rivalling the most crowded of the Asian states.

The geographic reality of Egypt, limiting the 'useful' area of the country to the narrow strip of green hugging the banks of the Nile, and flowing into the Delta in the north, has had two major effects on the distribution of the population. First, it has led to a high level of population density in the inhabited area. Second, within that inhabited area, the growth of cities in the last centuries, led by the spectacular growth of the Cairo metropolis, has further limited the cultivated area to some 2.4 per cent of the area of the country, representing 23 928 km^2 (Waterbury 1983, p. 42). The crowding of a large and increasing population within a restricted geographical area, bounded by inhospitable deserts to the east and west, and by the Mediterranean Sea to the north, has led to a great degree of homogeneity within the population of Egypt.

The aim of this chapter is to present a brief outline of the historical origins of the population of Egypt, discuss the status of the ethnic minorities and their numerical strength, and examine population trends in Egypt. Because of the limited number of ethnic minorities in Egypt, around 1 per cent of the total population, we shall concentrate our analysis upon the examination of trends within the population as a whole.

Historical origin and ethnic distribution of the population

The geographical position of Egypt, at the crossroads of three continents, as well as its position bordering the ancient trade routes in the Mediterranean, lead one to think that there cannot have been a racially unmixed, original population of Egypt. Rather, it appears that with the end of the last ice age, roaming groups of hunters and gatherers settled in the marshlands of the Nile basin, and developed a primitive, sedentary, agriculture. These groups belonged to a Caucasian race, but mixed with an African Negroid population very early in the history of the region. In addition, population movements from the Mesopotamian river basin and the shores of the eastern Mediterranean also contributed to the moulding of the population of what was to become Egypt, (US Department of the Army 1964, p. 43). By the beginning of the Neolithic Age, the main elements of the population of Egypt appear to have been present in the Nile Valley and the Delta. The original Egyptians were, probably, a mixture of negroid elements, 'brunet Mediterraneans', and 'Cro-Magnon' type, leading ultimately to the evolution of a white African population known as 'Hamitic' (CNRS 1977, p. 15).

However, since 1955, there have been theories that, ethnically, Egyptians are part of the Negro race, and that the civilization of Egypt, pre-dynastic and dynastic, should be included as part of the ancient Negro civilizations (CNRS 1977, p. 15). This position has been reinforced by the distinguished Africanist Leopold Sédar Senghor of Senegal, arguing in a lecture given at Cairo University in 1967, that the term 'Hamitic' was greatly lacking in clarity. Indeed, Senghor goes as far as stating that 'En verité, le mot "hamite" est un prétexte politique' (Senghor 1967, p. 68). Senghor goes on to argue that the Mediterranean civilizations appear to have been built by a mixture of the Negroid elements, present at first with the Caucasian Cro-Magnon race, moving progressively to the south, thus giving rise to an early Neolithic race of Mediterranean man, *Homo mediterraneus*. Later in the Neolithic, this race received admixtures of *Homo alpinus*, comprising Sumerians, Elamites, Hittites, Armenians, Assyrians, and Jews. This led to a gradual 'albinization' of the area (Senghor 1967, pp.69–72). By the end of the fourth millennium BC, the blending of races within the Nile Valley melting pot appeared complete (Posener 1962, p. 237).

The discussion of the ethnic origins of the Egyptian population has been further complicated by political factors, which stress the remarkable degree of continuity, evident in the resemblance of the modern Egyptian peasant, the fellah, with the bas-reliefs of the dynastic era, thus arguing in favour of an identifiable, integrated Egyptian 'nation' from the 'dawn of history' (Hamdan 1970, pp.47–8). According to this interpretation, the succeeding waves of invasions—Central Asian (Hyksos), Persian, Greek, Roman, Arab, and European—which bore down on Egypt for the last twenty-five centuries, did no more than establish a temporary ruling elite, ethnically and culturally separate from the bulk of the Egyptian population. The Arab invasion produced a slight effect ethnically, although it was much more important culturally. The Arab presence in Egypt started before Islam, with limited razzias on the eastern flank of the Nile Delta around the Biblical land of Goshen (now the governorate of Sharqiya), and also with some settlement of Arabs who married fellah women before Islam. The Islamic invasion of Egypt in AD 640 led to the presence of greater numbers of Arabs, though most of these Arabs did not remain in Egypt, but went on to pursue the conquest of north Africa and northern Sudan. Eventually, with Arabs who did settle in the eastern regions of the Delta and in upper Egypt, a large degree of intermarriage took place and the Arabs, originally not much different from the Egyptian population, disappeared into the 'melting pot' (cf. CNRS 1977, p. 17; Hamdan 1970, pp.51–4). The CNRS report on Egypt estimates that only some 6 per cent of the present population of Egypt has a clear Arab origin, while only 2 per cent has a Berber origin.

After the Arab invasion, Egypt witnessed the settlement of small groups of invaders such as the Mamelukes. The Mamelukes were brought into Egypt by the descendants of Saladin at the end of the twelfth century AD as military slaves. By the middle of the thirteenth century, the Mamelukes were ruling Egypt, holding power until the Ottoman invasion of 1517. The Mameluke system of government had a marginal effect upon the constitution of the Egyptian population. Thus, young boys from the Black Sea, Circassia, Central Europe, and from the northern shores of the Mediterranean were imported into Egypt as military slaves. The most able of these slaves took power, and imported the next generation of rulers. After the first generation, the Mamelukes who were settled in Egypt lost their elite status, and often intermarried with the local population. This phenomenon is most evident in Cairo and in the eastern governorate of Sharqiya where a Caucasian element sometimes comes to the surface, often associated with a patronymic denoting European origin such as Abaza (from the name of a Caucasian tribe), Gritly (Cretan), or even Sherkessy (Circassian). In the main, however, the melting pot can be seen to have led to an homogenization of the population throughout most of Egypt.

The great majority of the Egyptian population is thus similar in ethnic and racial characteristics. The people are generally small in stature, with a straight or concave nose, a slightly elongated head, large eyes, and straight or wavy hair, rarely curly, and usually black. Skin colour varies with the degree of exposure to the sun, from an olive tan in the northern Delta areas to a darker brown in the *Sa'id* or upper Egypt (US Department of the Army 1964, p. 44). Egyptians will refer to their own colour as *qamhi*, wheaten, which appears to be a light brown.

There are three ethnic minority groups which taken together account for around one per cent of the population. These are the Berber pockets of the oases in the Western Desert, the Nubians in the south of the Nile Valley, and the sedentarized and semi-sedentarized Arab Bedouin populations in the Eastern Desert and the Sinai Peninsula.

In the Western Desert, especially in the oases of Siwa and Kharga, there are pockets of Berber-speaking people, whose number does not exceed a few thousand, although it is difficult to number them. The Berber population of the oases is rapidly disappearing due to intermarriage with the fellah stock brought in to colonize the land reclaimed in the New Valley project. Also, it is difficult to classify the population of the oases as an ethnic minority. This is because, for many centuries, the original Berber stock has intermarried with Arab and Negro slaves who used to pass through the oases on the slave trade caravan route, Darb el Arba'in, linking western and central Africa to the north and Mediterranean shores of the continent (Encyclopaedia Britannica 1984, p. 450).

In the Eastern Desert, there are two ethnic minorities. In the south of the Eastern Desert, are the Hamitic Beja, who are physically similar to pre-dynastic Egyptians. The Beja are divided into two tribes, the Ababdah and the Bishari with the majority living in the Sudan. In the northern areas of the Eastern Desert and Sinai, there are some semi-nomadic Arab Bedouin tribes and a population numbering some 150 000 (CAPMAS 1985b, p. 13).

South of the Egyptian town of Aswan, from the first to around the fourth cataract, there are some 200 000 Nubians. Around one-quarter of these Nubians live in Egypt. The Nubians are tall and thin with caucasoid features, and are much darker than the average Egyptian. After the flooding of large areas of Nubia due to the construction of the High Dam, Egypt's Nubians were resettled North of Aswan by the village of Kom Ombo in 1963–4 (Adams 1977, pp.45 and 654). A large number of Nubians have emigrated to the cities of Cairo and Alexandria.

Finally, the urban conglomerates of Cairo and Alexandria have a small percentage of Levantine minorities, as well as small numbers of Armenians and Greeks who came to Egypt in the middle of the nineteenth and at the beginning of the twentieth centuries.

The great majority of the Egyptian population is homogeneous, however, with ethnic minorities forming small, marginal, pockets in the cities of Cairo and Alexandria as well as in the outlying districts of the Nile Valley and Delta. Ethnic integration and homogeneity are much more observable in the heartlands of the inhabited regions. It is only the frontier districts to the west, east and south where relatively large concentrations of ethnic minorities can be identified (Hamdan 1970, p. 61). In the next section, trends in the population of Egypt will be discussed without distinguishing among various ethnic groups.

Population trends

Since the second half of the twentieth century, two main trends have become apparent in the development of the population of Egypt. The first of these is the high rate of increase in population leading to an extremely high population density, the second is the rapid urbanization of the Egyptian population.

Population growth in Egypt since the nineteenth century

The first reasonably accurate census of the Egyptian population in modern times was in 1897, when the population was recorded at 9.734 million. Previous to that date, estimates of the population had varied from a high level of 8.5 to 9 million under the Roman occupation, to Volney's low level of 2.3 million at the end of the eighteenth century (Hamdan 1970, p. 325). Two further estimates, one based on the strength of the army, and a second based on the tax list for 1831 gave Egypt's population at the time as about 2.5 million (Abdel Hakim 1972, p. 17). In 1882, the British occupation forces recorded a population of some 7.8 million in a preliminary census (CNRS 1977, p. 158). However, the methodology was uncertain, due to the political unrest of the country at the time, and fifteen years later the 1897 census was undertaken (Abdel Hakim 1972, p. 18).

In the twentieth century, population censuses were carried out regularly every ten years from 1897 until 1947. Then, because of the 1956 Suez War and the ensuing political upheaval, the next census was undertaken in 1960. The following census was undertaken in 1966, then 1976, and the one after that in 1986, thus returning to ten-year intervals. The results of the population censuses are given in Table 5.1. These show the dramatic progression in the population of Egypt during the twentieth century. Starting with 11.287 million in 1907, the population reached 38.198 million in 1976, and some 50 million in 1986 according to the preliminary results of the most recent census. The increase in population during the twentieth century occurred faster than in the nineteenth.

The rate of increase

The rate of increase of the population of Egypt, the difference between the crude birth rate and the crude death rate, has fluctuated throughout the twentieth century. A high birth rate coupled with a high death rate, kept

TABLE 5.1. Egypt: population trends, 1907–88

Year	Population (thousands)	Birth rate (per thousand)	Death rate (per thousand)	Net rate of increase (per thousand)
1907	11287	NA	NA	15.8
1917	12751	40.1	29.4	10.8
1927	14218	44.0	25.2	18.8
1937	15933	43.5	27.2	16.3
1947	19022	43.7	21.4	22.3
1960	26085	42.9	16.9	26.0
1966	30083	40.9	15.8	25.1
1976	38198	36.6	11.8	24.8
1977[a]	38794	37.5	11.8	25.7
1978	39767	37.4	10.5	26.9
1979	40889	40.2	10.9	29.3
1980	42126	37.5	10.0	27.5
1981	43465	37.0	10.0	27.0
1982	44673	36.2	10.0	26.2
1983	45915	36.8	9.7	27.1
1984	47191	38.6	9.5	29.1
1985	48950	39.8	9.4	30.4
1986	49863	38.7	9.2	29.5
1987	51297	37.9	8.6	29.3
1988[b]	52919	37.5	8.6	28.9

Population data for Egypt can vary between sources and back years are revised in the light of recent information. Data for 1989–92 can be estimated approximately based on an annual growth rate of 2.7 per cent giving a figure in excess of 57 million for 1992 (Beshai 1992, personal communication).

[a] Population statistics after 1976 include Egyptian nationals living abroad.
[b] Estimate.
Source: Abdel Hakim (1972); CAPMAS (1990).

the natural rate of increase low in the early years of the century and in 1918, perhaps due to the effects of the First World War, the rate of increase fell below its 1907 level. The rate of increase then picked up, reaching the 20 per thousand mark (20.6) in 1930 for the first time. After 1930, the rate of increase of the population returned to its previous level below 20 per thousand, dropping to single figures (9.5 per thousand) in 1942 due again, in my opinion, to the War. After the Second World War, the population increased rapidly, from a rate of 16.2 per thousand in 1946, to 22.3 per thousand in 1947. The average rate of population increase per thousand was 26.4 in the 1950s, 25.55 in the 1960s, 26.68 during the 1970s and 28.33 during the 1980s. Preliminary results of the 1986 census put the natural rate of increase of the population of Egypt at 29.5 per thousand in that year (CAPMAS 1989, p. 24).

The population of Egypt seems to have passed through three stages. First, up to the 1930s, a stage of low growth due to the high natural birth and death rates. After the Second World War, improvements in health and the reduction in the child mortality rate contributed to an increase in the rate of population growth, but the rate of growth remained at about 25 per thousand, until the end of the 1960s. In the 1970s and the first half of the 1980s, an increase in the birth rate from 35.1 per thousand in 1970 to 40.2 in 1979, decreasing to 37 and 38 per thousand during the 1980s, as well as a decrease in the crude death rate, dropping to less than 10 per thousand in 1983 for the first time in the history of Egypt, appear to have added a new vigour to the rate of population growth.

This pattern of growth has led to a high proportion of the population of Egypt being under the age of 20 years (Table 5.2). The youth of the population and the consistency of a high birth rate despite the reduction in mortality rates, would suggest that the rate of population growth in Egypt may remain at 29–30 per thousand until the end of this century.

TABLE 5.2. Egypt: age distribution of the population, 1960, 1976, and 1986

Age group	1960 (%)	1976 (%)	1986 (%)
0–4	15.8	13.76	14.8
4–9	14.62	12.78	13.0
10–14	12.23	13.39	11.6
15–19	8.29	10.89	10.4
20–24	6.91	8.42	8.5
25–29	7.36	7.33	7.5
30–34	6.35	5.81	6.2
35–39	6.64	5.61	6.2
40–44	4.91	5.14	4.5
45–49	4.40	4.17	4.2
50–54	3.83	3.99	3.6
55–59	2.45	2.44	2.9
60–64	2.59	2.65	2.4
65+	3.47	3.59	3.8

Refers only to Egyptians living in Egypt. Figures for 1986 do not include population in collective residences (prisons, hospitals, and social care).

Source: CAPMAS (1990, pp. 26, 28, and 31).

Distribution of the population

Concomitantly with the rapid increase in the population of Egypt during the second half of the twentieth century, one can identify a very strong trend towards increasing urbanization. In 1987, it was estimated that the size of the urban population had overtaken that of the rural population; this was also predicted by Aliboni (1984, p. 168). Within urban areas, moreover, there has been a strong tendency for an increase in the size of the megalopoles of Cairo and Alexandria, at the expense of provincial cities (Table 5.3).

The following sections will deal with three topics: first, general trends in population migration to the urban conglomerations of Cairo and Alexandria; second, variations in density in rural areas; and third, variations in density in urban areas.

Population migration

The distribution of the Egyptian population has changed dramatically over the last century. In the 1907 census, some 82 per cent of the population lived in rural areas, but this figure had declined to 55.6 per cent by the 1986 census. It has been estimated that the rate of growth of the urban population is 4.5 per cent per annum, of which 1.5 per cent is due to migration from rural areas.

By 1966, one-fifth (19.9 per cent) of the urban population of Egypt consisted of lifetime migrants from rural areas. In Cairo alone, 51 per cent of lifetime migrants came from rural areas, while this proportion dropped to 49 per cent in Alexandria and 44 per cent in Suez (Abdel Hakim and Abdel Hamid 1982, p. 8, 18). In addition to rural – urban migration, there has been a net movement of migration from the smaller to the larger urban centres. Thus, a further one-fifth of Cairo's population in 1966 consisted of urban – urban migrants. This situation was aggravated by the destruction of the Suez Canal zone cities of Suez, Port Said, and Isma'iliya in the 1967 War, causing a further urban – urban stream of migration to Cairo. Since the 1970s, however, with the reconstruction of the Canal cities, there has been a reversal of this flow.

In summary, the most important aspect of internal migration is the movement from rural to urban areas. This movement takes place from two main regions: the southern Nile Delta, especially the governorates of Minufiya and Daqahliya which have contributed most to the migration to Cairo; and upper Egypt, where the governorates of Minya, Sohag, and Qena have contributed most to the migratory flow to the cities. The main recipient areas are the cities of Greater Cairo and Alexandria, although the Suez Canal cities have recently been net recipients of population as have frontier areas of recent development.

TABLE 5.3. Egypt: population density by governorate, 1976 and 1986

Governorate	1976	1986	Change (%)
	density per km²		
Cairo	23688	28359	+19.2
Alexandria[a]	7372	1089	–
Port Said	3644	5550	+52.3
Suez[b]	632	18	–
Isma'iliya	246	377	+53.2
Beheira[c]	537	222	–
Damietta	978	1258	+28.6
Kafr el-Sheik	409	524	+28.1
Gharbiya	1181	1478	+25.1
Daqahliya	789	1009	+27.8
Sharqiya	626	818	+30.6
Minufiya	1117	1454	+30.2
Qalyubiya	1679	2511	+24.8
Giza[d]	2284	3496	+53.1
Faiyum	625	845	+35.2
Beni Suef	840	1092	+30.0
Minya	908	1171	+29.0
Asyut	1093	1431	+30.9
Sohag	1244	1578	+26.8
Qena	924	1217	+31.7
Aswan	912	1181	+29.5
Red Sea[e]	0.3	0.44	+46.7
New Valley[e]	0.2	0.3	+50.0
Matruh[e]	0.5	0.8	+52.0
Sinai (N and S)[e]	–	3.3	–

The table does not include Egyptians living outside Egypt at the time of the census. Figures for 1976 exclude Egyptians living under occupation in the Sinai, hence precise population density figures for the Sinai in 1976 are not available.

[a] The new district of Ameriya in Alexandria is included in 1986 but not in 1976.
[b] The district of Ataqa is included in 1986 but not in 1976.
[c] The desert area of Wadi Natrun is included in 1986 but not in 1976.
[d] The Bahariya Oases are not included in 1976 or 1986.
[e] Major desert areas are included.

Source: CAPMAS (1985b, pp. 12–15); CAPMAS (1990, p. 22).

Rural population density

Rural Egypt is divided into two main areas: Lower Egypt comprising the area to the north of Cairo including the Nile Delta and some desert to the north by the shores of the Mediterranean; and Upper Egypt including the area from Giza, south of Cairo to the southern frontier with the Sudan. Each of these areas is itself divisible into three parts insofar as population density is concerned.

Lower Egypt has three main density regions. The southern tip of the Delta comprises the governorates of Minufiya, Qalyubiya, Daqahliya, and the southern part of Gharbiya. This is the most densely populated area in rural Egypt. The low-density area is found in the north of the Delta, which comprises the governorates of Kafr El-Sheikh, Sharqiya, and Beheira as well as the desert areas of Matruh. Medium density is found in the transitional area in the north-east Delta such as in the governorate of Damietta.

The frontier between the southern, highly populated areas of the Delta and the northern, relatively low density areas is the 7-m contour line. The soil north of the 7-m contour line is characterized by high salinity in the north, and high sandiness in the east and west. Cultivation problems, therefore, have led to migration from these areas and a low population density. In the more fertile land, in the southern regions of the Delta, the more fertile soil supports a higher population (Abdel Hakim 1972, p. 27). The migration from the southern Delta governorates to neighbouring Cairo has done much to soften the contrast in population density in northern and southern areas of the Delta.

In terms of population density, Upper Egypt can also be divided into three areas. A high density area lies in middle Egypt, roughly from the cities of Asyut to Nag Hammadi. An area of average density is found in northern areas of Upper Egypt, from the rural hinterland of Giza to Asyut, including the oasis province of Faiyum. The least dense area, characterized by high migration, as well as the flooding of large areas due to the waters retained by the High Dam, can be found south of Qena to the Sudanese frontier. Density divisions, however, are less clear in Upper Egypt than in the Delta (Hamdan 1970, p. 37).

Urban population density

Over the last fifty years, the urban rate of population growth has been around 4.5 per cent per annum. Of this 4.5 per cent, only 1.5 per cent is accounted for by the rate of rural – urban migration. The remainder is natural urban growth, thus showing that the urban population has a higher natural increase than the rural population. There is also an increase in the concentration of population in the larger cities to the detriment of smaller towns. Thus, whereas in 1947 towns with a population of 100 000 or more accounted for 54 per cent of the urban population, this proportion had risen to 75 per cent in 1976 (Abdel Hakim and Abdel Hamid 1982, Table 12). The province of Greater Cairo alone, comprising Cairo governorate, the cities of Giza, Shubra el Kheima, Shibin el Qanatir, El Hawamdiya, El Badrashein, and outlying large villages, amounts to 20.2 per cent of the resident population of Egypt (CAPMAS 1989, p. 15). Taken together, Greater Cairo and the city of Alexandria account for 26 per cent of the total population of Egypt, and some 60 per cent of the urban population. The four largest cities of Egypt: Cairo, Alexandria, Suez, and Port Said account for some 64 per cent of the urban population.

Distribution by age and sex and dependency ratio

The population of Egypt includes around 1.1 million more males than females (CAPMAS 1989, p. 27). The number of males per 100 females varies, however, from equal numbers in Aswan, to much higher proportions (152 males per 100 females) in the Red Sea, North and South Sinai governorates. The reason for the high variation in the male:female ratio is the rate of temporary male migration in search of higher income in Egypt's urban agglomerations and in neighbouring Arab countries. Areas of recent settlement, in the Sinai for instance, also display a high male:female ratio.

Over the last fifty years, there has been a high proportion of young people in the Egyptian population. The proportion of the population between 0 and 19 years was 47.6 per cent in 1937, 52.6 per cent in 1970, and 50.8 per cent in 1976 (CNRS 1977, p. 167; CAPMAS 1985b, p. 20). In the 1986 census, the age group between 0 and 19 years represented 49.8 per cent of the population. Together with this high percentage of youth in the population, there has been a slight decrease in the 'active' group, aged 20–64 years, from 48.5 per cent in 1937 to 44.2 per cent in 1970, 45.56 per cent in 1976, and 46 per cent in 1986. A large percentage of the population, however, remains concentrated in the 0–14 years age group, i.e. those not officially of working age, although this proportion has dropped slightly from 42.75 per cent in 1960 to 39.4 per cent in 1986 (Table 5.2).

The dependency ratio, defined as the number of people supported by those in work, is quite high. This is due to a variety of reasons. First, the age distribution of the population is such that the active population, aged 15–65 years, must support a large inactive population, i.e. those below 14 years and over 65 years, a total of 43.2 per cent of the population in 1986. Second, the high rate of unemployment, especially among women, but also among men (some 20 per cent in 1986), has worked towards increasing the dependency burden.

The figures, however, do not take into account the large number of children below 15 years, and persons over 65 years who are employed but are not declared as such to census collectors. In addition, the figures do not take into account housework by women, and the census collectors underestimate employment figures in the countryside where there is a general mobilization of the population, young or old, male or female, in peak employment times such as the cotton harvest, or when crises, such as the spread of the cotton boll worm, hit the countryside.

According to official statistics the dependency ratio was 28 per cent in 1976 and in 1986, but these figures should be treated with some care (CAPMAS 1985*b*, pp.20, 27; CAPMAS 1989, p. 23). If one included women of working age working in the home and children over 6 years in the countryside, it would seem that around 60 per cent of the population is economically active.

Employment

After appearing to have a labour surplus for many years, the situation in Egypt appears to be changing. There are labour shortages in the construction sector, where an estimated 40 per cent of the labour force has temporarily emigrated to neighbouring Arab countries.

The agricultural sector employs some 34 per cent of the labour force; the proportion declined to less than one-half the labour force for the first time in the early 1970s. In contrast, urban employment has increased and employment in services rose to 30 per cent of the labour force in the early 1980s (Arab Republic of Egypt 1984, p. 96). However, the increase in employment in the services sector should not be attributed entirely to urban concentration, for there is also a steady growth in co-operative employees, irrigation engineers, health services personnel, and government staff in rural services employment.

Conclusions

The population of Egypt is atypical of that of other Third World countries in one particular respect: it is relatively homogeneous and has no sizeable ethnic minorities. Indeed, the ethnic minorities which do exist in Egypt are either located outside the mainstream of the Egyptian economy, for example, the Berbers in the Western Desert oases and the Bedouin in the Eastern Desert and Sinai, or, as in the case of the Nubians in the south, are too small to pose a problem for the homogeneity of the population.

In other respects, the population of Egypt exhibits trends shared by many other countries in the Third World. The high rate of population growth, although it is by no means one of the highest in the Third World (compared for instance to Nigeria) is nevertheless quite high. The problem of a high rate of growth of the population is compounded in Egypt by geographical factors rendering 95 per cent of the land inhospitable, and by a high rate of urbanization. Both factors have contributed to a dramatic increase in population density over the last fifty years. Moreover, despite a slight rise in the total cropped area from 10.79 million feddans in 1960 to 12.16 million in 1985, the rising population in rural areas has led to a decrease in the cropped area per agricultural worker from 3.33 feddans in 1960 to 2.8 in 1985. The harvested area per caput of the whole population is around 0.25 feddans (Dethier 1989, p. 4–5).

In the near future, it is expected that the population trends outlined above will continue. This means, that the rate of growth of the population should continue at its present level, or indeed increase, as borne out by the relative youth of the population. A higher population density and increasing urbanization should follow.

The trend of employment away from agriculture and into services is expected to continue. The political instability in some of the countries which receive Egyptian migrant labour, and the possibility that labour-importing economies in the Gulf may have reached their peak level of absorption, indicate the strong possibility of a return of migrant labour. Tentative figures for returning labourers show that they are employed either in government and public sector (27 per cent), where their return increases disguised unemployment, or in construction and agriculture (Higher Specialized Councils 1984, p. 171). The return of migrant labour in the construction and agricultural sectors may have two effects. First, it should decrease child and female labour, and second, it may lead to unemployment in both sectors. This ought to lead to a reduction in agricultural and construction wage rates and increase the influx of labour into the urban informal sector

The influx of increasing numbers of the rural population to Egypt's major cities is expected to continue. The full effects of such an increase are yet to be seen.

References

Abdel Hakim, M.S. (1972). The population of Egypt: a demographic study. Reprint from *Bulletin of the Faculty of Arts, Cairo University*, (1967) 29, (1 and 2). Cairo University Press.

Abdel Hakim, M.S. and Abdel Hamid, W. (1982). *Some aspects of urbanization in Egypt*. Occasional Paper, No. 15. Centre for Middle Eastern and Islamic Studies, University of Durham.

Adams, W.Y. (1977). *Nubia: corridor to Africa*. Allen Lane, London.

Aliboni, R. (ed.) (1984). *Egypt's economic potential*. Croom Helm, London.

Arab Republic of Egypt (1984). *Minutes of the meeting of the Shura Council (Consultative Assembly) on 4 March, 1984*.

CAPMAS (Central Agency for Public Mobilization and Statistics) (1985a). *A statement on the population of the Arab Republic of Egypt*. Mimeographed Statement, October 1985. CAPMAS, Cairo.

CAPMAS (1985b). *Statistical yearbook*. CAPMAS, Cairo.

CAPMAS (1989). *Statistical yearbook*. CAPMAS, Cairo.

CAPMAS (1990). *Statistical yearbook*. CAPMAS, Cairo.

CNRS (Centre National de la Recherche Scientifique), Groupe de Recherches et d'Etudes sur le Proche-Orient (1977). *L'Egypte d'aujourd'hui: permanence et changements 1805–1976*. CNRS, Paris.

Dethier, J.J. (1989). *Trade, exchange rate, and agricultural pricing policy in Egypt*, Vol. I. World Bank, Washington DC.

Encyclopaedia Britannica (1984). *Egypt*. Vol. VI, p.450.

Hamdan, G. (1970). *Shakhsiat misr:dirasa fi'abqariat il makan* (Egypt's personality: a study in genius loci). Al Nahda Bookshop, Cairo.

Higher Specialized Councils (1984). *Taqrir al majlis al qawmi lil khadamat wal tanmiya'l Ijtima'iya*. Report of the National Council for Services and National Development. Presidency of the Republic, The Higher Specialized Councils, Cairo.

Posener, G., Sauneron, S., and Yoyotte, J. (1962). Race. In *A dictionary of Egyptian civilization*. pp.237–8. Tudor Publishing Co., New York.

Senghor, L. S. (1967). *Negritude, arabisme et francite*. Dar Al Kitab Allubnani, Beyrouth.

Department of the Army (1964). *United States army handbook for the United Arab Republic (Egypt)*. December 1964.

Waterbury, J. (1983). *The Egypt of Nassar and Sadat: the political economy of two regimes*. Princeton University Press.

6

Land tenure

A. RICHARDS

Introduction: the importance of land tenure

Ever since the eighteenth century, economists have argued that the distribution of land shapes the pattern of resource allocation, distribution, and growth in an agricultural economy. We have since accumulated a great deal of evidence which strongly suggests that the distribution of assets determines how the gains of growth are shared. Although the percentage of land in the value of total assets declines with economic development, historical evidence shows that the land tenure system has important consequences for the rate and distribution of human capital investment, which becomes increasingly important as development proceeds. Viewed historically, the distribution of land is fundamental to the distribution of income in the rural areas. Income distribution, in turn, may affect the pattern of industrial growth through final demand linkages. The structure of asset ownership and the pattern of growth and distribution are intimately connected (Johnston and Kilby 1975; Mellor 1976; De Janvry 1981; Adelman 1984).

Linkages between land tenure and resource allocation are not limited to demand feedbacks. Some argue that the poor information/limited communications so characteristic of less developed countries (LDCs) combine with concentrated ownership of land to generate market imperfections and inefficiencies. There is by now a very large body of literature which argues that small peasant farms and large estates face very different factor prices (Binswanger and Rozenzweig 1984; Griffin 1974). Such market segmentation gives rise to different combinations of inputs and outputs; one particular imperfection (in the labour market) provides the principal explanation for the oft-observed higher output per unit land on small farms (Berry and Cline 1978; Richards 1986). On the other hand, larger farmers may be able to adopt new technologies more readily than small producers, since the former often pay less for credit and/or can bear risks more easily than can the latter. Most economists agree that insecure property rights in land, especially land ceilings and/or rent controls, impose significant social losses. There is little doubt that the distribution of land and the rules of land tenure affect resource allocation in LDC agriculture.

Of course, the impact of land tenure on distribution and resource allocation is mediated by the political system. Not only do land reforms, land ceilings, etc. affect the allocation of land directly, but the owners of land may themselves wield considerable influence over agricultural policy. For example, large estate holders (the *pashas* of modern Egyptian history) dominated national politics before 1952. Although Nasser's land reforms eliminated them from the rural political economy (at least temporarily), the role of the 'second stratum' or rural middle class (owners of 10–50 feddans) was probably strengthened. Such farmers today exert considerable influence over agricultural policy formulation and, especially, implementation.

All of these issues come together in the debate over how the distribution of land affects growth and equity, or, as it is sometimes called, the 'agrarian question'. Here we may distinguish two broad perspectives. One maintains that agricultural development is *polarizing*: an ever increasing percentage of land will be held by large farmers, and a steadily rising fraction of the rural population will depend on wage-labour for their livelihood. This view is particularly associated with old-fashioned Marxism, but it has other adherents as well. The other view holds that because of limited economies of scale in crop production and animal husbandry, small peasants will be able to survive during the development process. We shall see some evidence to support both views in the Egyptian case; the current situation presents a mixture of 'polarizing' and 'preserving' features. Perhaps the best generalization would be that Egypt's agriculture began its development with a highly skewed land tenure system, then became (and remains) one in which small farms predominate in the best lands (the Old Lands of the Nile Delta and Valley[1]), but which now is undergoing technological and institutional changes which may be increasing the importance of large farms once again. Despite significant land reforms and fragmentation pressures, Egypt remains a 'bimodal' agrarian system, with a large number of small farmers producing alongside a relatively small number of large farmers (Johnston and Kilby 1975).

The historical legacy

Understanding the history of land tenure in Egypt is essential for two reasons. First, despite the significant changes introduced by the Nasser regime, the old order was only partially overturned. In contrast to countries such as Korea or Taiwan, the large landlords of the *ancien regime* in Egypt were weakened, but not eliminated. The terms of the land reform permitted them to retain some land, and various practices (such as sharing land among family members) permitted large estates to survive *de facto* if not *de jure* in some areas. Second, one must grasp the historical context to understand the strengths and weaknesses of Nasser's land reforms, which, in turn, constitute a fundamental determinant of the present distribution of farm assets.

We may distinguish four major historical periods in the modern history of Egyptian land tenure:

(1) the nineteenth century, when large estates were created, permanent irrigation was extended throughout the Nile Delta, and private property rights in land were consolidated throughout the country;

(2) the years between the First World War and the coup of 1952, when the land system was first consolidated in the 1920s and then entered a period of crisis during the Great Depression of the 1930s and the Second World War;

(3) the roughly two decades of Nasser's rule (1952–70), during which a series of land reforms were implemented;

(4) the era of the 'oil boom and bust', when the economy first experienced the well-known Dutch Disease phenomenon (roughly, 1974–85), and then a period of austerity and structural adjustment (1986–present).

A common theme throughout the macroeconomic gyrations of the final period has been a process of increased openess of the economy (*infitah*) and reduced government intervention in agriculture.

Two major historical forces created extensive private estates during the nineteenth century: Egypt's integration into the international economy as a major cotton exporter, and the attempts of Muhammad Ali and his successors to consolidate their power. The driving force was the ceaseless demand for revenue by the central government and the resistance of the peasantry to the ensuing exactions and depradations. The Egyptian Government needed revenue to extend the irrigation system (largely done with *corvee* labour), to improve the transportation infrastructure, and to finance foreign military ventures. The heavy taxation of land and peasant labour stimulated flight from the land and peasant revolts. Many peasants had their access to land confiscated as a result. New lands were also being opened up to cultivation for the first time and, although Muhammad Ali tried to administer the entire country himself, he was soon forced to decentralize. That is, he had to make grants of sequestered and newly reclaimed land to various family members and government officials. This was the origin of the *pasha* class, a tiny minority of Egyptians who held over 40 per cent of the cultivated area by the turn of the century. The holdings of these and other individuals were gradually turned into full private property in the latter half of the nineteenth century, a process which the British occupation completed. By 1900 full private property in land was the rule throughout Egypt.

The distribution of land ownership in 1900 is shown in Table 6.1. Most of these estates were exploited using a system known as the *'izbah* ('hamlet') system, in which resident permanent labourers (*tamaliyya*) were partially paid by granting them a (rotating) parcel of land on which to grow their

TABLE 6.1. Egypt: distribution of landownership, 1900–64

Year	Small properties (<5 feddans)				Medium properties (5–50 feddans)				Large properties (>50 feddans)			
	Number		Area		Number		Area		Number		Area	
	('000)	(%)	('000 feddans)	(%)	('000)	(%)	('000 feddans)	(%)	('000)	(%)	('000 feddans)	(%)
1900	761	83	1113	22	141	15	1757	34	12	1.3	2244	44
1906	1084	88	1293	21	134	11	1662	32	13	1.0	2476	47
1916	1480	91	1450	27	133	8	1645	30	12	0.7	2356	43
1936	2242	93	1837	31	146	6	1747	30	12	0.5	2254	39
1943	2376	94	1944	33	147	6	1774	30	12	0.5	2142	37
1952	2642	94	2122	35	148	5	1817	30	12	0.4	2042	34
Landreform												
1957	2718	94	2274	38	155	5	1915	32	12	0.4	1756	30
1961	2919	94	3172	52	171	5	1982	33	11	0.3	930	15
1964	2965	94	3353	55	168	5	1956	32	10	0.3	813	13

Percentages may not add to 100% because of rounding.
Source: Eshag and Kamal (1968).

subsistence crops of maize and beans. These crops rotated with wheat, clover, and cotton. The landlord or his agent specified the crop rotation, supplied seeds and fertilizers, controlled irrigation, and supervized labour on cotton, the major cash crop. The labour of these permanent workers was supplemented at peak season by outsiders, whether local villagers or migrant labourers (*tarahil*), who usually came from Upper Egypt. These workers were typically paid a cash wage above the wage which the *tamaliyya* received for harvest work.

Although this system has often been described as 'feudal' ('Amr 1958), one should eschew inappropriate parallels with European feudalism. There is much evidence that the *pashas* were economically rational and were interested mainly in the cash income of cotton production. At the same time, there is little doubt that rural social relations were characterized by landlords' exercise of arbitrary power and by self-protective deference, even obsequiousness, on the part of the peasants. The system resulted in significant social abuses; most damaging for long-term development was the hostility of many *pashas* to the education of their tenants' children. One of the principal goals of Nasser's land reform was the removal of such injustices.

As noted above, the agricultural labour force on large estates was two-tiered, composed of resident tenants and labour hired from the farm. Other major rural social groups were the land-holding small peasantry and the numerically small but highly influential rural middle class. A constant of modern Egyptian agrarian history is the steady improvement in the already strong position of the latter group. At the same time, the inelastic supply of land, the high percentage of land held in large estates, indebtedness, and population growth steadily swelled the ranks of landless agricultural workers. This rural proletariat may have amounted to 50–60 per cent of the rural population by 1950.[2]

This system produced mixed economic results. Although the large landlords were accused of being 'Ricardian drones', who conspicuously consumed their rents rather than investing them, the evidence supporting this charge is weak. Historical research indicates that many large landlords were actively engaged in managing agricultural production, responded to economic incentives, and invested in new technologies such as manufactured fertilizers, improved cotton varieties, and irrigation and drainage facilities. Some large landlords invested in industrial enterprises, such as those launched by the Bank Misr during the 1930s. The old *pasha* class are better viewed as wealthy agrarian capitalists than as 'feudalists' or 'Ricardian parasites' (Davis 1983; Richards 1982; Tignor 1984).

On the other hand, there is little doubt that the land tenure system of the era engendered some inefficiencies. Specifically, *pashas* and small peasants faced different factor prices, and often used different production techniques (for example, different crop rotations). Agricultural growth was sluggish, constrained by an increasingly tight supply of agricultural land, inadequate drainage, and, of course, the collapse of the international cotton market in the early 1930s. Agricultural output per worker deteriorated steadily throughout the period, and real agricultural wages stagnated. The poverty of the mass of the rural population sharply constricted the domestic market for simple industrial products. Human capital formation was exceedingly slow: in 1937 only 15 per cent of the entire population was literate, most of whom lived in cities. Finally, the equity picture was appalling: on the one hand a vast mass of illiterate, destitute peasants, afflicted with diseases like schistosomiasis and pellagra; on the other hand, a wealthy and sophisticated elite whose cosmopolitan consumption rivalled that of the European *haute bourgeoisie*. It is not surprising that the junior officers who seized power in 1952 made agrarian reform one of their principal goals.

Nasser's land reforms and the creation of the modern system

The Nasser regime gave land reform the highest priority for both political and economic reasons. Politically, they wished to deprive the family and friends of King Farouk, the deposed monarch, of all influence. They also sought the passive support of the second stratum of the rural middle class, and hoped to create political support among small peasants. They also wished to strengthen the central government: the Nasserists sought to substitute the state's power for that of the deposed *pashas*. Economically they pursued an import-substituting industrialization strategy by mobilizing an investable surplus from the agricultural sector. They believed that land reform would also widen the domestic market for local manufacturers. There was, in their rather Panglossian view, no trade-off between growth and equity, as their slogan, *al-adl w'al-kifayah* (justice and sufficiency) suggests. In practice, their policies stressed agricultural taxation rather than the creation of a wide rural domestic market (Richards and Waterbury 1990; Waterbury 1983).

A summary of the land reform laws and their dates is given in Table 6.2. It can be seen that the regime lowered the ceiling on land successively from 200 feddans in 1952, to 100 in 1961, and to 50 in 1969. The latter remains the legal limit for farm ownership on the Old Lands.

Several features of the reform stand out. First, all of the reforms taken together affected only 12.5 per cent of the cultivated area. Just under 342 000 families, amounting to perhaps 1.7 million persons (about 9 per cent of the rural population in 1970) received land. Second, these recipients were overwhelmingly ex-tenants, who alone were alleged to possess the necessary farming skills. Landless day labourers received little, if any, land: the old *'izbah* system, with its two-tiered labour force of permanent and casual workers, was reproduced by the land reforms. Third, much more

TABLE 6.2. Egypt: land distributed by agrarian reform laws, 1953–69

Origin of land	Area (feddans)	Families benefitting
1. Land Reform Law No. 178, 1952	365 147	146 496
2. Law No. 152, 1937 Law No. 44, 1962 (transfer of *waqf* lands)[a]	189 049	78 797
3. Second Land Reform Law No 127, 1961	100 745	45 823
4. Purchase of lands sequestered in 1956 (response to Suez invasion)	112 641	49 390
5. Law No. 15, 1963 (excluded foreigners from land ownership)	30 081	14 172
6. Other	98	49
Total	797 761	334 727

[a] Islamic law: land set aside ostensibly for charitable purposes and not taxed.
Source: Abdel-Fadil (1975, p. 10).

land changed hands as an indirect result of the reforms, because large landlords were allowed to sell land privately rather than have it confiscated by the state. The main beneficiaries of these distress sales were members of the rural middle class (Abdel-Fadil 1975; Radwan 1977).

The consequent pattern of distribution is shown in Table 6.1 (until 1964) and in Table 6.5 (for the effects of the 1969 law). Not only the land reforms, but also a fixed supply of land, rapid rural population growth, and Islamic inheritance law (which enjoins equal shares for all male heirs and half shares for all female) turned Egypt into a country of small farms. However, as the two tables show, significant inequalities in the distribution of land persisted; members of the rural middle class further consolidated their already strong position during the Nasserist period, while many landless peasants got nothing.

The old *pasha* class had fulfilled important marketing functions under the old regime. It was crucial that these functions be replaced, which was what the system of farm co-operatives was designed to do. Initially covering only those lands directly redistributed, in 1961 compulsory co-operatives were extended throughout the country. The principal features of the system were:

(1) the division of co-operative land into blocks or contiguous areas which usually contained the land of several peasant families;

(2) the planting of each block in the same crop, rotating every year in accordance with government decrees;

(3) the mandatory adherence of peasants to the government imposed crop rotation;

(4) most significantly, government monopsony or oligopsony over an extensive list of crops (some thirteen in the mid-1970s) and monopoly over crucial inputs like seeds and fertilizers.

By manipulating the crop/input terms of trade, the government used the co-operatives as the principal mechanism to tax agriculture (Dethier 1989). Beginning in the 1980s, crop production was slowly decontrolled; by 1991 the Egyptian Government retained its monopsony only over cotton and sugar cane, and there were plans to reduce government intervention still further. The government has also announced plans to retire from the input supply business.

The land reforms may have eliminated the more obvious differences in input opportunity costs facing different sized farms. First, the very largest farms were eliminated, and second, the compulsory co-operative system imposed crop rotations and supplied credit and fertilizers at uniform prices. Nevertheless, it remained (and remains) true that small farmers face a lower opportunity cost for the labour of female and child family members than do larger farms. The latter also appear to have hired a higher percentage of total labour than did small farms during the 1960s (Abdel-Fadil 1975). Further, farm inputs were subsidized, creating excess demand. The scarce inputs were rationed by political criteria: richer peasants stood at the front of the queue for credit, fertilizers, pesticides, etc. Richer farmers also used more farm machinery, while small farmers held two-thirds of the livestock herd.

The co-operative system and the land reforms did not eliminate differentiation by crop mix. The wealthier peasants could often purchase exemption from government imposed rotations, in particular, from the requirement in the Delta to plant (heavily taxed) cotton, freeing them to plant more remunerative crops like vegetables and fruits. On the other hand, small peasants continued to be more animal-intensive than larger farmers. We shall see below that this last tendency was a key to the continued survival of small farmers during the oil boom years.

The output and growth consequences of the reforms were probably positive. Unlike several other land reforms in the region (e.g. Iraq), there was no disorganization of production, since the co-operatives were able to enter into the marketing areas forcibly abandoned by the *pashas*. Agricultural growth rates were maintained, as more land was brought into cultivation and average yields either remained constant or increased. One must be careful not to attribute too much of this success to the land reforms; expanded water supplies from the Aswan High Dam were certainly at least as important. Nevertheless, it is notable that the equity gains of the land reform came at a very low, even nugatory, cost in foregone economic growth.

One should not, however, exaggerate these equity gains. The exclusion of casual labourers from the ranks of land recipients meant that the poorest peasants received no direct benefits. Although the Egyptian Government tried to help them by instituting a minimum wage, the regulation proved impossible to enforce, and in any case, would have had some undesirable employment consequences if it had been implemented. Further, since a larger percentage of the land was now in the hands of small farmers, who hired a smaller percentage of their total labour requirements, the demand for hired labour seemed to fall. Consequently, real agricultural wages fell during the 1950s, and then recovered during the 1960s, driven up by the rising demand for labour in industry, government services, and, especially, the construction of the Aswan High Dam. After Egypt's defeat in the Six Day War, June 1987, the economy plunged into depression, dragging down farm wages with it (Table 6.3). On the other hand, tenancy regulations greatly improved the welfare of tenants by making it virtually impossible to remove a tenant, and by requiring the owner to lease to the male heirs of a deceased holder of a written contract.[3] On both the equity and efficiency sides, the land reforms had positive effects, but they did not succeed in effecting as thorough a transformation of Egypt's agrarian structure as the post second World War reforms in East Asia.

Land tenure through the oil boom and bust

Modern Egyptian land tenure exhibits five fundamental features:

(1) a continued preponderance of small farms, especially in the Old Lands;

(2) the increasing importance of reclaimed land ('New Lands'), where larger farms predominate;

(3) a decline in the number of landless agricultural workers up until 1985, and (perhaps) an increase since then;

(4) a high and increasing role of off-farm economic activities for the rural labour force and for farmers' incomes;

(5) increasing mechanization, combined with a steady increase in the number of cattle owned and tended largely by small farmers.

Each of these features is connected to the others and may be seen as the outcome of three clusters of causes:

(1) continued pressure of the rural population upon the land, emigration and land reclamation notwithstanding;

TABLE 6.3. Egypt: real agricultural wages, 1938–74

Year	Average daily wage, adult males			
	Money wage Piastres[a]	Index	Cost-of-living index	Real wage index
1938	3.0	100	100	100
1939	3.5	117	101	116
1941	3.6	120	132	90
1942	5.0	167	198	83
1943	6.3	210	238	87
1944	9.3	310	262	117
1945	9.3	310	262	117
1946	9.5	317	297	107
1948	10.0	333	271	123
1949	10.0	333	259	130
1950	11.6	387	264	147
1951	12.6	420	263	160
1952	12.0	400	265	151
1953	12.0	400	269	150
1955	7.6	253	294	87
1956	10.0	333	342	97
1959	12.5	417	334	124
1960	12.5	417	337	123
1961	12.3	410	358	113
1962	14.0	450	367	122
1963	15.0	480	377	127
1964	19.0	609	438	138
1965	22.0	704	519	135
1966	25.0	801	468	170
1967	24.5	784	479	162
1968	24.5	784	499	156
1969	25.5	817	536	151
1970	25.0	801	576	138
1971	25.5	817	580	140
1972	27.5	880	613	143
1973	29.2	930	661	140
1974	32.2	1001	792	125

[a] 100 piastres = 1£E.
Source: Radwan (1977, p. 31).

(2) the economic consequences of the oil boom (and collapse in the mid and late 1980s);

(3) the changing impact of government policies.

First, the fixed supply of Nile Valley land, steady population growth, and Islamic inheritance law combined to reduce average farm size in Egypt. Although 900 000 feddans were officially reclaimed between 1952 and 1971, of which about 775 000 feddans were farm land, much land fell out of cultivation due to waterlogging and salinity problems, difficulties which were only repaired in the 1980s (if at all). In 1990, some 685 000 feddans of such land are considered productive. From 1971 to 1978, there was effectively no land reclamation (World Bank 1990); given village encroachment on cultivated land, the total cultivated area actually declined by approximately 900 000 feddans (perhaps 12 per cent of the total) (USDA 1988). In the 1980s the picture improved; not only were the drainage and salinity problems on reclaimed lands often improved, but an additional 570 000 feddans of land have been officially reclaimed. However, the World Bank estimates that only about one-half of these lands have actually achieved marginal productivity. The problem of assessing the quantity and quality of reclaimed land bedevils debates on land tenure in Egypt. However, one may hazard the generalization that the total effectively cultivated area in 1990 was still less than that in 1970.

This may not last, however. New technologies (drip irrigation, hydroponics, etc.) have helped to relax the constraint of soil infertility in the typically sandy soils of the New Lands. It is increasingly likely that water, not land, will be the binding constraint to horizontal expansion in Egyptian agriculture.[4] In the 1970s the Egyptian Government embarked on a policy of distributing the land of grossly inefficient state farms in reclaimed areas to fellahin and to secondary school and university graduates, among whom unemployment is a serious problem. Fellahin usually receive five feddans, while graduates receive between 10 and 30 feddans. Other areas of reclaimed land are sold at auction, usually to wealthy buyers who establish orchards. Such forces tend to increase the number and area of larger farms in the country.

Second, all aspects of the economy were profoundly affected by the regional oil boom (roughly 1975–85). Although Egypt's oil exports were always minor in comparative perspective (never exceeding 800 000 barrels per day, compared with a Saudi maximum in the late 1970s of over ten million barrels per day), oil replaced cotton as the principal commodity export in the late 1970s. Egyptian workers, including large numbers of rural young men, left the country in large numbers. Although the precise numbers are much disputed, a reasonable (and conservative) estimate for the number of Egyptians working abroad would be: 400 000 in 1975, one million in 1980, and two million in 1987 (Richards 1991). Some 30 per cent of all emigrants had previously been employed in agriculture. These migrants sent home funds which rose from $US123million in 1972 to over $US2billion in 1980 and which averaged $US3.2billion during the 1980s. Buoyed by such foreign exchange revenues, Egypt also borrowed heavily from abroad: foreign debt escalated from $US2billion in 1970 to $US22billion in 1980 and to about $US50billion in 1990.

The relaxation of the foreign exchange constraint stimulated an economic boom, in which per caput gross domestic product (GDP) grew at an annual compound rate of some 8 per cent from 1975 to 1982. In common with many Dutch Disease economies of the region (Gillis *et al.* 1987), Egyptian agriculture was shaped by the ensuing 'non-tradeables' boom. Labour was drawn out of agriculture to work abroad and in non-farm jobs, and there was a tendency for farm goods which enjoyed protection (whether created by transport costs, consumer tastes, or policy) from foreign competition, to expand much more rapidly than stagnating export goods like cotton and rice or import-competing crops like wheat. The demand for food, especially goods with high income elasticities like fruits, vegetables, and livestock, accelerated. The livestock boom had significant consequences for the viability of very small farms and for the agricultural labour market, because small peasants withdrew from hired labour to some extent to lavish more care on their animals.

The fundamental force transforming rural labour markets, however, was emigration and remittances. About one-third of all emigrants had previously worked in agriculture; perhaps one-third of the Egyptian farm labour force has participated in emigration for work abroad during the past fifteen years. This had both direct and indirect effects which tightened farm labour markets. First, emigration directly reduced the supply of adult male workers in agriculture; second, the remittances of rural migrants stimulated a boom in rural construction and other off-farm activities. By the mid-1980s, over one-third of the rural labour force was employed outside of agriculture, further reducing the supply of farm labour. Real farm wages soared: by 1985 they were roughly 350 per cent above 1970 levels. With the oil price collapse of the mid-1980s, however, the deceleration of emigration, combined with continued high labour force growth (2.7 per cent per year) and a contracting domestic economy, led to a reversal: by 1990 real farm wages had fallen back to approximately 200 per cent of 1970 levels (Table 6.4).

The final set of forces driving the observed trends in land tenure was the Egyptian Government's changing policy mix. The Government ceased to inhibit emigration in the early 1970s, the legal restrictions on farm size remained in force, and the state (slowly) modified its price policies, reducing the taxation of field crops (Alderman and Von Braun 1984; Dethier 1989). These policy changes began during the 1970s, but liberalization of the sector gathered steam in the 1980s.[5] By far the most important policy for land tenure was the quantitative import restrictions and tariff on livestock products (until 1988), a policy which complemented the market forces, described above, which raised the returns to animal husbandry. The Egyptian Government also continued to provide a variety of incentives for farm mechanization, such as credit and fuel subsidies. Finally, as noted above, land reclamation accelerated during the 1980s.

The growth and equity consequences of the current land tenure situation are examined in more detail below. During the 1970s, agricultural output probably actually fell; during the 1980s, by contrast, there was almost certainly a resumption of growth. It is doubtful, however, that land tenure played much part in either inhibiting or promoting growth. Other factors, such as the slow repair of the drainage network in the Nile Valley, the end of the Dutch Disease, the gradual pay-off of investments in agricultural research, and the improved policy environment were almost certainly more important than land tenure.

It is possible that small farm size and farm fragmentation inhibit the optimal

TABLE 6.4. Egypt: real agricultural wages, 1970–90

Year	Nominal wage (£E day^{-1})	Nominal wage index	Rural consumer price index (CPI)	Real wage index
1970	0.25	100	100	100
1971	0.26	104	101	103
1972	0.27	108	106	102
1973	0.29	116	115	101
1974	0.35	140	131	107
1975	0.47	188	137	147
1976	0.62	248	153	162
1977	0.76	304	168	181
1978	0.90	360	190	189
1979	1.07	428	194	221
1980	1.39	556	253	220
1981	1.81	724	287	252
1982	2.36	944	328	288
1983	3.09	1236	406	304
1984	3.72	1488	443	336
1985	4.36	1744	495	352
1986	4.84	1936	608	318
1987	4.80	1920	689	279
1988	5.13	2052	831	247
1989	5.52	2208	956[a]	231[a]
1990	5.47	2188	1099[a]	199[a]

[a] Assuming a rural CPI inflation rate of 15 per cent per year, a very conservative estimate (official CPI for these years not yet published).

Source: CPI data, CAPMAS (various years); wages data, Department of Agricultural Economics and Statistics, Ministry of Agriculture (unpublished data).

use of irrigation water. Egypt faces an impending water shortage, since the water supply is essentially fixed, while demand grows apace. Experience elsewhere in the world suggests that there are administrative economies of scale in pricing water. Some have suggested that the inequality of land holdings is a burden to solving the free-rider problem which the organization of water-users' associations creates in the Old Lands. However, this question has not been as intensively studied as it deserves; the truth is that we do not really know much about the impact of Egyptian land tenure on water-use efficiency.

To the extent that greater farm mechanization is necessary to attain the (extremely high) yield potential of Egyptian agriculture, it could be argued that current land ceilings inhibit such development (York et al. 1982). However, since mechanization has been subsidized, and since farm labour costs are falling, it is not obvious how important a force this would be in a deregulated price environment. On the other hand, it is noteworthy that in contrast with many other peasant agricultures of the world, there is remarkably little evidence that small farms have higher output per unit area or noticeably higher cropping intensities than do large farms. Evaluating this case is difficult, because the data often come from non-comparable sample surveys. Nevertheless, despite the variety of results, none shows a marked difference in output per feddan per crop by farm size. (Commander 1987; Radwan and Lee 1986; Richards et al. 1983)[6].

The same surveys typically show that tenancy is relatively widespread, ranging from 25 to 50 per cent of the farmed area. A recent study based on the 1982 Agricultural Census (Springborg 1990) argues that there has been a decline in the total area which is leased (from roughly 40 per cent in 1961 to 25 per cent in 1982). As noted above, such 'official renting', which in effect transfers property rights to the tenant, is likely to be less common than unofficial oral contracts, which often cover a single season. Some of this appears to be the traditional renting in of small parcels of land by very poor or landless peasants, However, renting out very small parcels of land by landowners whose main occupation is not farming is also a very common practice. As a result, large ownership holdings are often broken down, and extremely small parcels are consolidated. As farming becomes more profitable with deregulation, and to the extent that technological change (e.g. mechanization) is characterized by economies of scale, we would expect rentals to decline in favour of larger owners operating their own farms. The overwhelming majority of rentals are for cash; sharecropping is relatively rare in Egypt. There is little evidence that the current land tenure system is a source of inefficiency or of inadequate growth of farm output; how long this will remain true in the face of continued technological change and increasing water scarcity remains to be seen.

There is little doubt that rural poverty declined during the decade of the oil boom. The conclusion rests on two types of evidence: the historically unprecedented rise of real farm wages, 1970–85; and a comparison of the results of the Household Budget Surveys of 1974/5 and 1981/2. Although there are problems of comparability of the data, even critics such as Korayem (1987) agree that the number of poor rural households fell from approximately 1.8 million in 1974/5 to about 1.2 million in 1981/2.[7] Korayem estimates that the bottom 40 per cent of the Egyptian rural expenditure distribution increased its share of rural expenditure from 17.2 to 21.3 per cent between the two survey years. Despite the fact that small farmers held a somewhat smaller percentage of the total cultivated area in 1982 than in 1975 (Tables 6.5, 6.6), this effect was apparently swamped by the increasing importance of off-farm activities, rising real farm wages, and the livestock boom. Until the results of the 1989 survey are published, we cannot say much about the 1980s; however, the decline in real farm wages since 1985 is very disturbing. Poverty is almost certainly rising in rural Egypt, and inequality may also be increasing.

The current peasant differentiation picture may be summarized as follows. First, small peasants continue to hold most land, especially in the Old Lands. Roughly 50 per cent of Egyptian agricultural land is in farms of less than five feddans; when company land holdings (almost exclusively in the New Lands) are excluded, the figure rises to 57 per cent (Tables 6.5, 6.6; Springborg 1990). It has been argued recently that the agrarian reforms were being 'rolled back' under Sadat (Springborg 1990). This is only true in comparison with 1975, and, even here, some 75 per cent of the decline in total area of farms under five feddans was due to the decline of the 'postage stamp' farms of under one feddan. In view of the evidence cited in the preceding paragraph on poverty and distribution, it is plausible to conclude that during the period 1975–82 (the height of the oil boom) very small farmers in many cases abandoned farming in favour of urban construction work or emigration abroad.

Even in 1982, roughly one-third of all farms (but only some 6 per cent of the cultivated area) were in very small farms (less than one feddan). A major factor in the survival of such *zweigwirtschaft* (dwarf agriculture) has been such farmers' concentration on animal husbandry. As Table 6.7 shows, such very small farms are by far the most animal intensive. In one sample (Richards *et al.* 1983) over 60 per cent of all family labour time on very small farms was absorbed by livestock activities. This was a highly rational allocation of family labour time, given the very strong demand for such products together with the ease with which women and children can combine tending animals with other household activities. The increasing availability of off-farm employment activities also contributed to the survival of small and very small farms. Some studies show that between one-third and one-half of family income for such farmers comes from non-agricultural employment and remittances from relatives working outside the village. The national and regional boom, combined with government policy, shored up the position of Egyptian small farmers.

At the same time, the total land area of farms of 10 feddans and above

TABLE 6.5. Egypt: land distribution by size of farm in 1975

	Size of farm (feddans)						
	Small farms			Medium farms		Large farms	
	<1	1–<3	3–<5	5–<10	10–<50	>50	Total
Farm operators	1 124 286	1 160 147	354 841	148 459	65 059	131	2 852 923
Operators (%)	39.4	40.67	12.44	5.2	2.28	0.004	100.0
Area (feddans)	739 028	2 023 456	1 185 581	944 411	985 508	105 684	5 983 668
(%)	12.351	33.816	19.814	15.783	16.50	1.76	100.0
Average farm size (feddans)	0.66	1.7	3.3	6.4	15.1	806.7	2.1

These figures are for farms as units of production, unlike those of Table 6.1 which are for units of ownership.
Source: Department of Agricultural Economics and Statistics, Ministry of Agriculture (unpublished data).

TABLE 6.6. Egypt: distribution of land, 1982

Holding size (feddans)	Area (feddans)	(%)	Holdings (number)	(%)
0–1	399 357	6.0	796 394	32.3
1–5	3 084 173	46.5	1 427 446	57.8
5–10	1 098 224	16.6	173 248	7.0
10–50	1 206 156	18.2	67 350	2.7
50+	844 548	12.7	3 950	0.2
Total	6 632 464[a]		2 468 408	

[a] The total land area of the census is regarded as exaggerated by some authorities (USDA 1988); it reflects acceptance of the Egyptian Government's (sanguine) claims on the extent of reclaimed land.
Source: Ministry of Agriculture (1982).

roughly doubled between 1974–5 and 1982. In the latter year, less than 3 per cent of the number of farms contained just under one-third of the cultivated area (Tables 6.5, 6.6). Most of this is the result of land reclamation. There are undoubtedly economies of scale in desert farming: the heavy capital requirements of the drip and sprinkler irrigation systems which sandy soils require, and of fruit orchards, agronomically well-suited to such soils (and highly appropriate for drip irrigation), generate advantages for larger farms. Although many of these are corporate entities, large private farms are also common in reclaimed areas, as are the 'graduate farms' noted above. To

TABLE 6.7. Egypt: livestock density, 1982

Farm size	Cattle (head)	Water buffalo (head)	Land area (feddans)	Livestock density (head feddan^{-1})	
				I	II
0–1	372 627	352 763	399 357	0.933	1.816
1–5	1 518 270	864 213	3 084 173	0.492	0.772
5–10	328 827	112 821	1 098 224	0.299	0.402
10–50	184 737	36 014	1 206 156	0.153	0.183
50+	46 166	1 631	844 548	0.055	0.057

In addition, landless rural inhabitants owned 278 405 cattle and 255 253 water buffalo.
Livestock density: I, Cattle only; II, Cattle and water buffalo.
Source: Ministry of Agriculture (1982).

the extent that such areas continue to increase, we would expect to see an intensification of the bimodal agrarian structure which, land reforms notwithstanding, has long characterized Egyptian agriculture.

These tendencies may be reinforced by the Egyptian Government's determination to promote mechanization not merely to replace animal labour (as has largely been the case so far), but also to substitute for human labour, especially harvest work. Everywhere else in the world, farm mechanization has been associated with increasing farm size, rental markets for machine services notwithstanding. Despite the presence of well-developed tractor rental markets in Egypt, there is little reason to suppose that the Nile Valley's experience will be any different. Until the late 1980s, the Egyptian Government shored up small farms with its livestock policies while laying the basis for undermining them in the future with its mechanization subsidies. Now that both policies are to be cancelled, the consequences for future farm size is moot. The decline in farm wages and off-farm employment activities certainly bodes ill for the rural poor, including very small farmers. Although small farms are likely to remain the norm in Egypt for the immediate future, their ability to cope with the demands of the future is uncertain.

References

Abdel-Fadil, M. (1975). *Development income distribution and social change in rural Egypt (1952–1970): a study in the political economy of agrarian transition.* Cambridge University Press, Cambridge.

Adelman, I. (1984) Beyond export-led growth. *World Development*, 12, 937–49.

Alderman, H. and Von Braun, J. (1984). *The effects of the Egyptian food ration and subsidy system on income distribution and consumption.* International Food Policy Research Institute, Washington DC.

'Amr, I. (1958). *The land and the peasant: the agrarian question in Egypt.* (in Arabic). Dar al-Ma'arif, Cairo.

Berry, R.A. and Cline, W. (1978). *Agrarian structure and productivity in developing countries.* Johns Hopkins University Press, Baltimore.

Binswanger, H.P. and Rosenzweig, M.R. (1984). *Contractual arrangements, employment, and wages in rural labour markets in Asia.* Yale University Press, New Haven.

CAPMAS (Central Agency for Public Mobilization and Statistics) (various years). *Statistical yearbook.* CAPMAS, Cairo.

Commander, S. (1987). *The state and agricultural development in Egypt since 1973.* Ithaca Press, London.

Davis, E. (1983). *Challenging colonialism: Bank Misr and Egyptian industrialization, 1920–1941.* Princeton University Press, Princetown.

De Janvry, A. (1981). *The agrarian question and reformism in Latin America.* Johns Hopkins University Press, Baltimore.

Dethier, J. (1989). *Trade, exchange rate, and agricultural policy in Egypt.* World Bank, Washington, DC.

Dyer, G. (1989). Agrarian transition in Egypt: technological change in Egyptian agriculture and its impact on the relation between farm size and productivity. Unpublished M.Phil. thesis. University of Cambridge, UK.

Eshag, E. and Kamal, M. A. Agrarian reform in the United Arab Republic (Egypt). *Bulletin of the Oxford University Institute of Economics and Statistics*, **30**, 76.

Gillis, M. *et. al.* (1987). *Economics of development*, (2nd edn). W.W. Norton, New York.

Griffin, K. (1974). *The political economy of agrarian change: an essay on the green revolution*. Harvard University Press, Cambridge, Massachusetts.

Johnston, B. and Kilby, P. (1975). *Agriculture and structural transformation: economics strategies in late developing countries*. Oxford University Press, New York.

Korayem, K. (1987). *The impact of economic adjustment policies on the vulnerable families and children in Egypt*. Third World Forum and UNICEF, Cairo.

Mellor, J. (1976). *The new economics of growth: a strategy for India and the developing world*. Cornell University Press, Ithaca.

Ministry of Agriculture (1982). *Agricultural census, 1982*. Tables 2–A and 48–A. Ministry of Agriculture, Cairo.

Radwan, S. (1977). *Agrarian reform and rural poverty: Egypt, 1952–1975*. International Labour Office, Geneva.

Radwan, S. and Lee, E. (1986). *Agrarian change in Egypt: an anatomy of rural poverty*. Croom Helm, London.

Richards, A. (1982). *Egypt's agricultural development, 1800–1980: technical and social change*. Westview Press, Boulder.

Richards, A. (1986). *Development and modes of production in marxist economics: a critical evaluation*. Harwood Academic Publishers, New York.

Richards, A. (1991) Agricultural employment, wages and government policy during and after the oil boom. In *Labour and structural adjustment: Egypt in the 1990s*, (ed. H. Handoussa and G. Porter). American University in Cairo Press, Cairo and ILO, Geneva.

Richards, A. and Waterbury, J. (1990). *A political economy of the Middle East*. Westview Press, Boulder.

Richards, A., Martin, P.L., and Nagaar, R. (1983). Labor shortages in Egyptian agriculture. In *Migration, mechanization, and agricultural labour markets in Egypt*, (ed. A. Richards and P.L. Martin), pp.21–44. Westview Press, Boulder and the American University in Cairo Press, Cairo.

Springborg, R. (1990). Rolling back agrarian reforms in Egypt. *Middle East Report*, September – October, pp.27–33.

Tignor, R. (1984). *State, private enterprise, and economic change in Egypt, 1918–1952*. Princeton University Press, Princeton.

USDA (United States Department of Agriculture) (1988). *World agricultural trends and indicators, 1970–1988*. USDA-ERS, Statistical Bulletin No. 1781. USDA, Washington, DC.

Waterbury, J. (1983). *The Egypt of Nasser and Sadat: the political economy of two regimes*. Princeton University Press, Princeton.

World Bank (1989). *Arab Republic of Egypt: a study on poverty and the distribution of income*. Draft Report, (5 January). Country Economic Division, World Bank, Washington, DC.

World Bank (1990). *Arab Republic of Egypt: land reclamation subsector review*. Report (1 February). Country Department III, World Bank, Washington DC.

York, E.T. *et al.* (1982). *Egypt: strategies for accelerating agricultural development: a report of the presidential mission on agricultural development to Egypt*. Ministry of

Agriculture, Cairo, and Agency for International Development, Washington, DC. with the co-operation of the US International Agricultural Development Service (processed).

Notes

[1] 'Old Lands' refers in Egypt to the rich alluvial soil of the Nile Valley and Delta. 'New Lands', by contrast, are lands reclaimed from the edges of the desert. Land ceilings do not apply to the New Lands; the consequence of this law for the evolution of land tenure are taken up below.

[2] Since the numbers of landless are calculated as a residual (the difference between the agricultural population and the numbers of peasants with land), much uncertainty surrounds their exact number.

[3] Not surprisingly, such laws have induced landowners to switch to oral contracts, to which such stringent regulations do not reply.

[4] The relationship between the imperative of enhanced efficiency of water use and land tenure is discussed below.

[5] Dr Yusuf Wally, Minister of Agriculture since 1981, has long been one of the most forceful advocates of economic liberalization in the Cabinet.

[6] A recent study of this phenomenon finds considerable regional variation, with some governorates (weakly) showing the usual inverse relationship but others showing the reverse (Dyer 1989).

[7] Other sources, e.g. World Bank (1989) estimate the number of rural poor households in 1981/2 at about 1.0 million.

7

Food preferences and nutrition

H. ALDERMAN

Introduction

As is so often the case, an economist's notion of preferences differs from common usage. In consumer theory, observed behaviour says as much about prices and access to resources as it does about preferences. This is not merely an academic quibble, although theorists debate whether tastes are fixed or not (Stigler and Becker 1977). Both the level and the trend of food consumption patterns in Egypt reflect the influence of government pricing policies as well as habits. Consequently, it is necessary to delve into such policies if one is interested in understanding current consumption, or in speculating on what may occur were changes in price policy to occur. This chapter, then, will begin with a section discussing statistics of average levels of food consumption with reference to changes in recent years. It will then devote attention to subsidy and distribution policies before concluding with a discussion of indicators of nutritional status.

Diet composition

Although the beautiful friezes on the tombs in Saraqua and Thebes are as likely to depict barley as emmer and other wheats—the former remaining a major grain in Egypt through the last century (Rivlin 1961)—the current diet of the Egyptian people is based primarily on wheat and secondarily on maize. This is true whether referring to upper or lower Egypt, urban or rural areas. However, in the rural areas people consume more grain wheat and maize, while urban residents primarily purchase their bread ready-made (Table 7.1).

Within the broad pattern of a grain-based diet, a number of trends are apparent. Most significant is the marked rise in flour consumption in rural areas over the last decade. While there has been some decline in the utilization of grain wheat and maize in this period, as well as in sorghum consumption, the decline was far less than the increase in flour. Indeed, taken by itself the reported household consumption of flour in rural areas might appear dubious, while the corresponding entry in Table 7.1 might raise the eyebrow of the reader with a trusting as well as that of one with a skeptical nature. However, the levels of consumption observed in the household survey are consistent with trade and production data on the national as well as the governorate level (Alderman *et al.* 1982). In 1980, the Minstry of Supply distributed over 140 kg flour per caput through its marketing channels; over 200 kg per caput in the primarily rural governorates of Sohag, Qena, and Aswan in Upper Egypt. This remarkable level of consumption of marketed flour in rural areas and of ready-baked bread in urban areas has been fostered by subsidies on bread and flour as well as changes in urban employment and time allocation patterns.

While regional patterns in diet are not pronounced in Egypt, given the relatively homogenous population, there remains some diversity in the forms of breads preferred. In the south, leavened breads known as *sun* breads are consumed along with flat, platter-shaped loaves of a mixture of wheat and maize or sorghum. The percentage of maize in the dough is particularly high in middle Egypt and parts of the Nile Delta. There the prevalent type of bread is a large wafer-like bread known as *bettai* bread (May 1961). This bread may include up to 3 per cent fenugreek flour (fenugreek is a lysine-rich pulse which is also used in a gruel for postpartum women and convalescents). *Bettai* bread stores well and is baked at intervals of one week or longer.

When maize is used in baking, it is the white maize grown locally rather than imported yellow maize. The latter is used universally as an animal feed. Statistics on yellow maize utilization frequently report household consumption, but this is generally for poultry. White maize flour is often preferred to wheat flour as many villagers consider it more filling and healthy. In rural areas the culture continues to value home baking as indicated by the positive relationship between baking and income in villages, compared with the negative relationship usually apparent in cities. In both sectors the availability of bakeries and the length of the lines at these outlets influence baking and bread consumption (Alderman and Von Braun 1984).

Urban consumers generally buy a flat loaf called *baladi* bread, available from licensed government bakeries and vendors. While in recent years sizes, moisture contents, and flour extraction rates have varied according to regulations, the bread has been always cheap (Alderman *et al.* 1982). *Baladi* bread has a moisture content of 30 per cent and becomes stale quickly as it dries out. A similar flat loaf, called *shami* bread, made of a more refined flour, is also available in urban areas, as is a loaf reminiscent of French bread from which it derives its name of *afrangi* (or *fino*) bread. The price and content of these breads are also regulated by the Egyptian Government. In 1981, only 25 per cent of total urban bread sales by volume consisted of *shami* or *afrangi* breads.

Less common is a type of bread called *fateer*. This bread, which is usually rolled with shortening, is generally a breakfast food, although it is also

TABLE 7.1. Egypt: per caput food consumption in rural and urban areas

Food	Rural areas				Urban areas			
	1958/9	1964/5	1974/5	1981/2	1958/9	1964/5	1974/5	1981/2
	\multicolumn{8}{c}{Food consumption (kg caput^{-1} year^{-1})}							
Wheat grain	69.7	69.9	59.5	43.7	13.1	12.3	7.7	4.6
Maize	75.9	66.8	45.2	47.5	12.5	13.5	5.8	5.0
Rice	23.8	24.6	21.1	33.4	19.5	21.2	25.0	26.2
Wheat flour	16.1	22.5	40.7	143.9	33.4	8.0	25.4	44.5
Bread[a]	6.0	14.6	18.8	20.4	96.7	115.8	137.9	170.7
Beans	5.3	6.9	4.5	4.0	4.0	4.6	4.3	2.3
Lentils	3.2	4.5	3.1	3.6	3.3	4.2	4.1	1.8
Meat, poultry, and fish	12.5	13.2	13.6	30.5	19.4	19.9	19.0	35.3
Vegetable oils and fats[b]	2.7	5.0	7.8	5.8	6.6	8.7	8.1	7.9
Sugar	10.0	11.6	13.4	23.5	11.4	12.2	13.0	25.1
Calories (caput^{-1} day^{-1})	2729	2898	2590	3274	2252	2227	2433	3016

[a] bread is approximately 30 per cent moisture by weight.
[b] 1981/2 oil only.
Source: CAPMAS (1961, 1972, 1978); Alderman and Von Braun (1984).

FOOD PREFERENCES AND NUTRITION 117

available from urban vendors. If preferences held sway without the influence of prices, then clarified butter or ghee would be used *in lieu* of shortening.

Ghee is still used during festivals and by families with access to cheap EEC imports, but vegetable ghee or oil— generally cottonseed oil—is the more common cooking medium. Higher levels of consumption of oil are observed in the Delta.

Some regional patterns also exist in the consumption of rice and fish, both of which are more often consumed in the Delta region. However, much of the growth in rice consumption in recent years can be attributed to the increased access to rice in Upper Egypt through government marketing channels.

Fluid milk consumption is higher in urban areas. In the cities, pasteurized and sterilized milk, often imported from Europe, is more readily available. Furthermore, urban residents are more likely to have refrigerators. Rural milk consumption is predominantly from own production. Conversely, cheese consumption, generally of a highly salted white cheese—either young or aged—is higher in rural areas than urban.

Meat consumption has grown markedly in recent years in both rural and urban Egypt. In the rural areas, one-third of the consumption is from chickens, pigeons, rabbits, and guinea-pigs raised on the farm. In the urban areas, consumption is more or less evenly divided between chicken and meat, with fish consumption being about one-quarter of the total. The preference for boiled chicken has given Brazilian imports an edge over varieties primarily raised as broilers. Meat consumption also relies increasingly on frozen imports although fresh beef, buffalo, and camel are preferred. Swine are rarely raised, even among the Coptic minority. One exception is the community of garbage collectors called *zabaleen* who live on the outskirts of Cairo and feed food wastes to their animals.

There are meatless days mid-week when butchers are closed. This has little impact on those families at the top of the income distribution as they have refrigerators. It also has little impact on families on the lower tail of the income distribution. Most poor households consume meat once or twice a week, usually on Thursday nights or Friday noon.

Ministry of Supply officials indicate that meat consumption increases during the Moslem month of fasting, Ramadan. During this period, the consumption of sugar, refined flour, dried fruit, and nuts also increases. The Egyptian Government generaly makes special arrangements to import these fruits, especially dried apricots which in the form of apricot leather, called *qameradeen*, is traditional for the nightly breaking of the fast.

The Coptic community also observes a form of fasting during Lent and various fast days throughout the year. During these fasts no meat or dairy products are eaten. These observances, together with the Moslem observance of Eid ul-Azha, when a sacrifice of a goat is customary, all influence the timing and probable meat consumption through the year.

In addition to the foods indicated in Table 7.1, a considerable quantity of street (or prepared) foods are consumed in Egypt. Primary amongst them are *tamiya (falafel)* and *foul madames*. The former are fried bean cakes and the latter boiled faba beans (*Vicia faba*). These foods are not only consumed by workers in sandwiches, but are also brought to households for the morning meal. One still sees buckets lowered from upper floor balconies to be filled with beans from a vendor who has cooked them overnight. Data from the 1981 survey indicate that the amount of beans consumed in such form exceeds the amount actually prepared at home. Similarly, in urban areas a mixture of lentils, rice, pasta, and onions, called *koshari*, provides an inexpensive meal when bought from pushcarts and corner shops.

Finally, beverages form a major item of food expenditure and account for much of the appreciable sugar consumption in the Egyptian diet. While descriptions of life in Cairo and Alexandria a generation ago are replete with images of coffee shops, today the beverage is more likely to be tea, frequently boiled and always sweet. Also popular are infusions of *karkadé* (*Hibiscus sabdariffa*), fenugreek, anise or tamarind, as well as various syrups; these are usually served heavily sweetened. In larger cities, carrot, orange, strawberry, and mango juices, as well as sugar cane, are pressed when in season, and the locally packaged juices are widely, if sporadically, available. They are not, of course, as prevalent as fizzy drinks, usually brands familiar in Europe and the United States. These soft drinks are symbols of successful entrepreneurship and rampant consumerism, but they are, nevertheless becoming increasingly popular.

The impact of government policies on food consumption

Cairo is a city in which the building blocks of past civilization are reworked for modern needs; Pharaonic carvings become the floors of Mameluke tombs which in turn house contemporary Ciarenes, and lotus and papyrus motifs adorn tents from which the Koran is read long after the plants have disappeared from the banks of the Nile. Similarly, state granaries, market inspectors, and fixed prices of bread, which are recorded in the history of Egypt from the time of Joseph through Fatamid and Mameluke rulers and on to Muhammed Ali, still play a major role, although the role has been adapted for modern conditions. (Boas 1980; 1981; Rivlin 1961; Scobie 1981). In the past, various governments have played a central role in grain marketing and storage as a counter to the vagaries of the Nile's flood. Contemporary institutions such as the ration system and the system of flour marketing also have their origins in policies designed to stabilize prices and consumption. However, during the 1970s the system evolved into a general subsidy policy, as domestic prices for bread and flour were held constant in nominal terms during a period of inflation and currency devaluation. At the same time, the number of bakeries selling fixed-price bread increased in urban areas and outlets selling flour extended into rural regions. Consequently the volume and cost of the government marketing system increased. The growth of this consumer subsidy is shown in Table 7.2.

While a discussion of the macro-economic consequences are beyond the scope of this chapter (Scobie 1983), the magnitude of the transfer—over 6 per cent gross national product (GNP) in all but one year between 1974 and 1984—is an indication that the subsidy plays a major role in determining consumption patterns. The large increase in subsidy expenditure in 1974 can be attributed to the rise in the world price of grain in that year, while the increase in 1979 reflects domestic price stabilization policies subsequent to a currency devaluation. Similarly, the decline in 1983 reflects softening of international commodity prices, notably sugar and meat prices. However, the subsidy should not be considered merely as a passive response to external price shocks. The per caput volume of food delivered through the subsidy system grew rapidly in the 1970s, then it stabilized. Subsequent reductions were achieved by gradual increases in the price of commodities as well as reductions in the number of goods subsidized.

This consumer food subsidy system is complex. The cornerstone is the flour and bread distribution network, consisting of licensed bakeries and flour depots. As there are virtually no quantity restraints on purchases, the government, in effect, makes the domestic supply curve horizontal, with imports and aid making up the difference between growing demand and stagnating production. Since the price at which the flour is made available is below the import price, the distribution contributes appreciably to the level of consumption of wheat products and encourages shifts away from maize, sorghum, and rice.

Another prominent aspect of the system is the ration system which reaches over 92 per cent of the entire population, both urban and rural. Households are guaranteed a quota of sugar, oil, tea, and rice throughout the year. Beans and lentils have been available seasonally and were sold at low subsidized prices. Additional quotas were also generally available at fixed prices, which may or may not have been below import prices depending on market conditions.

A related marketing channel is a network of government-run shops and co-operatives which sell the above six commodities at fixed, generally subsidized, prices. The subsidy is, however, less than through the ration system and quotas are not assured. Rather, access is on a first come first served basis, which leads to long queues when a shipment of rice or sugar arrives. Frozen meat, poultry, and chicken are also sold at these co-operatives at prices below the domestic price of fresh meat and often below import costs. Per family quotas exist but are rarely enforced for these commodities. In addition, the Egyptian Government subsidizes the cost of pasta and of yellow maize. The subsidy on yellow maize is intended indirectly to keep the price of poultry down.

An indication of the role of these governmental marketing channels can be

TABLE 7.2. Egypt: indicies of government food subsidies, 1971–85

	1971	1972	1973	1974	1975	1976	1977	1978	1979	1980	1981	1982	1983	1984	1985
Nominal expenditures (£E million)	3	11	89	329	491	322	541	640	1020	1108	1689	1775	1645	2011	2053
Nominal expenditure as % of total government expenditures	0.2	0.7	5.5	16.5	16.9	8.5	13.8	15.4	18.2	16.4	21.4	15.3	14.7	15.6	NA
Nominal expenditures as % of Gross National Product	0.1	0.3	2.4	7.8	9.4	4.8	6.6	6.5	8.1	7.2	9.7	8.7	6.7	7.3	6.6
Index of per caput subsidies (1980=100) deflated	0.8	2.8	21.2	69.2	92.2	53.9	78.4	81.4	114.6	100.0	134.4	119.6	92.9	97.5	82.8

NA, not available.
Source: Alderman et al. (1982); Pinstrup-Andersen et al. (1987).

obtained from Tables 7.3 and 7.4. The first three sources of nutrients listed in the tables are government-regulated outlets. Clearly, they are the principal channels of food marketing even in rural areas. Open-market flour sales also reflect government subsidy policies in as much as these sales are generally legal resale of flour from depots. The tables also indicate the major role of grain in the generally adequate level of protein consumption (corrected for amino-acid composition by recording combinations of foods taken at each meal). Meat, fish, and chicken jointly contribute less than 20 per cent of protein, and in urban areas less than one-third of that protein comes from meats sold at the co-operative; in rural areas, a negligible portion of protein is from subsidized meats. Consequently, the policy of subsidizing meat products in the 1980s does not appear to have had either an appreciable nutritional impact, nor a justification in terms of a protein gap.

This cannot be said of the subsidy system as a whole. While average calorie and protein levels are high, they vary by income group. This, of course, is often observed (Reutlinger and Selowsky 1976), as are many of the commodity – income relationships implicit in Tables 7.3 and 7.4. The importance of the system for the nutrition of the poor stems from the fact that it provides an implicit income transfer which averages 18 and 13 per cent of the total expenditures of the poorest rural and urban income quartiles, respectively (Alderman and Von Braun 1984). This income transfer, coupled with income elasticities of 0.3 and 0.2 estimated for the poorest rural and urban income quartiles, respectively, contribute to the generally high levels of calorie consumption in the country. This impact on caloric intake is also augmented by the price response for many of the foods which are subsidized.

Nutritional concerns

Despite the high average level of food consumption and the moderately progressive nature of the food subsidy, malutrition levels in Egypt remain fairly high. Furthermore, infant mortality rates, often an indicator of under-nutrition, remain at a relatively high level for a middle-income country. The level in 1988 was 83 deaths per thousand (World Bank 1990). Note, however, that this is a marked improvement from the 104 deaths per thousand reported in 1982 (World Bank 1990).

A closer look at the incidence of malnutrition is useful in explaining the persistence of malnutrition in an environment which allows for high average levels of food consumption. A survey of nutrition conducted in 1978/9 found moderate rates of severe malnutrition (Nutrition Institute 1978). Only 0.8 per cent of the children were severely malnourished (defined as third degree malnutrition in the Gomez classification—under 60 per cent standard weight for age). Indeed, the survey found childhood obesity more prevalent than malnutrition. However, a simlar study in rural areas found 3.1 per cent severely

TABLE 7.3. Egypt: average daily calorie consumption by expenditure quartile

Method/source	Urban expenditure quartile				All urban house-holds	Rural expenditure quartile				All rural households
	1st	2nd	3rd	4th		1st	2nd	3rd	4th	
Calorie consumption (calories)										
24-hour recall	2343	2761	2915	3174	2798	2357	2574	2716	3149	2654
Food purchase	2420	2850	3072	3731	3016	2273	2892	3409	4571	3274
Source of calories (per cent)										
Ration system[a]	19	17	15	12	16	15	12	10	8	11
Cooperatives[b]	5	6	6	7	6	1	1	1	2	1
Flour and bread[c]	49	45	42	35	42	34	25	19	19	23
Additional share of open market flour	—	—	—	—	—	14	16	15	15	15
Sugar[d]	10	9	9	9	9	10	8	8	7	8
Rice[d]	8	8	9	8	8	9	13	12	16	13
Meat[d]	2	2	3	4	3	1	1	1	1	1
Chicken[d]	1	1	1	1	1	1	1	2	1	1
Fish[d]	1	1	1	1	1	<1	<1	<1	<1	<1
Production by household	—	—	—	—	—	8	13	13	14	12

Note: Expenditure quartiles were determined by ranking rural and urban households independently according to total reported expenditures per caput. The 1st quartile had the smallest expenditures; the 4th had the largest. Calorie consumption recorded by 24-hour recall is the food reported eaten in the preceding 24 hours converted to calories. The food purchase method of recording calorie consumption used the calorie content of the food purchased in one month by a household.

[a] These include both basic and additional rations.
[b] These figures include frozen meat.
[c] These figures are for bakeries and government flour shops only.
[d] These include all sources, including production by a household.

Source: Alderman and Von Braun (1984).

TABLE 7.4. Egypt: average daily protein consumption by expenditure quartile

Method/source	Urban expenditure quartile				All urban households	Rural expenditure quartile				All rural households
	1st	2nd	3rd	4th		1st	2nd	3rd	4th	
Protein consumption (grams)										
Purchase	72	88	96	114	91	70	90	107	125	95
24-hour recall	81	92	104	118	99	71	78	83	97	80
Amino-acid corrected	63	73	90	108	83	45	55	61	76	57
Source of protein (per cent)										
Ration system[a]	5	4	4	3	4	5	3	3	3	3
Cooperatives[b]	6	6	6	5	6	<1	1	1	1	1
Flour and bread[c]	58	51	47	41	49	41	31	23	22	28
Additional share of open market flour	–	–	–	–	–	17	20	18	17	18
Rice[d]	5	6	6	6	6	6	10	7	12	9
Beans and lentils[d]	3	4	3	4	3	4	2	4	4	4
Meat[d]	4	6	7	11	7	3	4	4	4	4
Chicken[d]	4	5	6	7	6	3	5	6	10	7
Fish[d]	3	4	7	5	5	1	2	3	3	2
Production by household	–	–	–	–	–	11	16	17	23	17

Note: Expenditure quartiles were determined by ranking rural and urban households independently according to total reported expenditures per caput. The 1st quartile had the smallest expenditures; the 4th had the largest. Protein consumption recorded by 24-hour recall is the food reported eaten in the preceding 24 hours converted to protein. The food purchase method of recording protein consumption used the protein content of the food purchased in one month by a household.

[a] These include both basic and additional rations.
[b] These figures include frozen meat.
[c] These figures are for bakeries and government flour shops only.
[d] These include all sources, including production by a household.

Source: Alderman and Von Braun (1984).

malnourished after correcting for sample weighing (El Lozy et al. 1980). The discrepancy between these results prompted the Institute of Nutrition to do a resurvey in 1980, visiting a subset of the 1978 sample in August and September. They found that in the 18-month interval, mean height for age as a percentage of the standard increased and the prevalence of stunting (chronic malnutrition) decreased (Nutrition Institute 1980). However, the prevalence of wasting, as measured by weight for height, increased, with a threefold increase in this measure in children with acute malnutrition.

An additional nutrition survey was conducted in 1986. This survey had a somewhat smaller sample frame than that of the 1978 survey and was undertaken only in the summer months. Surprisingly, the 1986 survey found only a slight reduction in the level of chronic malnutrition (weight for height) and an increase in acute malnutrition (height for age). Chronic malnutrition declined from 26.5 per cent to 21.4 per cent, while acute undernutrition increased from 2.9 to 7.0 per cent (CAPMAS and UNICEF 1988).

The improvement in long-term indicators of malnutrition concurrent with a deterioration in short-term measures is probably explained by seasonal patterns in diarrhoeal disease. If so, one can argue that the current problem of protein energy malutrition in Egypt is as much a sign of inadequate health and sanitation as a food problem. Progress in the treatment of the diarrhoeal diseases which contribute to malnutrition is reported in the 1980 nutrition survey. At that time nearly one-fifth of the disease episodes in rural Egypt were treated with oral rehydration salts. The amount of oral rehydration salts produced and distributed in the country more than doubled between 1980–2 and 1984–6 (NCDPP 1988). It was also shown that such oral rehydration salts reduce infant and child mortality significantly in controlled experiments. The use of such salts does not, however, reduce disease incidence.

Progress is also apparent with regard to pellagra, which was widespread in the 1950s (May 1961), but is no longer common. Indeed, in the absence of direct observations, Taylor (1979) used the reduction in reported cases of this debilitating and often fatal disease as evidence of the decline of maize consumption in rural areas.

Less progress is evident for anaemia, which is often observed in children and adult women (Nutrition Institute 1978, 1980). No change in average haemoglobin levels were observed between February 1978 and September 1980. While 38.4 per cent of preschoolers were anaemic in 1978, the 1986 survey found that 51.6 per cent of the children were anaemic. The more recent survey also indicated that 21.4 per cent of adult women were anaemic. The prevalence of anaemia reflects the frequency of pregnancies, high parasite loads, and a diet high in phytates and low in meat consumption, at least among the poor. Anyone familiar with anaemia levels in the developed countries cannot be optimistic that the problem will be resolved by changes in diet resulting from economic progress.

Further progress in nutritional status in Egypt is only likely to be achieved through improved health-care delivery and sanitation, although food policy changes may lead to some deterioration in nutrition. Given the current magnitude of subsidies, the only plausible direction for prices is upward. The speed at which subsidies are reduced, and the form that alternatives designed to protect the poor eventually take, will reflect political as well as economic conditions. Since the clinical and public health dimensions of nutritional problems in Egypt are as central as the food dimension, it is hoped that the policies that ensue in the expected period of retrenchment will be ones that effectively wed these dimensions.

Note on recent trends in health and nutrition

In the later half of the 1980s and the early 1990s the Egyptian economy went into a serious recession. While the severity of this downturn is indicated by growing unemployment and declining per caput incomes, no sharp deceleration of positive trends in health and nutrition were apparent by the end of the decade. For example, substantial increases in the use of oral rehydration salts for treatment of diarrhoeal diseases were recorded during this period. Similarly, major public health campaigns reduced the prevalence of diseases such as tuberculosis and bilharzia. Infant mortality, which had declined from 92.4 deaths per thousand births in 1973 to 74.8 in 1980, continued to decrease to 44.7 in 1987 (World Bank 1991). There is, however, significant regional disparity; in Upper Egypt governorates, infant deaths remained near—or, in the case of Qena, exceeded—100 per thousand births.

Despite increases in the real price of food, many food commodities remain subsidized. For example, the price of a loaf of bread was increased 150 per cent in mid-1989 and the size reduced. Neverthless, much of this increase served to recoup earlier erosion of the price due to inflation. In terms of the price of a calorie, or a kilogram of wheat equivalent, the price of bread remains below the cost of its ingredients. More important, since the fiscal costs of subsidies can also be viewed as income transfers to the population, Egypt has not achieved targeting of benefits to the poor, nor significant restructuring of the system, while it has reduced total transfers.

Nor is such targeting and restructuring an objective of Egypt's first Structural Adjustment Loan (SAL) agreed upon with the World Bank in 1991. No component of that loan addresses social policies, although a contemporaneous social fund was drafted precisely to deal with this issue. Indeed the resources for the social fund exceeded that in the SAL; the latter is for $US300million while the former is expected to exceed $US550million, including over $US400million in cofinancing from a number of donors. As with many similar funds (Alderman 1992), the emphasis in Egypt's social fund is on employment generation rather than on food or health policy. Only 2 per cent of the social fund is devoted to increasing the capacity for targeting subsidies and services. Another 8 per cent—not all of it additional—is earmarked for community development, including immunization, child care, maternal nutrition, and literacy. Over one-half is devoted to public works, with a priority in poverty areas, and another 20 per cent to expanding small enterprises.

References

Alderman, H. (1992). *Nutritional considerations in Bank lending for economic adjustment*. In *Proceedings of the 12th World Bank Agricultural Symposium*, (ed. J. Anderson and C. de Haen), pp.199–219. World Bank, Washington, DC.

Alderman, H. and Von Braun, J. (1984). *The effects of the Egyptian food ration and subsidy system on income distribution and consumption*, International Food Policy Research Institute Research Report, No. 45. International Food Policy Research Institute, Washington, DC.

Alderman, H., Von Braun, J., and Sakr, S. A. (1982). *Egypt's food subsidy and rationing system: a description*, International Food Policy Research Institute Research Report, No. 34. International Food Policy Research Institute, Washington, DC.

Boas, S. (1980). Grain riots and the 'moral economy': Cairo 1350–1517. *Journal of Interdisciplinary History*, **10**, 459–78.

Boas, S. (1981). Fatamid grain policy and the post of the Muhtasib. *International Journal of Middle East Studies*, **13**, 181–9.

CAPMAS (Central Agency for Public Mobilization and Statistics) (1961). *Family budget survey by sample in the Arab Republic of Egypt, 1958/59*. CAPMAS, Cairo.

CAPMAS (Central Agency for Public Mobilization and Statistics) (1972). *Family budget survey 1964/5*. CAPMAS, Cairo.

CAPMAS (Central Agency for Public Mobilization and Statistics) (1978). *Family budget survey 1974/5*. CAPMAS, Cairo.

CAPMAS and UNICEF (Central Agency for Public Mobilization and Statistics, and United Nations Children's Fund) (1988). *The state of Egyptian children*. UNICEF, Cairo.

El Lozy, M., Field, J., Ropes, G., and Burkhardt, R. (1980). *Childhood malnutrition in rural Egypt: results of the Ministry of Health's weighing exercise*. Health Care Delivery System Project, Monograph, No. 4. Massachusetts Institute of Technology, Cambridge.

May, J. (1961). *The ecology of malnutrition in the Far and Near East*. Haefner Press, New York.

NCDDP (National Control of Diarrheal Disease Project) (1988). Impact of the National Control of Diarrheal Disease Project on infant and child mortality in Dakahlia, Egypt. *The Lancet*, July 16, 145–8.

Nutrition Institute, Ministry of Health (1978). *Arab Republic of Egypt National Nutrition Survey, 1978*. United States Agency for International Development, Washington, DC.

Nutrition Institute, Ministry of Health (1980). *Arab Republic of Egypt, National Nutrition Survey*, II, United States Agency for International Development, Washington, DC.

Pinstrup-Anderson, P., Jaramillo, M., and Stewart, F. (1987). Recession, adjustment, and child welfare in the 1980s. The impact of government expenditure. In *Adjustment with a human face*, (ed. C. Giovanni, R. Jolly, and F. Stewart), pp.73–89. Oxford University Press, Oxford.
Reutlinger, S. and Selowsky, M. (1976). *Malnutrition and poverty*, Johns Hopkins University Press, Baltimore.
Rivlin, H. (1961). *Agricultural policy of Muhammed Ali*, Harvard University Press, Cambridge.
Scobie, G. (1981). *Government policy and food imports: the case of wheat in Egypt*. Research Report, No. 29. International Food Policy Research Institute, Washington, DC.
Scobie, G. (1983). *Food subsidies in Egypt: their impact on foreign exchange and trade*. International Food Policy Research Institute Research Report, No. 42. International Food Policy Research Institute, Washington, DC.
Stigler, G. and Becker, G. (1977). De gustibus non est disputandum. *American Economic Review*, **67**, 76–90.
Taylor, L. (1979). *Food subsidies in Egypt*. Unpublished mimeograph, Massachusetts Institute of Technology. Masachusetts Institute of Technology, Cambridge.
World Bank (1990). *World development report*. Oxford University Press, New York.
World Bank (1991). *Egypt: alleviating poverty during structural adjustment*. World Bank Country Study, World Bank, Washington, DC.

8

Government and the infrastructure in Egyptian agriculture

N. N. AYUBI

Introduction

Government has always had much to do with the prosperity or decline of Egyptian agriculture, and therefore of the country's whole economy. Traditionally, the influence of the Egyptian Government has been through its control of irrigation networks and its ownership of arable land. The state control over land was, however, weakened from the mid-nineteenth century, when private ownership of land was introduced, a process which, one century later, left much of the country's arable land in the possession of a few owners.

Following the 1952 Revolution, the Egyptian Government introduced a series of agrarian reforms and agricultural projects that were aimed not only at achieving higher equality in the society and higher productivity in agriculture, but also at ensuring the government's organizational and political control over the countryside.

From the mid-1950s, the provision of goods and services in the countryside was to operate under state directives through a consumer subsidy programme. Efficient allocation of resources was attempted by means of mandatory rules of production. Acreage allotments, compulsory quotas, and consolidation schemes were all used in an attempt to channel agricultural resources in the direction of the Egyptian Government's overall development plans (Radwan and Lee 1986, pp.158–60). The provision of credit, equipment, and agricultural inputs (seeds, fertilizers, insecticides) as well as the marketing of products were, to a large extent, controlled by the state through an extensive network of over 5000 agricultural co-operatives all over the country (CAPMAS 1983, pp. 67–9).

These agricultural policies, which delivered some gains in the 1950s and 1960s, remained basically unchanged until 1990, since when there have been changes and a general liberalization of government policy. The crisis in agriculture which was growing during the 1970s and 1980s, and whose impact was magnified by factors such as the adoption of the 'open door' policy (*infitah*) within Egypt, and the repercussions of the oil price boom at regional and international levels, reached a peak in the late 1980s. At this time, with worsening economic conditions, changes in policy were considered necessary.

This essay does not set out to examine Egypt's agricultural policies in detail, but rather aims to survey certain aspects of the Egyptian Government's crucial role in this field as it has been manifest in the last few decades, and as it has been reflected in some of the recent plans concerning the development of agriculture in the country.

Main elements of the governmental structure

The vital part played by the Egyptian Government in running the country's economic affairs continues, in spite of the shift during the mid-1970s towards liberalizing the economy by reducing governmental influences and controls. It is useful at this point to review briefly the functions of the main governmental institutions to see how the politico-administrative machinery operates in relation to the planning and management of economic activity generally (Ayubi 1982).

The Presidency

Due to political culture and legal tradition, the President is the dominant political and governmental authority in Egypt. Any major economic project must normally have the approval of the President before it can proceed with a reasonable prospect of success. Presidential powers include the right to propose, veto, and promulgate legislation. When the People's Assembly is not in session, the President may issue decrees that have the force of law (and which are ratified later by the Assembly). In some circumstances, and by investiture of the Assembly, he may, and does, issue decree laws whether or not the Assembly is in session (ARE 1971, Articles 108, 112, 147).

Linked to the Presidency is the Central Agency for Public Mobilization and Statistics (CAPMAS). It is the main data bank in Egypt, it is important as a source of information for any project, and it also has a controlling role since any sizeable statistical study conducted in Egypt has to secure the approval of the Agency in advance. CAPMAS is high in the governmental hierarchy and should, in principle, report directly to the President. Affiliated also to the Presidency is the Central Auditing Agency (CAA), which is the supreme control and evaluation agency for the public sector and for the economic activities of the State.

The unique office of the Socialist Public Prosecutor was incorporated for the first time in the 1971 Constitution. His jurisdiction is ambiguous with a mixture of legal, administrative, and also political 'control' functions, which can be likened to a parliamentary ombudsman, state prosecutor, and political commissar, all in one.

Other institutions directly associated with the President are the prestigious but basically advisory National Specialized Councils (NSC) introduced in 1974, whose declared purpose is to assist the President in drawing up national plans and policies by surveying the available resources and their potential, and advising on the optimum utilization of existing capabilities for the fulfilment of the objectives of the State.

The Cabinet

The President determines the general policies of the State, and the Cabinet, headed by the Prime Minister, supervises their implementation. To fulfil its functions, the Cabinet may issue decrees pertaining to economic or other matters. 'Decisions' may also be issued, in descending order, by the Prime Minister, and by individual ministers and governors. Most ministries have a number of public companies, and some also have public authorities and public organizations, belonging to them.

Ministries with related activities are often associated in a 'ministerial group'. Agricultural affairs are dealt with by at least three ministries (with varying titles and jurisdictions) for agriculture, irrigation, and land reclamation; other names may include 'food security', 'reconstruction and new communities', and 'popular development'.

The Cabinet is composed mainly of 'technocrats' with academic or professional backgrounds, particularly university professors, engineers, economists, and agronomists.

The planning apparatus

Although CAPMAS enjoys the role one would expect of a central planning agency in a country like Egypt, the formal planning function is actually entrusted to an ordinary Ministry of Planning that is not superior to any other ministry, either in terms of legal or functional status. This is a continuation from the Nasserist era. The Ministry of Planning really emerged after the preparation of the first—and to all intents and purposes the only—Five Year Development Plan (1959/60 to 1964/5). Historically, it was never very influential. Although planning units at all levels were supposed to produce follow-up reports every three months, the only outcome was an annual report by the Ministry of Planning. This was supposed to provide guidelines for the ensuing year, but usually appeared several months after the beginning of the year it was supposed to cover.

Since the first Five Year Plan, and due to several technical and political considerations, Egypt can be said to have had some development 'plans' but very little development planning. The hope of producing something more than just a public investment programme has not in any way materialized and there has been very little effective co-ordination between the Ministry of Planning and the planning units in ministries, public enterprises, and local government (Ayubi 1980, pp.226–37). The economic and political aftermath of the defeat in the 'Six Day War' of 1967 changed everything and the outcome was the preparation of annual public investment programmes. It was the annual state budget which functioned as the major tool of direction in the economic system, while the CAA continued to play a major role in the financial control of the public sector. Indeed, it is possible to argue that little actual planning has taken place at all in recent years, in spite of the publication of a Five Year Plan for 1978–1982, and another for 1982–1987, the latter divided into annual mini-plans (NBE 1985, pp.127–48).

Local administration

Major laws organizing local administration in the country were passed in 1960, 1971, 1975, and 1979 and the hierarchy of local administration levels in the country is as follows:

(1) *muhafaza* governorate or province;
(2) *markaz* county or a central or large town together with a number of satellite or related villages, a unit first introduced in 1975;
(3) *madina* city or town;
(4) *hayy* urban district, quarter, or neighbourhood, a unit also introduced for the first time in 1975;
(5) *qariya* village.

Egypt has twenty-six *muhafazas*, including four city governorates that combine the functions of regional and town administrations, namely, Cairo, Alexandria, Port Said, and Suez. Of the remaining governorates, nine are in Lower Egypt, eight in Upper Egypt, and five are 'frontier governorates'. Altogether there are twenty governorates that include rural areas within their boundaries. There are also about forty *madinas* (including—apart from city governorates—a dozen cities with more than 100 000 inhabitants at the time of the 1976 census, namely, Giza (1 233 000), Shubra el-Kheima (364 000), Tanta (285 000), Mahalla (293 000), Mansura (258 000), Asyut (214 000), Zagazig (203 000), Damanhur (189 000), Faiyum (167 000), Isma'iliya (146 000), Minya (146 000), and Aswan (144 000). The country also has about 4000 *qariyas*, of which fewer than one-fifth had local councils in 1976.

The most recent local government law of 1979 contained new elements relating to two major areas: the first involved the expansion of the developmental functions and authorities of local units; and the second involved a broadening of the executive and political powers of the *appointed* local officers (Ayubi 1984).

The law postulates that local governors will have the primary jurisdiction in creating and administering all 'public utilities' located within their boundaries, and in assuming all the powers previously enjoyed in the localities by the various ministries. Local units are also given extended rights to state-owned and local land, and to formulate regulations relating to land reclamation and land distribution, including the right to create a special fund in which the revenues from these activities can be deposited and utilized for purposes of low-cost housing and land reclamation within the governorate.

The law also gave local councils wide financial powers to lease or rent local properties within higher financial limits than had been allowed under the previous law, and doubled the borrowing limits from their previous levels for purposes of local investment. The councils were also given the right to approve the creation of free zones or joint ventures (with Arab or foreign capital), subject only to the approval of the General Authority for Investment and the Free Zones (GAIFZ). Local units have, in addition, the right to increase the rates of additional taxes on imports and exports, sales, production, and property within their boundaries, and can also appropriate 50 per cent of all increases in the local revenues of the governorates which exceed that specified in the budget. Further, they may retain the full value of the sale of state buildings and lands within towns that are under their jurisdiction.

In general, therefore, a fair degree of decentralization of resource allocation and development planning is intended to take place. In practice, however, '. . . There is only modest evidence that a shift of resources from the governorate to the *markaz* or village council is taking place. Rather, governorate-level technical staff often contend that lower levels of government do not have the technical or administrative capacity to plan and implement development projects, contentions previously advanced by the ministries to explain why the governorates should not be given greater authority.' (Mickelwait and Eilerts 1981, p. 44).

The infrastructure

Elements of the infrastructure and of the development plans that are particularly pertinent to agriculture are discussed in the following sections and are also reviewed in Ayubi (1991).

Irrigation

One reason for the strong position of Central Government in Egyptian society has been its crucial role historically in providing irrigation networks and other infrastructural facilities, without which the economy could not survive and prosper (Ayubi 1980).

In earlier times, irrigation depended on the extension of canals and the building of dykes to water the land during the annual flooding season (basin irrigation). Subsequently, irrigation became more flexible as perennial irrigation was extended and a complex system of canals, barrages and reservoirs, was built up. In 1816 Muhammad Ali initiated Egypt's modern irrigation system with the construction of two deep-water canals (the Mahmudiya and the Ibrahimiya) that provided summer water for the fields in the Delta. In 1843 work started on two barrages (*qanatir*) across the two Delta branches of the Nile, but they did not function satisfactorily, and were rebuilt in the early years of British occupation. The system was expanded and improved by the British, and subsequently by the Egyptians, with the building of five additional barrages; two in the Delta at Idfina (1915) and Zifta (1943), and three between Cairo and Aswan, at Asyut (1902), Isna (1908), and Naja'Hammadi (1930). Main canals drew water for distribution from behind these barrages, often through various gates, to secondary canals and smaller channels leading to the feeder system that reached nearly every field in the Nile Valley and Delta by the 1980s (Nyrop 1983, p. 126).

To distribute water over a period of time (because one-half of the Nile flow occurs in August and September) it was necessary to regulate distribution through a storage system. The first Aswan Dam was initiated in 1898, completed in 1902, and heightened in 1912 and again in 1934, thus allowing for the storage of floodwater for use during the rest of the year.

After the 1952 Revolution, a much larger project for water storage was sanctioned as early as 1953, and in 1960, with technical help from the Soviet Union, construction was started on the Aswan High Dam. The first phase of this huge operation was completed in 1964, the power plant began operation in 1968, and the whole dam was completed in 1970. By the early 1980s, the dam had made it possible to convert some 880 000 feddans from basin to perennial irrigation, and to reclaim about 650 000 feddans out of a target reclamation of 1.3 million feddans (NSC 1980*b*, p. 17ff.; EICA 1978, p. 6ff.). By controlling and regulating the flow of water stored behind the dam, it has been possible for Egypt so far to avoid the impact of the recent catastrophic droughts in Africa. Electrification has also proceeded with the installation of twelve turbines at the dam, with transmission lines to Cairo and to industrial plants around the country.

At the same time, there are a number of drawbacks to the High Dam, many of which were anticipated in advance of its construction (NSC 1980*a*, p. 14ff.). The most serious of the side effects is probably the problem of salinity.

Drainage

Drainage is one aspect of the irrigation system that was neglected for far too long, and with predictable results. With the expansion in perennial irrigation after the completion of the Aswan High Dam, but with no concurrent overall improvement in the drainage networks, the land became increasingly and seriously waterlogged; the problem had reached a significant level in the early 1980s, with some land becoming barren, and production in several areas declining quite noticeably.

The Egyptian Government had recognized the dangers of inadequate drainage and a thirty-year priority programme (1960–1990) was established to improve the drainage system by building tile drains under the fields, leading to larger drains taking the water back to the river—the drainage system would thus mirror the irrigation system (Nyrop 1983, p. 130). Covered drains were to be used as much as possible to lessen the dangers of bilharzia, and to allow cultivation over the drains (which would otherwise consume about 10 per cent of the limited arable land). The eventual goal was to install covered drains for about 4.5 million feddans—over two-thirds of the cultivated land. In the 1960s implementation of the drainage programme lagged because of the country's financial problems.

Since the 1970s, Egyptian experts and international aid organizations have considered the deterioration in the fertility of the soil through rising water tables and increasing soil salinity to be the single most important factor determining the current performance of the agricultural sector in Egypt. Until an efficient system of charging for irrigation water can be introduced in order to constrain the traditional habit of overwatering, the country will continue to depend on drainage projects to remove surplus water after it has been applied to the fields (NSC 1980*a*, p. 27–8).

In 1970, the World Bank initiated a series of loans to facilitate the installation of drains on about 2.8 million feddans (Nile Delta Drainage I and II in 1970 and 1977, and Upper Egypt I and II in 1973 and 1976). In the late 1970s, the United States Agency for International Development (USAID), the United Nations Development Programme (UNDP), and other international donors contributed to projects aimed at setting up Egyptian schemes to produce local cement and plastic drains for these programmes.

Many of the projects have involved restoration and deferred maintenance works, in order to prevent further production decline and restore capacity to previous levels, rather than to provide net increments in productive capacity (World Bank 1978, Vol. III, p.45). Although the drainage programme has been speeded up since the late 1970s, it is likely that waterlogging and salination have been increasing at a faster rate. Current and future projects are designed to expand the drainage system, rationalize water allocation decisions, and improve the system of water management in general.

Electrification

Electrification of the Egyptian countryside was adopted as a political target in 1969, and became an official public policy in 1971 with the completion of the Aswan High Dam. With technical help from the Soviet Union, a five-year project was prepared for introducing electricity generated through the High Dam into villages that did not have electric power for consumer or production purposes (Ministry of Electricity 1971). In the mid-1970s, electricity generated from the High Dam alone was equivalent to the electricity produced from all sources in the mid-1960s. Indeed, during the mid-1970s, the High Dam was producing over one-half of the country's total energy although its turbines were not all running at that time (NSC 1980*a*, p.40).

The countryside electrification project slowed down for a while in the late 1970s, but it picked up momentum again and, by the early 1980s, electricity had been made available in around 4500 villages and hamlets (Table 8.1). This was equivalent to about 80 per cent of the total number of villages in the country (*Mayo* 1981; CAPMAS 1983, p. 166–8).

Transport and telecommunications

As only 4 per cent of Egypt's total area is densely inhabited, much of the demand for transport is concentrated along the 1000 km corridor linking Alexandria, Cairo, and Aswan. The main method of transport is by road and the road network carries more than 80 per cent of total freight tonnage and around 75 per cent of passenger kilometres. There are 27 000 km of roads in the country, about 12 000 of which are paved. Favourable climatic conditions have kept Egyptian roads in relatively good condition, and the regional roads are nearly all usable for normal motor vehicles.

The flat terrain in the Delta and along the Nile Valley has enabled an extensive railway network of over 3900 km to be developed. However, since 1970, freight traffic on the railways has been declining due to a shortage of locomotives, withdrawal of trains from some routes and unreliable services—factors which have provoked customers to switch from rail to other modes of transportation (World Bank 1978, Vol. V., pp.15–6).

Egypt's telecommunications system, on the other hand, was inadequate, until major modernization projects were initiated in the 1980s. The national density of 1.34 telephones per 100 population in the late 1970s was significantly lower than that of other Middle Eastern countries, and of several other countries in the Third World (World Bank 1978, Vol. V., pp.24–5). Almost a thousand villages in Egypt have no access even to public telephone facilities. However, it is estimated that the few telephone sets that exist in rural Egypt seem to be utilized generally for the benefit of the consumer rather than of the producer, i.e. they are used for family reasons more than for commercial or administrative purposes (Kamal *et al.*, 1980 p. 3). By contrast, in Cairo, approximately 95 per cent of all calls during business hours are business-related.

Inland postal services remain limited in scope and poor in quality. Only 125 million inland-outgoing letters were sent in 1981–2 (compared to 80 million foreign-outgoing letters). Improvement of the postal services is being given

TABLE 8.1. Egypt: number of electrified and non-electrified villages by governorates, 1981/2

Governorate	Electrified	Non-electrified	Total
Damietta	62	–	62
Daqahliya	391	51	442
Sharqiya	465	–	465
Qalyubiya	185	–	185
Kafr el-Sheik	182	14	196
Gharbiya	313	–	313
Minufiya	300	–	300
Beheira	413	37	450
Isma'iliya	22	–	22
Giza	47	–	47
Beni Suef	207	11	218
Faiyum	150	7	157
Minya	337	–	337
Asyut	228	14	242
Sohag	240	25	265
Qena	185	4	189
Aswan	84	–	84
Red Sea	–	–	–
New Valley	72	37	109
Matruh	7	3	10
Sinai North	21	16	37
Sinai South	3	5	8

Source: CAPMAS (1983).

some emphasis at present, and there are some private post offices functioning. A system of mobile post offices has been introduced to serve localities without a mail service (CAPMAS 1983, pp.103–16).

Agriculture in the development plans

In the mid-1970s it was considered desirable to draw up a long-term and futuristic plan for the development of Egypt's economy up to the year 2000. It was envisaged that the usual five year plans could be fitted into this long-term plan. The plan was published in 1978 and it stipulated the promotion of productive economic activities for regions outside the existing heavily congested urban and rural areas. The purpose was to attract labour from the cities, to reduce rural–urban migration, and to provide areas capable of absorbing the expected population increase. It aimed for a more balanced population structure, and for improved living standards in various parts of the country.

Plan for the Year 2000

Within the long-term *Plan for the Year 2000*, three groups of desert and marginal areas were designated for reclamation, settlement, and development: proposals for these northern, western, and eastern zones (Ministry of Planning 1978*a*) included the following areas.

The Northern Development Belt (north Delta).

The preliminary outline for development in this area concentrated on one axis extending all the way along the Delta on the Mediterranean coast. Projects included:

(a) Agricultural reclamation of the northern Delta zone north of the existing general line of arable land (i.e. Foah, Sidi Salim, Hamul, Belqas, and Shirbin), and the construction therein of an extensive network of agricultural settlements;

(b) Exploitation of fish resources along the north coast of the Delta and the construction of summer resorts, the most important of these areas being Abu Qir bay and Rosetta, where the Rosetta branch of the Nile meets Lake Burullus;

(c) Agricultural reclamation of the Salhiya Valley, located in the Suez Canal zone south of the Cairo-Isma'iliya road, where there is sufficient surface water available to reclaim half a million feddans if a network of canals were to be provided;

(d) Agricultural projects in the Western Delta, extending from El Amiriya and Maryut to the Natrun Valley. Projects in these areas would depend on the possibility of utilizing available surface and subsurface water for irrigation, and would be a natural extension of the West Nubariya, Maryut, and Tahrir projects, and the farms of Giannaclis and the Natrun Valley. This area would see extensive cultivation, as well as the construction of new cities, such as Sadat City. Urbanization would be assisted through the extension of power lines from the Qattara Depression to the Delta, while roads and communications systems would be provided throughout the region.

The Western Development Belt (the Western Desert)

Western Desert development would be concentrated in two areas; the north-western (Mediterranean) coast between the Delta and Egypt's western boundary (with Libya), and the Qattara Depression and the south-western desert. North-western coast projects would concentrate on the development of pastures and the planting of barley, with irrigation water supplied by rainwater, by the old Roman wells in the area, and by existing canals in the valley which carry Mediterranean water for the cultivation of limited areas of fruits and vegetables. Animal husbandry would also be developed in this region, parts of which would be highly suitable for the development of seaside resorts, with a port and a free zone to be constructed at Marsa Matruh. Water resources would be developed all along the coast.

It is anticipated that if and when completed, electrical energy from the Qattara Depression project could produce approximately five times the amount of power generated by the High Dam at Aswan. In order to realize this scheme, a lake the size of the Delta in area would be constructed in the Qattara Depression. This vast body of water would possibly effect significant climatic changes such as increased rainfall and improved weather thus permitting some agriculture; the amount of subsurface water would also be likely to increase. Such changes would stimulate development areas in desert lands, in the coastal region, and around the Depression and the oases. The new lake could be an important source of salt and fish and it could provide a site for chemical industries based on the extraction of chlorine and sodium. New settlements would accompany the developments.

Digging the Toshki Canal to channel overflow waters from Lake Nasser into the Kharga region could enable 1.3 million feddans to be cultivated south of the Western Development Belt, although construction of this canal would depend on what water resources could be spared from projects in the Upper Nile area. In the meantime, 'modern villages' would be established on the western shores of Lake Nasser for people working in fishing, related industries, and tourism.

The Eastern Development Belt (Eastern Desert)

Along the Red Sea coast, development activities would be concentrated on tourism, especially at Ain Sokhna and Ghardaqa. The exploitation of petroleum and other mineral resources would also be possible.

To the west, opposite the Nile Valley, there would be reclamation for cultivation of small areas near the Valley (particularly near Qena, the Asyut Valley, and the eastern shore of Lake Nasser). There are other scattered areas along the eastern boundary of the Nile Valley from south of El-Saf to the north of Aswan, where land reclamation would require Nile water to be channelled through a pipeline system to be extended initially between Qena, Safaga, Qift, El-Qusair, Idfu, Ras Gharib, Aswan, and Brius.

The viability of these projects would depend heavily upon the provision of transport and services into these areas which would be extremely costly.

In this grand design for the future, there were hopes that between 1978 and the year 2000, around six million feddans would be reclaimed for cultivation. Immediate plans envisaged the reclamation of approximately 2.5 million feddans in the various regions (Table 8.2). Of necessity, such

TABLE 8.2. Egypt: lands to be reclaimed until the year 2000

Region	Reclamation area ('000 feddans)
North-west region	208
Suez Canal region	487
Lake Nasser region	1300
Other areas	1000
Total	2495

Source: Ministry of Planning (1978a).

reclamation would require large capital resources for activities such as levelling, flooding, leaching and washing, irrigation, drainage, electricity, mechanization, and the building of rural residential areas. Although it is well known that such activities yield very little short-run capital turnover on investments, the Egyptian authorities felt that the long-term advantages would be worth it: the increase in the acreage of cultivated land, the construction of modern communities for the fellahin, a gradual increase in agricultural production, and an accompanying increase in incomes, would all be positive results from such capital investment. Other advantages would be the growing use of modern technology and scientific methods in agriculture, the creation of new employment opportunities in agriculture, and, hopefully, a reduced population growth rate in the congested urban areas. This last objective was to be supported by the construction of six new cities in the desert (ARE 1979, 41ff.); work on some of these has already begun.

Five year development plans

Within the framework of the long-term *Plan for the Year 2000*, the *Five Year Development Plan for 1978–1982*—which was later adjusted to cover the period 1980–4 also gave high priority within its 'investment strategy' to: '. . . those projects which respond to the needs of the masses, i.e. food security, clothing, housing, strategic storage and distribution, and the absorptive capacity of ports' (Ministry of Planning 1978b, p.8).

The rural sector of Egypt was allocated investments of £E2624 million (25.8 per cent of the total investment of the Five Year Plan) for promoting village development and improving the socio-economic standards of the farmers. As a productive economic sector, 'Agriculture, irrigation and drainage' was allocated investments worth £E878.9million (22.8 per cent of all investments in the production sector and 8.6 per cent of the total investments).

The Five Year Plan allocated investments worth £E64million to animal, poultry, and fish production with a view to increasing the number of farmers involved in such activities.

A further Five Year Plan for the period 1982/3–1986/7 takes more account of Egypt's dependence on external sources, and is divided into almost self-contained annual plans. Although agriculture remains the main source (30.4 per cent) of gross domestic product (GDP) generated in the commodity sector, only a modest annual growth rate of 3.7 per cent was envisaged for it during the period of the Five Year Plan (NBE 1985, pp.132–5).

The Egyptian Government and rural development

As it became increasingly clear to the authorities that rural development lay within the jurisdiction of several ministries and other government bodies, it was decided that a co-ordinating system was needed in order to initiate an integrated programme for the development of the countryside. In 1973, the Agency for the Reconstruction and Development of the Egyptian Village (ARDEV) was established by Presidential Decree. ARDEV's function (under the leadership of a Deputy Minister for Local Government) was to draft a general plan for the development of the rural community within the framework of the general policies of the country. Following its approval by the Ministerial Committee for Local Government, this plan would be passed to the Cabinet by the Minister of State for Local Government. ARDEV also had to implement the policy and its programmes in collaboration with the ministries, the local government units, and with any other agencies concerned, while its own projects were to be co-ordinated with the rural development projects initiated and executed by the village councils within the confines of the local government system itself. A framework was thus established for implementing projects within the rural development plan, including—within a set time schedule—the necessary feasibility studies, evaluation, training programmes, preparation and execution, and follow up.

ARDEV's work plan began with a definition of reconstruction and development for the Egyptian village that implied a process of rebuilding all aspects of the rural community, both in physical as well as economic, cultural, and social terms. Certain experimental villages were chosen to provide the baseline for project planning in the long term, and to set the outline for the future developed rural community which was to achieve an 'industrialized agriculture' and an 'industrial farmer' responsive to 'the requirements of social change and accelerated development' (EICA 1978).

ARDEV was organized along the following lines. At the central level, and with the aim of establishing integrated rural development planning at the economic, social and physical levels, ARDEV was allowed to utilize, through different committees, the expertise of various development specialists. The Consultative Committee, headed by ARDEV's director, laid down general policy, plans and programmes, while more detailed studies were undertaken by sub-committees for Economic Development, Social Development, and Physical Development and were presented to the Consultative Committee. Two commissions were established for the integration and co-ordination of the various ministries' plans and programmes. The Ministerial Commission for Local Government, as stipulated in the Local Government Law, was to be responsible, with ARDEV, for the administration of all ministerial efforts in the area of rural development. The Commission for the Co-ordination of Rural Services included the senior undersecretaries of the various ministries and departments concerned with rural activities (Youth, Transport, Education, Health, Information and Culture, Agriculture and Irrigation, Planning, Housing and Reconstruction, Social Affairs, Local Government and Rural Electrification).

At the regional level, branches of ARDEV were set up within the organizational structure of the various governorates, districts, and villages, thus underlining the principle that rural institutions should assist in the planning and co-ordination of rural development efforts in the governorates. Branches of ARDEV were also supposed to implement the various rural development efforts at the district, village, and other levels of the local government system, and to report back to ARDEV. They were expected to participate with ARDEV headquarters in studies relating to progress in the villages and to the improvement of their services and utilities.

Field experimentation for the Village Development Plan

Field experimentation for this plan was based on a comprehensive survey of the 'first phase villages'. For this purpose, 17 villages were selected in 15 rural governorates. The villages represented various forms of economic activity and included agricultural and fishing villages, together with villages in rural, industrial, and touristic areas. Programmes to supply the minimum services and utilities required by the village development plan were started in the 17 villages late in 1972, and continued to 1974–5; other villages were included during the period 1975–8 (Table 8.3). The programmes included projects for social and economic development, for village planning and housing improvement, for construction of new roads, and for the 'building of model houses considered appropriate to a family's social and economic conditions' (EICA 1978).

In the house-building programme, the State bore 20 per cent of the cost of a dwelling, while the new occupant made an initial payment of 20 per cent of the final value. In most cases this was equal to the value of the compensation due to him for the demolition of his previous house. The balance of 60 per cent was payable over a twenty year period through interest-free annual instalments.

In the second phase, the experimentation and sampling base was expanded by selecting a village from each *markaz* (county) in the country, and providing basic services (agricultural, health, social, and educational) in addition to water and electricity. These development projects followed the same lines as those established in the 17 villages during the first phase. In a new housing system, costs were reduced by utilizing village labour and local contractors. Other development projects included the reclamation of marshy areas, or their transformation into fisheries, together with the improvement of sanitation facilities, domestic water supplies, and electricity.

Programmes in social development were intended to extend existing activities and maximize the 'social capital' provided by the 'combined units' that had been established in many villages in the 1950s to offer agricultural, medical, social, and educational services at the same site. Social development programmes also aimed to expand vocational training centres for boys and girls, with the aim of transforming villages into dynamic production centres and of creating new job opportunities that would increase the incomes of rural families, thus stemming the continuing migration of manpower to the towns. Cultural and recreational centres were also established for village women and girls, and youth centres were enlarged and strengthened. The planning of social programmes was linked to existing plans for economic development, such as schemes for agricultural mechanization, animal husbandry, and poultry breeding. It was also linked with the provision of training in the trades and crafts required in the villages.

Although from 1972 ARDEV initiated a number of physical, economic, and social development projects in various governorates (Table 8.3), by the end of the 1970s its activities had slowed down considerably, mainly due to lack of finances.

TABLE 8.3. Egypt: cost of ARDEV projects, 1972–8 (£E)

Years	Physical projects	Economic projects	Social projects	Others	Total cost of projects
Experimental villages (total 17)					
1972					847 000
1973 1974	282 300	162 561	80 773	78 996	604 630
1975					9 678
Other villages					
1975	113 089	892 964	83 724		1 089 777
1976	827 624	1 484 241	261 085		2 572 950
1977	430 510	1 380 418	271 072		2 082 000
1978	632 000	2 253 700	358 000	104 000	3 347 700

£E 1.00 = $US 1.45.
Source: EICA (1978).

The situation was reversed by the early 1980s when decentralization became the fastest growing part of the USAID portfolio in Egypt (Weinbaum 1985, pp. 20-2). The portfolio encompassed five closely related aid activities brought together in 1981, including programmes for the provision of government-administrered AID grants that were to be made available to village councils as funds for revolving credits for the financing of income-producing public loans, and a programme for basic village services that provided funds to some 20 rural governorates for infrastructural improvements at the village level. These included such undertakings as market roads, water and sanitation systems, the lining of canals, and drainage programmes. Over $US200million in grants and forgiven food loan repayments were set aside for such purposes with about one-half spent by the beginning of 1984. Over 3500 small projects had been completed or were being implemented. Another programme committed $US100million, of which 40 per cent was disbursed for the supply of American equipment to rural governorates, and funded such purchases as road maintenance equipment and water filtration plants. Another decentralization grant earmarked $US30million for enhancing local government institutional capacity in some provincial towns.

Reliance on ARDEV for disbursing American rural aid implies that much of that aid has to go to public sector projects, a practice that runs counter to American aid policy. However, a decentralization drive helps towards achieving another of the American administration's objectives—the transfer of decision-making away from the central bureaucracy (cf. Weinbaum 1985, p. 21). Moreover, decentralization projects are not necessarily inimical to the private sector, and a modest boom in private enterprise has in fact been claimed.

This huge infusion of American money and advice into Egypt's rural sector does involve risks. On an institutional level it makes ARDEV, not one of the most powerful units of the bureaucracy, become one of its richest. On a broader policy level, the United States has, by asserting its own priorities, obliged the Egyptian Government to sink its scarce investment resources into areas of development where the American funds were expected to be made available. America has thus established economic goals and strategies that limit the freedom of Egypt's planners to determine how they wish to allocate their resources. A frequently cited example is the diversion of local resources away from investment in new land development (Weinbaum 1985, pp.16-17).

Although a reasonable physical and organizational infrastructure exists in Egypt for the implementation of agricultural policies, the Egyptian Government's plans and projects in this field are rather too ambitious, given the relatively limited funds allocated to agricultural development. If the country is to confront its current agricultural dilemma and its growing food deficit with a measure of success it will need to work out an order of priorities.

There is little doubt that Egyptian agriculture needs to be improved and that the countryside needs to be developed; it will be regrettable if the main influence on policy-formulation in this field is the availability of funds and if the special priorities of the money suppliers become the blueprint for the country's agricultural policies.

References

ARE (Arab Republic of Egypt) (1971). *Dustur jumhuriyya misr al-'arabiyya [The constitution . . .]*, Ministry of Information, Cairo.
ARE (Arab Republic of Egypt) (1979), *Ta'mir misr [The reconstruction of Egypt]*, Dirasat Qawmiyya No. 3. Markaz al-Nil, Cairo.
Ayubi, N. (1980). *Bureaucracy and politics in contemporary Egypt*. Ithaca Press, London.
Ayubi, N. (1982). Organization for development: the politico-administrative framework of economic activity in Egypt under Sadat. *Public Administration and Development*, **4**, 279–94.
Ayubi, N. (1984). Local government and rural development in Egypt in the 1970s. *Cahiers Africains d'Administration Publique*, **23**, 61–74.
Ayubi, N. (1991). *The State and public policies in Egypt since Sadat*. Ithaca Press, Reading.
CAPMAS (Central Agency for Public Mobilization and Statistics) (1983). *Statistical yearbook of the Arab Republic of Egypt 1952–1982*, CAPMAS, Cairo.
EICA (Egyptian International Centre for Agriculture) (1978). *Integral rural development and the role of the Agency for the Building and Development of the Egyptian Village*, Mimeograph. EICA, Cairo.
Kamal, A. *et al.* (1980). *Communication needs for rural development*, Cairo University-MIT Technology Planning Programme, Report No. 11. Cairo University, Cairo.
Mayo (1981). September issues, (issued weekly) Cairo.
Mickelwait, D.R. and Eilerts, G. (1981). Law to practice: the status of implementing decentralization in Egypt. In *Building capacity for decentralization in Egypt: some perspectives*, IRD Working Paper No. 10, (ed. T. Walker). Development Alternatives Inc., Washington DC.
Ministry of Electricity, Egypt (1971). *Khittat kahrabat al-rif [The rural electrification plan]*, Public Authority for Rural Electrification, Cairo.
Ministry of Planning, Egypt (1978a). *The road to the year 2000*. Ministry of Planning, Cairo.
Ministry of Planning, Egypt (1978b). *The Five Year Plan 1978–1982*. Ministry of Planning, Cairo.
NBE (National Bank of Egypt) (1985). *Economic Bulletin*, **38**, 127–48.
NSC (National Specialized Councils) (1980a). *Al-Sadd al-ali wa atharuhu [The Aswan High Dam and its impact]*, NSC Series No. 4. NSC, Cairo.
NSC (National Specialized Councils) (1980b). *Al-Tawassu'al-zirap'i al-ufuqu [Horizontal expansion in agriculture]*, NSC Series No. 8. NSC, Cairo.
Nyrop, R. F. (ed.) (1983). *Egypt: a country study*, (4th edn.). The American University, Washington DC.
Radwan, S. and Lee, E. (1986). *Agrarian change in Egypt: an anatomy of rural poverty*. Croom Helm/ILO, London.

Weinbaum, M. G. (1985). Foreign aid politics and rural sector development: U.S. economic assistance to Egypt, 1975–84. Unpublished paper presented at the International Political Science Association's 13th World Congress, Paris.

World Bank (1978). *Arab Republic of Egypt: economic management in a period of transition*, 6 Vols. Report No. 1818-EGT, World Bank, Washington DC; also available in book form: Ikram, K. *et al.* (1980). Johns Hopkins University Press, Baltimore.

9

The role and impact of government intervention in Egyptian agriculture

T. A. MOURSI

Introduction

Since millennia past, government intervention has been a distinctive feature of Egyptian agriculture. The distribution and control of irrigation water was perhaps the *raison d'etre* of early Egyptian governments. The unique system of basin irrigation that was practised required systematic, orderly, and timely flooding, during the flood period of the Nile River, of the different basins in which the agricultural land was divided. This enabled the soil to acquire and retain adequate moisture to guarantee good growth of the ensuing crops. There was a need for a strong central authority for the proper administration and operation of this important and delicate task.

Water control and distribution is still one of the Egyptian Government's main concerns that purportedly—at least from the viewpoint of state bureaucracy—necessitates an administrative role in the agricultural sector; other economic, social, and political concerns have amplified and spread that role over the years to cover many other activities in that sector. Direct and indirect government intervention in agriculture is regarded as an instrument of economic growth and consequently has always been a major component of national and development policies in Egypt. The relatively limited amounts of usable land and irrigation water and the rapid rate of population growth, combined with the Egyptian government's announced objective of providing food and raw materials—usually at low prices—for the increasing population and domestic industry, has enticed state intervention in the direction and management of agricultural resources, ostensibly to secure efficient allocation of those resources. The government has also intervened in agriculture to effect a change in social welfare, and political objectives including equity of wealth and income distribution, social justice, alleviation of poverty, food and national security, and political stability.

The purpose of this chapter is to present a general schematic review of the main policies and control instruments of government intervention in agriculture, their impact, and also the recent government policy changes. The first section presents the historical underpinnings of the current agricultural incursion programme. The second section reviews the major instruments of government intervention policy in Egyptian agricultural production and food distribution since the revolution. The socio-economic effects of the intervention are discussed in section three, and section four presents a brief overview of the direction of recent changes and liberalization of the agricultural policy.

Government control and measures in the 1940s

The agrarian revolution started by Muhammad Ali during the first half of the nineteenth century was characterized by extensive state incursions. These included the introduction of new cotton varieties, development of irrigation projects, price and quantity rationing of basic staples, and mandatory purchases of major crops to stabilize domestic prices and to generate export tax revenue to finance Ali's regime. By the 1940s, however, this revolution seemed to have spent itself. Indeed it was in the 1940s, and especially during the war years, that the severity of many problems and their unyielding nature to traditional reform programmes were revealed. The Egyptian agrarian system seemed to suffer many ailments, some resulting from failure of the agrarian transformation to overcome the externalities generated by the agricultural revolution. There was, first and foremost, a flagrantly inequitable distribution of agricultural land, while at the same time the dominant class of landowners was sufficiently powerful to hold down the wages of farm labourers. 'Sticky' wages, land rents, and the tenant/owner relationship favoured landowners. In addition, land productivity was decreasing steadily and this was reflected in declining crop yields. However, the adverse effects of multiple cropping and inadequate drainage facilities on soil fertility factors had not been adequately assessed. In combination with a rapidly expanding population and a slow rate of reclamation of new lands, the decline in agrarian productivity led to a decrease in the per caput share of agricultural production. The stage was set for greater effort on the part of the government to save an ailing system.

The institutional and legislative measures taken by the State in the agrarian field in the 1940s were not something new. In the 1940s, however, a new phase of state intervention in agriculture—the principal source of national income—began. With the eruption of the Second World War, it was deemed necessary by the government to institute a number of restrictive programmes including regional acreage allotment restrictions for major food and feed crops, production ceilings, forced deliveries, mandatory pricing (such as Law No. 18 passed in 1942 to encourage maize production), and supply restrictions (e.g. sugar rationing). However, these programmes were primarily crisis-precipitated, to deal with the exigiencies of the war. They lacked the necessary co-ordination because the responsibilities for their development and administration were scattered among a number of state agencies. Meanwhile, the institutional restrictions on a specific crop or group of crops, and their impact on other crops and on the whole production and consumption structure, was ignored. There was then an artificial dichotomy between decisions related to production and consumption. While the cessation of war hostilities witnessed a selective relaxation of most intervention programmes, it may be of interest to note that the roots of many of the country's current government agrarian control policies had their origin in the 1940s programmes.

Government intervention in agriculture since the revolution

It is customary, when dealing with government intervention in Egyptian agriculture, to divide the period since the July 1952 revolution into at least two main sub-periods. During the first period, from 1952 until prior to the 1973 October War with Israel, the country was professed to embrace socialist ideologies. In the second period, starting with the *infitah* (the opening) policy launched by the late President Sadat after the October War, a 'more' liberal stance was adopted with the intention of developing stronger ties with the West. Under President Mubarak, the liberal philosophy seems to persist although presumably resting on a somewhat stronger technocratic—*vis-à-vis* charismatic—foundation in comparison with the Sadat era. The focus of this section is a review of the main policies and instruments of state incursion in agriculture. However, the period from 1952 until recently is not subjected to the usual historical dichotomy, mainly because of the absence of a fundamental stylized change in the taxonomic nature and role of government agrarian control policies during that period. This helps us to focus on the main instruments and impact of the policies without being entangled in related sociopolitical details, which are important in their own right, yet beyond the scope of this review.

Agrarian reform

The Agrarian Land Reform Law No. 78 of 1952 was an important exercise of government action that had profound and far-reaching effects on agriculture and on the agrarian community. The most important elements of that law and its amendments are the limitations it sets on land ownership and the regulation of the owner/tenant relationship. The implementation of the law has resulted in significant changes in land distribution and the size of holdings. While large ownerships and holdings disappeared, the number of owners in the <1 feddan group increased and so did the percentage of holdings, the area held, and the average size of holding in that group. There was also an appreciable increase

TABLE 9.1. Number of holdings, area, and average size of holdings distributed among different holding groups

	Size of holding group (feddans)						
	<1	1–<3	3–<5	5–<10	10–<50	>50	Total
1950							
Number of holdings	210 634	411 140	160 483	120 363	80 242	20 061	1 002 923
%	21.00	41.00	16.00	12.00	8.00	2.00	
Area (feddans)	122 878	737 270	552 935	798 710	1 474 542	2 396 130	6 082 465
%	2.00	12.00	9.00	13.00	24.00	39.00	
Average size of holding (feddans)	0.58	1.79	3.46	6.64	18.38	119.44	
1961							
Number of holdings	434 219	672 505	274 317	170 019	80 516	10 384	1 641 960
%	26.44	40.95	16.70	10.40	4.90	0.63	
Area (feddans)	211 155	1 153 237	990 029	1 100 669	1 431 886	1 335 863	6 222 839
%	3.40	18.50	15.90	17.70	23.00	21.50	
Average size of holding (feddans)	0.49	1.85	4.00	6.47	17.78	128.60	
1974							
Number of holdings	1 124 286	1 160 147	354 841	148 459	65 059	131	2 852 923
%	39.41	40.67	12.44	5.20	2.28	0.0046	
Area (feddans)	739 028	2 023 456	1 185 581	944 411	985 508	105 684	5 983 668
%	12.35	33.82	19.84	15.78	16.47	1.77	
Average size of holding (feddans)	0.66	1.74	3.34	6.36	15.15	82.93	

Source: Data for 1950 and 1961, Ministry of Agriculture (1961); data for 1974, Ministry of Agriculture (1979).

in the percentages of the area held by the 1–3 and the 3–5 feddan groups and in the average size of holding in the latter group (Table 9.1). Meanwhile, the average size of holdings in the 5–10 feddan and the 10–50 feddan groups and the percentages of their holdings gradually decreased (Tables 9.1 and 9.2). Some 364 469 families did benefit from the law by acquiring land, including 150 000 landless farm labourers (Hagrass 1970).

In its attempt to redistribute rural income in favour of landless farmers and of land owners with limited income, the law regulated the owner/tenant relationship such that cash rent was limited to seven times the value of the land tax. It also stipulated a fifty – fifty share of the crop with an equal division of costs in the case of sharecropping. The law prohibited subletting in order to abolish the middlemen class that had plagued tenancy, and it made a land lease inheritable when at least one inheritor was a farmer. The distribution of land among the landless and low income farmers and the redistribution of income between owners and tenants in favour of the last group helped to raise the income of a large potential consumer group that would, later on, swell the country's food demand and consumption.

One other element of the Agrarian Land Reform Law was that all land reform beneficiaries were obliged to join the state-directed agricultural co-operatives which were to regulate crop rotation and extend input and credit facilities to members, provided that participants abided by the various programmes imposed by the system, e.g. land consolidation, crop rotation, and marketing arrangements beginning with cotton in the 1961/2 season (Ikram 1980). The same kind of co-operatives were soon to extend to the rest of the country's farmers. They were to become not only a primary channel for government services and agricultural production improvements, but also a powerful state instrument to control the functioning of the agrarian system, and directly influence the course of rural development (Radwan and Lee 1986). It is important to note, however, that while the co-operatives succeeded in fulfilling the first task, they failed to accomplish the latter task effectively, mainly because the increase in the number of co-operatives was not accompanied by a corresponding rise in their capital (Moursi 1986).

Moreover, the Agrarian Land Reform Law tried, without success, to improve conditions for farm labourers and to set a floor for wages. At the time, however, labour market conditions did not favour the implementation of that part of the law because there was an abundance of workers, in relation to available jobs, and labourers were therefore willing to work for relatively low wages. In fact the level of farm wages fell by about 7 per cent in 1953 because of the fall in prices of the cotton crop (Hagrass 1970).

On the issue of fragmentation of land holdings, the Islamic Law of Inheritance and the deeply rooted traditions stood as stumbling blocks preventing the fixing of a 5-feddan lower limit on land holding as was stipulated by the Agrarian Land Reform Law. In fact this segment of the legislation has never been implemented and continued to be practically void (Hagrass 1970).

TABLE 9.2. Distribution of landownership before the land reform in 1952 and after the law passed in 1964 and 1984

	<5	5–<10	10–<20	20–<50	50–<100	100–<200	>200	Total
1952								
Landowners ('000s)	2642	79	47	22	6	3	2	2801
%	94.3	2.8	1.7	0.8	0.2	0.1	0.1	100
Area owned ('000s feddans)	2122	526	638	654	430	437	1177	5984
%	35.5	8.8	10.7	10.9	7.2	7.3	19.7	100
1964[a]								
Landowners ('000s)	2965	78	61	29	10			3143
%	94.3	2.5	2.0	0.9	0.3			100
Area owned ('000s feddans)	3353	614	527	815	813			6122
%	54.8	10.0	8.6	13.3	13.3			100
1984								
Landowners ('000s)	3288	86	46	23	6	2[b]		3451
%	95.3	2.5	1.3	0.7	0.2	0.1		100
Area owned ('000s feddans)	2904	576	589	621	407	335		5432
%	53.5	10.6	10.8	11.4	7.5	6.2		100

[a] Does not include government ownership of fallow land that was being reclaimed for distribution.
[b] Includes organizations, companies, and individuals; state lands, desert, prairie, and land under distribution are not included.
Source: Data for 1952 and 1964, Hagrass (1970); data for 1984, CAPMAS (1988).

In association with agrarian reform, the Egyptian Government embarked on an ambitious public land reclamation programme. About 1 293 900 feddans were reclaimed by the public sector between 1952 and 1986/7, and most of this land (821 400 feddans) was reclaimed from 1952 to 1967/8. Up to 1987, 214 166 feddans of the reclaimed land were distributed among farmers of limited income. This figure does not include land distributed by the Egyptian Organization for Desert Development. During the same period, some 864 521 feddans of agrarian reform land were distributed (CAPMAS 1986).

It has been more than thirty-nine years since the first Agrarian Land Reform Law was promulgated, and many changes have taken place since then. Land returns have increased disproportionately relative to land rents, cash rent has become the dominant form representing 83.5 per cent of all kinds of rents in 1982, and for all practical purposes tenants have become the actual land owners. It has become extremely difficult, if not impossible, for an owner to exercise any reasonable power over his/her land; he/she cannot retrieve or sell it without the tenant's consent or without paying him/her unduly high compensation. The owner/tenant relationship has tipped too much in favour of tenants, but legislation now being considered in the People's Assembly is designed to effect a proper, just, and equitable redistribution of rural income. If the seven-times land tax formula is to be maintained, one main element of the impending legislation is to raise rents by increasing the land tax. Alternatively the proportion could be raised. Sharecropping would be a better and fairer way of managing tenancy. One other important element is to give owners more control over their land by limiting land leases to three years, renewable by mutual consent.

Crop production and procurement

By the mid-1960s, it was apparent that Egyptian planners had become aware that the development of a comprehensive agrarian development policy was a necessary prerequisite for the establishment of an efficient socialist economy. In addition, the population explosion of the previous three decades, together with many endogenous and exogenous factors—an increase in rural wages, redistribution of wealth and income in favour of small landowners and landless farmers, and an increase in government spending—led to significant increases in consumer demand and in the prices of major agricultural staples. This emphasized the political importance of food in the Egyptian economy. The need for preserving internal stability by maintaining stable and cheap food prices motivated the Government to sustain and occasionally intensify its intervention in the agricultural input and output markets as well as in food distribution. Thus, an all inclusive Agricultural Act, known as Law No. 53 for 1966, was promulgated with details of various government objectives and instruments of government organization and incursion in agriculture. The strong drive of the State to control the agrarian sector was manifested by the requirement that farmers be issued 'holding cards' containing a detailed description of their production activities.

By virtue of that law, until very recently the major features of the Egyptian Government's agricultural control policy had remained about the same since the 1960s. Over nearly thirty years, the institutional intervention instruments and programmes were subject to relatively minor or no changes. On the input side of production, they involved administrative controls and incentive programmes including acreage allotment especially for the main crops cotton and sugar cane, and distribution and fixed allocation of key inputs (e.g. chemical fertilizers, pesticides, and certified seeds). All fertilizers were distributed to producers by the Principal Bank for Development and Agricultural Credit (PBDAC) in accordance with government-determined norms, specified as the product of a normative quantity of fertilizers required per feddan for each crop and area under that crop. Fertilizer prices, both at factory and retail levels, were also fixed by the government at levels below the corresponding international prices (Moursi 1986). In addition, the Government was involved in the introduction, breeding, and distribution of high-yielding crop varieties as well as the allocation of animal feed at low mandatory prices via village co-operatives.

From time to time the government continued to allocate limited areas for specific crops or varieties of crops, either to ensure adequate production of a specific commodity for local consumption and/or export purposes (e.g. rice and cotton) or to prevent over-production (e.g. cotton at times). Although evasions are not uncommon, area allotments undoubtedly have a significant impact on the pattern of agricultural production as they affect the areas planted with other crops that are not subject to restrictions, through crop rotation linkages.

On the output side, state incursion included farm-gate price and quota restrictions. Fixed price procurement laws require farmers to deliver to the Egyptian Government a defined quota per feddan of a specific crop at a fixed price. With some crops, cotton and sugar cane, all production is procured by the Government at announced fixed prices. The fixed price quota procurement programme does not preclude regional or geographical variability in the per feddan quota of the same crop. Markets for crops subject to fixed price delivery quota, except for cotton and sugar cane which are procured *in toto*, are characterized by a dual price structure, an announced fixed price, and a free market price. Usually the institutional prices are on average below the free prices and their international equivalents (Moursi 1980). When delivery is voluntary (e.g. wheat in 1976 and subsequent years) or when there is over delivery (e.g. rice in certain years), the government mandatory price serves as the lower floor for the free market local price of the relevant commodity. During 1970–81, the announced fixed farm-gate prices for winter onions exceeded the free market prices because the government wanted to procure as much of the crop as possible to increase its export volume.

Other major instruments that affected the production of agricultural and food commodities indirectly were trade policies such as tariff and trade restrictions, exchange rate management, and fiscal policy. Most of these controls are being lifted gradually or relaxed in line with Egypt's new liberalization and structural adjustment policy.

Government subsidies of agricultural inputs

In its attempt to maintain high crop production and productivity levels, the government has tried to make available to farmers, at reduced prices, important agricultural inputs such as fertilizers, pesticides, certified seeds, and animal feed. (Implicit subsidies on other important inputs also exist, e.g. fuel, lubricants, and diesel, which are sold below free market prices). In spite of the substantial increase in the cost of locally produced and imported fertilizers, prices to farmers, until recently, had not changed much since 1960, while at the same time prices of agricultural outputs have sustained manifold increases. During the period from 1960 to 1973, except for 1965–6, the sale of fertilizers implied a tax on farmers (Moursi 1980). From 1973 until 1987/8, fertilizer sales to farmers included an increasing subsidy that by 1987/8 had reached approximately £E183million (Tables 9.3 and 9.4). The increase in volume of fertilizer subsidies, which are borne out by the Government's

TABLE 9.3. Taxes on the sale of chemical fertilizers for the years 1961–86/7

Year	£Emillion	Year	£Emillion
1961	1.6	1975	−77.2
1962	1.8	1976	−35.1
1963	1.8	1977	−13.9
1964	1.0	1978	−32.0
1965	−1.1	1979	−76.2
1966	−0.7	1980	−41.7
1967	1.0	1980/1	−107.1
1968	3.4	1981/2	−143.9
1969	6.2	1982/3	−115.0
1970	10.2	1983/4	−121.4
1971	13.7	1984/5	−136.7
1972	19.7	1985/6	−130.2
1973	−1.5	1986/7	−130.9
1974	−52.4	1987/8	−182.9

Source: Compiled from unpublished data from the Fertilizers Stabilization Fund for 1961–72 and from the Public Agency for the Agriculture Stabilization Fund for 1973–87/8.

Agricultural Stabilization Fund, was partly due to the general increase in prices of domestic and imported fertilizers and in part to the gradual rise in fertilizer consumption (NSC 1989). Soil conditioners such as agricultural gypsum were also subsidized by about £E1million annually.

Due to the rising cost of cotton pest control operations, the Government began to subsidize this activity, beginning with the 1971/2 cotton planting season. The pesticides subsidy was later extended to a few other crops and was to rise gradually from about £E12million in 1971/2 to more than £E70million in 1987/8 (Table 9.4). In addition, the government's general seed policy involves the annual production and distribution of certified seeds for all the cotton planted area and for an identified proportion of the areas under other major field crops. The programme is supported by an annual government subsidy of more than £E1.5million.

Because of the importance of animal feed to the livestock sector, manufactured feed supplies are allocated among farmers by the Ministry of Agriculture, distributed through village co-operatives, and are sold at below international prices. It should be pointed out that the Egyptian Government pays farmers very low prices for the ingredients involved in animal feed manufacturing. Yellow maize, a basic ingredient in the manufacture of poultry feed, is also substantially subsidized; the subsidy accorded to that commodity has increased gradually from about £E54million in 1980/1 to about £E241million in 1984/5 before it began to fall down again to a meagre £E6.8million in 1987/8 (Table 9.4).

Other input subsidies extended to farmers take the form of nominal subsidies on agricultural credit provided by the PBDAC. The PBDAC provides farmers with short, medium, and long-term loans at fixed interest rates of 4.5, 9.5, and 4.5 per cent, respectively, compared to 14.5 per cent on loans provided by commercial banks; the loans granted are also subject to a low service commission (3 per cent) and to only 5 per cent interest in the case of reinstated delinquent borrowers for overdue claims (Moursi 1986). The deficit generated from the discrepancy between the fixed and the market interest rate, which the PBDAC has to pay on its loans from the banking system, represents an additional subsidy paid to farmers that is usually covered from the government budget. In general, there has been a steady increase in nominal loans provided by the PBDAC from 1980 to 1986 and accordingly of nominal subsidies on agricultural credit of about £E95million in 1987/8 (Table 9.4).

Fertilizer subsidies have usually represented the greater bulk of all input subsidies, followed by yellow maize, pesticides, and agricultural credit subsidies. Total nominal input subsidies as a percentage of the value of agricultural production varied from 7.5 per cent in 1981/2 to 2.6 per cent in 1986/7 (Table 9.4). All subsidized inputs are provided on credit by the PBDAC as short-term loans. Additional subsidies extended to the agricultural sector are the substantial subsidies accorded to agricultural research and development, education,

TABLE 9.4. Nominal input subsidies

Year	Fertilizers[a]	Cotton pesticides	Agricultural credit	Yellow maize subsidy	Additional input subsidies	Total subsidies	%[b]
			£Emillion				
1980/1	113 647	46 000	213 615	54 404	8320	245 086	4.7
1981/2	192 945	48 000	32 153	147 309	9582	430 388	7.5
1982/3	136 407	68 929	37 724	133 912	9974	386 945	4.8
1983/4	115 595	69 730	43 144	164 739	14 095	407 474	4.9
1984/5	111 117	66 548	49 708	241 098	11 325	479 826	5.3
1985/6	121 874	68 200	53 326	107 426	10 887	361 713	3.2
1986/7	124 551	69 371	95 269	27 266	9298	325 757	2.6
1987/8	182 940	70 200	94 830	6788	11 656	366 414	2.8

[a] Differences with corresponding figures in Table 9.3 are due to the difference in data source.
[b] Input subsidies as a percentage of the value of agricultural production.
Source: Compiled from unpublished data assembled by the Agricultural Stabilization Fund.

and extension, the management and operation of irrigation facilities, the construction of irrigation works including the digging of canals, the building of dams and reservoirs, the installment of lifting and pumping stations, and also the free supply of irrigation water.

Government intervention and food consumption

The current consumption subsidy programme that was originated in the 1940s evolved from policies aimed at ensuring access to basic commodities, especially food staples, by all income groups. It involved such programmes as price wedging between purchase and selling prices, the absorbing of marketing and transport costs, the placing of price ceilings and marketing margins on privately traded commodities, the fixing of prices on goods produced by public sector companies, and the rationing of basic staples in the consumption market (Alderman et al. 1982).

During the 1970s and 1980s, food subsidies, administered mainly by the General Authority for Supply Commodities, became a major constituent of the government policy. Nominal food subsidies which amounted only to £E2million in 1945 increased steadily to reach £E329.1million in 1974, £E1108million in 1980/1, and £E1862.6million in 1984/5. Subsidies for wheat and flour alone for 1974 and 1980/5 were £E216.4million and £E776million, respectively, representing approximately 66 per cent and 70 per cent, respectively, of total food subsidies and 52.8 per cent and 49.4 per cent, respectively, of total direct subsidies related to the agricultural and food policy (Table 9.5). 'The expansion of subsidies was largely a response to . . . the 1972 to 1974 grain-price explosion' in an attempt by the Egyptian Government to maintain food prices at their pre-crisis levels (Taylor 1979). Food subsidies have come to form a substantial portion of the income of consumers especially in the low income groups. A detailed review of the different programmes and description of the distribution system at the height of their full implementation are presented by Alderman et al. (1982), Von Braun (1980, 1981), and Von Braun and De Haen (1983). The remainder of this section is largely based on their description.

The principal food commodities subsidized by the Egyptian Government are wheat, rice, maize, broad beans (*Vicia faba*), lentils, cooking oil, meat, poultry, fish, tea, and sugar. These commodities are all the more important to low income groups who spend two-thirds of their per caput expenditure on food, of which 50 per cent is spent on cereals and cereal products in rural communities, compared with 35 per cent in urban areas (Von Braun 1981). While bread is distributed mainly in urban areas, wheat flour is distributed essentially in rural areas through controlled government stores. Subsidized wheat and flour are made available to all consumers without restrictions. Most of the subsidized wheat flour is sold to bakeries which are regulated by the Ministry of Supply (MOS). Loaf weight, moisture content, and prices are

TABLE 9.5. Direct subsidies related to agricultural and food policy

£Emillion

	1974	1975	1976	1977	1978	1979	1980	1981	1982	1983	1984
1. General Authority for the Supply of Commodites	329.1	490.9	321.5	313.4	449.6	1001.9	1108.0	2190.5	1336.5	1208.8	1862.6
2. Wheat and flour alone	216.4	260.9	171.6	149.1	222.8	647.0	776.0	736.0	NA	NA	NA
3. Wheat and flour subsidies as a percentage of 1	65.7	53.1	53.3	47.6	49.6	64.6	70.0	33.6	NA	NA	NA
4. Total subsidies	410.0	622.1	433.5	649.6	710.0	1352.0	1571.5	2909.2	2053.7	1986.6	2478.7
5. Wheat and flour as a percentage of 4	52.8	41.94	39.6	22.9	31.4	47.9	49.4	25.3	NA	NA	NA

NA, not available.
Source: Compiled from Dethier (1987) from data provided by the Ministry of Finance.

fixed. Most bakeries are privately run and are under the constant inspection of inspectors of the MOS to guarantee that the product meets the government requirements and that flour designated to bread making is not diverted to the making of pastry which is not under price control. Public sector bakeries have become increasingly important in recent years. Through such bakeries, the Government is trying to increase the capacity of mechanized bread making while ensuring at the same time high quality and health standards and avoiding rising labour costs.

Rice, cooking oil, tea, and sugar are rationed commodities. Monthly rations are provided at low subsidized ration prices to about 90 per cent of the population through ration cards. The commodity quotas are determined by the MOS and vary depending on the region and its rural and urban nature (Von Braun 1981). The principal outlets for the MOS commodities are private groceries that are registered with the MOS and who are licensed to receive rations from the public wholesale companies to distribute to consumers. The consuming household holding a ration book must register with a licensed grocer who records his monthly purchase of the rationed commodities. Besides the highly subsidized rations, there is, in most cases, an additional ration of limited quantity that is made available at a less subsidized price. The additional ration along with most of the subsidized food commodities are distributed via a network of consumers' co-operatives and state-owned stores. In most areas that are not covered by government retail stores, the main subsidized food distribution channels are the consumers' co-operatives. A consumer may obtain membership upon payment of a nominal fee, but in practice members are not usually active and there are seldom any profits left for distribution.

Besides the main rations and the additional rations of the above commodities, there are usually additional supplies that are sold on the open market at uncontrolled free (or black) market prices determined by supply availability and demand. Free prices are likely to vary from region to region according to market forces. Apart from the multi-tier price structure induced by the government's food subsidy and rationing policy, the Government set alternative prices for rationed or subsidized commodities that are to be used as inputs (e.g. wheat and sugar) in the production of certain products.

Finally, government intervention in the food consumption market has extended to involve the imposition of price ceilings on most fruit and vegetables that are sold in private stores. Price ceilings may vary from region to region and are revised weekly by the MOS authorities. Despite government efforts, however, enforcement of the ceilings set by the Egyptian Government has proved very difficult and open market prices which usually transcend the government ceiling prices have almost consistently prevailed.

The impact of government intervention in agriculture

The extensive and comprehensive system of state controls described above has substantial effects on the agrarian sector and on the rest of the economy. While there appears to be some consensus among economists and policy makers about the limitations and implications of the intervention policies—for example the inconsistency of the different control instruments—there are usually disparate views on the economic and social viability of intervention policies, the effects of various agricultural controls, and the practicability and workability of alternatives. This section presents a cursory review of different views on the effects of selected agricultural controls; these views emerged from independent studies and therefore occasionally may lack homogeneity.

The impact of the procurement policy

Several studies were carried out to analyse the impact of the fixed price/quota delivery programme as an instrument taxing the agricultural sector. While a few crops were subsidized, major food and export crops were heavily taxed by the price disincentive programme. Moursi (1980), in his study on the net effect of the government pricing policy on the prices of seven major field crops—cotton, wheat, rice, maize, beans, winter onion, and sugar cane—during the period 1965–77, showed that the selected crops were taxed heavily (cotton and rice, the major export crops, were the most taxed) and that there was accordingly a net annual average resource transfer from the agricultural sector in the order of £E695million, after taking into account all direct and indirect subsidies to agriculture. Von Braun and De Haen (1983) indicated that during the period 1965–80, agricultural output prices were distorted in different ways; wheat, maize, rice, beans, and lentils were usually taxed, while meat and milk were protected and subsidized through subsidized feed. Similar results were later obtained by Dethier (1987) who found that, during 1970–84, cotton and rice were taxed considerably and that the net effect of direct and indirect intervention on agricultural inputs and outputs resulted in a state appropriation of approximately 35 per cent of agricultural GDP (Dethier 1989). Emara et al. (1989) pointed out that the net producer subsidy equivalent (PSE) measured during 1980–7 for wheat, maize, rice, cotton, and sugar cane was negative—indicating an effective tax on agriculture—and became increasingly negative during 1984–7. The study also showed that keeping domestic farm-gate prices, especially of cotton, rice, and sugar cane, below world prices, while at the same time maintaining trade control policies that did not allow farmers to export maize and wheat when the world prices exceeded domestic prices, contributed to the negative overall effect.

The heavy tax burden associated with the forced delivery programme did not, however, significantly affect the cropping pattern in the short-term. This was probably due to the influence of non-price and institutional variables (e.g. acreage allotment) on farmers' behaviour (Cuddihy 1980). In the long-term, however, cropping patterns were more responsive, shifting in favour of untaxed or subsidized crops (e.g. *bersim*, fruits, and vegetables). Because the mandatory price and quota system is limited to a particular subset of export and staple crops, a heavier tax burden was eventually borne by smaller farmers (Cuddihy 1980). In addition, the implicit tax rate on controlled crops varied with set institutional restrictions, thus the tax burden was not shared evenly by farmers growing those crops. The distortive effects of the production tax were amplified further by agrarian input subsidies that were received indiscriminately by all agricultural producers.

Despite the rigid system of government controls and implied resource transfer, agricultural supply in general appears to be responsive to changes in prices and other factors (Moursi 1986). El Saadani and Abu Rawash (1989) showed that farmers respond positively—albeit with a lag that might exceed one crop season—to increases in procurement prices, and that other factors considered by farmers in response to a rise in procurement prices include revenue from other competing crops and crop rotations.

In spite of the strong positive collinearity between procurement and the cost of production (Beshai *et al.* 1989), the farmers' supply responsiveness seems to have helped them profit from higher farm-gate prices. Losses suffered by producers were reduced steadily in recent years as a result of the gradual increase in the fixed procurement prices of controlled and partially controlled crops to the extent that producers of sugar cane, rice, and maize benefitted from an implicit subsidy in 1985. Adjustment for exchange rate evaluation, however, showed that the reduction in price bias against producers was notably less than it appeared (Beshai and Moursi 1987). The overall impact of the government pricing policy on agricultural output has been a decline in the average annual growth rate in the 1980s from 2.6 per cent during 1964–70 and 3.5 per cent during 1970–80 to 2.6 per cent during 1980–3 and −0.2 per cent in 1983/4 (Dethier 1987).

The impact of the input subsidies programme

The comprehensive system of input subsidies described above was designed to expand agricultural production, improve productivity, and support farm income, especially of farmers most adversely affected by the forced delivery programme. Cuddihy (1980) maintained that despite the price responsiveness of farmers, lower factor prices did not lead to higher input intensities (presumably due to the distortive nature of the controls as well as institutional and quota restrictions). To the contrary, there was an excess supply of specific inputs (e.g. farm machinery) and secondary (black) markets emerged for inputs such as fertilizers and pesticides in excess demand (Cuddihy 1980).

Regional distortions, however, crippled market forces so that simultaneous excess supply and demand of certain inputs often prevailed. Moreover, the interaction of relative input price and quantity distortions with procurement controls contributed to the poor productive efficiency of Egyptian farmers (Chiao 1985; Moursi 1991).

Although various researchers have unequivocally emphasized the distortive and negative effects of the input subsidy programme, some studies warned of unpropitious effects that might result from an abrupt cancellation of the programme. The tenor of the findings of those studies seems to suggest caution with a gradual removal of selected distortive subsidies.

For instance, El Saadani and Abu Rawash (1989) indicated that decreasing subsidies on cotton production inputs would have a negative psychological effect on farmers and their use of fertilizers and pesticides, affecting productivity accordingly. They also suggested that removal of the subsidies would entail high economic cost; input subsidies represent only a very small percentage of the value of agricultural production (2.6 per cent and 2.8 per cent in 1986/7 and 1987/8, respectively) so that reducing or cancelling these subsidies would save the Egyptian Government around £E350million annually, but would result in higher input prices and production costs and lower producers' income, unless accompanied by higher output prices (reflecting both annual production cost increases and those resulting from input price increases). Consequently, they concluded that if input subsidies had to be reduced or cancelled, as it is being anticipated, transition from subsidies to economic pricing should be gradual and producers should be alerted beforehand.

The literature provides justification for input subsidies in some special cases. Shalit and Binswanger (1985) pointed out that under certain circumstances (e.g. if the government, regardless of economic viability, chose to increase food self sufficiency) the only theoretical case for a permanent fertilizer subsidy would be in the existence of a non-optimal output tax and if the crop(s) were highly fertilizer responsive (e.g. irrigated rice and maize). These conditions apparently prevail to a very high degree in Egypt. Fertilizers play a very important part in Egyptian agricultural production. The World Development Report (World Bank 1989) showed that the country used 3193 g of plant nutrient per hectare of arable land in 1986, placing Egypt highest among the class of middle and upper-middle income countries with respect to fertilizer consumption that year. It is imperative, therefore, that great caution should be taken relative to any change in the volume of subsidy extended to this essential input.

The impact of food subsidies

The rising subsidy bill increased government spending dramatically and consequently its current deficit has attracted considerable attention. Elaborate studies on Egyptian agricultural and food subsidies have been undertaken by national and foreign economists and institutions, and as food subsidies are much larger they have been studied more than input subsidies. Food subsidies affect the majority of Egypt's population, including the more vocal and much more concentrated urban group. In addition, because they represent a significant portion of the income of low income groups, food subsidies have an important social aspect deserving particular consideration. Different studies have reached different conclusions regarding the role and socio-economic effects of food subsidies; in spite of these differences, there is some putative agreement about the importance of the subsidies themselves and of policies that may affect them directly or indirectly. The following few paragraphs will deal with the social, economic, and political impacts of food subsidies.

Abdel Fadil (1989) summarized early studies on the social impact of reducing or cancelling food subsidies. He showed that higher income groups tended to benefit proportionately more from some subsidies—particularly if the subsidy rate was invariant across income groups (e.g. the wheat subsidy). He concluded that an overall reduction of food subsidies would have a regressive impact on the economy as a whole, with the poor, especially urban low and middle income classes characterized by high marginal propensity to spend on subsidized food items, 'bearing the brunt of real income loss'. Hence, the inequitable distribution of food subsidies provoked the widening gap between lower and upper income groups and urban and rural consumers.

With respect to the agricultural sector, Cuddihy (1980) stipulated that 'since some farmers do not grow all their food, they benefitted from the low food prices, but typically it was the urban consumer who gained most'. Moreover, Von Braun and De Haen (1983) indicated that when the agricultural pricing programme was combined with income transfers from food subsidies, both large and small farm households suffered from the price and subsidy policies. It should not be assumed, however, that eliminating food subsidies would directly lift rural hardship. For example, removing food subsidies would increase food prices eliciting wages also to rise; and if the food price increase was offset by raising investment spending, there would still be a loss in rural incomes (Taylor 1979).

Perhaps the most important economic aspect of the food subsidy programme is its impact on consumption and income. The regressive nature of the food subsidy programme may have been, at least in part, counterbalanced by some of the benefits of the ration component of the government food intervention programme. Alderman and Von Braun (1984) emphasized that while rations had only a slight effect on consumption levels, they had significant effects on income; furthermore, the income transfers they produced, while untargeted, reduced income inequality by allowing higher proportional subsidies to the poor.

The effects of food subsidies on macro-economic variables and on structural imbalances in the economy have also been analysed. Scobie (1981, 1983) found that actual subsidies, including those on food, did not appear to affect investment. He also explained that although food subsidies, and deficit government financing, may affect both the exchange and inflation rates, there was not sufficient evidence to support the argument that food subsidies would lead to a classic trade off between the requirements of equity and of growth. Scobie, however, warned that food subsidies could have serious destabilizing structural effects on the economy. For instance, stable food consumption based on large-scale food subsidies may transfer instability in world prices and foreign exchange rates into the domestic economy. Transmission of this instability may have a more detrimental impact on lower and middle income groups as terms of trade turn against the domestic economy.

The studies referred to above analysed some important negative aspects of the Egyptian food subsidy system. However, it was the Shura Council (Senate) that probably presented the most critical review of the economic, social, and political effects of that system. Its summary highlighted the following: allocative inefficiency due to distortions in price mechanism, decrease of local production and increased imports of subsidized commodities, discriminate benefitting of all income groups, misuse of subsidized commodities, prevalence of black markets, arbitrage, the appearance of a class of middlemen with untaxed incomes, political risks associated with any attempt to rationalize or abolish the system, increase of consumption and greater dependence on foreign aid to finance imports of essential commodities, sectoral imbalances (e.g. rural versus urban or agricultural versus service sectors), and the failure of the system to reduce income irregularity. Nevertheless, despite the serious limitations of the food subsidy system, there is little doubt that the programme had some significant merit. As Taylor (1979) concedes '. . . the subsidies were inevitable on macro-economic grounds'. Few studies endorsed the complete abolition of the programme, on the premise that it is not always necessary for the negative impact of a system to outweigh its positive impact. Rationalization rather than maintaining or abolishing the system seems a more viable and more economically efficient alternative.

Liberalization and recent changes in government policy

In the late 1980s, when the economic policy that had been implemented by the Egyptian Government for the previous thirty years failed to improve worsening economic conditions and to achieve sustained development, a change in economic policy was deemed in order. Since 1990, in consultation with the World Bank and the International Monetary Fund, the Egyptian

Government has begun to put in motion an all embracing macro-economic reform, structural adjustment, and stabilization programme. This provides a greater role for market forces and private initiative, and includes price and interest rate management, exchange rate and trade policy reform, public and private sector development, and privatization.

Since early 1990, there has been a gradual liberalization of the distribution of important products produced by public enterprises, as well as a phasing out of their subsidies. It is contemplated that all goods produced by the industrial public sector will be subject to the liberalization process with the exception of a few products which will remain controlled subject to periodic review. The reform package implied significant changes in the government crop procurement and agricultural input policies.

With respect to the fixed cotton prices paid to farmers, the Egyptian Government has, since 1990, been pursuing a programme to adjust procurement prices gradually so that they approach the international price level. The procurement price is planned to be 66 per cent of the international level for the 1992 crop. The mandatory rice delivery quotas are to be phased out in three years, by 1992, and input subsidies for fertilizers, pesticides, and livestock feed are also to be phased out in three years.

With the progress of the liberalization process, however, it is expected that other government controls will be eliminated or at least ameliorated and that privatization and private sector development will continue to limit government activities in fields such as land reclamation, poultry, meat, and animal feed production, and vegetable and fruit production.

The proposed economic adjustments will certainly have their adverse effects. In order to overcome these effects, and to gain support for the reform programme a 'social fund' will be created with contributions from the World Bank, bilateral and multilateral donors, and perhaps non-government organizations. The fund, which will be a temporary institution lasting three to four years will help to protect vulnerable groups that are likely to suffer most from the implementation of the economic reform programme especially labourers and fixed income groups. The areas of the fund's activities are still being considered, but it is expected to cover areas such as employment development, socio-economic development, labour intensive public works, and direct social assistance.

The potential effects of the changes in the government's agricultural intervention policy are beyond the scope of this review. The above, however, suggests that the policy change should have important socio-economic effects on the performance of different sectors, on the various social groups, and on the overall economy. Economic and policy research on the socio-economic effects of changes in the government incursion programme is therefore of paramount importance, especially if the Egyptian Government continues to adhere to the philosophy and the commitments of the adjustment programme.

Conclusion

Historical endogenous and international constraints that crippled the performance and growth of Egyptian agriculture and rural development prompted the Egyptian government to intervene directly and indirectly in agricultural production and food distribution. By the mid-1960s the burden of adopting and implementing an 'efficient and equitable' agrarian strategy had gradually shifted almost completely to the State. To achieve its announced objectives, the government introduced a comprehensive set of elaborate economic and institutional intervention instruments that allowed government control of the agrarian sector. The intervention programme was responsible for the generation and the mobilization of a sizeable surplus that was transferred to satisfy the government's increasingly unquenchable resource needs. The gains to the government associated with resource transfers are channelled, within the institutional system, primarily for maintaining domestic stability, through support of the government budgetary improvement of the balance of payments and supporting the urban programme for cheap food prices.

The agrarian incursion instruments, however, were *ad hoc* and crisis-orientated, lacking consistency and other necessary prerequisites to generate agricultural development; indeed not only did the agricultural intervention programme fail to instigate overall agrarian development, it also contributed positively to the languishing performance and stagnation of agriculture. The interaction of socio-economic, institutional, and political factors associated with government intervention in agriculture contributed to the failure of agrarian policy. A significant element contributing to that failure, during the last three decades, related to the inability of the institutional framework to resolve equitably the struggle between the conflicting interests of the poorer tenants and landless farmers on the one side, and large landowners and government bureaucracy on the other. The difficulties emerged once the minor gains to the peasantry, associated with the initial comprehensive implementation of the reform programme, were reversed in favour of the bigger farmers and the Egyptian Government. These were able to put pressure on the peasantry and take advantage of institutional controls and of the agrarian system (Radwan and Lee 1986).

In the late 1980s, when it became evident that the agricultural reform and related institutional programme put in motion more than thirty years ago could not achieve their objective, there was a call to reject the direct physical intervention in agricultural production and food distribution and to dismiss the controls as a necessary prerequisite for agrarian development. The call was provoked and stimulated by regional and international political changes supporting more liberal economic regimes. More important, the call was also endorsed by the government, on one hand as part of an overall domestic liberalization programme, and on the other hand as part of a reform agreement with international donor organizations.

While there are uncontestable merits for the deregulation and liberalization of the agrarian sector, abrupt dismantling of the agricultural intervention programme in Egypt may introduce other hazards. Several studies, which have suggested the need for radical agricultural reform, have favoured gradualism in the removal of the intervention instruments. In 1991 there appears to be some criticism of the details of the Egyptian Government's future agricultural reform programme. There is a dire need for methodological socio-economic research which will evaluate eclectically the alternative agrarian development strategies.

References

Abdel Fadil, M. (1989). The social dimensions of economic policy in Egypt. Unpublished paper presented at a seminar on the Social Market Economy, March 13–15. Cairo, 1989.

Alderman, H. and Von Braun, J. (1984). *The effects of the Egyptian food rationing and subsidy system on income distribution and consumption*. International Food Policy Research Institute, Research Report No. 45. IFPRI, Washington, DC.

Alderman, H., Von Braun, J., and Sakr, S.A. (1982). *Egypt's food subsidy and rationing system*. International Food Policy Research Institute, Research Report No. 34. IFPRI, Washington, DC.

Beshai, A.A. and Moursi, T.A. (1987). Farm-gate price bias against agricultural producers in Egypt. Unpublished paper, Department of Economics, American University in Cairo, Cairo.

Beshai, A.A., Nassar, A., and Fletcher, L. (1989). *Agricultural input subsidies in Egypt: magnitude and consequences 1980/1981 through 1987/1988*. Working Paper presented at the Conference on Agricultural Policy Reform in Egypt: Current Status and Future Strategy. Ministry of Agriculture and Land Reclamation, Egypt, June 24–25, 1989.

CAPMAS (Central Agency for Public Mobilization and Statistics) (1986). *Statistical yearbook*. CAPMAS, Cairo.

CAPMAS (Central Agency for Public Mobilization and Statistics) (1988). *Statistical yearbook*. CAPMAS, Cairo.

Chiao, Y., (1985). Frontier production function approaches for measuring efficiency of Egyptian farmers. Unpublished Ph.D. thesis. University of California, Davis.

Cuddihy, W. (1980). *Agricultural price management in Egypt*. World Bank Staff Working Paper, No. 388. World Bank, Washington, DC.

Dethier, J.J. (1987). *Agricultural prices in Egypt: issues, policies and perspectives*. Unpublished paper presented at FAO Workshop on Agriculture Price and Marketing Policies in Egypt, Cairo, April 11–16, 1987.

Dethier, J.J. (1989). *Trade, exchange rate and agricultural pricing policies in Egypt. Vol. 1. The country study*. World Bank Comparative Studies: The Political Economy of Agricultural Pricing Policies, World Bank, Washington, DC.

El Saadani, R. and Abu Rawash, A. (1989). *Economic effects of increasing procurement prices of Egyptian cotton*. Working Paper, presented at the conference on

Agricultural Policy Reform in Egypt: Current Status and Future Strategy. Ministry of Agriculture and Land Reclamation, Egypt, June 24–25, 1989.

Emara, A., Abbas, H., Nassar, A. and Gardner, G.R. (1989). *Egypt's producer subsidy equivalent: measurement of government's intervention in agriculture*. Working Paper presented at the Conference on Agricultural Policy Reform in Egypt: Current Status and Future Strategy. Ministry of Agriculture and Land Reclamation, Egypt, June 24–25, 1989.

Hagrass, S. (1970). *Agrarian reform: history, philosophy, programme*. Ein Shams Book Shop, Cairo. (in Arabic).

Ikram, K. (1980). *Egypt: economic management in a period of transition*. World Bank Country Economic Report. The Johns Hopkins University Press, Baltimore.

Ministry of Agriculture (1961). *Agricultural Census, 1961*. Ministry of Agriculture, Cairo. (in Arabic).

Ministry of Agriculture (1979). Distribution of the numbers of land holders and area of holdings according to indicated groups. *Annual Bulletin of the Institute of Agricultural Economics*, pp.20–25. Ministry of Agriculture, Cairo. (in Arabic).

Moursi, T.A. (1980). Agricultural pricing policy in Egypt: an empirical study 1965–1977. Unpublished M.A. thesis. The American University in Cairo, Cairo.

Moursi, T.A. (1986). Government intervention and the impact on agriculture: the case of Egypt. Unpublished Ph.D. thesis. University of California, Berkeley.

Moursi, T.A. (1991). *Measurement of productive efficiency of wheat producers in El-Menoufia Governorate: a frontier cost function approach*. Unpublished paper presented at the Second Conference of the Department of Economics, Faculty of Economics and Political Science, Cairo University.

NSC (National Specialized Councils) (1989). *The present and future fertilizers policy*. National Specialized Councils, Cairo. (in Arabic).

Radwan, S. and Lee, E. (1986). *Agrarian change in Egypt: an anatomy of rural poverty*. Croom Helm, London.

Scobie, G.M. (1981). *Government policy and food imports: the case of wheat in Egypt*. International Food Policy Research Institute, Research Report No. 29. IFPRI, Washington, DC.

Scobie, G.M. (1983). *Food subsidies in Egypt: their impact on foreign exchange and trade*. International Food Policy Research Institute, Research Report No. 40, International Food Policy Research Institute, Washington, DC.

Shalit, H. and Binswanger, H. (1985). *Is there a theoretical case of fertilizer subsidies?* World Bank Discussion Paper, RAU 27. World Bank, Washington, DC.

Taylor, L. (1979). *Macro models for developing countries*. Economics Handbook Series. McGraw-Hill, Maidenhead, UK.

Von Braun, J. (1980). *Agricultural sector analysis and food supply in Egypt*. Unpublished interior report on the Joint Project of the Institute of National Planning, Cairo, and the Institute of Agricultural Economics of the University of Goettingen, German Federal Republic.

Von Braun, J. (1981). *A demand system for Egypt: estimation results and scenario analysis for alternative food price policies*. Institute of Agricultural Economics, University of Goettingen, German Federal Republic.

Von Braun, J. and De Haen, H. (1983). *The effects of food price and subsidy policies on Egyptian agriculture*. International Food Policy Research Institute, Research Report No. 42. IFPRI, Washington, DC.

World Bank (1989). *World Development Report 1989: financial systems and development, world development indicators*. Oxford University Press, Oxford.

Further reading

Financial Times (1985). Financial Times Survey: Egypt. June 5, 1985.
Financial Times (1990). Financial Times Survey: Egypt. April 4, 1990.
Financial Times (1992). Financial Times Survey: Egypt. January 21, 1992.
Financial Times (1993). Financial Times Survey: Egypt. April 22, 1993.
Menas Associates' (1990–). *Egypt Focus: monthly news and analysis of economic, political and financial developments in Egypt*. Means Associates' Ltd, London.

10

Agriculture in the national economy

J.M. ANTLE

Introduction

Agriculture played a central role in the national economy of Egypt in the Pharaohs' time and continues to do so today. Like other developing countries, a large share of Egypt's national income is generated in the agricultural sector; the state of agriculture is integrally related to the welfare of the Egyptian people and to the national economy. However, in many ways Egypt is unlike other developing countries due to its unique position in geography and history. Egypt's agriculture along the Nile River and in the Nile Delta has, historically, been one of the most productive in the world, and Egypt's position as a crossroads for trade has linked its agriculture to developing world markets since the industrial revolution.

Agriculture is central to understanding the major economic problems facing Egypt. Among the most important are the rapidly growing population and related food and employment needs, and the foreign exchange requirements for the financing of investments in the non-agricultural sectors. Agriculture is central to the former set of issues because about one-half of the population resides in the rural sector and a substantial proportion of employment is in agriculture. The country's access to foreign exchange has long been tied to Egypt's export of food and fibre commodities, especially cotton.

One can analyse the role of Egyptian agriculture in the national economy in terms of the same forces of change that shape developing economies throughout the world. However, one can best understand the special character of Egyptian agriculture and its relation to the national economy by understanding the role it has played in Egypt's economic history. This chapter begins with a brief review of some major events in Egypt's economic history. The remainder of the chapter relates the recent developments in Egyptian agriculture to those of the national economy.

Key events in Egyptian agricultural history

During the sixteenth, seventeenth, and eighteenth centuries, Egyptian agriculture was an important part of the economic base of the Ottoman Empire.

Egypt's major crops—wheat, rice, lentils, and beans—provided staples for the Empire, while taxes on agriculture were a major source of revenue. The system of tax collection was based on the Mamelukes, or tax collectors, who became corrupt, usurped tax revenue, and raised and maintained armies. The Mamelukes established an essentially feudal system based on the exploitation of the peasant farmers. In the early 1800s Muhammad Ali became the dominant Mameluke and the *de facto* ruler of the country. In order to establish economic independence from the Ottoman Empire, he sought to expand trade with Europe.

The industrial revolution that was taking place in Europe at that time provided lucrative markets for agricultural commodities such as cotton and sugar. Muhammad Ali thus began the introduction of long staple cotton in Egypt, along with a series of changes in the technical and economic organization of Egyptian agriculture. Among these was a strict set of regulations for farmers growing cotton; these regulations would provide a precursor for the cotton cultivation rules enforced in modern times. Ali also sought to concentrate all decision making in his central government. When this failed, he decentralized decision-making into the hands of village sheikhs and *pashas*, a class of wealthy landowners. This system was the precursor of the modern-day government co-operatives found in each village, and to this day they are still dominated by the wealthier land-owning peasant class.

Another major contribution of Muhammad Ali was to begin the transition, eventually completed by the British, from basin irrigation (using the annual flooding of the Nile Delta) to perennial irrigation. This transition was needed in order to increase the cropped area, and to increase the production of cotton for export to the European market.

Ali's defeat by the Turks in 1840 marked the beginning of the forty years of Turkish rule of Egypt. Without protection from foreign competition, Ali's infant textile industries failed, but Ali's plan for irrigation development was continued by the Turks. The period 1860–80 saw further investments in irrigation for increased production of cotton and sugar cane to fill the gap in supply created by the the United States Civil War. From 1830 to 1860, Egypt's trade surplus averaged about £E4million; by 1880 it had risen to over £E20million. Under the period of British rule which began in 1882 and ended in 1914, the trade surplus grew to £E60million. In addition to irrigation investments, railway building began in 1853, with some 1500 km of track by the 1870s and nearly 3000 km by the early 1900s. The opening of the Suez Canal in 1869 and the opening of ports at Suez and Port Said made the Egyptian agricultural economy more accessible to international trade.

The period of British rule saw other major investments in agriculture. They brought to completion the transformation of the irrigation system which was begun under Muhammad Ali. The Aswan Dam, completed in 1902, allowed much larger areas of the Nile Delta and Upper Egypt to be cultivated throughout the summer. The improvement in irrigation allowed a shift from the traditional three-year rotation to a two-year rotation. These changes allowed a rapid expansion in sugar and cotton production, but the gains in production through area expansion and intensification came at the cost of a decline in productivity. Perennial irrigation led to a rise in the water table in the Delta, due to the lack of adequate drainage, and also a deterioration of soil quality due to the intensification of cropping. In response to these problems, farmers attempted to use more fertilizers and pesticides to improve yields. Problems associated with irrigation continued to be a major constraint to the growth in productivity of Egyptian agriculture, and they remain major problems today.

The 1940s saw several important changes in Egyptian agriculture. The shrinking of European markets for cotton and the demand for cereals to feed British soldiers led to a major shift from cotton back to more traditional cereal crops. Grain and fertilizer shortages led to the nationalization of those markets, which provided a precedent for the organization of agriculture in the post-war period.

The Free Officers Coup in 1952 brought Nasser into power and led to major agrarian reforms. Land holdings of over 200 feddans were redistributed to landless peasants and smallholders, and village-level agricultural co-operatives were established for crop marketing and input distribution. The co-operatives were used as instruments for the execution of a wide variety of agricultural policies, including forced sale of crops to the Egyptian Government at below world-market prices (an implicit taxation of agriculture), and the distribution of subsidized inputs (such as fertilizers and seed) to farmers. In addition, the Government began a programme of centralized crop rotations. To ensure that farmers produced crops that would be marketed through the government co-operatives, they were required to grow certain acreages of crops such as cotton and rice.

Recent developments in Egyptian agriculture

The legacy of economic imperialism in Egypt since the sixteenth century had a profound impact on Egyptian agriculture. This legacy left rural Egypt in a semi-feudal state in the early 1950s. The Mameluke tradition had evolved into a system of relatively few large landowners (*pashas*), a relatively small group of land-owning peasants, and a large number of landless or near landless labourers who were economically dependent on the landowners for employment. The 1952 revolution was undertaken by a military clique of officers who had socio-economic roots in the land-owning peasantry and their urban counterparts. In order to build a political base, they set out to discredit the *pashas*, their political rivals. Not surprisingly, the resulting system of agricultural reforms strengthened the economic and political power of the richer land-owning peasant class, and led to some economic improvements for the poorer peasants.

The legacies of Muhammad Ali and his successor regimes also manifested themselves in the role that the Egyptian Government played in the reforms of the 1952 revolution. Following the tradition of the Mameluke system, the Egyptian Government used agriculture as a source of resources to be extracted to finance the military, social works projects, and economic development in other sectors. In particular, the Government sought to acquire foreign exchange through agricultural exports, and to support a cheap food policy to subsidize industrialization and thwart political instability in urban areas. As in all such programmes, the key problem was how to achieve their aims without reducing farmers' production incentives.

The Egyptian solution to the problem, as in many other countries in the post-war period, was to force farmers to sell crops to the Government at prices below those in the international markets. These products were used to subsidize food for urban consumers, and were also sold internationally to acquire foreign exchange. To guarantee adequate production of the taxed crops, the centralized crop rotation was instituted in conjunction with the forced delivery of crops to the village co-operatives. Input and credit subsidies were combined with the centralized rotation and crop marketing system to help ensure adequate production and to encourage adoption of modern production technology.

A key problem restricting the growth of production in Egypt is the limited fixed acreage of available arable land. Long ago, with irrigation development, all the land suitable for the production of most crops was brought into production. Recently, the Egyptian Government has attempted to reclaim desert lands for agriculture, but these efforts have been largely high-cost and low-output. Thus Egypt, like many of the smaller Asian countries, can only hope to achieve significant increases in production through growth in yields.

In pursuit of higher production through higher yields, Egyptian agriculture joined the 'green revolution' of the 1960s and 1970s by adopting improved varieties of wheat and rice and increasing the use of nitrogenous fertilizers. Insecticides were introduced to deal with pest problems, especially in cotton, and small diesel and gasoline powered pumps began to replace the traditional animal-powered water wheels for irrigation. Tractors were introduced but, due to the government monopoly of tractor importation and distribution, only relatively few large tractors were made available until the late 1970s and early 1980s.

As a result of these policies, nitrogen fertilizer consumption increased from 437 000 tonnes (t) in 1970 to over 800 000 t in the early 1980s. Yields of major crops, especially wheat, maize, and rice, increased in the late 1960s and early 1970s, and have continued to increase through the 1980s as indicated in Table 10.1. and 10.2. Cotton yields were erratic in the 1960s and 1970s, increased in the late 1970s, but have shown a steady decline from 1982 to 1989. Sugar cane

TABLE 10.1. Egypt: area, production, and yield of cotton lint and maize, 1950–91

Year	Area ('000 feddans)	Cotton production ('000 t)	Yield (t feddan^{-1})	Area ('000 feddans)	Maize production ('000 t)	Yield (t feddan^{-1})
1950–4	1765	371	0.21	1746	1568	0.90
1955–9	1791	393	0.22	1851	1624	0.88
1960–4	1751	443	0.25	1727	1823	1.06
1965–9	1694	478	0.28	1510	2269	1.50
1970–4	1552	492	0.32	1595	2462	1.54
1975–9	1286	427	0.33	1879	2922	1.56
1980	1245	529	0.42	1910	3231	1.70
1981	1179	499	0.42	1907	3308	1.73
1982	1064	461	0.43	1936	3347	1.73
1983	1012	421	0.42	1800	3509	1.95
1984	988	401	0.41	1974	3698	1.87
1985	1081	435	0.40	1914	3699	1.93
1986	1055	403	0.38	1530	3608	2.36
1987	990	351	0.35	1879	3619	1.93
1988	1014	306	0.30	1964	4088	2.08
1989	1005	288	0.29	2007	4529	2.26
1990	993	306	0.31	2024	4798	2.37
1991	857	294	0.34	1676	4400	2.63

Source: Gardner and Parker (1985); Parker (1992, personal communication, based on unpublished data, Economic Research Service, USDA, collated from sources including MOALR, CAPMAS, and FAS Agricultural Counsellor, Cairo).

TABLE 10.2 Egypt: area, production, and yield of wheat and paddy rice, 1950–91

Year	Wheat Area ('000 feddans)	Wheat production ('000 t)	Wheat Yield (t feddan^{-1})	Paddy rice Area ('000 feddans)	Paddy rice production ('000 t)	Paddy rice Yield (t feddan^{-1})
1950–4	1571	1318	0.84	519	830	1.60
1955–9	1501	1464	0.98	654	1385	2.12
1960–4	1387	1504	1.08	799	1784	2.23
1965–9	1268	1362	1.07	1033	2177	2.11
1970–4	1302	1716	1.32	1095	2432	2.22
1975–9	1354	1894	1.40	1045	2369	2.27
1980	1326	1796	1.35	971	2384	2.46
1981	1400	1938	1.38	957	2236	2.34
1982	1324	2017	1.52	1026	2438	2.38
1983	1357	1996	1.47	1007	2442	2.43
1984	1169	1815	1.55	981	2330	2.38
1985	1186	1874	1.58	926	2312	2.50
1986	1267	1929	1.52	1009	2445	2.42
1987	1379	2722	1.97	983	2279	2.32
1988	1421	2839	2.00	837	2132	2.55
1989	1457	3183	2.18	983	2679	2.73
1990	1955	4268	2.18	1037	3168	3.05
1991	2214	4482	2.02	1100	3447	3.13

Source: Gardner and Parker (1985); Parker (1992, personal communication, based on unpublished data, Economic Research Service, USDA, collated from sources including MOALR, CAPMAS, and FAS Agricultural Counsellor, Cairo).

yields averaged 38.7 t feddan^{-1} in the 1960s, 35.9 t feddan^{-1} in the 1970s, and declined again to 33.1 t feddan^{-1} in the early 1980s (Gardner and Parker 1985). In recent years yields have increased (Chapter 14, Table 14.3). The yields of fruits and vegetables have shown a pattern similar to that of cereals, increasing as nitrogen fertilizer became more available.

Despite technological advances, Egyptian agriculture is not an example of completely successful agricultural development. Agricultural productivity has barely managed to keep pace with population growth, but the increase in productivity has not resulted in significant increases in real income for the rural people. In the last 10 years per caput food production has increased by 23 per cent whereas total food production has increased by almost 60 per cent (Table 10.3).

Thus, the Egyptian system of government intervention and management of agriculture failed to achieve the dramatic growth in productivity seen in some other areas of the world with favourable conditions, for example, the Indian Punjab. Part of this failure can be attributed to the government's pricing and marketing system which has distorted farmers' incentives to produce. In addition, the government's management of the input distribution system failed in certain cases to provide the technology needed by farmers. For example, in the 1970s, the government's fertilizer distribution system was effectively constraining the availability of fertilizers, and the rules for its allocation across crops were restricting its potential contribution to productivity. Furthermore, the government monopoly over tractor distribution resulted in only large tractors being available and these were not suitable for many of the small-scale

TABLE 10.3. Egypt: indices of agriculture and food production, 1979–90

Year	Total food	Total agriculture	Crops	Per caput food	Per caput agriculture
1979	98.5	98.3	98.3	101.1	100.8
1980	98.4	99.3	100.5	98.4	99.3
1981	103.1	102.5	101.2	100.5	99.9
1982	111.1	108.4	105.4	105.6	102.9
1983	118.1	113.0	105.5	109.3	104.6
1984	120.9	115.5	104.9	109.0	104.1
1985	130.6	125.0	112.1	114.8	109.9
1986	138.6	131.3	118.5	118.8	112.6
1987	146.1	136.5	123.2	122.3	114.2
1988	152.1	140.7	126.5	124.2	114.9
1989	153.1	140.9	127.5	122.2	112.4
1990	157.9	145.9	131.1	123.2	113.8

Index: 1979–81 = 100.
Source: FAO (1991).

farms. Consequently, as labour migrated from agriculture to other sectors of the economy in the 1970s, mechanical substitutes for human labour were largely unavailable. It appears, therefore, that the government's policies have not always contributed to higher agricultural productivity.

Agricultural contribution to economic growth

Agriculture plays an important part in the early stages of a country's economic development and this was certainly so in the case of Egypt in the nineteenth and early twentieth centuries. Agricultural land and water supplies constitute the major natural resources, and the majority of physical capital investment is associated with land clearing and levelling, irrigation, and livestock. Since the 1952 revolution, however, Egypt has embarked on an aggressive economic development programme which has necessarily led to the transfer of resources, especially labour, from the agricultural sector to other sectors of the economy.

Agriculture can make a significant contribution to economic growth by providing resources for other sectors to draw upon, but if the agricultural sector is not growing, economic growth will be possible only by sacrificing agricultural output. When the agricultural sector is growing, resources can be transferred to other sectors while per caput food and fibre production grow. In this way, agriculture can make a double contribution to overall economic growth.

In very broad terms, economic growth and development can be defined as the process of capital accumulation, with the term capital encompassing all assets—physical, human, social, and institutional—that create income and contribute to human welfare. Thus in order to evaluate Egyptian agriculture's contribution to the national economy's economic growth, it is necessary to evaluate how it has contributed to the overall process of investment in the various kinds of capital.

Because a large share of the productive resources are employed in agriculture, it is logical that the Egyptian Government should resort to the agricultural sector as a tax base for the financing of public sector investments and other government activities. As noted above, Egyptian agricultural policy has, at least since the Mamelukes, been designed to tax agriculture, and this tradition has continued in modern times. The taxation is accomplished through several avenues: implicitly, through an over-valued official exchange rate; and explicitly, by forced deliveries of crops to marketing agencies at prices below world prices. These taxes are partially offset by government subsidies on inputs, especially fertilizers and pesticides, fuels, and machinery. Estimates of the net tax rate on agriculture due to these various policies differ, but evidence suggests that it may have been 30–60 per cent or higher in certain cases (Ikram

TABLE 10.4. Egypt: gross domestic product (GDP) at factor cost and percentage of GDP resulting from agriculture, 1969–89

Year	Total GDP (£Emillion)[a]	Value of GDP from agriculture (£Emillion)[a]	GDP from agriculture (%)
1969	2446	730	29.8
1972	3002	933	31.1
1975	5056	1468	29.0
1978	9021	2286	25.3
1981	16 552	3326	20.1
1984	27 401	5494	20.1
1987	42 031	8640	20.6
1988	52 054	10 052	19.3
1989[b]	61 158	11 345	18.6

[a] Current £Emillion.
[b] Estimate.
Source: World Bank (1991).

1980). Clearly, agriculture has made a major contribution to the government's attempt to modernize and industrialize the economy by providing a substantial tax base for public sector investment.

There are fundamental economic forces involved in the transition from a largely agricultural economy to a largely non-agricultural economy. As economic growth yields higher incomes, the demand for non-agricultural goods grows at a faster rate than the demand for agricultural goods. Producers of non-agricultural goods respond to the increased demands by bidding away resources (land, labour, and capital) from agricultural production and devoting them to non-agricultural production. Higher wages in non-agricultural employment are a major reason for the migration of labour from agricultural to industry.

Table 10.4, which shows the gross domestic product (GDP) for the nation as a whole and the percentage resulting from agriculture, demonstrates that this pattern has indeed been observed in Egypt. As is typical of growing low-income economies, a relatively large share of GDP (19.3 per cent in 1988) is generated in agriculture. However, this share has declined continuously from about 34 per cent in the mid-1950s (Ikram 1980). Table 10.5 shows similar data for the United States, Japan, and India over the same period. Clearly, the lower the per caput income in a country, the higher the contribution made by agriculture to GDP. It is also worth noting that, in 1950, agriculture produced about 35 per cent of GDP in Egypt and Japan, but by 1980, agriculture produced only 4 per cent of GDP in Japan compared with about 18 per cent in Egypt. This difference reflects the fact that Egypt's economy did not grow

TABLE 10.5. Percentage of gross domestic product (GDP) produced from agriculture in different countries

Year	USA	Japan	Egypt	India
GDP (%)				
1950	7	36	35	51
1960	4	13	28	47
1970	3	7	29	45
1980	3	4	18	38
1988	NA	3	19	33
1989[a]	NA	NA	19	30
GNP per caput ($US)				
1980	12 000	9870	500	240
1989[a]	21 000	23 810	630	340

NA, data not available or non-existent.
[a] Estimate.
Source: World Bank (1980, 1991).

and modernize as rapidly as did Japan's. Hence, resources were transferred less rapidly from the agricultural to non-agricultural sectors.

Table 10.6 shows the total population and rural population for 1965 until 1990. The data show that population grew rapidly during this period, at an average annual rate of about 2.5 per cent. The proportion of agricultural labour in the total labour force has declined from 55 per cent in 1965 to 40.5 per cent in 1990. Compared with other industries, agricultural employment has increased absolutely but declined in relative terms. As Table 10.7 shows, manufacturing employment grew rapidly during the 1960s and 1970s, tripling in absolute terms and doubling as a percentage of total employment.

These data are typical of countries whose per caput incomes are in the same range as those of Egypt, but the share of population in agriculture is much higher than in high-income countries. For example, in 1900, the United States had about 45 per cent of its labour force employed in agriculture compared with about 65 per cent in Japan. In 1960 the corresponding figures were about 10 and 25 per cent, respectively, and in 1970 they had declined to less than 10 per cent for both countries.

Agriculture and trade

Trade in agricultural products has been a major component of the country's economy since Egypt began to supply cotton to Europe during the Industrial Revolution, and cotton exports remain central to Egypt's ability to earn foreign exchange to finance imports. Trade allows for a greater degree of

TABLE 10.6. Egypt: population, total labour force, agricultural labour force, 1965–90

Year	Population (millions)		Economically-active population (millions)		Agricultural employment as a percentage of	
	Total	Agricultural	Total	Agricultural	Total population	Economically-active popuulation
1965	20.39	16.18	8.13	4.47	15.23	55.0
1970	33.05	17.17	9.17	4.76	14.42	52.0
1975	36.29	17.72	10.04	4.90	13.51	48.8
1980	40.88	18.66	11.24	5.13	12.55	45.7
1985	46.51	20.02	12.79	5.50	11.83	43.1
1988	50.05	20.77	13.80	5.73	11.45	41.5
1989	51.23	21.01	14.15	5.80	11.32	41.0
1990	52.43	21.23	14.51	5.88	11.21	40.5

Source: FAO (1991).

TABLE 10.7. Egypt: distribution of total workforce by economic activity, 1961, 1968, and 1974

Activity	1961		1968		1974	
	('000s)	(%)	('000s)	(%)	('000s)	(%)
Agriculture	3747.2	56.53	4202.8	52.90	4198.3	46.25
Mining	5.2	00.07	9.0	00.13	20.2	00.23
Manufacturing	587.0	8.85	1117.8	14.07	1355.7	14.93
Structural and building	132.4	1.99	191.1	2.41	232.9	2.56
Electric and gas	16.4	0.24	49.8	0.64	40.4	0.44
Trade	603.6	9.13	706.2	8.88	1031.4	11.36
Transportation	266.7	4.02	293.8	3.69	396.6	4.36
Services	1177.4	17.76	1195.5	15.04	1461.5	16.14
Other activities	–		100.6	1.26	85.6	00.94
Undetermined	93.1	1.41	77.9	0.98	253.3	2.79
Total	6629.0	100.00	7944.5	100.00	9075.9	100.00

Source: CAPMAS (1961, 1968, 1974).

specialization in production; trade, specialization, and technological progress all lead to increases in productivity—increases in the amount of real products that can be obtained from a given resource base. Increased productivity provides both the resources and the incentives for the capital investment which is the basis for economic development.

Economic theory tells us that each region specializes in the products in which it has a comparative advantage, that is, those products which it can produce relatively more efficiently than its trading partners. The key factor in the theory of comparative advantage is the cost of producing one product relative to another, not the absolute cost of production. Historically, Egypt's favourable conditions in the Nile Delta have given it a comparative advantage in a number of major food and fibre crops, especially wheat, rice, maize, beans, and long staple cotton. In modern times, cotton has been Egypt's primary agricultural export commodity.

As Table 10.8 shows, agriculture contributed 71.2 per cent of total export value in 1966, but this share declined continuously to a low of about 21.6 per cent in 1982, since when it has remained at around that level. On the import side, agricultural products accounted for between 27.2 and 48.3 per cent of the value of all imports during the period 1966–90. In absolute terms, the volume of imports has grown dramatically during the past two decades in both agricultural and non-agricultural products, and much faster than exports. Since 1984 the export of agricultural products has declined continuously.

Table 10.9 shows the production, trade, and supply for domestic use of selected commodities in Egypt between 1970 and 1991. Imports of grain, particularly wheat, have increased dramatically in the 1970s and 1980s, whereas the degree of self-sufficiency in exported commodities such as rice has declined since 1970. Table 10.10 provides a more comprehensive summary of Egypt's major agricultural exports by quantity since 1975. Table 10.11 shows exports by value for 1988–91. Cotton, rice, potatoes, onions, oranges, and sugar have been the major exports, although some products such as refined sugar and groundnuts have fluctuated in quantity in the 1980s. Wheat and maize have been major imports and large amounts of both crops have been imported in connection with the United States Public Law 480 programme in the late 1970s and early 1980s.

The decline in agriculture's share in exports is explained partly by the growth in GDP and the decline in the share of agriculture in GDP. The decline also reflects the growing domestic demand for home-produced agricultural products which has reduced the amount available for export. Another factor has been government policy, which has decreased farmers' incentives for production. The resulting failure to produce major export crops at their potential level has limited Egypt's ability to exploit its comparative advantage in international markets.

Furthermore, Egypt has the potential to export crops other than those traditionally exported, for example, fruits and vegetables to other parts

TABLE 10.8. Egypt: value of agricultural trade, 1966–90

Year	Imports ($US10 000)			Exports ($US10 000)		
	Total merchandise trade	Total agricultural products	Agricultural imports (% total)	Total merchandise trade	Total agricultural products	Agricultural exports (% total)
1966	107 054	32 549	30.4	59 911	42 643	71.2
1968	66 600	24 855	37.3	61 991	43 482	70.1
1970	70 016	20 953	27.2	76 140	51 249	67.3
1973	91 442	31 134	34.0	111 671	71 715	64.2
1975	393 452	141 377	35.9	140 218	78 208	55.8
1978	672 785	199 689	29.7	173 745	66 378	38.2
1980	486 145	235 018	48.3	304 688	67 731	22.2
1982	908 061	321 672	35.4	312 111	67 292	21.6
1984	1 076 904	394 154	36.6	314 085	75 610	24.1
1986	1 150 549	334 759	29.1	293 511	66 938	22.8
1988	865 731	323 240	37.3	212 041	51 376	24.2
1990	920 200	326 382	35.5	223 400	48 066	21.5

Source: FAO (various years).

TABLE 10.9. Egypt: production, trade, and supply for domestic use of selected commodities, 1970–91

Commodity/supply	Units	1970	1975	1980	1985	1988	1990	1991
Wheat								
Production	'000 t	1519	2034	1796	1874	2839	4268	4482
Imports	'000 t	1233	3645	5423	7238	7200	7000	6237
Supply	'000 t	2752	5679	7219	9112	10 039	11 268	10 719
Import share	%	45	64	75	79	72	62	58
Maize								
Production	'000 t	2397	2782	3231	3699	4287	4598	4400
Imports	'000 t	73	511	988	1912	1491	1874	1400
Supply	'000 t	2470	3293	4219	5611	5778	6472	5800
Import share	%	3	16	23	34	26	29	24
Rice, milled								
Production	'000 t	1745	1624	1597	1549	1428	1903	2309
Imports	'000 t	0	4	7	7	26	10	2
Exports	'000 t	654	104	194	16	100	85	159
Supply	'000 t	1091	1524	1410	1540	1354	1828	2152
Self sufficiency	%	160	107	113	101	105	104	107
Soyabeans								
Production	'000 t	1	5	92	133	165	NA	NA
Imports	'000 t	1	35	18	51	120	NA	NA
Supply	'000 t	2	40	110	184	285	NA	NA
Import share	%	50	88	16	28	42	NA	NA

NA, Data not available.
Source: Gardner and Parker (1985); Parker (1992, personal communication, based on unpublished data, Economic Research Service, USDA, collated from sources including MOALR, CAPMAS, and Agricultural Counsellor, Cairo.

TABLE 10.10. Egypt: major agricultural exports by quantity, 1975–91

Commodity (t)	1975	1980	1985	1988	1989	1990	1991
Cotton lint	202 500	179 661	146 686	79 924	58 407	35 000	24 000
Milled paddy rice	104 310	184 000	16 000	71 151	29 921	82 000	158 000
Oranges	210 317	109 513	161 121	97 072	215 000	259 000	175 000
Potatoes	47 565	143 860	127 923	166 206	155 510	180 000	210 000
Onions, dry	70 042	48 340	NA	50 091	50 527	28 000	35 000
Garlic	NA	NA	NA	3487	2104	2000	10 300
Tomatoes	2188	2057	NA	15 147	14 960	15 000	10 000
Watermelons	NA	NA	NA	10 840	7659	18 700	19 800
Refined sugar	40 393	8883	NA	300	12 166	0	184
Molasses	NA	NA	NA	157 465	158 052	158 052	87 000
Groundnuts, in shell	7979	10 976	2484	345	1818	2300	1900
Cottonseed cake	38 237	6371	NA	NA	NA	NA	NA
Cheese (whole cow milk)	NA	NA	NA	1314	5777	6000	6700
Beans, green	NA	NA	NA	10 127	9413	12 000	14 400
Vegetables, dehydrated	NA	NA	NA	5651	5386	6300	6700

NA, Data not available.
Source: Gardner and Parker (1985); Parker (1992, unpublished data, Economic Research Service, USDA, collated from sources including FAO, CAPMAS, and Agricultural Counsellor, Cairo).

TABLE 10.11. Egypt: major agricultural exports by value, 1988–91

Commodity ($US '000s)	1988	1989	1990	1991
Cotton lint	287 039	274 502	170 000	98 000
Milled paddy rice	17 190	7550	19 000	48 700
Oranges	49 283	72 992	132 000	108 000
Potatoes	31 505	26 884	53 000	87 000
Onions, dry	12 442	9692	8800	9900
Garlic	1414	704	2 100	18 700
Tomatoes	2927	4227	5200	6400
Watermelons	3150	2737	6700	7700
Refined sugar	39	3610	0	90
Molasses	11 561	8363	8363	6100
Groundnuts, in shell	288	1404	3100	1870
Cheese (whole cow milk)	2462	3797	5000	7500
Beans, green	2881	3936	5100	6400
Vegetables, dehydrated	8948	7515	9300	9800

Source: Parker (1992, unpublished data, Economic Research Service, USDA, collated from sources including FAO, CAPMAS, and Agricultural Counsellor, Cairo).

of North Africa, the Middle East, and Europe. However, to enter these markets successfully, Egypt must improve significantly both the quality of its produce and its ability to market fresh produce. Substantial investments in marketing infrastructure are needed and these have only recently been part of the government's agricultural strategy.

In the future, Egypt's agriculture could also contribute to the country's trade balance in livestock production. Egypt is among the group of developing countries with a high use of cereals for livestock feed (Sarma 1986). Due to the very high income elasticity for meat and poultry products, the demand for these products grows very rapidly as per caput income grows. Egypt could reduce its food imports by devoting more of its agricultural resources to the production of feed grains and livestock.

Conclusion

Throughout history, agriculture and agricultural trade have played central roles in the Egyptian economy. Since the 1950s, Egyptian agriculture has undergone a significant degree of transition from the traditional agriculture of the nineteenth and early twentieth centuries, towards a more modern agriculture. During this transition, the growth in productivity has kept pace with a rapidly growing population, and the agricultural sector has provided a pool of resources which has, in large part, helped to finance growth in other sectors of the economy. Egypt's arable land is essentially fixed in quantity, so further growth in productivity will require significantly greater investments in modern, yield-increasing technology. Agricultural policies which provide the technology, incentives, institutions, and markets needed to support a modern agriculture will be the key to future agricultural growth. It remains to be seen whether the major agricultural reforms set in motion at the beginning of the 1990s will provide a favourable environment in which that growth can occur.

References

CAPMAS (Central Agency for Public Mobilization and Statistics) (1961, 1968, 1974). *Statistical yearbook*. CAPMAS, Cairo.
FAO (Food and Agriculture Organisation) (1991). *Production yearbook 1990*. FAO, Rome.
FAO (various years). *Trade yearbook*. FAO, Rome.
Gardner, G. P. and Parker, J. B. (1985). *Agricultural statistics of Egypt, 1970–84*. Economic Research Service, Statistical Bulletin No. 732. USDA, Washington, DC.
Ikram, K. (1980). *Egypt: economic management in a period of transition*. Johns Hopkins University Press, Baltimore.
Sarma, J.S. (1986). *Cereal feed use in the Third World: past trends and projections to 2000*. International Food Policy Research Institute, Research Report No. 57. International Food Policy Research Institute, Washington, DC.
World Bank (1980). *World tables*, (2nd edn). Johns Hopkins University Press, Baltimore.
World Bank (1991). *World tables 1991*. World Bank, Washington, DC.

11

The history of agricultural development [a]

T. RUF

Introduction

Anyone travelling today in the Nile Valley and meeting a farmer coming back from his fields with his hoe on his shoulder might think that he is seeing a farmer straight out of Egyptian antiquity; he might well imagine that Egyptian agriculture has not changed since time immemorial. In his defence it must be said that none of the many tourist guides he will consult during his journey will help him to form an opinion. They nearly all present the Egyptian fellah as someone living by traditions and techniques that are a thousand years old. If they are to be believed, nothing has changed in Egypt since the days of Rameses.

This myth of the fellah as a direct descendant of the peasant from the land of Amon is still so rooted in people's minds that it must be denounced. There is now practically nothing in common between today's agriculture and that of antiquity or the Middle Ages. Techniques, crops, agricultural productivity, and even the very landscape have been transformed. These deep-rooted developments in agriculture stem, without any possible doubt, from the ways in which the Nile waters have been used. Whereas for thousands of years farming communities carried out extensive agriculture when the waters were low, today's farmers irrigate their fields through a very dense system of irrigation canals. Theirs is a very intensive and productive type of agriculture which was still quite unheard of at the start of the nineteenth century.

This chapter provides an overview of the recent history of Egyptian agriculture. For a fuller treatment, beyond the scope of this chapter, the points made should be developed further and considerable nuance introduced. There is, however, a basic pattern of Egyptian agriculture which gives a certain unity to the country.

---

[a] Translated from French by D. and M. Hanley, Department of French Studies, University of Reading.

The significance of the changing use of the river Nile

From the time when the banks of the Nile were first built up downstream from Aswan by the first Pharaohs who unified Egypt until the fall of the Mameluke dynasty after French intervention in 1798, there had been some five thousand floods of varying strength, frequency, and effect—favourable or unfavourable.

For 5000 years the survival of inhabitants of the Nile Valley and Delta depended on how well they coped with the rapid rise of the waters between July and September. Not only was a certain amount of genius required to invent and deploy a system for spreading out the flood waters, moderating their force, and distributing their benefits over the land, but there also had to be some technical and political organization so as to keep the *nili* system of dams and canals in good repair. The use of the river has always been central to the objectives of central political and military authorities, and they have been able to organize the State around the Nile's vital function. This is an indispensable key to understanding the turbulent history of the country which was sometimes ruled by absolute aristocratic power and sometimes broken up into small independent fiefdoms in permanent conflict with each other. The latter description best fits the country as it was by the end of the eighteenth century.

After the period of French intervention and the failed restoration of the old regime, a new political-cum-military organization was set up in 1805 under the leadership of an Albanian officer, Muhammad Ali. Although nominally under the suzerainty of the Ottoman Empire, he re-established central authority and cleared the way for the restoration of the dams and canals which controlled the floods. The new regime intended to carry out its policies independently of Constantinople, while at the same time getting rid of the old feudal lords, who were finally murdered in 1811.

Muhammad Ali needed income from trade in order to begin his building works, guarantee his independence, and extend his authority to nearby regions. To this end he gave a considerable boost to the new crop, long staple cotton, which was discovered by Jumel (Gali 1889; Gregoire 1862; Rivlin 1961). Long staple cotton seemed an ideal product, being a non-perishable product much sought-after by the European textile industry, easy to export, and not on any list of products barred from trade outside the Ottoman Empire.

However, long staple cotton was not just a winter crop that could be grown when the waters were low; it demanded a lot of heat and water, and had a very long growing cycle centred on spring and summer. As a result the water reserves of the soil and the occasional shower (too infrequent to be of much use) were no longer adequate. This out-of-season crop needed to be irrigated at the start of its cycle when the Nile was at its lowest, and protected at the end of its cycle when the floods were likely to ruin the harvest.

Mindful of this twin problem, Muhammad Ali improved the central area of the Delta where he had at his disposal considerable landed estates, taken over after the murder of the Mamelukes (Gregoire 1862). In order to irrigate the cotton, he deepened the canals which distributed the flood waters so that they could take water from the river when it was low (Barois 1887). Then on the edge of the fields he installed vast numbers of machines to lift the water from the bottom of the canals into the furrows. These *saqqiyas* were wooden wheels with pots attached to their rims; the pots went down into the canal for the water which was then emptied out sideways as the pots reached the highest point of the wheel's revolution. *Saqqiyas* were to play an essential role in the development of twentieth century agriculture. The metal type of pot developed in the 1920s and 1930s proved to be extremely efficient and these are still widely used today (Ruf 1986*a*).

Around 1830 the river's history underwent a sudden change. At this time the main idea was to optimize the distribution of the low waters of the Nile. Therefore, in the mid-nineteenth century the building of the Muhammad Ali Dam began at the narrow point of the Nile Delta. The intention was to raise the water level so as to feed big irrigation canals which were to be built. Unfortunately the dam was poorly built and it cracked when put into use; only in 1939, with the building of a second more solid dam was it possible to raise the low-water level by four metres (Besançon 1957).

The Nile Valley was developed progressively from the end of the nineteenth to the middle of the twentieth century. A series of dams were built to raise the water level and distribute the water needed to irrigate summer crops or year-round crops like sugar cane.

Having concentrated on raising different stretches of the Nile and reinforcing the dams, the irrigation service began, from 1885, to set up a system of irrigation water distribution using water towers to feed tertiary canals: these were fed with water for six days then left dry for twelve. It was in fact impossible to supply all the canals at once because of the very small volume available at low water (Barois 1887). This system is still used today in the cotton areas of the Nile Delta and Valley. In the north of the Delta where rice is grown the system is four days water, four days dry.

During the twentieth century the main concern of development engineers was how to increase the volume of water available. As early as 1902, the first Aswan Dam made it possible to store 1 milliard (1 milliard = 1000 million m^3) of water amounting to about one per cent of the river's annual flow through Aswan. The dam was raised twice, and a dam on the White Nile in the Sudan was added. By 1940, storage capacity was around 9 per cent of the annual flow. Although the volume at low water had been boosted, as evidenced by the increase in *sefi* crops (irrigated from February to August), water shortage remained one of the main factors limiting agricultural productivity. The building of the Aswan High Dam solved this basic problem, but led indirectly to others; these included a rise in the water table and an increase in salinity stemming from the delay in improving drainage.

The last Nile floods on Egyptian territory were in 1964. Today the river is kept at a regular volume all year. There are no more fears of the waters rising devastatingly, and every farmer in the land is certain of having water at all times of the year, provided that he has equipment for pumping water out of the supply canals and that he is not unlucky enough to be the last man in a chain of canal users where some users ahead of him take more than their share of water.

Population growth and the revolution in land ownership

When the first attempts at perennial irrigation were made around 1830, Egypt had barely four million inhabitants. With an arable surface of some 20 000 km^2 and a population density of 200 inhabitants per 'useful' km^2, Egypt was already one of the most densely populated areas in the world. Most of the people lived in agricultural communities, and were obliged to pay taxes to the State and to perform compulsory labour to maintain the earthworks. Today there are some 50 million Egyptians living on an arable surface of nearly 30 000 km^2. The rural population density can be calculated at 1000 persons per agricultural km^2, because half of the population lives in towns (Table 11.1).

The rural population is composed mainly of small peasant families, some working farms of one or two feddans, others having only one *kirat* of land (one twenty-fourth of a feddan). In 150 years, population and ownership structures have changed completely. In the nineteenth century, as a result of Muhammad Ali's tax and land reforms, with the spread of cotton production and rural trading, with the effects of conscription, and the opening up of new horizons for a great many peasants, rural society gradually changed its attitudes and its mode of organization. The rural community, originally formed of large extended families, broke up into smaller groups and collective identities weakened. Tax was levied only on individuals and finally, at the end of the process of change, the reform of land law gave peasants legal status as small property owners; for the first time their land rights were subject to the same law as those of the large landowners (Gali 1889; Rivlin 1961).

Under Muhammad Ali, the large landowners had been able to amass their hugh estates. The Viceroy of Egypt hit on the idea of rewarding his army officers and civil servants for services rendered, by granting them properties confiscated from the vanished Mamelukes, or land from villages which had fallen behind with their tax payments. The twentieth century saw a clash between two sectors of landowners—the big estates with peasants and agricultural labourers living in precarious tenure and status, and the small landowners whose insecurity derived from being indebted to various moneylenders.

The twentieth century began badly for the small peasant landowners, for they encountered two phenomena which questioned both their system of agricultural production and their system of social reproduction. The first of

TABLE 11.1. Egypt: population, area under cultivation, and area harvested, 1800–1976

Year	Total population (millions)	Rural population (millions)	Area under cultivation (million feddans)	Area harvested (million feddans)	Population density (persons per km^2)
1800	2.3	NA	NA	NA	NA
1820	2.5	NA	NA	NA	NA
1844	4.4	NA	NA	NA	NA
1882	6.8	NA	NA	NA	NA
1897	9.7	7.8	5.0	6.8	462
1907	11.2	9.1	5.4	7.7	494
1917	12.7	10.0	5.3	7.7	570
1927	14.2	10.9	5.5	8.7	614
1937	15.9	11.9	5.3	8.4	714
1947	19.0	13.2	5.8	9.2	780
1960	26.0	16.3	6.1	9.9	1014
1966	30.1	17.9	6.4	10.5	1119
1976	36.6	21.4	6.8	11.2	1281

NA, data not available or non-existent
Source: Data for 1897, 1907, 1917, 1927, 1937, 1947–8, and 1960, Al Sarki (1964); 1844, Rivlin (1961); 1937, 1947, 1960, 1966–76, CAPMAS (1982); 1882, 1897, 1917, 1927, 1937, and 1947, Encyclopaedia Britannica (1964).

these, the salting up of land due to inadequate drainage exacerbated the second, which was debt. The social and economic crisis which was taking a grip on the country seemed too much for the Anglo-Egyptian Government. The Kitchener Act of 1913, to protect small property, forbade the expropriation of peasants' land for non-payment of debts, kept the peace within society, and kept this sector of agriculture alive. However the Act prevented peasants from obtaining agricultural credit for want of collateral, until the Nasser government brought in a system of supervision for agriculture.

Population growth meant the breaking-up of small property and it also resulted in a more difficult relationship between the big landowners and the agricultural labourers or sharecroppers. The 1952 Revolution was aimed particularly at the landowning oligarchy, and one of the first measures of the new authorities was a land reform bill.

In reality, the bill did not dismantle the big owners' sector which still survives today. Only big owners with ties to the previous regime were expropriated and these estates, some of which were very large, amounted to only 13 per cent of the arable surface of the country. This land was given to landless labourers and sharecroppers (9 per cent of farming families) in a system of strict state supervision; state co-operatives brought together the new but indebted landowners who were to pay off the cost of the land in 30 annual payments (Radwan 1977). The Egyptian Government arranged the rotation of crops, looked after marketing, gave out seed and fertilizer in advance of the harvest, and took the lead in the struggle against cotton pests; this crop had been the mainstay of agricultural policy for over a quarter of a century.

Developing systems of production

When crops were grown after the floods had receded, their rotation depended almost solely on how far the floods had spread. Wheat was grown in most flooded areas but, where there had been insufficient water, barley was grown because it was more tolerant of dryness. Cotton became associated with a particular type of site, but the practice of growing it in the same place continually was soon given up and the site was then changed every year. Gradually the idea of planning a succession of crops was established. Cotton was grown every three, four, or five years, and in the meantime winter crops such as wheat, barley, and *bersim (Trifolium alexandrinum)* were grown. Sites were left fallow in summer so as to keep all available water for the cotton fields (Gregoire 1862; Gali 1889).

With more effective control of the flood waters and crop protection, peasants grew more and more crops, particularly maize, in the *nili* season. At this time it was easier to irrigate because the canals were at their highest levels. It is symbolic of the changes in society and in landownership that, at the

TABLE 11.2a. Egypt: area of winter crops (*chetwi*) in the nineteenth and twentieth centuries (thousands of feddans)

Year	Total area	*Bersim*	Wheat	Barley	Beans	Flax	Lentils	Others
1844	3582	286	914	872	839	306	168	197
1880	3972	941	1241	520	776	10	150	334
1950–4	4480	2160	1580	120	330	5	85	200
1955–9	4700	2360	1500	135	350	NA	80	275
1960–4	4760	2450	1380	130	365	NA	75	360
1965–9	4780	2630	1270	110	350	NA	65	355
1970–4	4900	2800	1300	80	280	NA	80	360
1975–9	NA	2804	NA	NA	290	NA	NA	NA

NA, data not available
Source: Data for 1844, Rivlin (1961); 1880, Gali (1889); 1950–74, El Tobgy (1976); 1974–9, Ramah (1982).

TABLE 11.2b. Egypt: area of summer crops (*sefi*) and perennial crops in the nineteenth and twentieth centuries (thousands of feddans)

Year	Total area	Cotton	Rice	Maize	Sorghum	Others	Orchards	Sugar cane
1844	507	224	98	0	0	185	NA	12
1880	1207	866	78	69	155	45	NA	70
1950–4	2280	1760	500	30	385	200	95	95
1955–9	3180	1790	640	55	390	295	110	110
1960–4	3600	1750	790	270	415	395	150	120
1965–9	4720	1680	1020	1070	460	460	205	145
1970–4	4870	1550	1090	1240	465	515	250	200

NA, data not available
Source: Data for 1844, Rivlin (1961); 1880, Gali (1889); 1950–74, El Tobgy (1976); 1974, Ramah (1982).

TABLE 11.2c. Egypt: area of autumn (*nili*) crops in the nineteenth and twentieth centuries (thousands of feddans)

Year	Total area	Maize	Sorghum	Others
1844	899	NA	NA	NA
1880	785	596	139	50
1950–54	1860	1720	50	80
1955–59	1970	1800	60	115
1960–64	1670	1460	55	155
1965–69	680	430	45	170
1970–74	620	350	35	215

NA, data not available
Source: Data for 1844, Rivlin (1961); 1880, Gali (1889); 1950–74, El Tobgy (1976); 1974, Ramah (1982).

end of the nineteenth century, maize became the staple food of the peasantry and barley declined (Tables 11.2a–c).

Cotton, which had been introduced originally by Muhammad Ali, was now grown in all sites in the Delta which had year-round irrigation, and it was now grown every two years. The Civil War in the United States, which deprived the Western textile industry of its main supplier, undoubtedly boosted the growth of cotton. Buyers turned to Egypt and prices went up fivefold in the 1860s. This was when the country and its farmers became integrated into the world market and, as a consequence, became indebted (Lorca 1979).

The State was caught in a spiral of investment, some of it productive as in the case of hydraulic infrastructure, some of it prestige; when it became bankrupt the country came under the economic control of the West and, in 1882, under British military and political tutelage. The hydraulic and agricultural policies which were pursued in order to get rid of the huge public debt were focused entirely on cotton. Despite fiscal pressure which kept down the prices paid to producers, most of the country's farmers still grew cotton, sometimes forced to do so by landowners, but not always. In a money economy, the small grower had to face costs: water did not flow onto his fields by gravity alone, but had to be pumped by *saqqiyas*. Very often several small farmers shared this cost. They also had to pay land taxes which at the end of the nineteenth century still accounted for 40 per cent of government income (De Chamberet 1909)[1].

Thus cotton became the country's main crop in the areas which had been converted to year-round irrigation. For over 50 years the main cropping system was based on a biennial rotation: cotton in the first year followed by subsistence crops, such as winter wheat or *bersim* followed by *nili* maize.

One can understand the development of the crop rotation in terms of creating a cropping system that eliminates idle periods when the land is unused due to lack of irrigation. At the start of the nineteenth century one annual crop was grown after the floods; by the 1970s and 1980s the rotation had developed to include six different crops grown in a three-year span. This meant a remarkable intensification of agriculture which has caused new problems for farmers. Tables 11.3 and 11.4 illustrate the development of crop rotation in the Nile Delta (Daqahliya governorate) and the Nile Valley (El Minya governorate)

During this period of development, at the start of the nineteenth century, the main agricultural problems still besetting Egypt today became evident—namely, land exhaustion, salinity, and pest attack.

The more intensive agriculture resulting from improvements in water distribution was not immediately accompanied by changes in fertilization. In fact no one had worried about fertilization for thousands of years, because the Nile mud contained basic fertilizer. There are two vital elements to be taken into account in order to understand the crisis in Egyptian agriculture during the first half of the twentieth century. Firstly, the value of Nile mud as a fertilizer was something of a myth, for it could only compensate for the loss of nitrate when a single crop yielding only about 400 kg feddan^{-1} was grown. Secondly, from the end of the nineteenth century onwards, the new hydraulic infrastructure prevented the floods from remaining for long in the Nile Delta, thus considerably limiting any fertilizing input from the mud. At this time the main fertilizer used, *sebakh koufri*, came from old inhabited sites. Soon, however, the best sites became exhausted, and the effect of the residues on the soil was not only of negligible value but was also destructive, because of the high content of harmful salts. Chilean nitrate was imported from 1903 onwards but was used only marginally for a long time (Société Sultanienne d'Agriculture 1920; Mosseri 1928). It took more than 50 years for mineral fertilizer to become widely used.

TABLE 11.3. Crop rotation in Daqahliya governorate in 1844, 1885, and 1980

Year	Area ('000 feddans)			Percentage of agricultural area		
	1844	1885	1980	1844	1885	1980
Chetwi crops						
Bersim	20	100	390	5.7	26.3	63.5
Wheat	72	98	156	20.6	25.8	25.4
Barley	77	69	0	22.0	18.2	0
Beans	77	29	12	22.0	7.6	1.9
Flax	30	0	11	8.6	0	1.8
Others	10	24	29	2.8	6.3	4.7
Total *chetwi*[1]	286	320	598	81.7	84.2	97.2
Fallow	64	60	0	18.3	15.8	0
Sugar cane	0	0	2	0	0	0.3
Orchards	0	0	15	0	0	2.4
Total[2]	64	60	17	18.3	15.8	2.7
Agricultural area[3]	350	380	615	100	100	100
Nili crops						
Maize	100	65	41	28.6	17.1	6.7
Sorghum	0	4	0	0.0	1.1	0.0
Others	0	0	30	0.0	0.0	4.9
Total *nili*[4]	100	69	71	28.6	18.2	11.5
Sefi crops						
Cotton	30	154	200	8.6	40.5	32.5
Rice	38	54	274	10.9	14.2	44.6
Maize	0	12	58	0.0	3.2	9.4
Sorghum	0	11	0	0.0	2.9	0.0
Others	30	0	0	8.6	0.0	0.0
Total *sefi*[5]	98	231	532	28.1	60.8	86.5
Total area of cropping[6]	484	620	1218			
cropping rate[7]	1.4	1.6	2.0			

Agricultural area[3] = Total[1] + Total[2].
Total area of cropping[6] = Agricultural area[3] + Total[4] + Total[5] − fallow area.
Cropping rate[7] = Total area of cropping[6]/Agricultural area[3].
Source: Data for 1844, Rivlin (1961); 1885, Gali (1889); 1980, Directorate of Agriculture, Daqahliya Province.

TABLE 11.4. Crop rotation in El Minya governorate in 1844, 1885, and 1980

Year	Area ('000 feddans)			Percentage of agricultural area		
	1844	1885	1980	1844	1885	1980
Chetwi crops						
Bersim	11	47	120	3.8	13.3	27.8
Wheat	90	100	85	30.9	28.2	19.7
Barley	60	36	0	20.6	10.2	0
Beans	80	104	75	27.5	29.4	17.3
Flax	30	1	0	10.3	0.3	0
Others	18	38	45	6.2	10.7	10.5
Total *chetwi*[1]	289	326	325	99.3	92.1	75.3
Fallow	0	0	45	0	0	10.4
Sugar cane	2	28	42	0.7	7.9	9.7
Orchards	0	0	20	0	0	4.6
Total[2]	2	28	107	0.7	7.9	24.7
Agricultural area[3]	291	354	432	100	100	100
Nili crops						
Maize	30	9	35	10.3	2.5	8.1
Sorghum	30	20	0	10.3	5.6	0
Others	0	0	25	0	0	5.8
Total *nili*[4]	60	29	60	20.6	8.1	13.9
Sefi crops						
Cotton	0	0	120	0	0	27.8
Rice	0	0	0	0	0	0
Maize	0	5	190	0	1.4	44.0
Sorghum	0	24	0	0	6.8	0
Others	0	7	50	0	2.0	11.6
Total *sefi*[5]	0	36	360	0	10.2	83.4
Total area of cropping[6]	351	419	807			
Cropping rate[7]	1.2	1.2	1.9			

Agricultural area[3] = Total[1] + Total[2].
Total area of cropping[6] = Agricultural area[3] + Total[4] + Total[5] − fallow area.
Cropping rate[7] = Total area of cropping[6]/Agricultural area[3].
Source: Data for 1844, Rivlin (1962); 1885, Gali (1889); 1980, Directorate of Agriculture, El Minya Province.

Declining fertility does not alone explain the general decline in agricultural yield between 1905 and 1920. Rising ground water was also a factor. Until the crisis, the only real concern had been to increase irrigation opportunities. There was no drainage system. Gradually agricultural engineers and land managers became aware of the harmful effects of shallow stagnant water and the accompanying process of salting-over (Audebeau 1913). In succeeding years the irrigation service installed a huge system of open drains.

Finally in these conditions which were so unfavourable to plant growth, parasites appeared and multiplied in a very short time, destroying harvests. The leafworm (*Spodoptera literalis*) and the pink bollworm (*Pectinophora gossypiella*) did untold damage to cotton. In order to protect crops, the Egyptian Government was obliged to take legislative action, calling up peasant children to fight the worms by hand (Dudgeon 1918; Société Sultanienne d'Agriculture 1920). This method is still in use today; chemical methods are only used when other methods have failed to stem the tide of pests.

In this way the bases of a peasant system of agriculture for the twentieth century were laid down. This system bore the marks of a century of political history and international economic history, and of an ecosystem that was becoming more and more artificial. The system was also characterized by the ups and downs of Egyptian agronomy which was still at an early stage of development, and by rather rough-and-ready ways of adapting to the difficult conditions of production. Finally, the system was affected by the increase in social tension and especially by pressure on land (Ruf 1984).

The link between agriculture and livestock production: the key to understanding the development of peasant agriculture

Today, Egypt's main crop is neither cotton nor wheat nor even maize; it is the Alexandrian clover (*Trifolium alexandrinum*) known as *bersim* (Tables 11.2a–c). From November to April or May, farmers grow *bersim* in about two-thirds of their fields. This explains the importance of livestock, and why farmers are so keen to keep one or two head of cattle (or female buffaloes or ewes or goats) on their plots, even though they are sometimes tiny. The 1913 Act for the protection of the cotton fields prevents farmers growing fodder crops between May and October. In spite of this ban, cotton pests develop on other plants, and especially *bersim*. In summer, farmers feed their animals with the remains of other crops, such as stubble from cereals which they sometimes eke out with green leaves from the maize. The problem of animal feed has become so acute that cereal straw can now fetch a higher price than the grain.

Why then are farmers so stubborn? Why do they persist in keeping a cow or a buffalo at all costs, despite the technical difficulties and the fact that it is taking up space which could be used for crops to feed humans? Once more history sheds light on the behaviour of Egyptian peasants.

Firstly, we must note that the widespread links between agriculture and livestock production seem to be quite recent; they derive from the mid-nineteenth century period of State interventionism when it was necessary to have animals near irrigated fields in order to work the *saqqiyas*. The links grew stronger at the end of the nineteenth century as peasant families became more individual entities; each one recognized by its land, house, and means for undertaking irrigated agricultural production, i.e. a *saqqiya*, a swing-plough and enough animal power to drive them. The number of cows and buffaloes registered was roughly equivalent to the number of peasant families: about a million (Table 11.5). After the crisis of 1900–20, the rise in animal production kept up with the rise in the number of small family farms. Today there are some 2.7 million cows, 2.5 million buffalo, and over 5.5 million sheep and goats (CAPMAS 1990). The density of animal units[2] per feddan is very high, working out at one animal unit per feddan over the country as a whole and 1.4 animal units per feddan in the centre of the Delta, where integrated farming is most developed.

The importance in social terms of owning animals, especially cows or buffaloes, does not wholly explain why small farms persist in integrating agriculture and livestock. Livestock perform vital functions in production units of this type, but animal traction, which goes back to the beginnings of integrated farming, is no longer the main one. Nowadays there is a tendency to mechanize pumping and any kind of soil cultivation because of the time gained in a system where crops follow on without rest periods[3].

Another very old function of the link between agriculture and livestock is fertilization. Surprisingly, its importance has not been recognized in Egypt, probably because the myths about fertile Nile mud have distracted attention from the much less spectacular task of dung-spreading. The tripling of the animal population between 1920 and 1980 has led to a similar increase in the supply of dung, which is the main source of fertilizer today. According to Ministry of Agriculture data, each feddan can be given some 12 t of manure (a mix of earth and excreta) which would provide about 30–40 kg N, 20–40 kg P_2O_5, and about 100 kg K_2O. Growing *bersim* provides an additional 100 kg N fixed by symbiosis. The system of state co-operatives set up by Nasser enables farmers to supplement this natural fertilizer with chemicals, mainly urea and composite fertilizers. These supplements provide an average of 60 kg N, 12 kg P_2O_5 and 0.3 kg K_2O feddan^{-1} (El Tobgy 1976, supplemented by the author's fieldwork, 1981–2). These data cover some very different situations, but they do show the importance of livestock in what has been described as the 'reproduction of the cultivated ecosystem'.

The food value of livestock is also important in helping to provide a protein balance in the diet of peasant families. The economic value is also essential;

TABLE 11.5. Livestock production in Egypt, 1885–1980

Year	Cattle	Buffalo	Sheep	Goats	Donkeys	Camels	Arable area (million feddans)	Animal units[a] per feddan
			(million head)					
1885	0.25	0.25	NA	NA	0.2	0.1	5.0	0.15
1905	0.75	0.7	NA	NA	0.5	0.1	5.4	0.35
1915	0.5	0.5	0.8	0.4	0.6	0.1	5.3	0.32
1935	0.95	0.9	1.7	0.7	0.8	0.15	5.3	0.55
1945	1.2	1.0	1.5	0.7	0.8	0.15	5.5	0.59
1955	1.35	1.3	1.3	0.7	1.0	0.15	6.1	0.62
1970	2.0	2.0	2.0	1.1	1.3	0.12	6.5	0.85
1975	2.1	2.2	2.0	1.4	1.4	0.1	6.8	0.87
1980	2.4	2.4	2.8	1.1	1.4	0.1	6.5	0.99

NA, data not available

[a] Measure equivalent to the energy (or fodder) needs of an adult cow with an annual milk production of 800 kg.

Source: Data for 1885, Gali (1889); 1903–17, Société Sultanienne d'Agriculture (1920); 1937–55, Enclyclopaedia Britannica (1964); 1952, 1960, 1970, and 1975, Ramah (1982); 1960, 1970, 1971, 1978, 1980, Fitch and Soliman (1981); 1970–4, CAPMAS (1982).

livestock provide a living form of savings together with a significant income, which is often managed by the women.

Complaints are often heard about the large share of the crop rotation taken up by *bersim* or the low level of productivity (in commercial terms) of traditional livestock breeding. The suggestion that 'unproductive' sites used for fodder crops should be reduced would jeopardize the keystone of the peasant production system. On the contrary, it is highly likely that the improvement of living conditions in the countryside will involve the solving of problems posed by integrated farming: the fodder system and animal feeding will have to be improved, as will zootechnical performance. If such steps were taken the result might be a huge leap forward in productivity right across the whole system of agricultural production (Ruf 1986*b*).

Recent history: changes and continuity under Nasser, Sadat, and Mubarak

Seen in historical perspective, the Nasser period must be regarded more as continuing the tendencies of the early twentieth century than as breaking sharply with the agriculture of the 1940s. Land reform brought changes in living conditions for large groups of the peasantry but did not change land distribution fundamentally. This underlying continuity is seen in hydraulic policy, where the building of the Aswan High Dam marked the end of the engineering programmes begun in the nineteenth century and designed to gain total control over the Nile. The setting-up in 1962 of compulsory state co-operatives for all farmers under the guise of 'Arab socialism' might suggest a change in agricultural and economic policy. From 1964 the Egyptian Government made it compulsory for farmers to rotate crops over about 50 feddans, justifying this measure by the need for better crop planning and a more effective struggle against cotton pests[4]. In return the co-operatives provided inputs such as seed and fertilizer on credit, to be paid for by law at harvesting. This policy continued the authoritarian tradition of twentieth century agricultural policy, which has been centred on cotton growing. It has much in common with similar measures taken in the nineteenth century by Muhammad Ali to get cotton growing started on the central delta estates and it fills out the legal framework set up in the crisis years 1900–1920.

In terms of progress, the results of this policy were disappointing. Not only did the dream of a 'scientifically planned' agriculture fail to lead to an industrial type of agriculture, but the growth in agricultural production still lagged behind population growth. Twenty years on from the beginnings of widespread state supervision of agriculture the country's dependence on food imports has become one of the biggest in the world; some 70 per cent of the wheat consumed is imported (Table 11.6 and Chapter 10, Table 10.9). Rice production which has developed with the available water supply is just able

TABLE 11.6. Wheat in Egypt, 1930–84

Year	Area (million feddans)	Production (million ardebs[a])	Yield (ardebs per feddan)	Consumption (million ardebs)	Imports (million ardebs)	Production: consumption ratio
1930–4	1.6	NA	NA	NA	NA	NA
1935–9	1.4	8.3	5.9	8.2	0.1	1.0
1940–4	1.6	7.9	4.9	8.0	NA	1.0
1945–9	1.6	7.5	4.7	9.2	NA	0.8
1950–4	1.5	8.7	5.8	12.7	3.5	0.7
1955–9	1.7	9.7	5.7	16.2	5.4	0.6
1960–4	1.5	9.8	6.5	17.5	8.0	0.6
1965–9	1.4	9.0	6.4	19.5	14.4	0.5
1970–4	1.3	11.0	8.5	26.0	15.5	0.4
1975–9	1.25	12.5	10.0	35.0[b]	22.5[b]	0.35[b]
1980–4	1.2[b]	13.0[b]	11.0[b]	45.0[b]	32.0[b]	0.30[b]

NA, data not available
[a] 1 ardeb of wheat = 150 kg.
[b] Extrapolation.
Source: Data for 1935–59, Al Sarki (1964); 1950–74, El Tobgy (1976); 1961–78, Lebas and Levy (1979); 1950–78 EMCIP (1978) and USAID (1976).

TABLE 11.7. Rice in Egypt, 1930–84

Year	Area (million feddans)	Production (million t)	Yield (t feddan^{-1})	Consumption (million t)	Exports (million t)	Production: consumption ratio
1930–4	0.5	NA	NA	NA	NA	NA
1935–9	NA	0.48	NA	NA	NA	NA
1940–4	NA	0.50	NA	NA	NA	NA
1945–9	NA	0.75	NA	NA	NA	NA
1950–4	0.5	0.80	1.6	NA	NA	NA
1955–9	0.64	1.05	1.6	NA	NA	NA
1960–4	0.79	1.80	2.3	1.5	0.4	1.2
1965–9	1.02	2.20	2.2	1.7	0.6	1.3
1970–4	1.09	2.50	2.3	2.0	0.3	1.3
1975–9	1.00	2.40	2.4	NA	NA	NA
1980–4[a]	1.00	2.40	2.4	NA	NA	1.0

NA, data not available
[a] Estimate.
Source: Data for 1935–59, Al-Sarki (1964); 1950–74, El Tobgy (1976); 1961–78, Lebas and Levy (1979); 1950–78, EMCIP (1978) and USAID (1976).

TABLE 11.8. Cotton in Egypt, 1900–85

Year	Area (million feddans)	Production (million kantars)	Yield (kantars feddan^{-1})
1900	1.2	6.0	5.0
1905	1.5	5.0	3.3
1910	1.5	5.0	3.3
1915	1.4	5.0	3.6
1920	1.5	5.0	3.3
1925	1.6	7.0	4.4
1930	1.5	7.0	4.7
1935	1.5	8.0	5.3
1940	1.0	5.0	5.0
1945	1.0	5.0	5.0
1950	1.8	7.5	4.2
1955	1.7	7.0	4.1
1960	1.8	9.0	5.0
1965	1.7	9.5	5.6
1970	1.5	9.5	6.3
1975	1.3	8.5	6.8
1980	1.1	10.0	9.1
1985	1.0	NA	NA

NA, data not available
1 kantar = 45 kg fibre.
Source: Data for 1900–18, Société Sultanienne d'Agriculture (1920); 1900–39, Besancon (1957); 1900–61, Al Sarki (1964); 1910–76, Ikram (1980); 1960–76, El Tobgy (1976); 1973–80, CAPMAS (1982).

to meet domestic demand (Table 11.7 and Table 10.9), and even cotton is in decline because of the difficulty in disposing of production on world markets (Table 11.8 and Tables 10.1, 10.9). Generally, peasants have tried hardest to increase production of the two crops which are not supervized by the state, namely *bersim* and maize (Table 11.9 and Table 10.1), in other words the subsistence crops of which they can be sure.

Moreover, the policy of developing desert land, which Nasser began, has run into problems. The model state farms occupying tens of thousands of feddans in Liberation Province, to the south-west of Alexandria, have failed; farmers were reluctant to give up rich land in order to settle in high-risk areas where farming is believed to be difficult.

Political change comes slowly in agriculture. Thus ten years after Nasser's death little had happened, with the exception of many exemptions from strict State supervision, granted to big landowners who wanted some freedom in their choice of crops. As the co-operative system was having trouble recovering its debts, the neo-liberal Government of President Sadat undertook a

TABLE 11.9. Maize in Egypt, 1935–74

Year	Area (million feddans)	Production (million ardebs)	Yield (ardebs feddan^{-1})	Human consumption (million ardebs)	Animal consumption (million ardebs)	Production: consumption ratio	Imports (million ardebs)
1035–39	1.5	11.7	7.8	NA	NA	NA	NA
1940–44	1.8	10.3	5.7	NA	NA	NA	NA
1945–49	1.5	10.3	6.9	NA	NA	NA	NA
1950–54	1.75	11.2	6.4	NA	NA	NA	NA
1955–59	1.6	11.5	7.2	NA	NA	NA	NA
1960–64	1.6	13.0	8.1	11.5	2.5	0.9	1.5
1965–69	1.5	16.5	11.0	13.5	2.5	1.0	0.75
1970–74	1.6	18.0	11.3	14.5	3.2	1.0	2.0

NA, data not available
1 ardeb of maize = 140 kg.
Source: Data for 1930–59, Al Sarki (1964); 1950–74, El-Tobgy (1976); 1961–74, Lebas and Levy (1979).

reform of agricultural credit in 1979–80. This made debtors deal directly with the banks. If they had not repaid their loan at the year's end, the bank would not allow the co-operative to supply them with inputs for the following year. Some farmers were thus cut off from the supply and credit system which had always been run by the administration. This raised the fundamental long-term problem of how the State might expect to systematize cotton growing if some of the growers no longer had the advantages of the state co-operative. A solution to the problem is all the more urgent as farmers are tempted by other crops (market gardening or horticulture) which give much better returns than cotton.

The agricultural situation today will certainly lead to the revision of the rules for the game played by the protagonists of the agricultural economy, namely the State, small farmers, and agribusinessmen. The State no longer depends on cotton which has been the main cash crop. Its main income comes from oil exports, Suez Canal dues, and tourism. Furthermore, the nation's accounts are swelled by remittances from the many Egyptians living abroad. The area in cotton production began to be reduced fifteen years ago, but there is great inertia in the State cotton system, which employs hundreds of thousands of people. It is hard for the Egyptian Government to liberalize the peasant economy, because there is a high risk that the whole textile sector will collapse. It seems that the cotton trap, which John Ninet denounced during the boom years of 1860–5, is still hampering the political choices that have to be made for the year 2000.

The small farmers who are regimented by the state co-operative system, but who are also to some extent protected by it, are facing enormous difficulties, due for the most part to the tiny size of their farms. The fragmentation of plots among heirs has meant that they no longer have enough land to survive. As in other countries, migrating to the capital city seems one way out, but often proves to be an illusion. Those who stay will probably intensify their system of growing, especially if they can stop growing cotton, which is too time-consuming. Moreover, official research in agriculture is moving towards intensification by testing very early varieties of cotton. In some non-cotton areas farmers are already growing seven crops in three years; by replacing cotton with soya beans, they can grow *bersim* before it and maize after. In the author's view small farmers should aim at better integration of crop production with livestock rearing, including the development of fodder production systems, genetic improvements, organic manure, perhaps even the production of biogas. It should also involve better mastery of the techniques needed for intensive growing, namely small-scale mechanization, which has in the case of Egypt the advantage of shortening the time-gap between one crop and the next.

Finally the agribusinessmen, who are as mistrustful of the State as they are of the landless peasantry, have gone in for systems of production which will make them fairly independent of political and social risks. Thus, the big orchards in the Delta employ very little direct labour. The owner sells the harvest to a wholesaler when it is still on the trees and he recruits his own day-labourers. The other operations are mechanized wherever possible. There is no doubt that these 'gentlemen-farmers' all fear the return to power of a Nasser-style regime which would cut back their profit margins or even carry out a new land reform.

Notes

[1] Fiscal pressure decreased in later years, and today land taxes are insignificant.
[2] The animal unit is a conventional measure equivalent to the energy (or fodder) needs of an adult cow with an annual milk production of 800 kg (this is the average production for Egypt).
[3] Government energy policy also encourages the use of machinery: agricultural diesel is heavily subsidized.
[4] Cotton production plunged sharply in 1961. It was put down to the inefficiency of anti-pest measures. The market-gardening areas around the big cities are not subject to compulsory crop rotation.

References

Al Sarki, M.Y. (1964). *La monoculture du coton en Egypte et le développement économique*. Droz, Genève.
Audebeau, C. (1913). *Observations faites en 1912 dans le centre du delta*. Commission des domaines de l'Etat égyptien, Cairo.
Barois, J. (1887). *L'irrigation en Egypte*. Paris. (réédité en 1911).
Besançon, J. (1957). *L'homme et le Nil*. Gallimard NRF, Paris.
CAPMAS (Central Agency for Public Mobilization and Statistics) (1982). *Statistics yearbook*. CAPMAS, Cairo.
CAPMAS (1990) *Statistics yearbook*. CAPMAS. Cairo.
De Chamberet, R. (1909). *Enquète sur la condition du fellah égyptien au triple point de vue de la vie agricole, de l'éducation, de l'hygiène et de l'assistance publique*. Darantière, Dijon.
Dudgeon, G.C. (1918). The maintenance of the quality of Egyptian cotton. *Bulletin of the Imperial Institute*, **16**, 160–70.
El Tobgy, A.L. (1976). *Contemporary Egyptian agriculture*, (2nd edn). Ford Foundation, Cairo.
EMCIP (Egyptian Major Cereals Improvement Project) (1978). *Present status and future of wheat and barley improvement in Egypt*. EMCIP, Cairo.
Encyclopaedia Britannica (1964). Encyclopaedia Britannica, London.
Fitch, J. and Soliman, I. (1981). *The livestock economy in Egypt*. University of California Economic Paper, No. 29, Cairo.
Gali, K. (1889). Essai sur l'agriculture de l'Egypte. Thèse de l'Institut agricole de Beauvais. H. Jouve, Paris.
Grégoire, M. (1862). De la culture du coton en Egypte: historique, état actuel, avenir. *Mémoires de l'Institut égyptien*, **1**, 437–86.
Ikram, K. (1980). *Egypt: economic management in a period of transition*. World Bank Country Economic Report, Johns Hopkins University Press, Baltimore.

Lebas, L. and Levy, M. (1979). *Politiques nationales et techniques agricoles, le cas de l'Egypte*. Institut National de la Recherche Agronomique (INRA), Paris.

Lorca, A. (1979). *John Ninet, lettres d'Egypte, 1879–1882*. Centre National de la Recherche Scientifique (CNRS), Paris.

Mosseri, V.M. (1928). La fertilité de l'Egypte. *L'Egypte Contemporaine*, **91/2**, 93–126.

Radwan, S. (1977). *Agrarian reform and rural poverty, Egypt 1952–1975*. International Labor Office (ILO), Geneva.

Ramah. A. (1982). *Forage problems in Egypt*. Centre d'Etudes et de Documentation Universitaire Scientifiques et Techniques (CEDUST), Service Culturel Ambassade de France, Cairo.

Rivlin, H. (1961). *The agricultural policy of Mohammed Ali in Egypt*. Cambridge University Press, Cambridge.

Ruf. T. (1984). La coexistence de systèmes de production différents dans une region du delta du Nil: intérêt de l'approche historique pour le diagnostic regional et l'action de développment. *Les Cahiers de la Recherche-Développement*, **3–4**, 30–41.

Ruf, T. (1986a). La sakkia égyptienne, interface entre l'aménagement hydro-agricole et le systèmes de production paysans. CIRAD-DSA Séminaire, 16–19 décembre, 1986. Aménagements hydro-agricoles et systèmes de production. Collection DSA No. 6, Vol. II, pp.375–81.

Ruf. T. (1986b). L'intégration de l'elevage dans les petites exploitations du delta du Nil, approche historique des fonctions de l'élevage bovin: traction, fertilisation, épargne. Communication présentée au séminaire 'Relations agriculture-élevage' CIRAD/DSA, Montpellier, 10–13 septembre 1985. *Les Cahiers de la Recherche-Développement*, **9–10**, 100–106.

Société Sultanienne d'Agriculture (1920). *Mémento Agricole égyptien*. Institut Français d'Archéologie Orientale (IFAO), Cairo.

USAID/Ministry of Agriculture (1976). *Egypt: major constraints to increasing agricultural productivity*. Foreign Agriculture Economic Report No. 120, USAID, Washington.

Further reading

Ayrout, H. (1952). *Fellahs d'Egypte*, (2nd edn) Sphinx, Cairo.

Berque, J. (1957). *Histoire sociale d'un village égyptien au XXe siecle*. Mouton, Paris.

Brehier, L. (1900). *L'Egypte de 1798 à 1900*. Combet, Paris.

Charles Roux, F. (1936). *Histoire de la nation égyptienne*, Vol. 6, *L'Egypte de 1801 à 1882*. Hanotaux, Paris.

Ministry of Agriculture (1969). *Agricultural Cooperative Societies Law*, No. 51, February 1969. Ministry of Agriculture, Cairo.

Ministry of Agriculture (1980). *The New Agricultural Cooperative Law*, No. 122, 1980. Ministry of Agriculture, Cairo.

Ruf, T. (1986c). Deux siècles d'interventions hydrauliques et cotonnières dans la vallée du Nil. In *Dynamique des systèmes agraires, l'exercise du développement*, pp.279–310. Institut Français de Recherche Scientifique pour le Developpement en Cooperation (ORSTOM), Paris.

12

Systems of agricultural production in the Northern Littoral Region

M. J. MARTIN

Introduction

The Mediterranean coastal zone of Egypt, known as the Northern Littoral Region can be divided into two sections:

1. The north-west coastal zone, which comprises the arid zone west of Alexandria along the Mediterranean coast and includes El Amiriya, Burg el Arab, El Alamein, Sidi Abdel Rahman, El Daba, Fuka, Baqquish, Marsa Matruh, Sidi Barrani, and Salum at the Libyan border. The Qattara Depression and Siwa mark the southern regional boundaries inland, and the region is administered by the Marsa Matruh governorate.

2. The north-east coastal zone, which comprises the area extending to the east of Alexandria, bordering the Delta area, and reaching beyond Port Said towards the Israeli border. It includes the littoral lakes: Maryut, Idku, Burullus, Bardawil, and Manzala which border the northern Delta area.

The land area bordering the north-east coastal zone is composed of marine alluvial soils which are recent, level, heavy in texture, and mostly saline. The agriculture of this area is covered in the following chapter on the Delta Region.

The north-west coastal zone

Land area and soils

The north-west coastal zone has a land area of 2.4 million ha (6 million feddans) of which 95 per cent is rangeland.

Soils intergrade between desert and calcic brown soils, and these are found in a narrow strip, interrupted by sand dunes and limestone along the coast from Lake Maryut to Libya. Salt crusts and marshes are found in many places. In general, soils are less than 25 cm deep, they have a loam or sandy loam texture, a pH of 7.5–8.5, and good water retention characteristics. The surface soil is light in texture due to the wind-blown sand, which may completely cover the soil surface. The calcium carbonate level is high, averaging 20 per cent, and the organic matter content is low. The soils of the coastal zone border on desert and regosols inland. Wind-borne sand accumulates into dunes 30 m or more in height and these continue for long distances.

Climate

The region has an arid Mediterranean climate. Harsh drought conditions and high radiation prevail but these are tempered by the maritime influence on the atmospheric relative humidity and temperature in the coastal zone. Temperature and rainfall data are shown in Tables 12.1 and 12.2.

TABLE 12.1. Northern Littoral Region: mean minimum and maximum temperatures

Station	January Minimum (°C)	Maximum (°C)	July Minimum (°C)	Maximum (°C)
Alexandria	9	18	20	35
Marsa Matruh	8	18	20	3
Siwa	4	20	20	38

TABLE 12.2. North-west coastal zone: mean annual rainfall

Station	Mean annual rainfall (mm)
Alexandria	192
El Daba	143
Marsa Matruh	145
Sidi Barrani	148
Salum	102

Water conservation

The limited water resources of the region are mainly committed to intensification of horticultural production to meet human requirements.

Fresh groundwater for agriculture is available for extraction along the coast, either from dug wells, collecting galleries, or drilled wells depending on the depth, nature, and location of the water. Details of the types of construction are given in Table 12.3.

The methods of utilizing runoff from wadis for artificial winter watering practices include flooding, water spreading, terracing, and the construction of small dams and dykes to divert the water to the surrounding cultivable areas for irrigating crops and enriching aquifers. Between 1961 and 1983, 162 km of medium and large dykes were constructed across the wadis (El-Shafei 1984). Sheet runoff water is utilized for crop production by constructing small dykes parallel to the contour lines and collecting the water in man-made cisterns for future use.

A small number of drilled wells exist. At Fuka, where groundwater is contained in the upper limestone bed, it has been established that 2000 m^3 day^{-1} could be withdrawn from properly located wells without affecting other wells in the area or seriously depleting the aquifers. Cisterns are normally situated in rocky sites, from which sheet runoff water is channelled into dyke/gallery/cistern complexes which harness the flow of some 218 wadis. Dyke lengths vary from less than 600 m to over 15 km; collecting galleries are most extensive at El Qasr close to Marsa Matruh, where ten collecting galleries have a total length of 11.5 km and a daily discharge rate of 60 l per linear metre. A cistern with a capacity of 1000 m^3 requires a catchment area

TABLE 12.3. Methods of water conservation in the north-west coastal zone

Type of construction	Volume (m^3)	Cost (£E)	Area served (feddans)
Small earth and stone dykes for use with sheet runoff	510 287	1 020 000	10 000
Divert and enlarge earth dykes on wadis (by bulldozer)	532 000	160 000	5000
Construct dykes in terraced formation, using soil and stone without mortar	61 500	922 500	3075
Construct bunds, divert water flow over a larger area for conservation (project on 36 wadis)	541 500	391 000	2800
Total	1 645 787	2 493 500	20 875

Source: Aboukhaled (1984).

of 120 feddans. Between 1975 and 1985 the Egyptian General Development Organisation repaired many ancient Roman cisterns which were silted up and also built new ones with silt trap screens (Table 12.4). These cisterns provide water storage in excess of 1 400 000 m³ for domestic use, livestock watering, and horticultural production. It is assumed that each cistern will provide irrigation for 5 feddans over 5 years, i.e. 1 feddan year^{-1}.

TABLE 12.4. Number and capacity of cisterns in the north-west coastal zone

Type	Number	Capacity (m³)
Roman cisterns	4056	1 058 988
New cisterns	2221	356 808

Source: FAO (1985a).

The traditional method of water distribution is by open surface U-section concrete channels (25 cm x 25 cm in section), or by asbestos or steel pipelines up to 3 km long, terminating in small earth channels in the field area, which in turn feed round basins for trees or small rectangular basins or furrows for vegetables and other crops. Water loss is reduced after the first irrigation due to the formation of a crust in the calcareous soils. However, there is some seepage through cracks and up to 40–50 per cent water loss can be expected in earth channels through seepage. Some progressive farmers use movable irrigation hoses in their fields for 'hose basin irrigation', and in some areas contour banks have been installed across wadis to control surface runoff on the arable areas.

Dug wells are up to 30 m deep and 1–4 m in diameter. The water is raised mainly by bucket or *shaduf* at a rate of 2 m³ day^{-1}. It was estimated in the mid-1980s that a well dug to a depth of 10 m deep could irrigate 3 feddans at a cost of £E0.22 per m³, and there were then about 900 dug wells in the region. Submersible electric pumps are being introduced in dug wells. They are powered by a generator or by mains electricity where available.

Between 1959 and 1965 the General Desert Development Authority installed 1010 low-cost windmills between Burg El Arab and Salum. Each cost £E350. Today there are almost 300 still working, but they are gradually being replaced by small kerosene and diesel powered pumps.

A windmill can discharge irrigation water at a rate of 5 m³ day^{-1} into small concrete-lined reservoirs 3–4 m³ in capacity. Water is distributed from these reservoirs by bucket or small earth canals to trees or vegetable gardens nearby. Tractor-drawn water bowsers are used to carry water to outlying sites, but transport costs can be prohibitive.

Drip irrigation is being introduced on a few orchard sites and it is most effective with improved, imported cultivars of crops such as Spanish and Italian olives. Drip irrigation is not used to its best advantage with traditional varieties which have established root systems and do not respond to irrigation.

On the north-west coast, the Ministry of Works and Water Resources is currently implementing a USAID-financed feasibility study to utilize treated seawater for the irrigation of salt-tolerant fodder crops such as alfalfa for animal production. The study started with a 50-feddan pilot project in 1988 but has the ultimate goal of irrigating half a million feddans (Abu Zeid 1988).

Population

Prior to 1960, the north-west coastal zone was isolated from the mainstream of development activity in Egypt. Since then the Egyptian Government has put increasing emphasis on developing agriculture and tourism within the area.

The inhabitants of the region are 250 000 Bedouin (from the arabic *al-badu* or *badwiyyin* meaning desert dwellers) (Cairo Today 1985, 1986) Traditionally these were nomadic, living in tents in family units of seven to eight persons, and having a primary income derived from livestock production primarily through the sale of sheep, goats, and wool, but with manure, skins, and goat and camel hair providing further income.

The Awlad Ali (literally 'the sons of Ali') who inhabit the whole of the northern coast of Egypt, from Salum to Alexandria, are descended from both Arab Beni Suleim and Berber tribes. They migrated from Eastern Cyrenica (Libya) about 200 years ago, and until quite recently continued to move freely across the border to trade and to visit their Libyan kinsmen. As they pushed forward into the semi-desert of north-west Egypt, they subdued smaller tribes that already lived in the area and integrated them into their own tribal system. These dependant groups were called *morabiteen* (literally 'the bound' or 'the tied') because they were tied to an Awlad Ali tribe (the *hurr* or free tribes).

Today the difference between the two groups is no longer as significant as it used to be, since the Egyptian Government controls a vast majority of the area over which the tribes once held sway. However, despite recent extensive administrative development within the governorate, traditional forms of organization continue to play an important part.

The five largest tribes of Awlad Ali and the integrated *morabiteen* groups are subdivided into clans with several thousand members each. The leaders of the clans are *omdas* and sheikhs who are very powerful men with traditional and modern offices. Their traditional duties include resolving conflicts between two clans along the lines of tribal law. Their modern function includes acting as middlemen between their tribes and the government administration in Marsa Matruh.

Over the past thirty years, 75 per cent of Bedouin have settled. They now live in stone houses, cultivate orchards of olives, figs, and almonds, and drive the ubiquitous pick-up truck which has largely displaced the camel as the primary form of transport. Settlement has come about with assistance from the Egyptian Government in the form of self-help programmes, financial incentives, food aid, subsidized concentrates for animal feed, and co-operative credits.

Livestock husbandry

During the past fifteen years, the low cost of free open grazing and the high demand for sheep meat from the local and export markets at relatively high prices have encouraged flock owners to increase their flock numbers. The present livestock population has been estimated to be three times greater than the carrying capacity of the available natural vegetation.

In the absence of grazing rights, overgrazing has resulted in the deterioration of the natural vegetation, causing flocks to become more dependent on supplementary feeding. During the dry season and drought years there are more frequent migrations to the Delta and the Nile Valley. The price of live animals drops by varying degrees depending on the severity of a drought period, but during the following good years, and before the natural vegetation regains its former condition, the animal population starts another escalation cycle. Flock owners retain their animals in order to build up their flock numbers, and once again the market price soars. Thus, within one or two good seasons, the sheep and goat population once more exceeds the carrying capacity of the available natural vegetation.

Mechanization, the increased need for grain, and the expansion of grain production by ploughing along steep wadi beds have destroyed the natural vegetation cover and contributed to the process of land degradation.

The original natural vegetation cover, dominated by thorn, shrubs, and trees, stretching inland to a distance of 10–30 km in the 125–150 mm rainfall zone, has been largely eradicated through shifting livestock grazing combined with opportunist barley cropping and firewood collection.

A Department of Agriculture survey in the mid-1980s estimated livestock numbers at about 107 500 sheep, 26 500 goats, and 12 800 camels in the Northern Littoral Region of Egypt. These numbers are estimated to be three times the carrying capacity of the available land due to the great demand for sheep meat in Egypt and from abroad. Flocks of sheep and goats graze in the rainy season, from November to March. Thereafter they are fed conserved fodder and when the feed supply is exhausted they migrate to the Nile Valley or Delta.

Sheep
Barki fat-tailed coarse-wooled sheep were introduced to North Africa by Phoenician colonists and are the only surviving breed of sheep in the desert area. They are kept for meat and wool production, but there is no systematic

breeding programme; sires are selected phenotypically on the basis of size and growth rate. The weight of a mature ewe is 40–45 kg and that of a ram 50 kg. The ewes and rams are run together all the year round, but the main breeding season is June, with lambing in November/December.

Ewes have a lambing percentage of 70–80 per cent and can produce 40 l of milk (5.1 per cent fat) in an 18-week lactation. Lambs have an average birthweight of 3–5 kg, and weigh 20–25 kg when weaned at 4 months. In a non-drought year the mortality rate between birth and weaning is about 10 per cent. The annual yield of manufacturing quality wool is between 2.0 and 2.5 kg per head year^{-1} and this is used to make blankets and carpets.

In the absence of any systematic fattening system, sheep are slaughtered at any age or weight depending on demand and the prevailing market prices which invariably exceed the official fixed price. Sheep are also smuggled 'on the hoof' across the border into Libya. Dealers from the Marsa Matruh/Salum area put to good use their old tribal links with their Libyan kinsmen, and will pay above the Marsa Matruh and Salum market price because they can get higher prices in Libya.

There has been a scheme to distribute improved Barki rams obtained through selection and breeding at El-Hamman (Ministry of Reclamation Project No. 2270), and this has been sponsored by FAO as part of the World Food Programme.

Goats

Unimproved Baladi goats are very small flat-bodied animals with well developed digestive systems; they are probably descendants of the Mediterranean derivatives of the wild *Capra prisca*. Goats are herded with sheep in flocks of 40–50 animals and are kept to supply the Bedouin with milk, meat, and hair. The main breeding season for Baladi goats is June, with kidding in November/December. Females can produce 50 l of milk (4.8 per cent fat) in a 5-month lactation and they have a kidding percentage of 130 per cent. Their milk yield is low compared with that of improved goat breeds kept in similar conditions in other countries.

As part of a programme of development assisted by FAO (El-Shafei 1984), work has been undertaken to cross the indigenous breed with Damascus goats to increase milk production whilst maintaining tolerance to the harsh local environmental conditions.

Camels

Single-humped camels (dromedaries) graze the natural vegetation of thorny scrub in flocks of 15–40 animals. The Bedouin have no systematic breeding or fattening system for camel production but their camels are mated from 4 years old and then every 2/3 years, calving five times up to the age of 20 years. Camels are used primarily as pack animals but also for some draught work. They are worked from 6 years old and can be used until they reach 20 years of age. Camels moult once a year in June/August and the camel hair, yielding almost 2–3 kg animal^{-1}, is collected and made into scarves, carpets, and headcovers. Occasionally young camels are slaughtered during festivals.

Camels can produce up to 9 l milk (5.5–6.6 per cent butterfat) if they are milked twice daily, i.e. 2500–3650 kg in an 18-month lactation. The milk is either drunk fresh or processed into yoghurt, butter, or soft cheese, generally for domestic consumption.

The Marsa Matruh Central Co-operative has sixty branches along the northern coastal belt. This co-operative distributes subsidized government concentrate and markets the sheep wool and hair produced in the region.

Veterinary services

Only minimal attention has been given to animal health, but there are twelve veterinary centres in the region, each staffed by at least one qualified veterinarian and two assistants who operate on a small budget with limited facilities. The service falls short of meeting the basic requirements for control of ecto- and endoparasites.

Crop production

An opportunist barley crop is taken after the first rains, but dry planting is carried out prior to the rains as the rainfall distribution is not uniform. The area planted varies considerably from year to year.

Cultivation practices have serious limitations as there is no systematic rotation and the same land areas are cropped year after year. However, a form of fallow barley results from the fact that the total area cropped with barley varies from year to year according to rainfall variations, drought, and seed shortages after a drought year. Barley cultivation is very extensive. More efficient land use would limit cultivation to those areas where an average yield of barley could be obtained in all but drought years. The present system allows land to lie fallow for several years because a good barley crop can be obtained only in years of high rainfall. Fallow land is wasted, because the natural vegetation has been destroyed by cultivation and grazing is impossible; it is also susceptible to wind erosion.

Cereal root disease, seed-borne pathogens, and rat infestation are all common in the region. Barley is also susceptible to wind blast because its growing point is not protected. Therefore the strong dust-laden khamsin winds affect production and can cause crop failure. In drought years, a crop likely to fail is either grazed or pulled out, conserved, and fed as hay.

In 1972 it was estimated that about 400 000 feddans were sown with barley (Aboukhaled 1984). In the 1987/8 winter season the Matruh governorate stated that seed had been distributed by the co-operative movement to plant in excess of one million feddans of barley between El Amiriya and Salum, including reclaimed lands at West Nubariya, Maryut, and El Amiriya

(MENA 1988). The actual area of barley grown was probably considerably less. Nowadays the Egyptian Government is actively encouraging farmers to switch to wheat for human consumption.

Barley and Emmer wheat were the first cereals to be domesticated in the Middle East, at least 9000 years ago. From these evolved the six-row barley grown in the north-west coastal zone today.

Barley for seed is purchased by the co-operative movement from the local market for resale to farmers. Varietal control, seed cleaning, and seed treatment are not practised.

Traditionally, cultivation was carried out with donkeys and camels used as draught animals, with locally made wooden ploughs. No fertilizer was applied. Now, more than 80 per cent of cultivation is mechanized. The implements used are a simple tractor with three-point linkage-mounted light disc harrows or locally made chisel ploughs. These implements are used to stir the surface soil, after which barley is sown by hand at a sowing rate of 75–100 kg ha^{-1}. The seed is incorporated to varying depths with a second pass of the same or similar implement.

Some cultivation equipment used in the region is owned by contractors or farmers, but 70 per cent of the cultivation equipment used is hired from the Maryut Agricultural Company located at the eastern end of the coastal zone. The allocation of tractors and implements is co-ordinated by the Marsa Matruh Central Co-operative and is organized at the local level by the co-operatives. Although the system is rather inefficient, it does meet local requirements to a large extent.

Harvesting is from April onwards and yields of grain (200–400 kg ha^{-1}) and straw (up to 800 kg ha^{-1}) vary according to rainfall and available crop nutrients. When grain yields of barley are expressed per unit area and per mm of annual rainfall, production in Egypt (2.0 kg ha^{-1} mm^{-1}) is lower than in Libya (3.5 kg) and Australia (5.5 kg), where conditions are similar.

Traditional harvesting is by hand and involves pulling the plant complete with root system out of the ground. As tribal grazing rights include the harvested crop areas, farmers ensure that they remove as much crop residue as possible. This method of harvesting leaves the top 4 mm of unstable powdery surface soil very prone to wind erosion in this coastal region.

Threshing is carried out with donkey-drawn sledges or tractor wheels and winnowing is done by hand with wooden forks. The barley grain is used for human and animal consumption and the straw is used for stockfeed during the dry summer months and drought periods. Grain for immediate human consumption is stored in sacks in the family house or tent, while chaffed straw is transported to the farmer's dwelling on the backs of camels and donkeys where it is stored in heaps. The grain for storage is stored in pits (1.5 m in diameter) dug 30 cm below ground level. These pits are lined with chaffed straw and the stored grain is covered with soil built up 1 m above ground for protection against the elements. The grain may be kept for two years or more, during which time insect damage may occur, and the almost airtight storage may reduce seed viability.

As a result of an FAO research review and planning mission to Egypt (FAO 1985a), it was established that 95 per cent of the total land area of the north-west coastal zone is rangeland of which 192 000 feddans are suitable for arable rotational cropping and a further 288 000 feddans have potential for green/dry pasture establishment in the 125–150 mm rainfall zone near the coast.

In an attempt to improve crop productivity and stabilize the available land, it was suggested that barley should be direct-drilled with a tined-seed drill at a sowing rate of 35–40 kg ha^{-1}. Sowing at a constant pre-set depth with a minimum of soil disturbance would help to conserve soil moisture. An application of triple superphosphate at a rate of 250 kg ha^{-1} would help to increase yields to a level comparable with those attained in Australia and Libya.

In order to increase the quantity of forage available for animal production, several demonstration areas in the north-west coastal zone have been sown with improved cultivars of the annual legume, medic (*Medicago littandis*) imported from Australia. Cultivars such as cv. Harbinger, which is similar to indigenous unimproved cultivars found in the region, are drilled at 5–6 kg seed ha^{-1} and given triple superphosphate at an application rate of 100 kg ha^{-1}.

Once established, medic is self-generating. It fixes atmospheric nitrogen and can produce a large quantity of fresh/dry forage, with the added bonus of improving soil fertility. On the established demonstration fields, medic is being grown in a 2-year rotation with barley.

In and around these demonstration areas the tribal grazing – cropping rights are being displaced and the practice of hand-plucking the barley crop complete with roots is gradually being replaced. Cutting the crop with a sickle, scythe, or even a tractor-drawn mowing machine leaves the root material in place to anchor the soil and control desertification.

Vegetable production

It is estimated that in the Northern Littoral Region, between 1000 and 4000 feddans of vegetables are grown annually under irrigation (Aboukhaled 1984), the actual area being dependent on the stored water available. Cultural techniques are labour-intensive and largely primitive, improved seed varieties and modern techniques gaining only slow acceptance.

The major field crops grown are watermelons, onions, broad beans (*Vicia faba*), and tomatoes. Broad beans are a winter crop, grown as a green shelled bean around Marsa Matruh and as a dry crop in the more isolated areas, where they are grown on a relatively large scale because they are tolerant of salinity and low water levels. Onions have good salt tolerance and will grow with minimal water. They are often intercropped with fruit trees, mainly olives. Tomatoes are grown under irrigation in the summer but they are dry-farmed in the winter in areas where there is sufficient runoff water. Watermelons are dry-farmed and are the main summer crop. They are a valuable alternative to drinking water in desert areas, where parched seeds are chewed.

Small-scale garden plots of cucumbers, peppers, lettuces, cantaloupe melons, mint, garden rocket (*Eruca sativa*), squash, radish, parsley, and other herbs are also grown, normally for family consumption.

Fruit production

Olives, figs, dates, and almonds are grown by traditional methods, but farmers prefer olives because they do not attract animals and the crop is easy to market.

Olives

The old local varieties of olive tree which are already established and have extensive rooting systems have limited genetic potential for improved yield and once established do not respond to irrigation. Propagation, which is by grafting, is slow and expensive. Very little pruning is undertaken as there are no skilled pruners. When pruning is carried out, it is often done too early, only diseased and dead branches are hacked off, and the branch stumps are left exposed to sunlight.

Dry farming is typical for olives, but the young trees are watered for the first 3–4 years until the rooting system is established. In orchards near coastal sand dunes, where fresh groundwater is available from dug wells, and in areas surrounded by synclinal basins, supplementary irrigation is applied when necessary all year round. Strong dust-laden khamsin winds from the desert have a harmful effect on olive production, but there has been no planting of *Casuarina* or similar windbreaks as in the Delta.

The olive is the most important tree crop and the fruits are either pressed for oil extraction or pickled. The traditional local cultivars are Kalamata and Hamid, used for pickling, and cv. Shemlali is grown for oil production. Several improved Italian and Spanish olive cultivars have been imported and they respond well to irrigation. Drought or water stress is most damaging during the stone hardening and fruit swelling stages. One or two well-timed irrigations at these stages can increase the yield of improved cultivars significantly. Irrigated imported cultivars yield 55 kg fresh olive fruits tree^{-1}, compared with 18 kg tree^{-1} for dry-farmed local cultivars. There is considerable pest and disease damage and the growers do not generally recognize the importance of pest control. They are inclined to consider it uneconomic on their poor crops.

Harvesting often takes place before the olives are ripe and this has a harmful effect on the quality and quantity (only 14–18 per cent) of oil. Further deterioration due to 'sweating' is caused by mechanical damage from the palm wood crates used for transport from the field and subsequent storage prior to processing.

There are two mechanical and eleven primitive oil mills in the region, but their output is low and delays are incurred. Commercial-scale pickling is done in Alexandria.

Until recently indebtedness was common among producers as there was no governmental or co-operative credit scheme available in the region. Producers were dependent on the merchants for their seasonal loans, and to local retailers for their food and clothing needs.

Figs

Figs were first grown in Egypt before 4000 BC. Nowadays they are propagated vegetatively from cuttings but they tend to be pruned inadequately. Growers often cut back only random diseased branches without pruning sufficiently to encourage fruiting growth. Traditionally, figs are sold fresh at the local markets; the main cultivar is Sultana. Usually figs are eaten fresh or they are stewed as the fruit is often bruised during transport in palm wood baskets.

Processing of figs is still insignificant, but if cultivars suitable for drying were available they would increase overall productivity significantly. To develop fig production in the area the Ministry of Agriculture and Land Reclamation is establishing over 700 fig orchards. Each unit covers 20 feddans and has a dwelling to encourage further settlement of Bedouin. The scheme is being financed by the United Nations World Food Programme to encourage settlement in the region. It forms part of an aid programme valued at £E7.5million in farm inputs, food, and cash distributed to about 30 000 farmers in 1988. This form of support is intended to supplement farmers' income whilst they establish their new farms (Abu Zeid 1988).

Dates

Until recently, only hard varieties of dates were grown and these were sold fresh at local markets. In 1988, in a move to develop date production, a pilot model date-processing plant was established by FAO at the Siwa Oasis. It is planned that this will be followed by similar plants in the region, established through self-help ventures to exploit the full potential of the crop.

Almonds

Traditionally, almonds were of limited importance. The Bedouin were not inclined to grow almonds on a large scale, because the trees had to be protected against theft all through the fruiting season; for this reason the crop was often pickled green. With the establishment of a settled Bedouin community, this is no longer the case. Almond orchards are being established, and the trees are often intercropped with other trees.

The north-east coastal zone

Fish production

The eastern Mediterranean is generally poor in fish resources, with the exception of the waters off the Nile Delta. The traditional fishing grounds are the littoral lakes and the inshore waters on the continental shelf off the Nile Delta (Table 12.5). The littoral lakes of Egypt, which provide a rich environment for estuarine and marine fish, have been the main source of fish for a long time. Historically, production from the littoral lakes has constituted 65 per cent of the total national catch (Table 12.6) despite the fact that the fishing area does not exceed 15 per cent of the total water surface area of Egypt. The lakes are subject to heavy fishing pressure, the annual production having changed little over the years since the first catch data were recorded in 1928. These littoral lakes are among the most productive standing water bodies in the world because of their inherent shallow depth and the vast quantities of nutrient rich water which flow into them from the Nile Delta irrigation and drainage system.

The brackish lakes with a salinity of up to 5 g ml^{-1} comprise Manzala, Burullus, and Idku which are connected to the sea, and Maryut which has no sea connection. They have a total area of 2300 km^2 (500 000 feddans) and traditional lagoon/lake fishing is from small sailing or rowing boats, and on foot. The characteristics of the brackish northern littoral lakes are shown in Table 12.5.

The catch from the brackish lakes would be much higher if their area had not been reduced by land reclamation projects. Lake Maryut suffers severe oil and industrial pollution from the refineries and factories of Alexandria but

TABLE 12.5. Characteristics of the brackish northern littoral lakes

Lake	Area (km^2)	Depth (m)	Production (t km^{-2})	Fish species
Maryut	160	No data	26	Tilapia (70%),
Idku	133	1.0	31	mullet (15–20%), and eels
Burullus	560	0.5–1.0 (4–5 in sea channel)	19	Mullet and eel
Manzala	1450	4–5	19	Tilapia (80%), mullet (20%), and shrimp
El Mallaha	100	No data		Mullet caught with fine nets

TABLE 12.6. Egypt: fish catch in 1984 by area

Area	t year^{-1}
Mediterranean	20 000
Manzala	35 500
Idku	2660
Maryut	8050
Burullus	15 000
Bardawil	5000
Lake Nasser	20 000
Nile and branches	25 000

Source: Institute of Oceanography, personal communication.

a commercial fish farm is established on the landward side of the causeway carrying the 'desert road' to Cairo.

The saline lakes such as Bardawil are characterized by salinity levels nearly equal to that of the sea. The fish catch per km^2 is lower than that attained in the brackish lakes but this is offset by the higher unit price commanded by salt-water fish.

The principal species found in the lakes include tilapia, particularly *Tilapia nilotica, Tzilli, Tgalaea*, mullet, shrimp, eels, and a variety of other sea and salt-tolerant freshwater fish. The tilapias are the most common fish, constituting 80 per cent of the catch. Species of marine origin such as mullet and shrimp usually enter the saline lakes for protection and feeding, but migrate to the sea for spawning. Statutory regulations governing fishing on the lakes have existed for more than forty years, and cover mesh size, vessel types, seasons, fishing locations, and minimum size of fishes. However, enforcement of these regulations is lax and the controls are generally ignored; many of the fish caught are of an illegal size.

There are about 100 000 full-time lake fishermen using some 19 000 rowing or sailing boats. They have no means of preserving their catch on board, so they must deliver it to the point of sale, market, transport vessel, or storage facility. Most of the catch is sold fresh but salted fish is in high demand in some areas; few fish are frozen or canned. The present catch could be increased considerably if the existing resources were managed rationally. Regulations on mesh size and an increase in the size limits of mullet and tilapia in the Delta lakes could double the catch.

The river Nile and its network of irrigation and navigation canals produce 15 per cent of the total catch which reaches up to 25 000 t annually. The present catch could be more than doubled if specific procedures were adopted. This should include the improvement of fishing gear and methods, prohibition of illegal fishing, stocking with selected indigenous species whose spawning grounds were eliminated by the construction and operation of the Aswan High Dam, and water quality improvement through pollution control. There is also potential for the utilization of the vast stretches of freshwater irrigation canals for domestic and commercial aquaculture.

Sea fishing

The Egyptian Mediterranean coast from Salum to Rafaa is 1100 km in length and the continental shelf runs its entire length. It is up to 200 m in depth and is composed largely of silty mud and muddy sands. In front of the Delta it has a maximum width of 70 km. The coast can be divided into three sections:

1. The western zone which extends for about 600 km from Salum to Alexandria and at Marsa Matruh, El Daba, and Salum there are small fishing communities of 50–80 fishermen who alternate between fishing and farming (El-Shafei 1984).

2. The central zone which includes the main fishing centres of Alexandria, Abu Qir, El Madia, Rashid, Idku, Damietta, and Port Said. There are also many small fishing communities in Tabiat el Sheik, Burullus, Gamasa, and Kobry el Saffara.

3. The eastern zone which stretches about 200 km from Port Said to El Arish.

Most of the sea fishing operations are in the central zone between Alexandria and Port Said, where trawling takes place at depths between 10 and 100 m. The rest of the shelf is scarcely exploited. Since 1964 the fish catch from the sea has declined as a direct result of the construction of the Aswan High Dam. This effectively stopped the annual flow of nutrients carried out by the Nile to the sea. Prior to that, the peak outflow of the Nile stream was in the *nili* season, between August and December. In September, the daily discharge was estimated to be 33 million m^3 of freshwater rich in biogenic material, mud and, silt, estimated at 140 million t per annum. The nutrient retention upstream as a result of the dam reduced the aquatic productivity in the eastern Mediterranean and the catch fell drastically.

In addition to demersal species, pelagic fish such as sardines were one of the main catch components in the *nili* season before the completion of the Aswan High Dam in the mid-1960s. Thereafter, the annual sardine catch dropped dramatically from an amount in excess of 18 000 t to only 500 t. Sardine fishing has never recovered and the annual catch in the late 1980s was about 2000 t.

The annual prawn catch has also suffered a severe reduction from 10 000 t to around 3800 t, on account of the High Dam construction and also intensive fishing of immature prawn in the Delta lakes of Manzala, Idku, and Burullus. The prawn larvae enter the lake for shelter and feeding and then migrate back to the seas to mature and spawn. During this lake phase, young immature prawn are usually subject to intensive fishing, amounting to more than 1000 t per year in Lake Manzala alone.

Between 1962 and 1982 the annual catch from the Mediterranean dropped from about 38 000 t to less than 12 000 t (Table 12.7); its contribution to the total national fish catch fell from 30.5 to 8.5 per cent. There is now a small improvement in production due to the exploitation of a larger area of the sea.

TABLE 12.7. Egypt: national fish catch, 1962–82

Year	Total catch (t year^{-1})	Mediterranean waters
1962	124 159	37 832
1966	88 000	15 048
1973	93 500	9600
1975	106 574	5384
1977	104 541	6683
1980	140 397	17 466
1982	137 208	11 708

Total fish production in Egypt is difficult to quantify because of the disparity of available data. Unpublished data for 1987 (GAFRD, General Authority for Fisheries Resource Development, personal communication) gives annual fish production in Egypt at 220 000–300 000 t, production from the lakes at 130 000 t (50 000 t from Lake Manzala), production from the Mediterranean at 22 000 t, and annual fish consumption at 8.6 kg per caput.
Source: Institute of Oceanography (1982); Institute of Oceanography, personal communication.

The decline in fish production has resulted from a number of factors. The expansion of Egypt's oil and petrochemical industries has caused extensive pollution, which has significantly reduced the existence and proliferation of sea organisms and has therefore led to a deterioration in the marine environment.

The exodus of skilled fishermen to work as expatriates in the Gulf States, Greece, and Libya has resulted in over-fishing by unskilled fishermen and the use of illegal fishing gear and net sizes. Between 1979 and 1982, the number of sea-going powered boats increased by 71 per cent from 1265 to 2166 for a number of reasons:

1. The reduction in the catch from inshore sea fishing and the littoral lakes forced fishermen to fish further offshore in motorized boats.

2. Some fishermen who worked as expatriates overseas returned home and invested their savings in new boats and equipment.
3. People from outside the fishing community built fishing boats because it was considered a good investment.

The trend is towards a larger fishing fleet with a diminishing catch compensated for by a high unit price (FAO 1985b). There are at least 50 000 full-time fishermen (90 per cent of the total). The fishing industry is centred on the confluence of the Nile, and is run through 60 service co-operatives in the main fishing area between Alexandria and Port Said. Governmental control is minimal but the frontier guard corps of the Ministry of Defence monitor the area and issue fishing ground permits. These require information on the trip duration and the names of the fishermen on board (Undersecretariat of Aquatic Resources, Fisheries Department, personal communication).

For inshore fishing, fishermen operate in wooden boats of a typical Mediterranean design. Boats in the semi-industrial fleet are 20–30 m in length while the artisan fleet consists of small deck boats 8–13 m in length. There is no governmental control over the number of boats built.

On the littoral lakes, fishing methods include trammel, cast net, fixed net, land line, long line, gill net, and fish traps. For sea fishing in the western zone, fishermen use cast nets, hand lines, long lines, trammel, and gillnets. In the central zone, trawling (Italian trawls), purse seining, night fishing with lights, gill nets or ring nets (sardines), and veranda nets (mullet) are most common. Long lines are used for other commercial fish and, historically, fishing is to a maximum depth of 100 m.

Port Said and Marsa Matruh have berthing facilities but elsewhere the catch is shipped in smaller boats for landing. Boat yard and slipway facilities are available at Alexandria, Damietta, and Port Said. By percentage of total catch landed, Damietta, which lands 36 per cent of the total catch, is the main centre, followed by Port Said (29 per cent), and Alexandria (20 per cent). These figures include an estimated 10 per cent of the catch landed which is thought to be consumed by the fishing communities (FAO 1985b).

Fish marketing

In the main centres, fish are marketed fresh through fishmongers to whom the fishermen are often indebted for the purchases of fishing gear, boat maintenance, and repair work. Historically there has been no co-operative involvement in this area, similar to the agricultural co-operatives movement which supplies inputs for the compulsory crops which are then set against the farmers' revenue from their harvested crop.

A large percentage of the fish landed are wasted through lack of chilling facilities and bad handling in wooden fish boxes. Lack of ice can restrict summer fishing, especially away from the main centres where freezing facilities are negligible.

The Ministry of Supply operating through the Fish Marketing Company (FMC) organizes the importation, storage, distribution, and retail sale of all imported fish, fish produced from Lake Nasser, high dam production, and a small quantity of fish from the Mediterranean catch. This is done through the *gammayas* (government food shops found throughout Egypt) where the prices are subsidized. In 1985, the FMC handled 85 000 t of imported fish and 27 000 t of locally produced fish (Institute of Oceanography, personal communication).

Development prospects for the Mediterranean fisheries are uncertain. A limited increase in the fish production could be obtained by exploiting the eastern and western parts of the sea, fishing on the continental shelf, and more effective management of the presently exploited demersal stocks.

Fish culture

Fish culture is the most promising way of increasing the fish supply substantially in order to help meet the protein requirements of the Egyptian population. Although small-scale fish farming has been practised in Egypt for a long period, the plan to develop diverse commercial fish farming has only been in existence for the past ten years (FAO 1981).

The presence of endemic fish species which could be raised in fresh and brackish water and the availability of vast areas along the fringes of the north Delta lakes (*howash*), which are of marginal value for agricultural use, are good reasons for the promotion of fish culture. It is also possible that the utilization of drainage water from the existing extensive irrigation system would permit easy control of water flow for fish culture, provided that the water is free from pollution and pesticides.

In the early 1900s fish culture was practised for stocking inland lakes, such as Maryut, which have no link to the sea. Mex Fish Farm west of Alexandria produced mullet fry and young glass eels to restock Lake Maryut. Today the fish farming programme is well established and expanding rapidly. It covers some 60 000 feddans; 12 000 feddans is owned by the government and the remaining area is made up of about 1000 private farms.

Commercial fish farming takes the form of pond culture using mullet and carp either in mono- or polyculture, although tilapia are always present in ponds. *Hosha* culture, where fish are confined to an enclosed section of water, is most extensive in Lake Manzala and Lake Burullus, where 70 000–100 000 feddans are utilized in units of between 25 and 200 feddans. The system is either extensive where primitive seed fish are taken from a neighbouring lake, or intensive where developed enclosures are stocked with artificially produced and fertilized fish stock. To date, the integrated system introduced by the General Authority for Fisheries Resources Development (GAFRD) in 1982/3 has been confined to mirror carp produced in rice fields. In Kafr el Sheik governorate, 52 000 feddans of rice fields were stocked with mirror carp yearlings and the programme was very successful. The average production of fish was 60–100 kg feddan^{-1}, and a total yield of about 4000 t of marketable size carp were produced in three months. In addition, rice production was increased by 5 per cent. The programme was reintroduced in 1987/8 in the same region with the distribution of free carp seedling.

In its Food Security Programme for the area, Alexandria governorate initiated a duck-cum-fish farming project in the canals behind Nouza (Alexandria) Airport in an isolated part of Lake Maryut. This project has been developed further with the help of USAID and the Catholic Relief Agency who have supported the reclamation of a small area of the lake and constructed fish ponds to produce tilapia, mullet, carp, Nile perch, and ducks.

Future

Further development work is planned to satisfy the needs of the growing settled community in the Northern Littoral Region. It can be summarized as follows:

1. Improved utilization of the available surface and ground water, including the installation of new and the rehabilitation of existing windmills.

2. Increased barley production on the plateau area through the use of improved agricultural practices, seed selection and cleaning, and the use of artificial fertilizers.

3. The problems of desertification and feed shortages are being tackled by regenerating the vegetative cover with plantings of improved, self-perpetuating legumes, palatable shrubs, and trees. Windbreaks are being established around orchards and vegetable gardens.

4. The utilization of extensive available saline water resources for the production of tolerant fodder crops.

5. The improvement of sheep and goat flocks through the selection and distribution of improved sires.

The current five-year fishing improvement programme of the GAFRD includes all the criteria necessary to further develop Egypt's fishing industry. These are as follows:

1. To enforce existing legislation and protective measures which have been largely ignored to date.

2. To maintain lake–sea connections.

3. To expand the artificial propagation of shrip, carp fingerling, and mullet fry to cater for the expansion of fish farming, to stock inland lakes, and introduce cage culture on the network of irrigation and drainage canals.

4. To encourage farmers to use their rice fields for fish culture.

5. To provide funds for the fishermens' co-operatives to introduce medium and long-term loans for fishermen.

References

Aboukhaled, A. 1984). *Improvements in the irrigation practices and methods in the north western coastal zone of the Arab Republic of Egypt. A consultancy report.* Report No. FAO-AGO-EGY/82/015.
Abu Zeid, M. (1988). *Egyptian Gazette*, 7 February. MENA (Middle East News Agency), Cairo.
Cairo Today (1985). *Bedouin in a changing world.* Vol. 6, No. 9.
Cairo Today (1986). *The Bedouin.*
El-Shafei, S. (1984). *Agricultural development in the north west coastal zone. Report on water harvesting.* Report No. FAO-AGO-TCP/EGY/2205(MF). FAO, Rome.
FAO (1981). *Pilot project for intensified aquaculture production (freshwater fishes), Egypt.* Report Nos. FAO-FI-TCP/EGY/8904 and FAO-FI-TCP/EGY/8904. FAO, Rome.
FAO (1985a) *A review of the organization of agricultural and fisheries research, technology and development in Egypt. Report of an FAO Agricultural Research Review and Planning Mission to Egypt, April-May, 1985.* Report Fiche No. 86M00953. FAO, Rome.
FAO (1985b). *A review of the agricultural and fisheries research, technology and development in Egypt. Annex VI Research in fisheries.* Report Fiche No. 86M00953. FAO, Rome.
Institute of Oceanography (1982). *1982 yearbook.* Institute of Oceanography, Alexandria.
MENA (Middle East News Agency) (1988). *Egyptian Gazette*, January. MENA, Cairo.

13

Systems of agricultural production in the Delta

P. N. WARD

Introduction

The Nile branches just north of Cairo to form a triangular area which extends to the Mediterranean coast; both branches, the Rosetta (western) and the Damietta (eastern) are approximately 240 km long. The Nile Delta is the irrigated area between and around these branches; it consists of some 25 000 km^2 and includes nine of the governorates of Egypt (Fig. 13.1). The region is flat, extending from Cairo at 24 m, to Damietta at 1.5 m a.s.l. The cultivated area amounts to some 3.59 million feddans (1.51 million ha), and includes 60 per cent of the area under cultivation in Egypt (El-Tobgy 1974).

The Nile Delta is one of the oldest agricultural regions in the world. Much of its area has been under continuous cultivation for more than 5000 years and, until the twentieth century, the economy was agrarian with some exports of cotton, rice, and a few minor crops.

The area is typified by highly fertile alluvial soils, intense sunlight and a climate that allows year-round cultivation of a wide range of crops. These natural conditions, together with the availability of water from the Nile, means that high levels of crop production are obtained, in many cases above world averages (FAO 1991). Considering these very favourable conditions, productivity could be even higher than at present were it not for the poor drainage in some areas, losses due to pests and diseases, lack of timeliness in farming operations, irrigation water management, government policies, and lack of investment (FAO 1987).

Climate

The climate is Mediterranean with hot rainless summers (mid-March to mid-October) and mild, frost-free winters with some rain. Mean annual maximum and minimum temperatures are 28 and 14 °C, respectively. Summer maximum temperatures are in the range 32–5 °C with a minimum of 20 °C; mean winter temperatures show a maximum of 20 °C and a minimum of 7 °C.

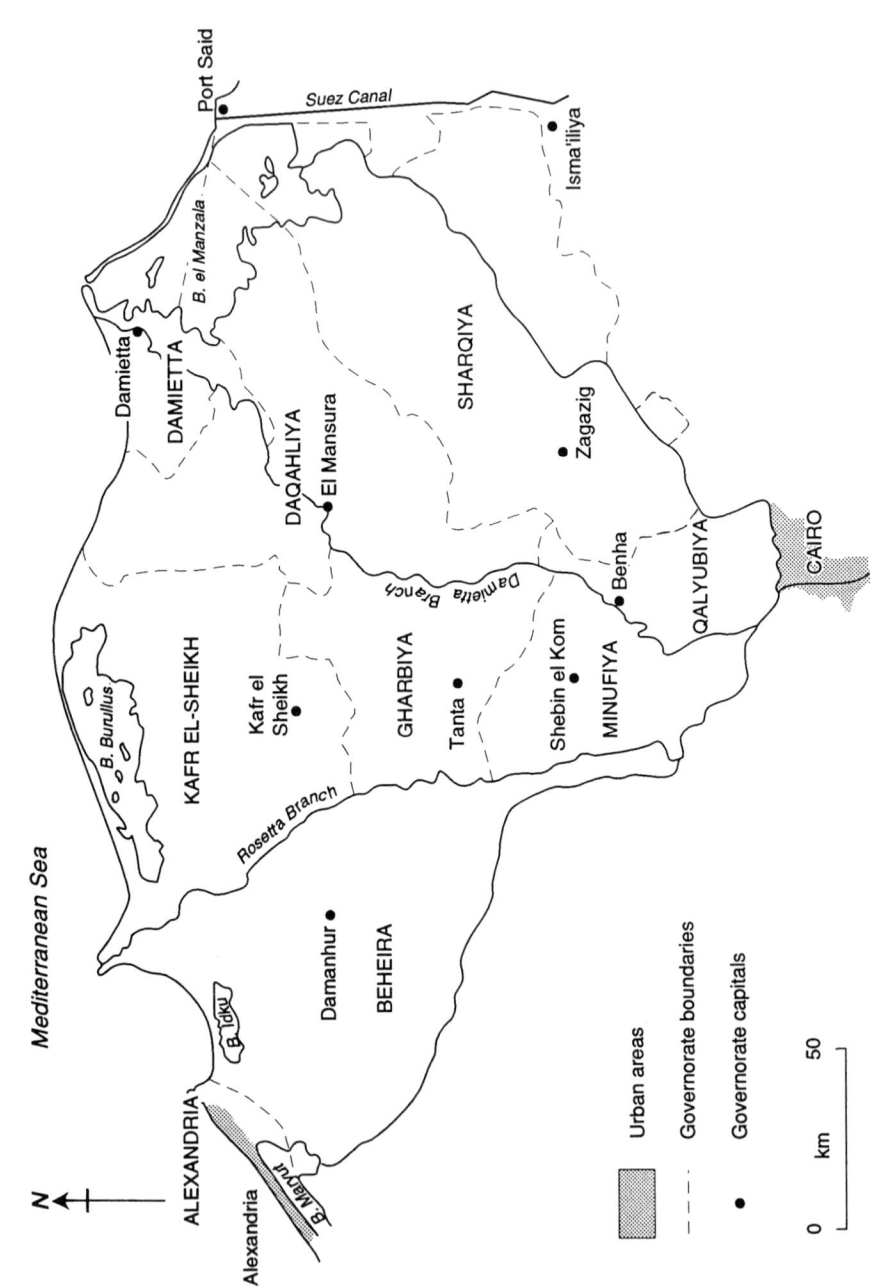

FIG 13.1. Egypt: the governorates of the Delta region

Humidity is on average 60–80 per cent, with a pronounced peak in August. Annual rainfall is on average 150 mm over the area, ranging from 24 mm in Cairo to 192 mm in Alexandria. All of the rain falls in the winter months, but even in winter, cloudy days are few. In agricultural terms, most of the area can be considered rainless and agriculture is almost completely dependent on irrigation from the Nile.

Soils

The alluvial soils in the Delta were formed from the dark grey clay which has resulted from the annual flooding of the Nile. The river brought suspended matter from its source areas, and deposited it on its flood plains. The clay cap, averaging 10 m thick in most of the Delta, was thus formed over coarser deposits of an earlier period. The clay cap is absent in the marginal areas of the Delta. Here most soils are deep, black, and heavy to medium in texture, but becoming gravelly in the most marginal areas.

Delta soils were classified by the Ministry of Agriculture in the late 1950s and early 1960s; this classification was based on productive capability rather than on physical characteristics (El Tobgy 1974). Soils were classified into six groups: Classes I–IV covered cultivated land ranked by productivity, Class V was barren land and Class VI was non-agricultural land.

Irrigation

Traditionally, most crops in the Delta were autumn-sown, but when the Delta Barrage came into use in 1863, it enabled the water level to be raised so that irrigation canals supplying the Delta could be fed; this allowed summer crops to be grown (Money-Kryle 1957). The construction of the Aswan Dam and further barrages permitted year-round irrigation. Since 1969, the Aswan High Dam has regulated the flow of the Nile in Egypt and the whole Delta system is now closed; water can only enter the sea via locks and the littoral lakes outlets.

Irrigation water is brought to every village free of charge by the Ministry of Public Works and Water Resources. The flow of water into the canals is fixed according to the area served and the official requirements for each region and cropping season as laid down by the Ministry of Agriculture. The size of each field outlet is fixed for each farmer depending on the area of the property; it has been claimed (Adams 1983), however, that farmers in many areas circumvent this restriction on the quantity of water reaching their land. Canals are normally cleared out and maintained annually in January.

Irrigation rotations are enforced by alternating the water levels in the canals to give wet and dry periods. The common patterns are:

(a) February–April: 5 days high level wet, 5 days low level wet, 5 days dry;

(b) April-December: 7 days wet, 7 days dry;

(c) rice and vegetable areas: 4 days wet, 4 days dry.

In most areas the farmer has to raise the water an average of 30–40 cm from the common irrigation channel to his own land. This is traditionally carried out by draught animal power using an Egyptian water-wheel (*saqqiya*), although motor-driven wheels and to a larger extent pumps, are becoming more common.

Delta farmers continue to irrigate in their traditional manner, which is to over-irrigate and allow the excess water to drain away. Water is normally available in sufficient quantity and free of charge, so there has been no incentive to modify this practice. As drainage systems are inadequate (FAO 1987), over-irrigation is considered to be one of the main causes of an increase in the level of the water table since the introduction of year-round irrigation (El-Tobgy 1974).

This high water table has a direct effect on plants, reducing the oxygen supply to their roots and increasing evaporation from the soil surface causing salinization in some areas; the combined effect is a reduction in crop yield by an estimated 5–10 per cent (FAO 1987).

The people

The peasant farmers of the Delta are known as fellahin. They are the successors of the ancient pharaonic people and have experienced very little racial mixing (Money-Kryle 1957). The normal style of clothing is the *gellabiya*, a loose tunic, and the staple diet is maize bread (although the highly-subsidized wheat flour or pitta bread is now taking its place), together with vegetables such as okra, tomatoes, *molokhiya* (*Corchorus olitorius*), cucumbers, and dairy products.

Some fifty per cent of the total rural and urban population of the Delta is still involved in agriculture. In some areas, shortages of labour are said to occur at peak times, but in general, the potential labour force is under-employed especially in winter, when some of the poorer farmers and farm-workers seek temporary employment outside agriculture, often as building labourers in the towns.

Houses in the Delta region often have two or three rooms on the ground floor, one of which is usually used for the cattle. Families are large and the houses tend to be crowded, but as the sons of the family marry, it is common for another storey to be added to the houses. Fuel consisting of cotton and maize stalks or even dried dung is stored on the roof. Floods, lack of land, and administrative constraints have caused most people to live in large villages and small towns with populations ranging between 1000 and 20 000 but averaging about 3000–4000. Thus, most Delta farmers live in an urban environment and have to travel from their houses in the village to their plots of land every day. Population densities within the villages are often as high as those in Cairo.

With a high population density located in a flat area, Egypt has been able to supply basic services to much of the rural population. All of the larger villages and many of the smaller ones have access via paved roads, and have the basic services including electricity, water supplies, and health services.

Cropping patterns

Knowledge of the main crop rotation patterns is essential as a basis to the understanding of the agricultural systems in the Delta. As irrigation water is available all the year round, continuous cropping is in practice, with an average of two crops per year. There are various rotations but all use winter and summer crops with some *nili* crops. Winter crops cover the season from November to May, the major crops being Egyptian clover, wheat, broad beans, and vegetables; summer crops include cotton, rice, maize, and vegetables and are grown between May and October. *Nili* crops, which are now of much less importance, were formerly the autumn-sown crops following the Nile floods; the term now refers to autumn-sown vegetable and maize crops. The land is not fallowed, and the most common rotation is a three-year cotton rotation, for example:

Year	Winter crop	Summer crop
1	Egyptian clover (1–2 cuts)	Cotton
2	Egyptian clover (3–4 cuts)	Maize or rice
3	Wheat (or barley)	Maize or rice

Rice is the more common summer crop in the northern Delta, maize in the southern Delta. Pulses can replace some of the clover and wheat in winter. Nearer the urban centres, vegetables are more common and can give three or four crops per year; intercropping is common. On the more fertile soils, a two year rotation is sometimes used, for example:

Year	Winter crop	Summer crop
1	Egyptian clover (1–2 cuts)	Cotton
2	Egyptian clover (3–4 cuts) or wheat	Maize or rice

234   THE AGRICULTURE OF EGYPT

The main crop rotations are summarized in Fig 13.2, and the agricultural year is shown to commence in November with the sowing of the winter crops. The main field crops require 4–6 months to harvesting, whereas most vegetable crops require only 3–4 months.

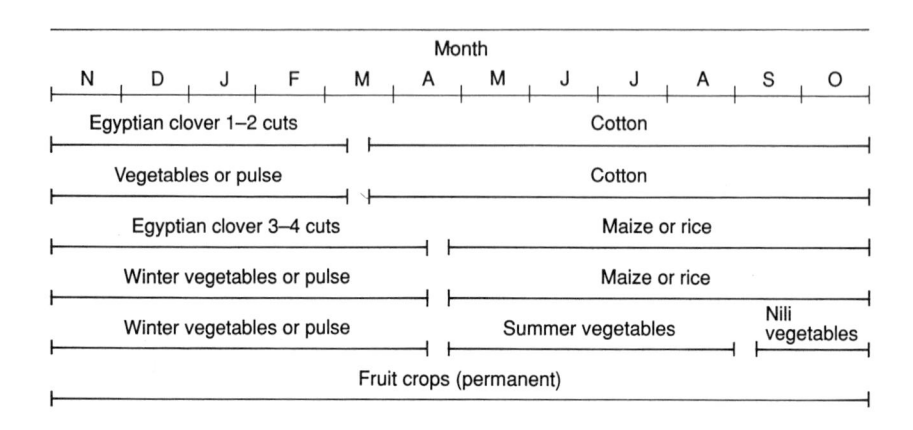

Month											
N	D	J	F	M	A	M	J	J	A	S	O
Egyptian clover 1–2 cuts					Cotton						
Vegetables or pulse					Cotton						
Egyptian clover 3–4 cuts						Maize or rice					
Winter vegetables or pulse						Maize or rice					
Winter vegetables or pulse						Summer vegetables					Nili vegetables
Fruit crops (permanent)											

FIG 13.2.   Pattern of cropping used in the main rotations in the Nile Delta

Government policy

A short summary of the agricultural policy of the Egyptian Government is included here because of its significant influence on Delta agriculture. A more detailed discussion of the government's intervention in Egyptian agriculture is given in Chapter 9.

Agricultural production in the Delta, as elsewhere in Egypt, was affected by changes after the 1952 revolution. During this period the government decided to use agricultural exports as a source of foreign exchange earnings to finance industrial development and military expenditure (Dethier 1989). Output was increased and food prices kept down by subsidies. Price control was brought in for the major farm inputs and products, and production quotas were established to increase exports and minimize imports. Systems were set up for the provision of inputs and the collection of the compulsory crops.

Government companies were set up to produce poultry, meat, and milk under a subsidized price system. From the 1950s to the early 1970s some areas on marginal soils along the western edge of the Delta were brought into production. These areas are now known as the 'old new lands'.

The centralized control of Egyptian agriculture as a whole remained the major influence in Delta agriculture until the late 1980s but since then reforms aimed at liberalizing the agricultural process have been in progress. By 1992, cotton and sugar cane were the only crops still regulated by State control. However, the changes taking place in Egyptian agriculture are ongoing and their desired effects are still to be fully realized. The description of Delta agriculture in this chapter focusses for the most part on the pre-reform period of the late 1980s.

Land tenure

Land reform was carried out and, under the last Act in 1969, the land-holding of an individual was limited to 50 feddans, that of a family to 100 feddans. Landholdings above the maximum were redistributed in quantities of 1–5 feddans to the former tenants and farm workers. The new owners were given full titles to this land when their mortgages were paid off (Saab 1967), and the final title deeds were to be distributed in 1988 (MENA 1987a). This land was not, however, allotted in one place; usually it was split into three different lots, with one lot corresponding to each block of the commonly used three year rotation.

The local inheritance custom of dividing property between all heirs (with females receiving half as much as males) has resulted in further fragmentation. It is estimated that only 0.2 per cent of Delta farmers now have more than 50 feddans while 94 per cent of farmers have less than 5 feddans and 58 per cent have less than one feddan. Average farm size is 2 feddans with an estimated 0.13 feddans of cultivated land per caput. This is decreasing as the population increases and agricultural land in the Delta is lost to urban development.

In an attempt to overcome the problems associated with the fragmentation of landholdings, the government undertook a programme of 'consolidation of land utilization' in the smallholdings not affected by the land reform acts. This amounted to an area of some five-sixths of the cultivated area. Under this scheme, a block system similar to that used in the agrarian reform operates; all holdings in a village are grouped into blocks to allow the three-course rotation (or two-course, where used) to be operated. This scheme was carried out without affecting land ownership with the aim of facilitating mechanized ploughing, cultivation, and spraying, as well as improving water management. It also made it easier for the Egyptian Government to control the areas of crops grown. Unlike their counterparts on the agrarian reform land, farmers, especially those with very small holdings, may have all their land in one block; they have to grow the rotational crops allotted to their area and sell or purchase feed or fodder as necessary.

Co-operatives and credit

In an attempt to avoid the fall in farm productivity found in most countries following land reform, membership of co-operatives was made mandatory for the beneficiaries of the agrarian reform; these organizations were to provide the credit, inputs, and services formerly provided by the original land-owners (El-Tobgy 1974). These co-operatives do not have the same role as in other countries; they are primarily a means of executing government policy rather than meeting the needs of individual producers. Their original primary functions were to ensure that the government had a supply of the major crops, to enforce crop rotations, and to co-ordinate the supply of inputs. The local level co-operatives are under Ministry of Agriculture control via a diverse system of co-operative societies, unions, and supervisory agencies.

Since 1977, the provision of credit has been taken over by the village banks, which are now also responsible for inputs (usually supplied on credit) such as seed, fertilizer, feedstuffs, and pesticides.

Every farmer must have an agricultural holding card by law under the 1966 Agricultural Act. This card should be in the name of the person actually farming the land and so may be the tenant, share-cropper, or head of the family, rather than the actual owner. The card will include the details of the holding, including the area of orchards. It is used to specify information, computed by the local co-operative, on the season and area to plant given crops. Cropping areas are decided centrally each year by the Ministry of Agriculture. The areas are then allotted to the governorates and down via districts to the village co-operatives.

The village bank allocates the relevant (subsidized) inputs on (subsidized) credit in accordance with the Ministry of Agriculture's instructions on levels of usage for a particular crop in a particular area. After harvest, the farmer has to submit specified levels of his output to the Egyptian Government via the co-operative and he risks being fined for non-compliance. A record is maintained of all transactions with the bank and the co-operative for each family unit. Thus the bank is assured of the repayment of its loans and the government is able to enforce its crop quota policy.

The prices of inputs including credit, certified seed, fertilizers, cotton pesticides, concentrates, equipment purchase and hire are all subsidized; diesel fuel is also subsidized on a national basis for all uses. These subsidies are however, less than the government collects by paying low prices for the controlled crops. The levels of inputs, which are often insufficient, are related to the yield required by the government, and this is based at the level of the worst farmers. For this reason a 'free' market also exists for the supply of inputs.

Prior to the current reforms, which are bringing an end to crop procurement policies, the co-operatives bought the entire cotton crop for the State Cotton Organization, a proportion of the wheat and bean crops for the Ministry of Supply, and a proportion of the rice crop. Quotas were in force in the appropriate areas for sesame, groundnuts, and potatoes and some of the horticultural crops for export were also marketed via specialist co-operatives.

Co-operatives have a manager hired by the Ministry of Agriculture, but he has minimum scope to manage and is only allowed to follow Ministry directives (Saab 1967). In theory, the manager is supervized by a board of directors drawn from local farmers, some 80 per cent of whom must have holdings of less than 5 feddans. A combination of illiteracy on the part of the board members and the fact that financial control is ultimately with the Ministry of Agriculture (through the agricultural reform authorities) means that these boards have very little effect.

The majority of the credit given by the village bank is in kind, in the form of inputs for the requisition crops, i.e. working capital. Only 5 per cent of loans are longer term, i.e. for capital items assisting the longer term development of agriculture (MOA et al. 1982).

Pricing

The prices of requisition crops are normally kept well below international levels to maximize foreign currency earnings, keep consumer prices down, and act as an indirect tax on agriculture. For this reason, farmers prefer to grow crops such as vegetables, fruit, and clover which can be sold on the open market whenever possible. The co-operative/village bank control mechanism is very important for the Government to ensure the production of requisition crops.

Despite favourable environmental conditions, low prices and low profitability in most agricultural sectors have meant that farmers have not been encouraged to save and invest in new technology to raise production levels. In the early 1990s a trend in relaxation of price and quota controls has started to occur following pressure from Western countries to boost the Egyptian economy by moving away from control planning to a free market.

Extension and research

The Egyptian Government runs both a centralized research service and an agricultural extension service through the governorates and there are agents at village level. Research is also carried out by the local universities, principally in the form of students' theses, although Shazly (1985) reports a lack of co-ordination between the universities and the research service. In general, the government has given priority to the control of agriculture and infrastructural improvements such as irrigation, drainage, and land reclamation with relative neglect of the advisory and development services (MOA et al. 1982). Evidence of this neglect is provided by the low adoption rates of high-yielding varieties and the large difference between the yields obtained by the average farmer and those obtained on some research stations.

The shortage of finance is not the only factor affecting the advisory and development services; as important are the problems of staff training and motivation. Most of the present staff working in the Ministry of Agriculture organizations are either secondary school leavers or graduates from one of the universities in the Delta area; both groups lack training and experience in farm management and practical agriculture. Practical 'hands-on' ability is vital in small farmer extension schemes and without a good knowledge of farm management, extension workers are unable to convince farmers about the benefits of new farming methods (Adams 1982).

From the time they enter secondary school, even the staff who come from agricultural backgrounds are not expected to take much interest in farming and they rarely do any manual work. By the time they have finished their education, they have little desire to assist farmers in any practical way as it is considered beneath them to do so, even when the rest of the family is involved in practical farming (Adams 1983). The net result is that very little technical advice is conveyed to farmers (Shazly 1985) and the majority of experienced, even if illiterate, Delta farmers have no confidence in their local advisers.

The other factor mentioned, poor staff motivation, is not confined to agriculture, but is common throughout the public sector in Egypt; one government report quoted that the average state employee works for twenty-five minutes per day (MENA 1987b).

Agricultural areas of the Delta

Agriculture in most of the Delta follows a system similar to that outlined above, the main differences being the crops grown in the different areas.

The northern Delta rice zone

The northern Delta rice zone includes the Alexandria, Kafr El Sheik, and Damietta governorates, the northern parts of Beheira, and parts of Gharbiya, Daqahliya, and Sharqiya governorates. This area stretches from Alexandria in the west nearly to Port Said in the east, and covers an area of 1.7 million feddans.

Soils vary from sandy marine soils in the north, which are poor (Classes IV and V), through the central area of brown soils with 60 per cent clay, where some areas are below sea-level, to the southern areas where soils tend to be very dark brown alluvial deposits with a 50–60 per cent clay content. Problems of salinity, alkalinity, and a high water table are common especially in the northern areas. Most of the soils are Class III or IV and are of poorer quality than those in other parts of the Delta. The area is low-lying so the Ministry of Agriculture's efforts to improve the soils have been by sub-soiling and the application of gypsum rather than by drainage. All land grows an average of two crops per year.

As rice is salt-tolerant, some 73 per cent of the nation's rice is grown in this area together with 33 per cent of the cotton and 25 per cent of the wheat. The main winter crops are clover (51 per cent of the area) and wheat; vegetables are also grown, tomatoes being the most important.

In the summer, rice is the major crop but maize, cotton, and horticultural crops such as tomatoes, watermelons, cucumbers, and potatoes are also grown. Autumn (*nili*) crops are not grown very much because the main summer crop is rice, and this is not harvested until October. The major fruit crop is oranges, but other citrus fruits are also grown together with grapes, figs, guavas, pears, and dates.

The peak labour demand occurs in June, while the peak water demand, when water use is at a rate seven times that of the lowest period of the year, is in July (mainly for rice).

The livestock population is relatively low with an estimated 0.75 animal units per feddan; feeding is based on Egyptian clover and rice straw.

The southern Delta

The southern Delta includes the southern parts of the Beheira, Gharbiya, Daqahliya, and the Minufiya and Sharqiya governorates. The area is a continuation of the northern Delta area, but the soils are of a higher quality. The area covers one million feddans with 60 per cent of soils in Classes I and II. Drainage is generally better than in the north and, in much of the area, tiled drains have been installed under the World Bank Nile Drainage Scheme (FAO 1987).

Some 12 per cent of the country's rice is grown here and cotton is also an important crop (17 per cent of the cropped area). Maize is grown and, in winter, Egyptian clover is grown on 60 per cent of the land. Tomatoes are the main vegetable and they account for 30 per cent of the vegetable-growing area. Potatoes, cucumbers, watermelons, and French beans are also grown. Citrus are the main fruit crops, accounting for some 36 000 feddans, but grapes and bananas are also grown. As in most of the Delta, the peak labour demand occurs in June.

Livestock numbers are high, around 1.4 animal units per feddan and, despite the slightly higher clover area, there is a greater forage deficit than in the northern Delta zone.

The Cairo vegetable zone

The Cairo vegetable zone includes parts of Minufiya and Sharqiya governorates. This area is the most fertile in the Delta and is typified by vegetable production for Cairo; some 17 per cent of the area is occupied by vegetables.

Most of the soils in the area are Classes I and II (70 per cent), with only 5 per cent in Class IV; most of the Class III soils are sands and sandy loams and are in the eastern part of the zone. Besides vegetables, the major crops are clover in the winter and maize in the summer; cotton and rice are not so important. Tomatoes are the most important vegetable, but cabbage, squash, lettuce, peas, carrots, potatoes, cucumbers, and runner beans are also grown. There is a large area of fruit grown, mainly citrus, but mangoes, bananas, watermelon, and other melons are also produced.

Virtually all fruit and vegetables are sold through the wholesale markets in Cairo. More labour is required in this agricultural area due to the number of horticultural crops grown; wages tend to be higher due to the proximity to Cairo. The high labour requirement and the location of the Cairo vegetable zone has led to urban development encroaching on to some of the agricultural land.

Livestock production is less important than in the southern Delta as a whole, but the supply of feed is better in early summer because only a small amount of cotton is grown and therefore more of the clover can be kept on for later cuts. Maize leaves and stover are used in the summer and vegetable by-products are also available.

The old-new lands

The western border zone

The western border zone includes the marginal areas of the western part of Beheira governorate and comprises some 153 000 feddans of soils in Classes III and IV. Soils consist of a layer of 0.25–1.00 m over sand and shell fragments and in the south they become more sandy and calcareous. Various crops are grown including cotton, rice, maize, and a small amount of wheat. Fruit and vegetables are also grown and the major crop is watermelon. Only a small area of permanent crops is grown, consisting mainly of citrus and grapes. Cropping tends to be less intensive, averaging 1.8 crops per year.

Individual units are larger, so fewer animals are kept per feddan for transport and subsistence. The supplies of roughage are better than in most areas.

The eastern border zone

The eastern border zone is the eastern part of Sharqiya governorate. It is similar to the western zone, but the soil is of better quality with more alluvial deposits and a dark brown clay. In the south of this zone loam covers sand. There are some saline and barren areas and drainage is often poor. Clover is the main crop in winter, followed by wheat. Rice and maize are the main summer crops, followed by cotton. Some fruit and vegetables are grown, tomatoes being the major vegetable and oranges, dates, and mangoes the main fruit crops. Livestock numbers are high and feed is insufficient even in winter.

Crop production

Crop production plays the dominant part in Delta agricultural systems. Cultivation used to be almost exclusively by draught animals, but from the mid-1970s to the mid-1980s the number of tractors in use in the Delta doubled to some 28 000 units. In 1989, the number had increased to about 35 000 units (estimated from FAO, various years). Undoubtedly, this process was assisted by the remittances of family members working in the Gulf states (MOA *et al.* 1982) and to a lesser extent by the government's mechanization policies.

Most primary cultivation is now carried out mechanically and most of the animal draught activity is expended in lifting water (90 per cent of the animal power used). The normal cultivation implement is the locally made, light chisel plough. The final seed-bed preparation is ridging-up or levelling depending on the crop grown. Levelling is normally carried out with a light, tractor-mounted scraper blade, which has largely replaced the traditional animal-drawn log. Seed rates are generally high, because it is usually broadcast and dressed seed is rare.

Weed control, ditch-cleaning, and irrigation are all usually carried out by hand using the *fas*, a type of short-handled hoe with a large blade. Hoeing to control weeds usually works well for the smallholder on his plot, but problems are encountered on the larger public-sector farms.

The five major Delta crops still predominate, with the summer crops of maize, rice, and cotton grown on 95 per cent of the arable area. In winter the two major crops, Egyptian clover and wheat, account for about 84 per cent of the arable area.

Cotton

Although the cotton crop is now declining in importance, it is still the Delta's most important crop and is by far the major export crop. It is the crop with the largest research input and the most organized marketing system; cotton breeding has been particularly successful.

Egypt only grows the higher quality, long staple varieties of cotton (*Gossypium barbadense*) to meet the higher value market for fine quality cotton. Although it only produces 4 per cent of total world cotton production, Egypt produces 30 per cent of the long and extra-long staple cotton grown in the world. Short staple varieties (*Gossypium hirsutum*) could be grown in the Delta to satisfy the local market. These would have the advantages of a shorter growing season and would allow the grower to take more cuts from the preceding clover crop, but the Egyptian Government fears contamination of the long staple varieties should short staple varieties be grown.

All seed is provided by the Ministry of Agriculture, which supervizes specialist growers and co-operatives, to grow it on contract. Cotton is usually planted in March after short-term clover (85 per cent of the crop) or sometimes after broad beans or other vegetable crops. The land is usually irrigated and ridged before planting. The seeds are sown in the sides of the ridges and triple superphosphate is applied as a fertilizer. After sowing, the area is irrigated with water above the ridges and this is repeated at monthly intervals.

Five weeks after sowing, plants are thinned and nitrogen is applied. Weeding is carried out by *fas* from two to five times as necessary so as to leave the plants on top of the ridge. Plants are checked regularly, often by girls, for the cotton leaf worm (*Spodoptera littoralis*) and the egg masses picked off. Spraying is carried out in cases where there are large concentrations of the leaf worm and also for infestations of bollworms. Other pests encountered are thrips, aphids, and spider mites, especially early in the season. The co-operative staff often only go into the field a few times and these visits are to oversee the cotton pest control activities, for which the co-operatives are responsible by law.

A dressing of nitrogen usually in the form of urea is applied again at flowering. The total application for the crop is 150 kg N ha^{-1}. The first picking is usually in late August or September and the second picking is three weeks later. Cotton picking is usually carried out by women and girls, but supervised by a man. The cotton is spread to dry in the sun. In October and November the leaves are grazed and the stalks are pulled and stored on the roofs of the houses for use as fuel. The main diseases are sore-shin root rot (*Rhizoctonia solani*) and wilt diseases (*Fusarium oxysporium*).

Since 1958, the Egyptian Government has had a single variety zoning policy to maintain purity; each farmer in a given area has to plant a certain variety. The finest quality varieties are usually allocated to the extreme north of the Delta while the other extra-long varieties are allocated to the rest of the northern and central Delta. Long staple varieties are grown in the southern Delta. Only one variety is processed in each gin to maintain purity.

Average yields of seed cotton were 2.47 t ha^{-1} in 1986 (MENA 1987*c*) producing a ginned lint cotton yield of 0.78 t ha^{-1}. Yields reported in 1990 (FAO 1991) were lower (2.30 t ha^{-1}) and estimates for 1991 were 2.11 t ha^{-1} (FAO AGROSTAT 1992)[1]. Lint cotton yields of up to 0.92 t ha^{-1} have been reported in field scale trials (MOA *et al.* 1982), and the percentage of lint is about 30–40 per cent.

The price paid to the farmer is approximately 30 per cent of the world price, so that although it is the country's largest crop in value and despite subsidized inputs, other enterprises, especially those which involve less state intervention, are more valuable to the farmer. Even the Ministry of Agriculture admits that farmers try to avoid growing cotton (Ingram 1983; Financial Times 1992). Furthermore, farmers do not always exert sufficient effort to ensure optimum management of the cotton crop and therefore ensure optimum yields. For example, it has been established that cotton sown in

March produces significantly greater yields than crops sown later because the pest problems are reduced. However the farmer has more to gain by taking a late cut of clover, even if the cut is light, and then accepting the lowered cotton yield.

The whole crop is marketed through the local co-operatives and the State Cotton Organization. Cotton is exported to many countries, with the main markets now in Eastern Block countries, Japan, France, and Italy. Raw lint cotton forms the bulk of the exports, but an increasing proportion is now processed through the expanding domestic textile industry. In addition, cotton is the country's main oil crop; one tonne of cotton seed usually yielding 200 kg of oil. The by-product from the oil extraction process, cotton-seed cake, is an important domestic protein source for concentrate feeds.

Maize

Maize is the main cereal crop grown in the Delta and it is traditionally used by the rural population for bread-making; it is also the only green fodder available in the summer. The increased availability of subsidized wheat flour and bread to the rural population has lessened its importance as a food and has increased its importance as a fodder crop (MOA *et al.* 1982). An estimated 1 714 500 feddans is grown in the Delta, particularly in the central and southern areas, where rice is not grown.

Maize is sown in May and June and farmers usually use the thinnings, and then the lower, middle and upper leaves, and later the tassels as fodder for their animals. The Ministry of Agriculture claims that this reduces grain yields by 30–35 per cent (El-Tobgy 1974), although others claim that the process has been developed by farmers in such a way as to reduce yield only slightly (Adams 1983). In areas near urban centres with a good market for meat and milk, many farmers grow maize solely as a forage crop. Plants are harvested individually at around sixty days. The seed for the next crop is sown between the rows of the previous crop just prior to harvest. This sowing and harvesting sequence is carried out on a continuous basis as soon as each day's requirement is harvested. In this way, after the first cut, there is always a supply of 60-day-old green forage for the animals.

As the crop is a dual-purpose subsistence and forage crop, the authorities' efforts to replace traditional types with high-yielding grain varieties have not been very effective; some of these hybrid varieties are also more susceptible to disease (El-Tobgy 1974). Leaf blotch and maize rust are the main diseases, and stem borers are the main pest, but these are not usually serious unless the crop is planted late. Farmyard manure from cattle and buffalo night yards or dove cotes is often used on the crop. If this is not available, up to 190 kg N ha^{-1} may be applied. Maize yields averaged 5.30 t ha^{-1} in 1990 (FAO 1991) and yields of up to 10 t ha^{-1} have been reported from research stations (Dessouki 1985). Estimates for 1991 were 5.78 t ha^{-1} (FAO AGROSTAT 1992) and a comparison of maize yields in Lower, Middle, and Upper Egypt in 1979 and 1989 is shown in Table 14.3.

All maize is consumed locally and, in the last 10 years between one and two million tonnes have been imported annually, much of it for incorporation in animal feeds (Parker 1992, personal communication). Many farmers use the grain for family consumption and animal feed and sell any surplus in the local village market.

Rice

Rice is the other major summer crop and it is grown in the northern Delta; the area increased greatly after the completion of the Aswan High Dam and 1 029 000 feddans were planted in 1989 (FAO 1990). The local consumer preference is for short-grain varieties and so these are the main varieties sown.

The Egyptian Government aims to supply one-third of the seed requirements with certified seed each year. Rice is sown in May or June, usually in nurseries in parts of clover fields; later it has to be transplanted. This process offers the advantage of extended use of the remaining clover in addition to saving water and increasing rice yields.

Rice is sown in the soil with a 5 cm covering of water. When the seedlings are about 5 cm high, the field is drained and the depth of the water increased as the plants grow. Each season, an average of 31 000 m^3 of water is used per feddan. During the growing season, an irrigation system of four days wet/four days dry is guaranteed for rice grown in the northern and central Delta rice blocks, where 96 per cent of the crop is grown. A small amount of rice is grown in the southern Delta, where the government-supplied water has to be supplemented by water pumped from private wells. During the last ten years the International Rice Research Institute has given assistance with the breeding and agronomy of the rice crop.

Harvesting is usually carried out by hand in October; combine harvesters are available in some areas, but the cost (£E200 feddan^{-1} in Beheira governorate in 1987) is considered too high by most farmers. Average paddy yields were 6.29 t ha^{-1} in 1990 (FAO 1991), with yields of up to 8.5 t ha^{-1} reported in trials (MOA et al. 1982), and estimates for 1991 at 7.28 t ha^{-1} (FAO AGROSTAT 1992). A comparison of rice yields in the Delta and other parts of Egypt is shown in Table 14.3.

In 1987, farmers in the Beheira governorate were required to sell 1.5 t of each feddan grown to the Egyptian Government. The price was £E240 t^{-1}. Yields were on average 2.5 t feddan^{-1} in the area, so an average of 1 t feddan^{-1} was available for the farmer to sell on the open market at £E550–600 t^{-1}.

Rice has become one of the basic Egyptian foods, so most of the crop is consumed locally with the surplus exported. In 1987, 160 000 t milled rice were exported whereas the total crop yield was 2.14 million t milled rice (3.2 million t paddy) (Parker 1992, personal communication). The general trend is a decrease in exports as internal consumption rises. In 1973, the production of milled rice was 1.5 million t (2.3 million t paddy), of which 298 000 t were exported. Rice exports are arranged by the State, mainly through the Nile Company for Crop Exports (MENA 1987e) to East European countries. Rice straw is used mainly for animal feed and bedding; a small amount is used in paper making.

Egyptian clover

Egyptian clover or *bersim* (*Trifolium alexandrinum*) is traditionally the mainstay of the crop rotation and the major feed crop for animal production. It is the most extensively grown of all the summer and winter crops in the Delta and has tended to increase in recent years reflecting the greater profitability of livestock production to the farmer. It is a vigorous true clover which is resistant to alkaline soils, is very palatable, and contains 22 per cent crude protein (Gohl 1981).

Clover is planted from September to November as soon as the preceding crop (usually cotton, rice, or maize) is off the land. Often there is little or no cultivation, the seed rate is usually 20–25 kg feddan^{-1} and some phosphate fertilizer may be used. At least four cuts can be taken between sowing and the following May ('full season' clover), but 85 per cent of the Delta cotton crop is sown after ('short season') clover in March; this gives only one or two cuts.

Clover is usually cut by hand and fed to cattle and buffalo. The first cut is taken at 60 days, the second at 45 days, and the third and later cuts at 30–45 days. Yields of fresh matter per cut vary between 5 and 8 t feddan^{-1} with an average of 6.5 t feddan^{-1}. Despite the importance of the crop, improved varieties are not available. The other major constraint on yield is over-irrigation.

Irrigation of Egyptian clover is prohibited after 10 May as it acts as a winter host for a major cotton pest, the cotton leaf worm; in any case, growth is limited by high temperatures at this time of year. Some 10–15 per cent of the crop is left for seed, which is harvested in May and June.

The abundance of fodder in the winter and the dearth in the summer months would seem to favour conservation, but very little hay is made because Egyptian clover is difficult to dry; the stems have a high water content, and the leaves tend to drop off (Gohl 1981).

Wheat

Some 60 per cent of Egypt's home-produced wheat is grown in the Delta, and the crop is grown following cotton, maize, or vegetables. An area of some 1.4 million feddans was grown in the Delta in 1992 (Parker 1992, personal communication). Before sowing in November, the land is ploughed and formed into small basins, which may be sown and then irrigated, or irrigated, ploughed again, and then sown. The sowing rate is usually high, averaging 60 kg feddan^{-1}. Fertilizer is used and rusts are the main disease problem. Wheat is harvested in May or June, usually by hand, and it is threshed by belt-driven threshers; the latter have largely replaced the traditional wooden sled or discs.

Long-strawed varieties of wheat are preferred because the straw is of more value than the grain. Straw is used as an animal feed in the summer and in brick-making. Until 1977, the farmer was obliged to sell the whole of his crop to the Egyptian Government, but this has now been reduced to a proportion of his crop (300 kg feddan^{-1} for wheat grown in Beheira governorate in 1987). Wheat is used in bread-making and in animal feed. The grain not sold to the Government is usually retained for family consumption and milled locally in the village. Average yields improved from 3.80 t ha^{-1} in 1986 (FAO 1989) to 5.21 t ha^{-1} in 1990 (FAO 1991) and an estimated 4.8 t ha^{-1} in 1991 (FAO AGROSTAT 1992). Yields of 5.5 t ha^{-1} have been reported (MOA et al. 1982), and yield data for Lower and Upper Egypt are compared in Table 14.3.

Although 72 per cent of Egypt's wheat (World Bank 1987) is imported, the area of wheat grown in the Delta decreased during the mid-1980s. The farmer avoided growing wheat if possible because the return per feddan was only £E170 compared with £E550–600 for Egyptian clover (McDougal 1987). Wheat is available cheaply on the world market and it is often subsidized under aid schemes. It is therefore advantageous to grow higher value crops such as vegetables and cotton (wheat is harvested too late to be followed by cotton) and sell them on world markets to buy wheat.

Other summer crops

Groundnuts

Some 28 560 feddans of groundnuts were grown in Egypt in 1990 (FAO 1991). In the Delta they are grown in Isma'iliya and Sharqiya governorates, where the crop is well adapted to the lighter, sandy soils. Groundnuts are planted in April and harvested in October. They are sold fresh, about 75 per cent being sold locally and 25 per cent exported. The average yield of unshelled nuts in 1990 was 3.00 t ha^{-1} (FAO 1991). The crop used to be government-controlled with a quota supplied to the co-operative set annually. Now it is among the crops freed from forced delivery in the Government's agricultural reform programme.

Sesame

A small area of sesame is grown in Isma'iliya and Sharqiya governorates. The crop is planted in April or May and harvested in September. Most of the crop is consumed fresh and the average yield of sesame seed is 0.93 t ha^{-1} (FAO 1991).

Summer forage crops

Very small areas of summer forages such as sorghum, Sudan grass and their hybrids are grown, usually on state farms or some private farms in the marginal areas.

Soyabeans

Soyabeans were first introduced in the 1970s and are grown in Kafr el Sheik to supply an oil processing plant. It is one of the few crops where the Egyptian Government has offered an incentive to growers by setting the price at above world levels. The return to the farmer was 30 per cent greater than for cotton in 1983 (Adams 1983) and farmers responded well. Soyabeans are planted in April and harvested in August. In 1990, the average yield was 2.14 t ha^{-1} (FAO 1991) and the estimated yield for 1991 is 2.55 t ha^{-1} (FAO AGROSTAT 1992). After oil extraction for local consumption, the soyabean meal is used mainly in poultry feeds.

Other winter crops

Barley

Barley is sometimes grown instead of wheat on the poorer soils, usually near the Mediterranean coast. Six-row varieties are grown for animal feed and two-row varieties are grown for malting. The average grain yield is 2.42 t ha^{-1} in 1990 (FAO 1991), with yields of up to 4.06 t ha^{-1} recorded in trials (MOA *et al.* 1982). The straw is used as an animal feed.

Flax

In early times, flax was the main fibre crop in Egypt. It is an early-maturing winter crop sown in November and harvested in March. It is a dual-purpose crop grown for both fibre and oil and it is grown by specialized growers on contract to retting and processing companies. FAO data for the whole of Egypt record an area of some 33 000–40 000 feddans in 1990 and El-Tobgy (1974) has recorded fibre yields of 5.97 t ha^{-1}. The fibre produced supplies the requirements of a domestic linen industry and the oil production meets about 50 per cent of the linseed oil demand for products including paint.

Sugar beet

A sugar beet factory was built in Kafr el Sheik in the mid-1980s and 40 000 feddans of beet were first grown in the 1987–8 season (MENA 1987*f*). A similar area continues to be grown yielding 41 t ha^{-1} in 1990 (FAO 1991) and an estimated 53.4 t ha^{-1} for 1991 (FAO AGROSTAT 1992). Sugar beet production is to help meet the local demand for sugar, which is rising as the population increases. In 1987, Egyptian sugar consumption was 30 kg caput^{-1} (MENA 1988a).

Vegetable crops

The climate of the Delta permits the cultivation of a wide range of temperate and tropical vegetables and the proximity of markets in the cities of Cairo and Alexandria (and more recently Europe and Arab countries), together with the availability of cheap labour for such labour-intensive crops, has led to a large increase in vegetable production. Indeed, while the area devoted to vegetables has doubled over the past twenty-five years or so, the government reported that vegetable production increased by 249 per cent between 1981 and 1987 (MENA 1987d), most of this being due both to the larger areas grown and the increased yields in the Delta.

The main vegetable-growing areas are close to the big cities of Cairo and Alexandria. Vegetables are grown mainly for domestic consumption, but a small and increasing proportion are exported, especially potatoes, but also tomatoes, garlic, broad beans (*Vicia faba*), and green peppers.

In most cases the seasons are not so well defined as in the case of other crops and there is considerable seasonal overlapping; exceptions which do fit into the common rotations are broad beans, lentils, and chickpeas in the winter, and onions, which are grown in the summer in the Delta.

Tomatoes

Tomatoes are the most important vegetable crop grown in the Delta; indeed, following the five major Delta crops, the tomato is the next most important crop both in terms of area grown and total value of the crop (El-Tobgy 1974). Two-thirds of the national crop is grown in the Delta amounting to an estimated 277 400 feddans in 1990 (FAO 1991), approximately one-third of the total vegetable-growing area in the Delta.

Tomatoes are grown all the year round, but yields are higher in the summer and autumn than for the winter crops. Average yields in all seasons were 24.6 t ha^{-1} in 1990 (FAO 1991). The varieties grown have now been introduced from abroad and have been found to be superior to the traditional varieties. Most of the crop is marketed fresh, although some is made into purée; both form important components of the local diet.

Potatoes

The main potato-producing area is in Beheira governorate with sizeable quantities also grown in Minufiya and Gharbiya. It is the second most important vegetable crop after the tomato and the most important vegetable export crop in the Delta.

Main crop potatoes are usually planted in January; this is brought forward to December or even late November when the crop is to be grown for export. An autumn crop is also planted from August to October for harvest in December and January for local consumption. The seed potatoes for the main crop are imported from Europe and locally produced seed potatoes from the previous main crop are used for the autumn crop. The main crop is higher yielding, partly owing to the superior seed tubers (El-Tobgy 1974).

Potatoes are usually planted after the land has been irrigated, ploughed, and levelled. The seed tubers are planted in furrows about 65 cm apart. Farmyard manure or artificial fertilizer is always used and growers are very aware of the response of the crop to increasing fertilizer application rates (up to 300 kg N and 150 kg P_2O_5 feddan^{-1}. The average yield in 1990 was 20.95 t ha^{-1} (FAO 1991).

The traditional export market was in early potatoes for Britain. These could be produced before competitors' crops were harvested. Now the market has been expanded to include exports of both early and main crop potatoes to Arab countries (Ingram 1983). The main organization involved in exports is the public sector Nile Company for Crop Exports, but some private companies are becoming involved in the trade. Potato exports amounted to 130 000 t in 1984 (Barkovky 1985).

Broad beans (Vicia faba)

Broad beans, a traditional crop, are often grown in the winter within the normal rotation. Approximately one-third of the national crop is grown in the Delta amounting to some 122 800 feddans in 1989 (FAO 1991). The largest areas of beans are grown in Beheira, Kafr el Sheik, and Sharqiya (El-Tobgy 1974). The crop is usually planted in November and harvested five months later in April. The broad beans grown are mainly local cultivars and their major disease problems are chocolate spot (*Botritis fabae*) and rust (*Uromyces fabae*).

In 1990 yields of dried beans averaged 2.64 t ha^{-1} (FAO 1991) and similar yields (2.5 t ha^{-1}) are estimated for 1991 (FAO AGROSTAT 1992). Yields of broad beans in Lower, Middle, and Upper Egypt in 1979 and 1989 are shown in Table 14.3 The crop used to be subject to the government quota system in the main growing areas. The beans are usually harvested dry, and they are a traditional breakfast food and an important source of protein, especially for the rural and poorer urban people in the Delta.

Onions

In the Delta, onions are grown mainly in the summer for local consumption. This is in contrast to Middle and Upper Egypt where they are grown in the winter for export. Onions are grown as a single crop or, more usually, especially in the southern Delta, they are intercropped with cotton (El-Tobgy 1974), using the higher value onion crop to increase the returns to the farmer from the low-value cotton crop. Yields of intercropped and single crop onions have been recorded as 3.5 and 14.6 t ha^{-1}, respectively, (El-Tobgy 1974).

A comparison of yields in Lower, Middle, and Upper Egypt is given in Table 14.3.

Onion seed, usually a traditional local cultivar, can be planted in September and October and sometimes transplanted in December, although it is usually planted in February and March and harvested in June, July, and August. Diseases are a problem, thought to be exacerbated by year-round irrigation.

Lentils

Lentils, traditionally grown in Upper Egypt, are relatively new to the Delta, where some 2500 feddans are now grown as a winter crop. Yields are shown in Table 14.3.

Other vegetables

Several other vegetable crops are grown primarily for domestic consumption. These include turnip, courgette, okra, *molokhiya*, radish, parsley, and leek, which are grown during most of the year; cabbage, cauliflower, pea, French beans, lettuce, beetroot, carrots, and celery, which are grown mainly in the autumn and winter; and cucumber, aubergine, and green pepper, which are only grown in the summer. Garlic, chickpeas, spinach, broccoli, and pumpkin are also grown.

Fruit crops

As with vegetables, fruit production tends to be concentrated near the centres of population, especially in Beheira, Qalyubiya, Sharqiya, Minufiya, and Alexandria. The most important fruit crops are oranges, watermelons, and dates, although many other crops are also grown. Permanent fruit-crop growers tend to be the larger farmers or town dwellers, who are not dependent on subsistence crops and are able to afford the costs of establishment. Fruit production is one of the expanding Delta agricultural enterprises and the Egyptian Government claimed a 251 per cent increase in fruit production between 1981 and 1987 (MENA 1987*d*), with three-quarters of Egypt's fruit produced in the Delta.

Citrus

Oranges are by far the most important fruit crop in the Delta and are the most important Delta export crop after cotton. Oranges represent some 80 per cent of the citrus area and production. Two other citrus crops, mandarins and limes are also of some importance. Smaller quantities of lemons, grapefruit, and sour oranges are also grown.

The main citrus growing areas are in Beheira and Qalyubiya, followed by Sharqiya and Minufiya, where 60 per cent of Egypt's citrus crops are grown. The most serious citrus disease is gummosis (*Phytophthora citropthora*) and the most serious pests are fruit flies and scale insects (El-Tobgy 1974).

Trees usually receive farmyard manure after harvest. Productivity is not as high as in other citrus-growing countries and there is scope for better management and, particularly, improved pruning (Barkovky 1985). The quality of the orange crop in terms of flavour, colour and juiciness is generally good, but variable.

Usually some 20 per cent of the crop is exported, formerly only to Eastern Block countries, but now also to Arab countries. Several packing factories have been established, especially by the El Wadi Government Company to enable export. As internal consumption rises, a lower proportion of the crop is available for export.

Watermelons and melons

Watermelons and melons are grown in the summer. Beheira is the leading growing area followed by Sharqiya, Kafr el Sheik, and Isma'iliya. Some breeding work has been carried out and small quantities are exported to neighbouring countries. In 1990 average yields of watermelon and other melon including cantaloupes were 16.42 and 23.50 t ha^{-1}, respectively (FAO 1991). Estimates for 1992 were similar.

Dates

Dates are grown throughout the Delta, normally scattered between blocks of land rather than in plantations. The date palm is tolerant of high concentrations of salt in the soil and so there are some plantations near the Mediterranean coast, where no alternative crop can be grown. Traditional varieties account for most of the crop and propagation is carried out using the offshoots from productive trees. Khamedj, or inflorescence rot caused by the fungus *Mauginiella scaettae*, and date palm leaf spot *Graphiola phoenicis* are the main diseases. Most of the crop is consumed fresh locally, and the yield per tree is estimated at 70 kg. Total production is declining owing to poor drainage and very high salinity (Barkovky 1985).

Mangoes

The mango has become an important crop. The main producing areas are in the southern Delta, although Beheira is also important. Mangoes are usually intercropped with another crop such as orange or custard-apple early in life as they take 12–15 years to reach maximum production. Mangoes are harvested between July and October and several different cultivars, ranging from yellow to green in colour when mature, are grown. Virtually all the crop is consumed locally, although there are some exports. The most serious disease is powdery mildew.

Grapes

The leading grape-growing areas are in Beheira, Minufiya, and Daqahliya. Grapes are pruned in January and February and new cuttings are planted in

February and March. The two main grapes grown are Thompson Seedless (*banati*) and *roumi ahmer* (Barkovky 1985). Powdery and downy mildew are the main disease problems and yields averaged 11.6 t ha^{-1} in 1990 (FAO 1991). FAO estimates for 1991 were 14.2 t ha^{-1} (FAO AGROSTAT 1992). Most of the crop is consumed locally as fresh grapes, although some of the crop is dried and some is made into wine.

Other fruit crops

Bananas are grown mainly in the southern Delta in Qalyubiya and Minufiya. Most of the crop is a small type, *Musa cavendish* (known as *baladi* or *hindi*), the flavour of which is preferred locally. Some larger varieties of bananas are grown on a small scale, but production in general is declining.

The guava is traditionally one of the principal fruits eaten by the poor and is important for this reason.

Pears are one of the few temperate fruit crops grown in quantity, but they produce rather a hard fruit. There are several orchards on the western edge of the Delta and, in 1991, annual production in Egypt was estimated to be 76 000 t (FAO AGROSTAT 1992). The mild winters have precluded growing the temperate tree crops such as apples, which require a cold winter period. During the last few years, however, 7000 feddans of apple orchards adapted to local conditions have been established and over the last decade annual production in Egypt has doubled to an estimated 65 000 t (FAO AGROSTAT 1992).

Peaches, apricots, and nectarines are grown in small orchards in the western Delta and production increased from 11 000 t in 1985 (FAO 1986) to 35 000 t in 1989 (FAO 1990) Custard-apples, pomegranites, plums, and cherries are also grown, and some figs and olives are grown in the marginal areas. Generally pome and stone fruits are inferior in quality, but producers benefit from high returns as they are in high demand. Some strawberries are grown and, although the area is small, their high value and export potential give them some importance.

Marketing of fruit and vegetables

Most fruit and vegetables go through the very congested wholesale markets in Alexandria and Cairo. In some cases produce goes direct to processing factories or specialized government or private outlets. Therefore most smaller farmers sell to a local trader who will sell via these markets. The standard crates used are made from palm-rib. Post harvest losses of up to 35–40 per cent are claimed for delicate crops due to inefficient handling, packing, and storage operations (MOA *et al.* 1982). The latest government figure for wastage is 17 per cent for fruit and vegetables (MENA 1988*b*).

Animal production

Traditionally, animal production has been an integral part of the mixed farming system in which arable farming predominated; animals were generally regarded as necessary accessories to crop production. More recently however, animal production has begun to develop in its own right, especially the modern poultry industry and cattle and buffalo production.

Cattle and buffalo

General characteristics

Cattle and buffalo are the principal livestock in the Delta. Their numbers amounted to 1 401 000 and 1 384 000, respectively, in 1986 (MOA, personal communication) and they contributed 80–90 per cent of local milk and meat production (Shazly 1985). The present-day local Baladi cattle found in the Delta are considered to be stabilized crossbreds between Shorthorn and Zebu cattle. The crossbreeding may have occurred in Egypt itself, or more likely, in Syria (Williamson and Payne 1978).

Baladi cattle are of medium size and fine-bodied with long legs. The short glossy coat is normally reddish in colour, but can vary from very light to a dark brown. The head is long and narrow with small horns and virtually no hump. The dewlap is noticeable and cows typically have long narrow teats extending from a small to medium size udder.

The buffalo is thought to have been introduced to the Delta from Asia in the middle of the seventh century (Itriby 1974). It has a bigger frame than the local cattle (averaging 600 kg as opposed to 350 kg), with stronger legs and larger hooves. The horns are much larger and broader than those of cattle. At birth buffaloes have a good coat of soft hair, but this becomes sparser as the buffalo grows, and the body of some housed buffalo appears to be nearly bare. Normal coat colour is dark brown, although variation from dun to light grey to black occurs. The local type found in the Delta is the Beheiri, which is a better milk producer than the type found further south in Egypt.

The role of cattle and buffalo in the Delta

Since around 300 BC, animals have been used to drive the *saqqiya* (Egyptian water wheels) which lift irrigation water. Until recently, animals were also the principal source of non-human power in farming; they were used for cultivation, threshing, and transport, and farming families used the milk for home consumption.

Animals were also important for their contribution to soil fertility. Their manure was invaluable, but also significant was the leguminous Egyptian clover, the principal fodder crop grown to support them. In combination with the annual flooding of the Nile, animal manure maintained the fertility of the Delta soils and thus supported the high rural population density.

During the last twenty-five years, however, developments have occurred which have affected this system. The increased availability of fertilizers has meant that Egyptian clover is not so necessary as a restorative crop within the rotation. In addition, the need for animals for cultivation has diminished as a result of the great increase in the use of tractors.

One factor remaining constant is the low rainfall; all forms of agriculture are dependent on irrigation. Land is either capable of supporting arable cropping or it remains desert; there are virtually no marginal lands which are only suitable for grazing. Thus keeping cattle and buffalo is largely in direct competition with crop production. Indeed, given that livestock are no longer so necessary for crop production and that Egypt's net agricultural trade balance is in deficit by about $US3000million (World Bank 1987) largely due to the import of basic foodstuffs, it might be supposed that the role of livestock would be diminishing and that animal production would become increasingly based on crop residues and by-products.

The opposite has occurred; both the numbers of animals and the areas sown to Egyptian clover have increased during recent years. This is because the average Delta farmer with his small landholding is very dependent on his animals as a source of milk, which is the major source of animal protein in his family's diet. Moreover, the increased demand for beef from the urban population has made calf sales more valuable. Therefore, cattle and buffalo are still very important to the Delta farmer, even though the reasons for keeping them have altered. Now he keeps them primarily for milk production, secondly for draught purposes (mainly for lifting water), and thirdly for calf sales (Hopkins 1980).

The Egyptian Government's policy of less stringent control over the farm-gate prices for milk and meat compared with major crops has favoured cattle and buffalo production. Although Egypt imports some 40 per cent of its meat (World Bank 1987), it is claimed that controls preventing the import of more frozen meat at a price lower than that of producing meat in the Delta, also favours local cattle and buffalo production (McDougal 1987).

General management

Cattle and buffalo are considered together because they are managed similarly by the farmer. Most comments on management relate to smallholder production (large-scale cattle and buffalo production is discussed in a later section). Ninety per cent of these smallholders' animals are small herds of one or two animals kept by the farmer to provide milk for consumption. There are some large herds on State farms and some recently established private enterprises in the marginal border zones close to large urban centres.

Cattle and buffalo are kept in the traditional manner. From November to May, they are fed on Egyptian clover. Animals are usually tethered at the side of the field, where they are fed clover which has been cut and carried to the animals to avoid poaching and destruction of the sward; they are taken back to the house at night. Fattening animals are generally kept inside all the time.

In summer, when most of the land is in arable crops, the animals have to survive on various crop by-products, which are mainly home-produced or purchased locally. Some maize grain and beans may be fed to working animals. In the summer, the animals may be left near the house all day or they may be taken to the fields during the daytime as in winter. Cattle and buffalo are usually taken to drink frequently. The buffalo are allowed to wallow, especially in the summer as they are less able to withstand heat than cattle. These tasks, along with feeding and watching the animals, are usually carried out by children, especially girls, because fewer girls than boys attend school in rural areas.

The women of the family are generally in control of the milk and dairy products. Liquid milk is not generally consumed and most of the milk is turned into dairy products for home consumption; together with pulses these form an important protein source for the farming family. The cream is turned into *samna* (ghee), and the skimmed milk is used to make the local salty cheese; fermented milk is also made from whole milk.

Custom dictates that selling surplus milk is a sign of a family's poverty and low status in the community. If surplus milk is to be sold, it should be as dairy products rather than as liquid milk because, traditionally, the latter is a sign of laziness on the part of the family's womenfolk (Saab 1967). Buffalo milk is preferred for the family's own consumption because its higher solids content makes it more suitable for the preparation of milk products.

Few problems are encountered with milk let-down and calves are usually sold to a dealer for slaughter or further fattening at a young age. If the calves were retained longer, the feed available, especially in the summer months, would be insufficient for both dam and calf, resulting in a decrease in the milk available for family consumption.

Calves remain with their dams for the first three or four days to ensure they receive colostrum. After this, concentrates or forages may be offered depending on their availability on the farm. Calves continue to receive some milk, but the amount of suckling is restricted more and more until the time of disposal. Early weaning takes place on some larger units, but only recently have efforts been made to introduce it to small producers.

Calf losses are as high as 20–30 per cent up to three months of age (Itriby 1974), although losses are fewer among small farmers, who generally give their animals better individual attention. Nevertheless, the threat of losing a calf, which is a significant proportion of the assets of a small farmer, further encourages an early sale. Replacement adult animals are purchased from a dealer as required.

Owing to the importance of cattle and buffalo in providing the family with food, a farmer too poor to afford the price of an animal will often form a partnership with a richer relative. It is usually preferable to keep these agreements within the family, but if this is not possible, one of the richer farmers or merchants in the village would be approached. The latter would purchase the animal, the farmer would be responsible for feeding it and would consume the milk produced. The value of the calf would be shared between them.

Feeding

During the period from November to May when Egyptian clover is plentiful, cattle and buffalo are generally well fed. For the remainder of the year, however, they have to survive on a diet of crop residues and by-products. Small quantities of maize leaves are usually the only source of green fodder. Other feeds such as straw, maize, stover, bran, some green parts of the rice plant, weeds, and cotton-seed cake are fed, but the feed supply is insufficient and because animals have to utilize their body reserves, large weight losses are very common.

Studies of the nutritional status of livestock show that on a year-round basis, both energy and protein are in deficit compared with the estimated theoretical requirements (Hathout 1986); on a seasonal basis, energy supply is always insufficient (90 per cent of the requirement in winter and 39 per cent in the summer). Protein is in deficit in summer (36 per cent of requirement), but is actually in excess in the winter because of the high content of clover in the diet.

Feed supply is the main limiting factor for cattle and buffalo production in the Delta; the feed available is mainly used to satisfy animals' maintenance requirements and consequently production levels are low. Resolving this problem is not straightforward, because an increase in the quantity of summer forage grown would mean a reduction in the food and cash crops grown. A reduction in animal numbers, so that less feed is used for maintenance and more for production, would be difficult as the majority of animals are in herds of one or two cows.

Increasing the feed supply with concentrates is another possibility and this has been taken up by the Egyptian Government, which supplies subsidized concentrate feeds. Cotton-seed cake produced in Egypt is the main protein source, and imported maize is the main energy source. There is insufficient subsidized concentrate to meet the demand and quotas are in force. The Government favours larger producers, who receive 81 per cent of the total concentrate allocation (Creek *et al.* 1983).

Theoretically, the larger producers should be more production-orientated; the smaller producers are more likely to use concentrates to maintain greater numbers of low-producing animals.

A further possibility is to make better use of crop by-products. The only crop residue available in a large quantity is rice straw which is palatable, but contains only 3.3 per cent crude protein and 5.4 MJ metabolizable energy per kg (Kearl *et al.* 1979). Its nutritive value is not usually sufficient to cover even the maintenance requirement of cattle (Gohl 1981). Work by FAO showed that the tough and fibrous nature of rice straw made it very difficult to chop and there was no effective and economic method for grinding it (Creek *et al.* 1982). Upgrading of straw by ammonia treatment was uneconomic at the farm level (Creek *et al.* 1983).

The most practical method of increasing the use of rice straw for animal feed would be to adapt the feeding systems presently used so as to use more rice straw in winter while conserving more clover for use in summer and so leading to more balanced diets on a year-round basis. This would require resources and some effort in applied research, demonstration, and extension before farmers would accept such innovations.

Field observations show that buffalo can be more easily maintained on poor quality fodders than cattle, and they seem to be more efficient at digesting fibre than cattle, although experimental results are conflicting and all comparative results do not sustain this view (Williamson and Payne 1978).

Other by-products, such as maize stover, could be used more effectively and fruit and vegetable processing wastes or even water hyacinth (a major weed on canals), but the total quantities are small.

Fertility

Fertility levels are generally low. Buffalo cows usually calve for the first time at over three years of age and have a calving interval of more than 540 days. They have the disadvantage of a gestation period of 316 days (Sidky 1953) compared with 280 days in the cow. Oestrus tends to be short and irregular. Cattle tend to have slightly better indices and the calving interval averages 490 days.

The main cause of the poor fertility is undoubtedly the low level of nutrition, especially in the summer. Studies have shown that the lowest conception rates are in July, August, and September, when animals are under most nutritional stress (Itriby 1974). In the Alexandria University dairy farm, with more adequate summer feeding, a calving interval of 379 days has been achieved.

The small size of the herds also indirectly affects fertility levels; there is a lack of bulls as very few farmers can afford to keep them. There are also the normal problems of heat detection associated with tethered animals and small herd size.

Artificial insemination is used by some smallholders, but to date the service is rather fragmented and not well organized. Several larger units have their own inseminators. The lack of a milk recording service means that there are no fully proven local bulls. Buffalo semen is more difficult to freeze than cattle semen and the artificial insemination techniques used in cattle have to be modified.

Milk production

The local cattle have been selected for a combination of draught performance and ability to withstand long periods on low nutritional planes, so both milk and meat production are generally low. The average lactation yield was 674 kg cow^{-1} in 1990 (FAO 1991)[2] excluding the milk taken by the calf (usually for the first 60–80 days). Cattle are generally kept for six to eight lactations with the highest yields in the fifth lactation (Shazly 1985).

Selection and breeding within a population with such low milk production levels would be laborious and so Friesian cattle and semen have been imported to increase milk production. In the northern Delta, especially, pure Friesian cattle have been very successful for specialized milk production. Average herd lactation yields of 4500 kg cow^{-1} have been achieved on the Alexandria University dairy farm, which is run on commercial lines, and average yields of 3770 kg have been reported on a smallholder scheme (Ilaco/Euroconsult 1987). Friesian cattle now represent 10 per cent of the cattle population and are found almost totally on state and some large producers' farms (Shazly 1985). For the average smallholder with limited financial and feed resources, the Baladi is still the main breed, although a minority of producers have Friesian x Baladi or even pure Friesian cattle.

Although cattle are important in the large milk-producing herds, the buffalo is still responsible for 65 per cent of milk production (Shazly 1985). Buffalo generally produce greater milk yields than the indigenous cattle under smallholder conditions; the lactation yield averages 1300 kg per animal when the stocking rate is about 0.75 animals per feddan in winter. Average lactation length is around 325 days followed by an average dry period of 215 days. Herds kept on higher planes of nutrition have averaged up to 3600 kg per lactation. Milk quality is high with butterfat usually ranging from 6 to 8 per cent (El-Tobgy 1974).

With a higher base yield and no readily available sources of high quality stock, genetic improvement is largely based on the breeding and selection in the Ministry of Agriculture herds.

Although, traditionally, smallholders sold very little surplus milk in liquid form, some small farmers close to the urban centres have become specialist milk producers. These producers normally use buffaloes and utilize maize for summer forage; winter milk production is based on Egyptian clover. Many can now obtain subsidized concentrates through their local co-operative. They supply the milk to private sector dealers, who in turn distribute it via street vendors on bicycles.

Some milk is sold to the public-sector milk factories. Producers selling milk receive a lower price than selling to a dealer, but are entitled to a quota of subsidized concentrate according to the quantity of milk sold to the factory. Most of the milk sold in this way comes from cattle on State farms and the large private-sector farms. There are a few schemes for milk collection from smallholders close to the factories, but their production is a very small proportion of the total handled by the milk factories and the majority do not sell through the government system.

Meat production

As stated earlier, most male calves are disposed of by the breeder at an early age. Buffalo calves with a birth weight of 40 kg are usually sold off for veal (*bitello*) after 40–60 days and at a liveweight of approximately 80 kg (Shazly 1985). The main reason is to save milk, but the farmer gets a good price due to the demand for veal. The meat of male buffaloes older than two years tends to be tough and adult male buffalo have a poor temperament. Cow calves are usually kept slightly longer because of their lower birth weights (24–27 kg) and slightly better growth rates.

A minority of calves are bought by dealers and beef producers to rear on to higher weights. For buffalo this is usually up to 18 months of age, at which stage deterioration of meat quality occurs.

The authorities have taken measures to encourage fatteners to retain cattle for beef. As calf mortality is a problem, the Egyptian Government has set up a livestock insurance scheme jointly administered by the veterinary and extension departments of the Ministry of Agriculture. Subsidized concentrates are also available for producers with more than five animals, who are obliged to have insurance, veterinary inspection, and to deliver the meat to the Government when requested. Although the meat price is 33 per cent lower than the market price, the subsidized concentrates means that it is more profitable than using the open market. As the supply of subsidized concentrates is limited, large producers (public sector, governorate 'Food security projects', and private producers) tend to use concentrates and sell to the public sector while the smaller beef producers tend to fatten their animals on Egyptian clover and sell to the private sector.

Another important source of meat is from the culled breeding females. Hides of calves are used for shoes, handbags, and suede leather articles, while adult hides are used for belts and heavy leather items.

Large-scale cattle and buffalo production units

Since the 1970s, the Egyptian Government has attempted to increase milk and meat production by stimulating and subsidizing large-scale mechanized companies specializing in livestock production. Such companies receive priority in subsidized concentrate supplies and, unlike most smallholders, grow summer forages such as sorghum x Sudan grass hybrids.

Saab (1967) noted that on larger units, 'Buildings were lavish, but not well adapted . . . and that animals were kept in dirty, wet conditions with little if any bedding and insufficient green fodder.' Unfortunately the same conditions prevail today on many public-sector farms, with only a few notable exceptions. Losses of animals are much higher (FAO 1977) despite the preponderance of veterinarians on these units, and is largely due to poor husbandry, especially lack of hygiene, lack of individual attention to animals, and poor feeding.

Sheep and goats

Sheep and goats are of less importance in the Delta. Beef is the meat of choice in Egypt and mutton is not popular in the urban centres. The main demand is seasonal for sacrificial slaughtering in the Eid El Adha Moslem religious feast.

The sheep and goats that are present tend to be in nomadic flocks surviving on crop residues only. Some improved feeding may occur prior to the feasts. The local sheep breeds belong to the fat-tailed coarse wool type, the two present in the Delta being the Ossimi and the Rahmani.

Donkeys and camels

Donkeys and camels are the animals used for transport. Because most farmers live in villages and therefore have to travel relatively long distances to reach their scattered plots, the donkey remains important for transport of the farmer, his family, and his tools.

The Delta camel population is diminishing with the spread of improved roads and the use of pick-ups. Nowadays they are used mainly for the transport of crops between field and village. The local type is the Egyptian Delta camel (Williamson and Payne 1978).

Poultry production

Traditionally most of the poultry production has been carried out by small farmers, who keep flocks of 10–50 free-range chickens, mainly of native breeds, especially the Fayoumi (Shazly 1985). This traditional sector produces enough to provide the rural areas and small towns with eggs and poultry meat.

Since the 1960s, the Egyptian Government has encouraged the setting up of modern intensive units in the public sector and also, more recently, in the private sector, to meet the demands of the urban population. It was also hoped to substitute poultry meat for some of the beef in the urban population's diet (El-Tobgy 1974).

The reasoning behind the emphasis on modern units is that virtually all the feed has to be imported and the modern environmentally controlled units convert feed to meat and eggs twice as efficiently as the local flocks (Shazly 1985). Thus in contrast with other types of livestock production, several very modern, vertically integrated poultry enterprises are found in the Delta. These units use imported breeds and depend on imported parent stock, vaccines, and medicines, as well as imported feeds. Both public and private sector intensive units depend on government assistance in the form of subsidized inputs, tax relief, and a protected market.

Since the late 1980s, however, with the relaxation of government controls on agriculture, there has been a reduction in subsidies on imported maize used for poultry feed. Poultry production showed a dramatic decrease owing to increased cereal prices but is now increasing following increased domestic cereal production.

In 1980, the village sector still produced 83 per cent of the eggs and 50 per cent of the poultry meat, but as the modern intensive sector increased its output, village production of eggs and meat dropped to 60 per cent and 37 per cent, respectively (Shazly 1985). The country as a whole is now self-sufficient in egg production (MENA 1987g), but still produces only half of its poultry meat demand (MENA 1987h).

Turkeys, ducks, and geese are also kept, with the number of ducks in particular rising during the last few years. Pigeon meat is eaten on special occasions and pigeon lofts are specially constructed.

Other animals

Some pigs are kept but are of minor importance because of the predominantly Moslem population. Some rabbits are kept for meat. Developments in fish farming are dealt with in Chapter 12.

Animal health

The Ministry of Agriculture runs a veterinary service which is responsible for vaccination against the major infectious diseases and for providing free veterinary advice to livestock farmers. Most larger villages have veterinarians, and farmers in the Delta tend to use them more than in other areas (Hopkins 1980).

The main diseases vaccinated against are rinderpest, haemorrhagic septicaemia, foot-and-mouth disease, blackquarter, and anthrax. There is a slaughter and compensation scheme for tuberculosis and brucellosis, but funding is insufficient and, in practice, it is only carried out on government farms. Other problems are sheep pox, some tick-borne diseases, parasites, and conditions relating to fertility. Cattle are also sprayed for external parasites. Poultry diseases are very common especially in the intensive units, where Newcastle disease, fowl pox, fowl cholera, and fowl plague occur.

Besides his veterinary duties, the local veterinarian is involved in the cattle insurance scheme, the distribution of subsidized feedstuffs, and sometimes artificial insemination. As the local agricultural advisers are only concerned with crops, the local veterinarian is accustomed to being the only authority on livestock husbandry as well as on veterinary matters in the locality. Indeed in the Beheira, it was reported that an attempt was made, as part of a local rural development project, to bring in specialist staff to carry out non-veterinary work such as artificial insemination and advice on animal husbandry methods. Strong resistance was encountered from the local veterinarians, and this was cited as the main factor limiting livestock development in the area (Fawzi, personal communication).

Future

Despite the high productivity of the Nile Delta region, there is scope for considerable yield increases in the major crops grown. It has been suggested that this could be achieved by improving the agricultural advisory services in the Delta and that such an investment would increase food production in excess of the gains realized from investing similar amounts in reclamation to bring land into production (MOA *et al.* 1982); indeed Commander (1989) believed that the net marginal impact of new land reclamation on agricultural output would remain limited at least over the short and medium term.

The Egyptian Government, however, taking a longer term view has re-emphasized its aim to continue desert reclamation (MENA 1987*i*) and this is one of the areas targetted in the current government reform programme. As government resources are finite, this would imply that no sizeable switching of funds into research and extension for the Delta region is likely to occur in the near future; the amount of water supplied by the Nile may, however, result in a re-consideration of these policies.

Drought in the Nile source regions is causing a drop in the river's rate of flow and the level in Lake Nasser has fallen sharply in recent years (Anon 1988). It is therefore difficult to see how the desert reclamation programme can continue, when Egypt is already exceeding its agreement with the Sudan to take an annual maximum of 55.5 milliards (1 milliard = 1000 million m^3) of water from the Nile. There are worries that the drying out may not just be part of the normal pattern of drought and floods, but may be caused by long-term changes in climate and rainfall patterns.

Thus it would seem that increasing the productivity of Delta agriculture would have to be a priority in order for Egypt to increase its food production. If the Egyptian Government proceeds with its desert reclamation programme, and in 1992 it continues to be a policy priority, the move of agricultural policies away from 'control' towards incentives and advice would provide opportunities for government staff to be switched away from controlling what the farmer grows to advising him on the best crops to grow. Since the late 1980s the relaxation of government controls on crop production and marketing have begun, partly as a result of pressure from Western aid-donors, but also as part of an overall domestic liberalization programme and an escalating food import bill.

Notes

[1] FAO crop yield data, obtained from *FAO production yearbook* (1986–91). or, in 1992, directly from the AGROSTAT agricultural data system produced by the FAO Statistics Division, refer to the whole of Egypt, not the Delta specifically.
[2] FAO milk yield data are for the whole of Egypt, not specifically the Delta.

References

Adams, M. (1982). *Agricultural extension in developing countries*. Longman. Harlow.
Adams, R. (1983). *Development and structural change in rural Egypt*. Johns Hopkins University, New York.
Anon. (1988). The Nile. *The Economist*, **306**, (7539), 80–1.
Barkovky, H. (1985). *Research in horticultural crops. A review of the organisation of agriculture and fisheries research technology in Egypt. Annex I*. FAO, Rome.
Commander, S. (1989) Some issues in agricultural sector policy. In *Egypt under Mubarak*, SOAS Middle East Centre Studies, (ed. C. Trip and R. Owen), p.137. Routledge, London.
Creek, M., Barker, T., and Hargus, G. (1983). *An evaluation of the use of anhydrous ammonia to process wheat straw*. UNDP/FAO Beef Industry Development Project Field Document No. 8. FAO, Cairo.
Creek, M. et al. (1982). *Preliminary findings on the use of anhydrous ammonia to process rice straw*. UNDP/FAO Beef Industry Development Project Field Document No. 6. FAO, Cairo.
Dessouki, E. (1985). *Research in crop production. A review of the organisation of agriculture and fisheries research technology. Annex I*. FAO, Rome.
Dethier, J. (1989). *Trade, exchange rates and agricultural pricing policy in Egypt*. World Bank, Washington, DC.
El-Tobgy, H. (1974). *Contemporary Egyptian agriculture*. Ford Foundation, Beirut.
FAO (Food and Agriculture Organisation) (1977). *Water buffalo*. FAO, Rome.
FAO (1987). *Nile Delta drainage project*. Completion report. No. 9/87CP-EGY40CR. FAO/World Bank Co-operative Programme Investment Centre, Washington, DC.
FAO (1988, 1989, 1990, 1991). *FAO production yearbook*. FAO, Rome.
Financial Times (1992). Financial Times Survey: Egypt. January 21, 1992.
Gohl, B. (1981). *Tropical feeds*. FAO, Rome.
Hathout, M. (1986). Feed resources for livestock in Egypt. In *Proceedings of the 2nd Egyptian-British conference on animal and poultry production*. University College of North Wales, Bangor.
Hopkins (1980). *Economic management in a period of transition*. World Bank, Washington, DC.
Ikram, K. (1980). *Animal husbandry and the economy in two Egyptian villages*. American University in Cairo, Cairo.
Ilaco/Euroconsult (1987). *Damietta dairy programme*. Ilaco/Euroconsult, Arnheim.
Ingram, S. (1983). A growing potential for agricultural exports. *Cairo Today*, **4** (5), 20–3.
Itriby, E. (1974). The buffaloes of the Near East. In *Health and husbandry of the domestic buffalo*, (ed. Rees), pp. 651–61, FAO, Rome.
Kearl, L., Harris, L., Farid, M., and Wadeh, M. (1979). *Arab and Middle East tables*

of feed composition. International Feedstuffs Institute Research Report No. 30, International Feedstuffs Institute, Utah State University, Logan and Arab Centre for Studies of Arab Agriculture, Damascus.

McDougal, M. (1987). Desert or Delta? *Cairo Today* **8** (6), 24.

MENA (Middle East News Agency) (1987*a*) Development of agrarian reform authority. *The Egyptian Gazette*, **33 282**, 2.

MENA (1987*b*). From the Cairo press. *The Egyptian Gazette*, **33 239**, 3.

MENA (1987*c*). 750,000 feddans to be reclaimed. *The Egyptian Gazette*, **33 274**, 2.

MENA (1987*d*). Significant increase in farm output since 1981. *The Egyptian Gazette*, **33 264**, 3.

MENA (1987*e*). Crops worth £E8m to be exported. *The Egyptian Gazette*, **33 304**, 2.

MENA (1987*f*). Move to increase sugar production. *The Egyptian Gazette*, **33 286**, 2.

MENA (1987*g*). IBRD loan for small farmers. *The Egyptian Gazette*, **33 277**, 3.

MENA (1987*h*). Poultry industry. *The Egyptian Gazette*, **33 316**, 2.

MENA (1987*i*). Top priority to desert land reclamation. *The Egyptian Gazette*, **33 337**, 2.

MENA (1988*a*). Sugar. *The Egyptian Gazette*, **33 403**, 2.

MENA, (1988*b*). Crops. *The Egyptian Gazette*, **33 366**, 2.

MOA (Egyptian Ministry of Agriculture), USAID (United States Agency for International Development), and USDA (United States Department of Agriculture) (1982). *Strategies for accelerating agricultural development in Egypt.* USDA, Washington, DC.

Money-Kryle, A. (1957). *Agricultural development and research in Egypt.* Publication No. 3. American University, Beirut.

Saab, G. (1967). *The Egyptian agrarian reform 1952–62.* Oxford University Press, London.

Shazly, K. (1985). *Research in animal production. A review of the organisation of agriculture and fisheries research technology in Egypt.* Annex V. FAO, Rome.

Sidky, A. (1953). *The buffalo of Egypt.* Egyptian Ministry of Agriculture, Cairo.

Williamson, G. and Payne, W. (1978). *Animal husbandry in the tropics* (3rd edn). Longman, London.

World Bank (1987). *Report on Egypt 4136 EGT 1987.* World Bank, Washington, DC.

14

Systems of agricultural production in Middle and Upper Egypt

A. A. BESHAI

Introduction

It may be that Egypt is now seen as a country which is open to the rest of the world. Its geopolitical situation is well recognized, and Egypt's role with respect to other Middle Eastern countries and African countries is as significant as her links with Europe. In the last two decades, out-migration of workers with its economic and social implications has become a noticeable feature of the country. Indeed, in recent years, remittances from Egyptians working abroad have often been the largest item of credit on the capital account of the balance of payments. Most of the migrant workers are from Middle and Upper Egypt, defined as the areas covered by the governorates of Giza, Beni Suef, Faiyum, and Minya (Middle Egypt) Asyut, Sohag, Qena, and Aswan (Upper Egypt) (Fig. 14.1)

However, it should be noted that, historically speaking, Egypt is a country with natural barriers which have resulted in the comparative isolation of the country's inhabitants throughout history. These natural barriers are the sea in the north and the deserts to the east and west. To the south, passage by way of the Nile River is impeded by a series of cataracts.

This geographic isolation is to a large extent responsible for the characteristic conservatism of the Egyptian peasants; even today, one sees peasants in Upper Egypt using ancient methods of cultivation as well as ancient implements. The wealth and prosperity of Egypt has been largely dependent on the produce of the soil and this is particularly true of Upper Egypt. The peasants are first and foremost agriculturalists, as is evident from the Arabic word for peasant: *fellah*, which means one who digs or tills the soil.

Whilst the methods of cultivation are essentially primitive, the system of storing water and distributing it has been adopted relatively recently and has undergone several changes. The first dam constructed on the Nile at Aswan in Upper Egypt was completed in 1902 and was raised twice. Work commenced on the High Dam, also at Aswan, in January 1960 and was finally completed in September 1967. However, alongside the sophisticated system

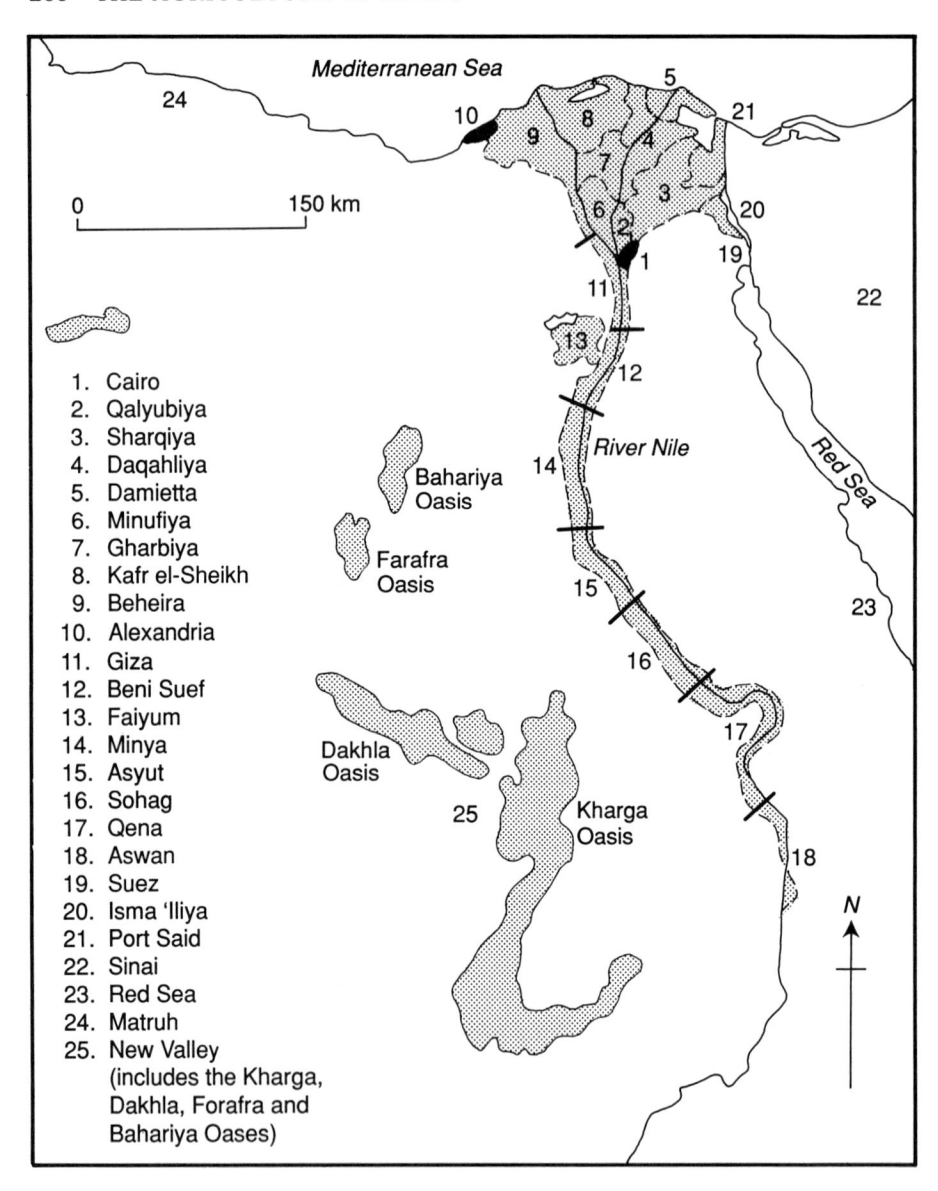

1. Cairo
2. Qalyubiya
3. Sharqiya
4. Daqahliya
5. Damietta
6. Minufiya
7. Gharbiya
8. Kafr el-Sheikh
9. Beheira
10. Alexandria
11. Giza
12. Beni Suef
13. Faiyum
14. Minya
15. Asyut
16. Sohag
17. Qena
18. Aswan
19. Suez
20. Isma 'Iliya
21. Port Said
22. Sinai
23. Red Sea
24. Matruh
25. New Valley (includes the Kharga, Dakhla, Forafra and Bahariya Oases)

FIG 14.1. Egypt: the governorates. Source: Abdallah (1965).

of water canalization, one still sees in Upper Egypt the very old methods of watering, namely the *shaduf* (a water-hoist), the water-wheel, and the Archimedean screw.

The cultivated lands lie along each of the banks of the river Nile; there is no 'delta' in Upper Egypt. Furthermore, the extent of the land that lies on either bank of the Nile fluctuates widely from one governorate to another. For example, cultivated land in Minya governorate is around 530 000 feddans while at the other extreme, Aswan governorate has only 128 000 feddans (Institute of Land and Water Research 1991, personal communication). There are no apparent boundaries to the cultivated land; there are no hedges to divide the fields. The fields are separated from each other only by ridges of earth and narrow trenches. As such, the very fragmented landholdings are not visible to the viewer. The implications of this fragmentation are important, and are examined in subsequent sections of this chapter. The plough, driven by oxen, is used because it is believed that it digs well into the ground. In the poorer parts of Upper Egypt, it is drawn by an incongruous couple—a camel and a donkey.

Ethnic identity of Upper Egyptians, the household, and division of labour

It is widely believed that the Upper Egyptians come closest to their ancestors, the Pharoahs; in Northern Egypt, Turks and other nationalities have intermingled with the inhabitants throughout history. However, the racial purity of the Upper Egyptians has not been maintained entirely through time; it has been affected by descendants of negroid elements who have settled and intermarried. This accounts for the presence of black or half-black people. The strong contrasts in the physical features of the country (green versus desert) are reflected in the psychology of the inhabitants, collectively and individually (Blackman 1927).

The household in Upper Egypt is not a subsistence type unit, self sufficient in providing its own labour and consuming its own produce. It has become 'a petty commodity-producing unit and the landless household is a provider of wage labour' (Hopkins 1987).

Men work outside the household as farmers, government employees, traders, or day labourers, and in farming households the men have the hardest and most strenuous work in the fields. The head of the household takes all the necessary decisions throughout the cropping cycle, sees to it that labour is hired, inputs are available on time and jobs are assigned (Hopkins 1987).

Women have an important role in the rural sector throughout Egypt, but in Middle and Upper Egypt it is rare to see women working in the fields. They manage the household and are responsible for animal care. Young girls work in the field up to the age of twelve or thirteen years. Their activities include cotton pest control, gleaning, weeding for lentils and chickpeas, and gathering dry roots and green fodder for the animals

(Hopkins 1987). However, with the massive spread of education, peasant children in Upper Egypt often go to school rather than work in the fields. There are usually primary schools in each village and secondary schools in the district.

In the home, the woman's activities include cooking, baking, looking after children, washing, and filling water containers. In the Upper Egyptian village, the household is a production unit and of specific importance is the woman's role in the preparation of white cheese, clarified butter, and poultry raising (Hopkins 1987).

At harvest time, all available labour is employed in the fields. The threshing of maize is done with a sledge-like machine, called a *norag*, drawn by oxen. Given the intense work at harvest time (and there is more than one harvest in a year), the disguised unemployment often claimed to occur in the countryside is a misnomer. There is seasonal unemployment, but it can be stated safely that, over the year, the marginal product of the country people is above zero. The high rate of unemployment in Egypt is essentially unemployment of the educated.

Population, land area, and land quality

According to the 1986 Census (CAPMAS 1987), some 35.5 per cent of the total population of Egypt, reported then as 48.5 million[1], are to be found in Middle and Upper Egypt. The distribution by governorate is shown in Table 14.1.

TABLE 14.1. The percentage distribution of population by governorate in Middle and Upper Egypt.

Governorate	%
Giza	7.7
Beni Suef	3.0
Faiyum	3.2
Minya	5.5
Asyut	4.6
Sohag	5.1
Qena	4.7
Aswan	1.7
Total	35.5

Source: CAPMAS (1987).

It may be of interest to relate the population to the cultivated land. For the eight governorates of Middle and Upper Egypt the cultivated land amounts to 1.8 million feddans, of a total of 7.2 million feddans. Thus, 35.5 per cent of the population is living on 25 per cent of the agricultural land and it should be noted that the land is of varying quality. The Institute for Land and Water Research divides land in Egypt into four main categories together with two other very low-quality land categories. Upper and Middle Egypt have rather high quality agricultural land as shown in Table 14.2. It is immediately apparent that Upper Egypt has a larger share of the first category land and a far bigger share of the second category land compared with lower Egypt.

TABLE 14.2. Land distribution by quality in Upper/Middle and Lower Egypt

Region	Land category (%)			
	1	2	3	4
Upper/Middle Egypt	6.0	59	20	6
Lower Egypt	3.4	12	31	8

However, land quality is not the only determinant of productivity. Productivity depends on a host of factors including seed quality, methods of cultivation, drainage, and weather conditions. For these reasons, variations in productivity are to be expected. Table 14.3 shows yields per feddan for different crops grown in Lower Egypt, Middle Egypt, and Upper Egypt, together with the average for the nation as a whole. The data also show the change in yield for each major crop over the last decade.

Wheat, broad beans, lentils, and onions are winter crops; the rest are summer crops. Yields of wheat for the country as a whole have increased over the last ten years. The main reason is the use of high quality seed and the same trend is true for maize. Beans seem to fare better in Middle and Upper Egypt than in Lower Egypt. Traditionally, Middle and Upper Egypt were the primary lentil-growing areas, but recently the use of improved seed varieties has succeeded in Lower Egypt. Onions always grew better in Middle and Upper Egypt because they require higher temperatures towards the end of the growing season. Traditionally, rice has been grown in Lower Egypt. In Middle Egypt it does not do so well for reasons of temperature and relatively lower water availability. Attempts to grow rice in Upper Egypt have not been successful. Sugar cane fares better in Middle and Upper Egypt because it requires high temperatures. In the

TABLE 14.3. Average yields of major crops grown in different regions of Egypt, 1979 and 1989 (t feddan^{-1})

	Wheat		Beans		Lentils		Onions		Maize		Rice		Sugar cane	
	1979	1989	1979	1989	1979	1989	1979	1989	1979	1989	1979	1989	1979	1989
Lower Egypt	1.38	2.09	0.79	1.12	0.37	0.88	5.1	8.0	1.74	2.41	2.4	2.7	32.3	35.2
Middle Egypt	1.40	2.21	0.91	1.27	0.85	0.80	6.6	9.9	1.55	2.55	2.3	2.5	38.9	40.3
Upper Egypt	1.19	1.98	1.18	1.38	0.42	0.64	10.1	11.4	1.65	2.46	NA	1.1	34.8	40.9
All Egypt	1.34	2.09	0.93	1.24	0.42	0.80	7.7	9.4	1.68	2.45	2.4	2.7	35.4	40.6

NA, not available.
Source: Compiled and computed from unpublished data, Economics Section, Ministry of Agriculture (1991).

case of all crops, there has been a considerable improvement in yield from 1979 to 1989.

The dynamics of agricultural production

Historical note

In order to trace the dynamics of agricultural production in Upper Egypt through time, it is important to refer to some historical facts. Before the time of Mohammad Ali in the early nineteenth century, Egypt used flood or basin irrigation. The Nile flooded in July, August, and September and with the low level of water in October and November, autumn-sown crops such as barley and clover were grown. No crops such as rice and maize were planted after the autumn-sown crops. Following the 1866 American Civil War and the rise in the price of cotton, attention was given to irrigation, and 30 000 water wheels were installed. Maize was also introduced early in this century. In Upper Egypt, sorghum was grown because its transpiration rate is very low and it stands heat well. After the erection of the Aswan Dam, summer crops increased gradually in Upper Egypt, but this was not true for cotton. In Upper and Middle Egypt the cotton area has continued to decline over the last three decades. Yields of cotton increased but the fixed price system did not make it sufficiently remunerative for farmers and they switched to vegetables and fruits as well as clover. Furthermore, with the open-door policy since 1975, there was a boom in poultry and animal production and this encouraged the farmer to grow sorghum and clover (Hattab 1992, personal communication).

Land tenure

The land tenure system in Upper Egypt is characterized by fragmentation of holdings. In Asyut governorate, land holdings of 5 feddans or less accounted for 94 per cent of the total land in 1987. In Beni-Suef, the respective percentage was 93.3 per cent; in Qena it was 93 per cent. Indeed, in Qena landholdings of one feddan or less accounted for 46 per cent of the total holdings.

The problems attendant upon the fragmentation of land holdings are many and include:

(1) Loss of land, owing to excessive delineation between holdings as well as innumerable irrigation and drainage canals;

(2) Reduction in crop yields due to differential treatment of pest control and irrigation among holdings;

(3) Reduction in crop yields due to the absence of economies of scale;

(4) Wasteful and inefficient use of water.

The importance of sugar cane in Upper and Middle Egypt

Sugar cane production assumes special significance in the agricultural production systems of Upper and Middle Egypt. Indeed, in Upper Egypt, the major crop is sugar cane.

When sugar cane was introduced in Upper Egypt in the 1860s, it was akin to cotton which was introduced in Lower Egypt in the 1820s. Indeed, sugar cane has often been referred to as the 'cotton of the South', or the 'cotton of Upper Egypt'. In spite of this, there has been a difference. For decades, all cotton was exported as raw cotton, whereas sugar cane was, of necessity, processed internally. Sugar became the major industrial output of Egypt.

Not only has sugar cane been analogous to cotton in terms of area and quantity of production, it also suffered the consequences of severe fluctuations in the world market. Whilst cotton growing in Egypt benefitted enormously from the American Civil War, the area of sugar cane cultivated rose to 90 000 feddans at the time of the Cuban War in 1898–9. Later, it fell to about 40 000 feddans in 1907–8. In the second half of the twentieth century, the area under sugar cane rose from around 100 000 feddans in the 1950s to around 175 000 feddans in the early 1970s, reaching some 250 000 feddans in the 1980s. The area under sugar cane cultivation by governorate is given for 1989 in Table 14.4.

In Lower Egypt, about 9000 feddans are under sugar cane. This amounts to some 3.4 per cent of the sugar cane area in Upper Egypt. Within Upper Egypt, by far the biggest producers are Qena and Aswan. This is understandable, because sugar cane thrives best in very hot climates and in Upper Egypt the temperature is 8 per cent higher on average than in Lower Egypt.

The preponderance of sugar cane in Qena and Aswan has meant that sugar cane dominates the crop rotation there. Sugar cane remains in the ground for four to five years and, in the past, this used to be ten years. Through experience the fellah learned that it was uneconomical to leave sugar cane for more than four to five years, because yields were reduced significantly after that time.

The Egyptian Ministry of Agriculture has established the Council of Sugar Products which takes £E1 per tonne of sugar cane produced by the farmer (it used to be 50 piastres). This money is put in a fund for sugar cane development and extension services.

Sugar cane, not unlike cotton, is considered a strategic crop. The recent liberalization policies of the Ministry of Agriculture have freed the majority of crops from forced delivery and/or fixed prices, but sugar cane and cotton

TABLE 14.4. Area of sugar cane cultivation by governorate in Upper and Middle Egypt, 1989

Governorate	Area ('000 feddans)
Giza	1.5
Beni Suef	0.8
Faiyum	0.2
Minya	30.2
Asyut	1.8
Sohag	15.2
Qena	157.7
Aswan	57.1
Total	264.5

Source: Compiled from unpublished data supplied by the Ministry of Agriculture, Egypt (1992).

still remain under prices fixed by the Egyptian Government. Although it is likely that cotton will be liberalized in the foreseeable future this is not likely to be the case for sugar cane. However, the Government's interference in sugar cane production and pricing should not be construed to mean that this will be adverse to producers.

Over the years, the returns from sugar cane production have not always been as remunerative as the fellah would have wished and, being a rational economic creature, he has tried to raise his income by concentrating on remunerative cash crops such as wheat, maize, sorghum, and *bersim* (*Trifolium alexandrinum*). Some vegetables and fruits are also grown in sugar cane areas. Qena, for example, has become a major producer of tomatoes and in the last ten years has been the source of tomatoes for Cairo in the winter months. Bananas have also gained in importance in the last seven years. For each of these two commodities, the area grown has reached 30 000 feddans in Upper Egypt and the two crops are the high-income source for many farmers. In 1991, when the output of tomatoes in other parts of Egypt was reduced and prices rose, it was estimated that in Upper Egypt the return per feddan reached £E5000 for tomatoes.

Apart from the fixing of the price of sugar cane, another problem for the grower is that the crop stays on the land for five years or so. In this respect, it is very unlike cotton or vegetables where the switch can be made every year or half-year. Indeed, sugar cane is a classic example of the type of crop favoured by the big landlord—or absentee landlord—because it does not need daily attention. Moreover, sugar cane is a good crop for the farmer who does not need his produce for his own subsistence, for example wheat or maize, or *bersim* required for the fellah's water-buffalo. The country squire who grew sugar cane could be an urban dweller and the crop presented no problem if the farmer were a big landowner. For the small farmer, however, and it has been shown that landholdings are particularly small as a result of the land reform law and the system of inheritance, sugar cane presents a problem. If a farmer has only one feddan and it is under sugar cane, he cannot grow wheat, maize, or any other crop such as *bersim* for his animals unless he hires additional land. This is often what he does. Moreover, if a farmer owns one feddan or less and his neighbours are growing sugar cane, he has no choice but to do so himself because of the shade factor. This accounts for the conglomerations of sugar cane production despite the fragmentation of landholdings.

To alleviate the problems of the farmer, the Council of Sugar Products utilizes the proceeds of its fund to attempt to raise the productivity of land under sugar cane. This is done by enabling the farmers to use sophisticated 150 horse-power tractors and sub-soilers to plough the land intensively after it has been under sugar cane production for a number of years. Land levelling is carried out using laser equipment and this can be done only in an area of not less than five feddans.

Added to this, the government has recently introduced a procedure for standardizing the age of sugar cane crops growing on specific plots of land. If the sugar cane on adjacent plots of land varies in age the plots are all cut at the end of the year and the land is levelled and planted uniformly with plants of the same age. This results in economies or cost reductions in terms of land preparation and irrigation.

In the last few years, there has been no specific law enforcing farmers to grow sugar cane. The recent increases in yields have helped to promote the crop and the remuneration from sugar cane to the farmers has been better than hitherto. The sugar cane grower now grows the crop by tradition and because he is affected by the factors that indirectly compel him to grow it. In a sense, farmers in Qena and Aswan do not have much choice. Each year the Ministry of Agriculture issues a directive about the crop rotation and this is usually followed. Given the climatic conditions and the preponderance of sugar cane, together with the other factors apertaining, the best they can do is to join the group in growing sugar cane. Growing wheat or maize in these areas is not that good in terms of yields. However, there are always farmers who appeal to the government to be exempted from growing sugar cane on the grounds that the land is not in good shape, for example in terms of drainage. In these cases the government would usually agree to their request and undertake to improve the land.

In terms of net returns, sugar cane comes only seventh or eighth after other rotations such as wheat/maize; beans/maize; rice/maize; and wheat/rice. The returns from sugar cane production are not always as high as farmers would wish but plans for freeing prices from central control are being considered and in 1992 the delivery prices for sugar cane have been raised. Furthermore, the world market in sugar is very volatile and no producer could withstand the caprice of price behaviour. As such it is not clear whether liberalizing the price of sugar, as has been done with other crops, would have a significant effect on production. Of late, the government seems to be considering the price behaviour of other agricultural products to ensure a fairly remunerative return to the sugar cane producer. Finally, the producer has the power to decline from growing the crop, for the law would subject him only to a fine of £E30 for doing so (Affifi 1992, personal communication).

Another factor for consideration is the very large water usage requirement of sugar cane. Sugar cane and rice together consume some 40 per cent of Egypt's share of the 55.5 milliards (1 milliard = 1000 million m^3) of Nile water. Sugar cane alone consumes some 15 per cent of the total. At a time when the availability of fresh water is beginning to pose a problem, and is being discussed at national and international levels, the Egyptian Government has introduced sugar beet. It succeeded well in northern Egypt and experiments at growing it have been carried out in Faiyum, Minya, Asyut, and as far south as Sohag. Early indications point to success with sugar beet and if the crop offers real potential, ways will be sought to drastically reduce the area under sugar cane cultivation, a practice which is an entrenched tradition. The Government could levy a 'sugar cane tax' which would be a masked tax on the use of water. The future seems to point in that direction but it remains to be seen how successful such a tax might be.

Causes and effects of the low standard of living in Upper Egypt

A significant characteristic of Upper Egypt, and to a lesser extent Middle Egypt, is out-migration. This out-migration can be directed to Cairo or outside Egypt; Egyptian migrant workers who work in other Arab countries come mostly from Upper Egypt.

The production of sugar cane, the major crop in Upper Egypt, does not need the constant presence of farmers. Planting takes place in March and harvesting is from mid-December to mid-April. For this reason farmers migrate to Lower Egypt each year for the four months between September and December. Their wives are left in charge of the farming activities but these are not at their peak. Some migrants work as ambulant petty traders in Cairo, while others work in cotton ginning factories as porters, foremen, and loaders. Some families eventually move and settle in Lower Egypt.

Poverty is not simply a matter of land, since some landless families are fairly prosperous (Hopkins 1987). Often wage employment outside the village forms the highest source of income. In Egypt the distinction between town and country is blurred because the rural habitat is not dispersed with isolated homes but concentrated in densely populated settlements. In two villages in Qena (Tafnis and Higaza), Radwan and Lee (1986) found that 23 per cent of the total income came from outside the village source.

In Middle and Upper Egypt wages are lower than in the big cities. Furthermore, agricultural wages are lower than non-agricultural wages. In Minya it has been estimated that the agricultural wage is only 68 per cent of the non-agricultural wage whereas in Egypt as a whole it is 75 per cent. In Minya, the average agricultural wage is 38 per cent less than the national average agricultural wage and the further south, the larger the discrepancy (Hani Naguib Elias 1989).

Non-farming jobs are principally of two types: work linked with agriculture such as camel drivers, winnowers, and pump guards; or work in the tertiary sector, for example, tailoring, masonary, weaving, pottery, trade in grains and livestock. Upper Egypt is important for pottery making and basket weaving.

The outlet for urban work is limited because Middle and Upper Egypt have had no more than 20 per cent of total industrial investment in the last 30 years. From north to south the type of industrial development changes. Minya and Asyut have a few industries based on agriculture, specifically on cotton, grains, and some sugar cane. These include cotton ginning, spinning and weaving, and flour-making. Recently, there have been agro-industries for vegetables, onions, and garlic.

In the zone stretching from Qena to Idfu, agriculture has made its mark on the type of industries. The predominant industries are based on sugar cane, and also on cotton. As a by-product of sugar production, paper manufacturing has been introduced at Idfu.

In the far south, in Kom Ombo and Aswan, paper and timber manufacture have been introduced, in addition to the well-established industry based on sugar production.

There is an interesting analogy between the development of agriculture and the development of manufacturing industry in Egypt. Agriculture in Egypt developed sequentially from north to south—from the Delta to Middle and Upper Egypt. The irrigation networks and industry follow the same pattern of development. Industrialization began in northern Egypt and moved slowly southwards, with the lion's share still remaining in the north. Indeed, some would argue that because of the agricultural output-mix of the region, industrial development took place in Upper Egypt almost against the will of the planners. No wonder, therefore, that Upper Egypt has earned the reputation of the 'dark south' or the 'oppressed south' (Hamdan 1982). The increasing population of Cairo is the result of both the national increase in population and the rural–urban migration. In Cairo, one person is born every minute and two arrive by train, primarily from Upper Egypt. It is the author's conviction that the only way out of the impasse in Upper Egypt is the development of the village as an agro-industrial complex. The success of

Egypt's small-scale industries which now account for over 40 per cent of the industrial labour force, could be capitalized on by encouraging such industries in rural Egypt. Apart from raising the living standards in these areas, there would be a deceleration in rural–urban migration.

Notes

[1] Population data for Egypt can vary between sources and back years are revised in the light of recent information (see also Table 5.1).

Acknowledgement

The author would like to acknowledge the help of Engineer Farouk Afifi, Chairman of the Council of Sugar Products, January 1992 for the information which he has contributed to the section on sugar cane and also to Dr Hillal Hattab, Emeritus Professor of Agriculture, Faculty of Agriculture, Cairo University for information contributed in the section on animal production.

References

Abdallah, H. (1965). *U.A.R. agriculture*. Foreign Relations Department, Ministry of Agriculture, Cairo.
Blackman, W.S. (1927) *The fellahin of Upper Egypt*. George G. Harrap & Company, London.
CAPMAS (Central Agency for Public Mobilization and Statistics) (1987). *Results of the 1986 population census*, Preliminary Report. CAPMAS, Cairo.
Hamdan, G. (1982). *Egypt's identity*. 'Shakhsiyaf Misr', Cairo (in Arabic).
Hani Naguib Elias (1989). Labour problems in Minia. Unpublished Ph.D. thesis, University of Assiut.
Hopkins, N.S. (1987). *Agrarian transformation in Egypt*. Westview Press, London.
Radwan, S. and Lee, E. (1986). *Agrarian change in Egypt: an anatomy of rural poverty*. Croom Helm.

15

The deserts of Egypt: desert development systems

A. BISHAY

Introduction

Egypt suffers from continued population growth and dwindling resources. With Egypt's population (57 million in 1992) concentrated in the Nile Valley, which constitutes less than 4 per cent of the total area of the country, development of her deserts to make them productive and habitable is an urgent need that is widely recognized. In order to solve the problems of food and housing for the continuously increasing population, the Egyptian Government has decided to expand horizontally (in the desert) as well as continuing the present successful vertical improvement of its cultivable land. According to the Government of Egypt, a minimum of 150 000 feddans per year will be reclaimed during the current Five Year Plan. Estimates for long-range desert reclamation plans range between 1.6 and 3.4 million feddans, with the figure of 2.3 million feddans adopted officially by the Government of Egypt.

According to Kassas (1989), inhabited Egypt is a 33 000 km^2 oasis amidst a desert area that extends over 1 000 000 km^2. Within the inhabited oasis are cultivated land, industrial centres, urban centres including the Cairo megalopolis, transport facilities, etc. The vast desert is hardly inhabited, and is often divided geographically into three principal sub-divisions: the Western Desert (671 000 km^2) extending from the Nile Valley to the Libyan border, the Eastern Desert (225 000 km^2) extending from the Nile Valley to the Red Sea, and the Sinai Peninsula (61 000 km^2). The desert of Egypt can also be divided into five climatic provinces; three hyperarid and two arid (Ayyad and Ghabbour 1986) (Fig. 15.1).

This chapter deals with the natural environment and resources of the three principal sub-divisions of the Egyptian desert; namely: the Western Desert, the Eastern Desert, and the Sinai Peninsula. Accordingly, the climate, water resources (rain, underground, Nile), land resources, as well as plant, animal and human resources will be described and evaluated. Human resources include nomadic, sedentary/semi-nomadic, and new settlement

DESERT DEVELOPMENT SYSTEMS 279

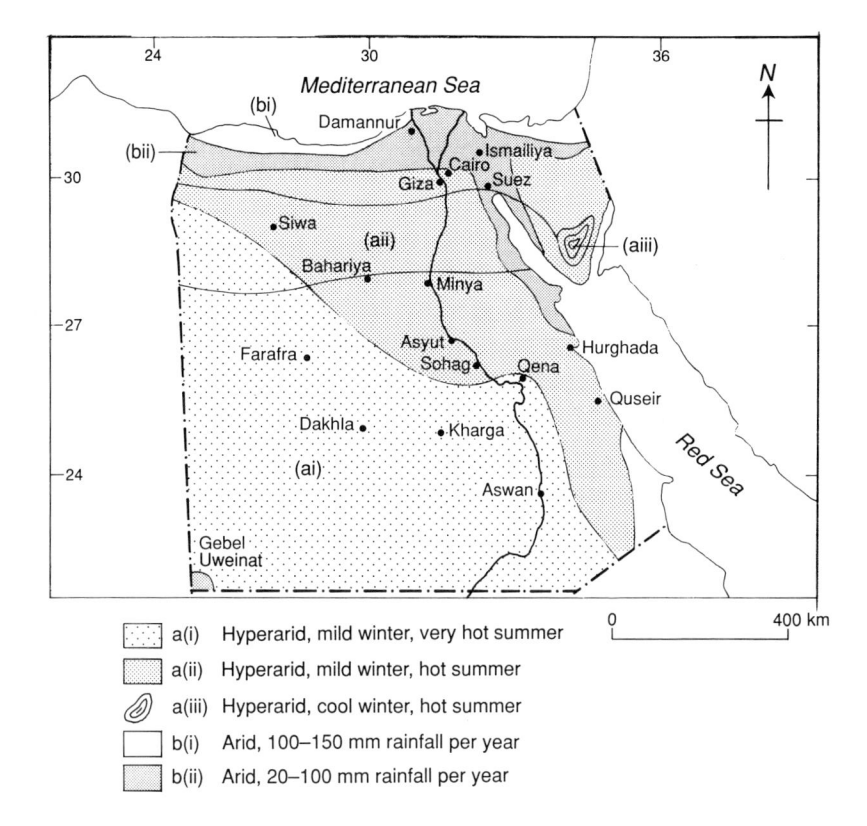

FIG 15.1. Egypt: climatic provinces in the deserts, adapted from UNESCO classifications. Source: Ayyad and Ghabour (1986) and reprinted with permission of Elsevier Science Publishers.

communities. The socio-economic base of indigenous desert communities (nomadic and sedentary) will be discussed with emphasis on their cropping and livestock activities. On the other hand, desert communities for new settlers and entrepreneurs will be dealt with in terms of the 'desert development systems' approach, covering biological, technological, and community aspects. Examples of agricultural new communities, and desert agri-business complexes will be cited and assessed. Finally, the future of desert development in Egypt will be discussed in the light of the Water Master Plan, the Land Master Plan, and socio-economic limitations. Sustainable desert development and its impact on applied research activities will also be considered.

Natural environment and resources

Climate

The temperature regime in the desert ecosystems of Egypt is generally governed by the latitudinal location and the maritime effect of the Mediterranean Sea and the Red Sea. The altitudinal effect is limited to a few highlands: the Sinai Mountains, the Red Sea coastal chain, and the Gebel Uweinat at the Libyan/Sudanese/Egyptian border. Five climate provinces have been identified in the deserts of Egypt (Ayyad and Ghabbour 1986):

(a) Three hyperarid provinces. These cover (i) the southern part of the Western Desert characterized by a mild winter (10–20 °C) and a very hot summer (>30 °C) and many rainless years; (ii) the northern part of the Western Desert, Gebel Uweinat, and the Eastern Desert, characterized by a mild winter, a hot summer (20–30 °C) and less than 30 mm annual rainfall; and (iii) the area around the summits of Sinai Mountains, characterized by a cool winter (0–10 °C) and a hot summer.

(b) Two arid provinces with a mild winter and hot summer and a rainy season extending from November to April, but mainly concentrated in December and January. These cover: (i) the coastal belt of the Mediterranean, with about 100–150 mm annual rainfall, and (ii) the more inland province along the Mediterranean, with about 20–100 mm annual rainfall.

In arid and hyperarid provinces, the temperature along the Red Sea coast varies between a mean minimum of about 10 °C for the coldest month, and a mean maximum of about 33 °C for the hottest month. The range of variation becomes greater further inland (from about 4 to 38 °C in the oases of the Western Desert). In continental locations, the temperature variation ranges between −4 °C (or less) for the coldest month (e.g. oases of the Western Desert) and 50°C for the hottest month (e.g. Kharga Oasis). In general, the coldest month is between December and February, and the hottest month is between June and August in the deserts of Egypt.

The relative humidity is affected mainly by the proximity of the location to the Mediterranean Sea and the Red Sea. The lowest records are generally obtained for inland locations during late spring (20 per cent), while highest records are obtained in early winter (50 per cent) for locations closer to the Mediterranean coast.

Wind circulation is mainly controlled by a permanent high-pressure belt: the Azores. Furthermore, seasonal high and low pressure systems alternate over the continental mass, the Red Sea, the Mediterranean Sea, and the Arabian Peninsula. In winter, the Sahara high pressure system dominates the circulation and the north winds bring cool dry air from the North African continental source region. In the summer, very hot dust-laden winds (the khamsin) often blow over most of Egypt causing a number of problems which include soil erosion, disruption of traffic due to reduced visibility, suffocation of cattle, and enormous quantities of dust deposition (calcite, quartz, amorphous silica).

In view of the interest in energy (mechanical or electrical) produced through windmills and wind turbines, a survey of wind speed[1] was made over most of Egypt. The survey showed that the highest speed is in the area of Ras Ghareb on the Red Sea coast and East Uweinat (6–7 m s^{-1}). Fig. 15.2 shows the different speed profiles obtained in the survey.

This information is valuable in connection with pumping underground water for agricultural, mining, touristic, and domestic purposes. A simple mechanical windmill was found very useful for pumping water for domestic purposes at the South Tahrir site of the American University in Cairo (AUC) Desert Development Center (DDC). On the other hand, a solar photovoltaic/wind (electric) hybrid system proved to be very efficient at the Sadat City site of the DDC. The maximum energy obtained from the sun is at around 12.00 h. At Sadat City (on the desert road between Cairo and Alexandria), the daily wind speed reaches its maximum at about 18.00 h. A battery system stores the solar and wind energies as d.c. power (direct current), which when converted to alternating current (a.c.) is used for pumping ground water (from a depth of 42 m) for irrigation purposes.

The evaporative power of the air in the hyperarid provinces of Egypt, as measured by the Piche evaporimeter, varies in January from 3.6 mm d^{-1} (Aswan) to 7.9 mm d^{-1} (Dakhla Oasis), and in June from 14 mm d^{-1} (Bahariya Oasis) to 24.3 mm d^{-1} (Dakhla Oasis). In arid provinces, the mean minimum evaporation rate during winter is, in general, within the same range as in the hyperarid provinces, whereas in summer, the mean maximum is notably lower (14.0 to 14.8 mm d^{-1})

The annual potential evapotranspiration is, in general, lower in arid than in hyperarid provinces. The lowest is that of Giza (1592 mm according to Penman's equation, and 1582 mm according to Turc's equation).

Water resources

Water is the most important ecological factor in the desert and represents the most critical resource in Egypt. Rainfall distribution in the deserts of Egypt has already been discussed under the section on 'Climate'. In general, rainfall is insignificant, only the northern stretches of the land and for a short distance inland receive an annual rainfall of up to about 200 mm, and this is used for sporadic cultivation along the coasts (Said 1979).

The Nile is Egypt's most important water source, at present supplying the country with almost all of its water requirements. The discharge of the Nile fluctuates from year to year, with an average annual discharge of 86 milliards (1 milliard = 1000 million m^3) an amount which puts this longest river in

282 THE AGRICULTURE OF EGYPT

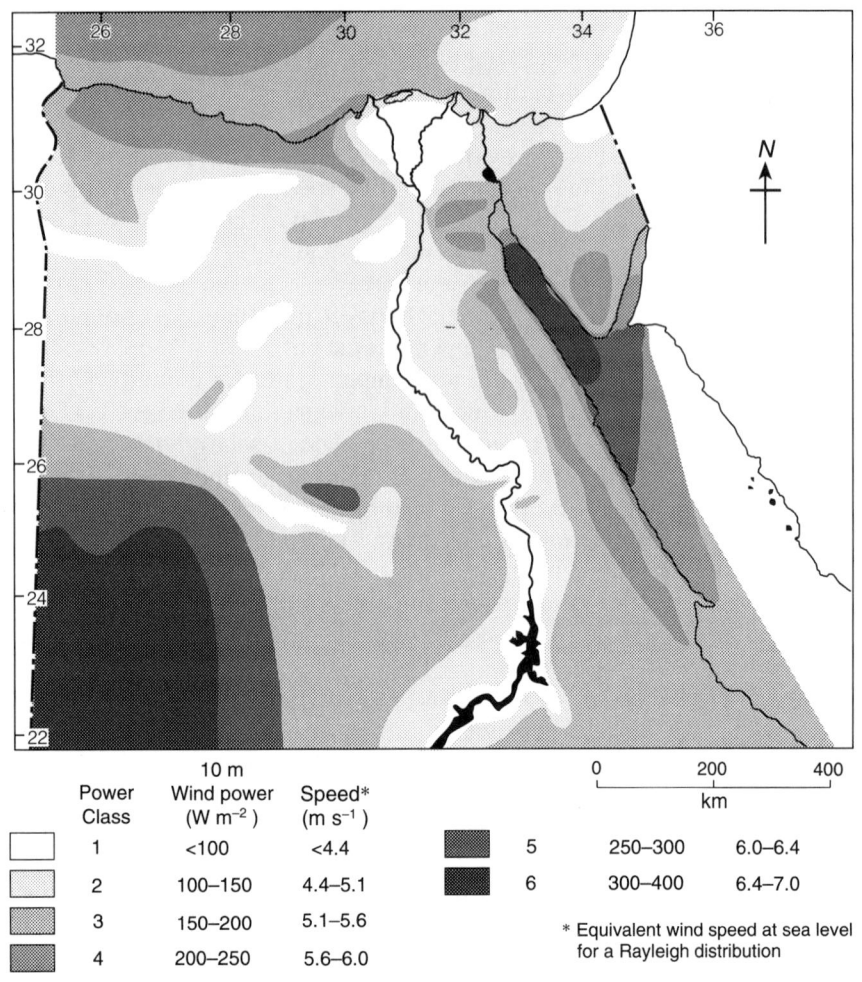

Power Class	10 m Wind power (W m^{-2})	Speed* (m s^{-1})
1	<100	<4.4
2	100–150	4.4–5.1
3	150–200	5.1–5.6
4	200–250	5.6–6.0
5	250–300	6.0–6.4
6	300–400	6.4–7.0

* Equivalent wind speed at sea level for a Rayleigh distribution

This map of annual wind power estimates was prepared by Battelle, Pacific Northwest Laboratories, Richland, Washington, under contract to Louis Berger International, Inc., for the Egyptian Renewable Energy Field Test Programme. The programme is being performed in cooperation with the Egyptian Electricity Authority of the Government of Egypt under the sponsorship of the United States Agency for International Development under contract number AID-263-0123-C-00-4069-00. The wind power estimates are based primarily on historical wind data available from weather-observing stations throughout Egypt. Extrapolations were made to areas where no data were available by considering large-scale terrain influences on the airflow. In general, the estimates represent values for locations that are well exposed to the prevailing winds. However, local terrain features can cause variability in the wind power over distance scales too small to be depicted on this map.

FIG 15.2. Egypt: annual average wind power estimates

the world at the bottom of the list of major rivers in terms of discharge. Egypt's annual share of the Nile water is 55.5 milliards (or 1760 m^3 s^{-1}), which is consumed around the year at the rate of 926 m^3 s^{-1} during the winter months and 2660 m^3 s^{-1} during the summer months. A number of studies have indicated that about 15 per cent of the Nile water reaching the lands of Egypt can be saved without involving dramatic changes in the habits of the Egyptian farmer and his methods of flood irrigation of the old land. This surplus water and an additional 2.0 milliards which should accrue to Egypt in the year 2000 after the completion of the Jonglei Canal to divert the Upper Nile in Southern Sudan and thus avoid the Sudd region, represents all the surplus that Egypt can obtain from the Nile. In addition, the present annual rate of re-use of agricultural drainage water is 4.6 milliards and will increase gradually to 7.0 milliards by the year 2000 (Abu Zeid 1990).

According to Abu Zeid (1990), the annual available ground water is 4.9 milliards, 2.6 milliards of which is now being used and an additional 2.3 milliards will be used annually in the future. This groundwater is another valuable source for use in land reclamation projects.

The main groundwater aquifer lies in the Western Desert and consists of a thick sequence of sandstones with clay lenses overlying basement rocks. The system is confined by a thick clay sequence in the oases and their vicinity. This produces artesian conditions and results in the free-flowing wells in the depression areas. The regional hydraulic gradient is from the southwest. Studies by El Baz (1979) and others indicate that this water has been stored from the old rainy ages and accordingly should be considered as fossil water, i.e. non-renewable.

Based on a mathematical model simulation as well as technical and economic feasibility studies, one milliard and five hundred thousand m^3 of this water can be used annually to irrigate about 143 000 feddans distributed among the different Western Desert oases: Kharga, Dakhla, Farafra, and Bahariya. This is in addition to the land which can depend on rainfall for its cultivation (the north-west coast and the Siwa Oasis).

For the last thirty five years, groundwater extraction in the Kharga and Dakhla depressions has exceeded natural recharge with resultant head decline and falling free flow discharges. The groundwater reservoir is immense, but the availability of water is essentially limited by the viability of the pumping head.

On the other hand, the groundwater in the southern part of the Western Desert is considered to be enough for cultivating about 189 000 feddans in East Oweinat and 50 000 feddans around the High Dam lake. However, although a solar/wind/diesel system has been used to pump water at East Oweinat, only 200 feddans have been reclaimed to date (Naghmoush 1989).

It should be noted here that most of the land reclaimed to date in the Western Desert near the western borders of the Nile Delta has depended mainly on Nile water from the Nasr Canal. This covers the areas of Tahrir,

Nubariya, etc. Plans are now underway to extend this canal towards the north-western coast. The Nasr Canal is already approaching the Alamein area, west of Alexandria. In addition, construction of the El Bustan Canal, will further increase the area to be reclaimed in the Western Desert. On the other hand, the Isma'iliya Canal with its two branches (Suez and Port Said) has been widened and deepened, while the Salhiya Canal, the Youth Canal, and the El Salam Canal are completed or are being constructed to supply the necessary irrigation water to the Eastern Desert and the Sinai (Radi 1989).

Another source of water for land reclamation is the water released during the winter closure in January for electricity generation and navigation. There is a plan to conserve this water in another lake and to use it for irrigation in future land reclamation projects.

Study of the Water Master Plan reveals that in the year 2000, Egypt will require 72.87 milliards of water; this represents the annual water supply expected to be available to Egypt from all resources. In the year 2015 when Egypt's population is expected to reach over 90 million, the situation is questionable. Some drastic measures and ingenious innovations are needed now.

Land resources

Geomorphology

The Nile, a perennial stream which makes its way over hundreds of kilometres of Egyptian desert to the Mediterranean, divides Egypt into two distinct geomorphological regions: the eastern dissected plateau and the western flat expanse, which forms an extension of the Libyan Desert. Although the land to the east of the Nile forms one geomorphological region, it is divided geographically into the Eastern Desert and the Sinai Peninsula, separated by the Gulf of Suez (Ayyad and Ghabbour 1986).

Eastern Desert   The Eastern Desert consists essentially of a backbone of high and rugged igneous mountains running parallel to the Red Sea coast from the Ethiopian Plateau northward up to Gebel Umm Tinassib (28° 30′N). These mountains do not form a continuous range, but rather a series of mountain groups with some detached masses and peaks, and are flanked to the north and west by intensively dissected sedimentary limestone plateaux (Said 1962).

The basement complex formations to the south of Qena separate the Red Sea coastal plain from the Nubian sandstone (mainly Cretaceous) fringing the Nile Valley.

The most pronounced geomorphological feature of the whole Eastern Desert of Egypt is its dissection by valleys and ravines. While eastward drainage of highlands to the Red Sea is by numerous independent wadis, the westward drainage to the Nile Valley mostly coalesces into a relatively small number of great trunk channels. The most notable of the wadi systems, dissecting the limestone plateau, is Wadi El Asyuty which pours into a depression joined to the Nile Valley. Wadi Qena is also one of the most notable features of the limestone desert; the north – south course of its stream is unique. The main wadi system dissecting the Nubian (sandstone) Desert is the Wadi Allaqi system (the most extensive in the Egyptian deserts).

Between the highlands and the shoreline, the coastal plain slopes gently. It varies in width, and is practically non-existent in certain parts of the Gulf of Suez. Near the hills, the inland desert plain is covered with coarse boulders; further away the surface sediments become less coarse. The shore-line comprises in some areas a number of bays and lagoons. The shallow water, separated by coral reefs, provides the habitat for mangrove vegetation. The main plain is covered by a series of silts, sands and gravel of fluviatile origin, often with a stony surface. In places, this stony surface is buried by blown sand or washed silt.

Western Desert The Western Desert is essentially a flat plateau with numerous closed-in depressions, except for the mountain mass of Gebel Uweinat (1907 m) on the extreme west of the Sudano-Egyptian border. To the north-east of Gebel Uweinat, a broad tract of high ground extends for more than 200 km, and slopes gradually northwards to the depression of the Dakhla and Kharga Oases. The Northern boundary of these oases is marked by a high escarpment which forms the southern edge of a great plateau of Eocene limestone (Said 1962). This plateau rises in places to over 500 m above sea level and forms the dominant feature of the major part of the Western Desert in Egypt. In this limestone plateau the great hollows containing the oases of Farafra and Bahariya are situated. To the north-west, this plateau slopes gradually towards Siwa and the great Qattara Depression, where the ground descends below sea level. To the north-east of Bahariya, the plateau rises again to form Gebel Qatrani overlooking the Nile-fed depression of Faiyum. Fluviatile palaeo-deposits cover the desert which extends west of the Nile Delta and embraces the district of the Wadi el Natrun depression, and the Cairo-Faiyum desert.

A prominent feature of the Western Desert is the parallel belts of sand dunes that extend in a north–south direction for hundreds of kilometres. Extensive flat expanses of drifted sand, especially in the south and west, have gained for the Western Desert the fame of being a sea of sand, but the total area covered by sand is in fact less than that occupied by bare rock. The absence of a soil mantle may be the single most significant geomorphic phenomenon.

Sinai Peninsula The Sinai Peninsula has, as a core near its southern end, an intricate complex of high, very rugged igneous and metamorphic mountains (Said 1962). The northern two-thirds of the Peninsula is occupied by a great northward-draining limestone plateau which rises from the Mediterranean coast and terminates in a high escarpment on the northern flanks of mountains which represent the highest peaks in Egypt: Gebel Katherina (2641 m), Gebel Umm Shomar (2586 m), and Gebel Serbal (2070 m). These mountains and their deep rocky gorges form one of the most rugged tracts in the country.

The higher part of the limestone plateau which flanks the igneous core to the north forms Gebel El-Tih. The central portion of the plateau drains to the Mediterranean by numerous affluents of Wadi El Arish. The eastern and western edges are dissected by numerous narrow wadis draining into the Gulfs of Aqaba and Suez.

In the northern part, the plateau surface is broken by the hill masses of Gebel Yi'allaq (1090 m), Gebel Halal (890 m), and Gebel Maghara (735 m). These are separated from the Mediterranean shoreline by a broad belt of sand dunes, some of which attain heights of more than 100 m above sea level.

As can be seen from the description of the geomorphology of the three main desert areas, the absence of a true soil cover is a prominent characteristic of the deserts of Egypt.

Mineral resources (Said 1979)

Egypt's mineral resources have been sought since time immemorial. Both ancient and modern Egyptians are known to have made enormous efforts exploring for minerals and using them in a most rational way. During the past twenty-five years, an extensive programme of mineral exploration, using sophisticated techniques, has been carried out. It has involved the employment of highly qualified scientific personnel, drilling, geochemical analysis, airborne and ground geophysics and remote sensing, tunnelling, and detailed mapping. The mineral wealth of the deserts of Egypt is reasonably well known; Egypt has a few occurrences of base metals, a limited quantity of iron ores, but is rich in phosphates, evaporites, salines, limestones, building materials, glass sand, kaolin, and other earthy materials. These could represent the base upon which industry in Egypt should be established.

Energy is another critical resource. Traditional energy resources are limited. Except for small quantities of coal known in Sinai, fossil fuels in the forms of oil and gas are known around the Gulf of Suez and the northern reaches of the Western Desert. Egypt is now self-sufficient in oil.

Foremost among the mineral resources of the deserts of Egypt is the reservoir of groundwater in the Western Desert, estimated to hold almost 230 000 km^3 of fresh water. The daily discharge is estimated to be 3 375 000 m^3. The consensus is that the recharge is minimal; the water has been held in the pore spaces of the Nubian sandstone reservoir since the pluvial of the Pleistocene and Holocene epochs.

Until an energy source becomes available on an economic basis, the future of the desert will necessarily be limited. Pumping water for agriculture, industrialization, and use of mineral resources all depend on the availability and cost of energy. Until such non-traditional or renewable sources of energy are developed and available at reasonable cost, the reclamation of the desert will be limited to centres around areas of some mineral resource, primarily water. Other reclaimed areas will have to depend on water transported from the Nile through lined canals and a number of pumping stations.

Soil microbiology

Counts of total viable bacteria in samples from the Siwa Oasis, Tahrir Province, Wadi Natrun, and the Kharga and Dakhla Oases were made by Naguib *et al.* (1971*a* – *c*). The samples ranged from bare sandy soil to salt-marsh soils under *Tamarix* spp., or wasteland under *Alhagi maurorum* or *Arthrocnemon glaucum*. The counts were made in nutrient broth and the bacteria were all Gram-positive, spore-forming rods. The strong relationship between numbers of micro-organisms in the soil and their chemical composition was clearly shown. The highly significant factors affecting the distribution of bacteria in the soils investigated were moisture, organic carbon, and nitrogen content, sulphates and sodium ion concentration. In general high counts were obtained in cultivated soils. High moisture levels favoured bacterial counts while high values of salinity, chlorides, and sulphates were thought to inhibit bacteria.

Fifty of the above isolates (out of 924) were found to be pectolytic (pectin hydrolysis—plant pathogens). The results of the identifications showed that these soils had forms similar to *Bacillus cereus*, *B. coagulans*, *B. licheniformis*, *B. megaterium*, and *B. subtilis*. The percentage of pectolytic bacteria was frequently very low (0–6 per cent) but could reach 25 per cent (Siwa) or 31 per cent (Tahrir Province) of the total bacterial count. These pectolytic bacteria are greatly affected by the presence of micro-organisms in the soils, and flourish in their presence. Soils with low levels of indigenous pectolytic bacteria did not favour the proliferation of additional inocula. Pectolytic bacteria which produce a more active enzyme flourish better in the presence of pectic substances in their mother soils. (Naguib *et al.* 1971*b*; Ayyad and Ghabbour 1986).

Fungal counts in the above-mentioned soil samples varied between 400 to 6000 fungi per grain of air-dry soil. High salinity inhibits fungal activity, and hence a large number of spores accumulate. The sand-dune soil had a relatively high number, indicating that fungi can develop in widely different soils. The majority of the fungi isolated belonged to the fungi *imperfecti*, with species of the genera *Aspergillus* and *Penicillium* most common.

Plant resources

Based on the climatic and geomorphological regional variations in the Egyptian deserts outlined in the previous sections, the following scheme of climatic-geomorphologic units were suggested (Ayyad and Ghabour 1986):

(1) Hyperarid Eastern Desert Region:
   (a) Red Sea Coastal Plain;

288 THE AGRICULTURE OF EGYPT

    (b) Limestone Plateaux;
    (c) Basement Complex Formations;
    (d) Red Sea Wadis;
    (e) Sinai Wadis;
    (f) Sinai Mountains;

(2) Hyperarid Western Desert Region;

(3) Arid Mediterranean Desert Region:
    (a) Sinai Plains;
    (b) Gulf of Suez Coastal Plain;
    (c) Wadis;
    (d) Gravel Desert;
    (e) Western Desert.

Each of these units includes a group of ecosystems varying in local physiographic features. In this section, the habitat and vegetation of the most common ecosystems will be described.

Hyperarid Eastern Desert Region

Red Sea Coastal Plain

Mangals
The shoreline morphology and climate of the Egyptian Red Sea coast, especially south of Hurghada, seems to favour the growth of mangal vegetation (Kassas 1957). Along the raised coral reefs there is a series of small bays that cut into the beach, and are partly land-locked by further coral reefs. These sheltered bays provide a favorable habitat for the growth of mangal vegetation (Zahran 1977).

Avicennia marina usually grows in pure stands, but may be found mixed with *Rhizophora mucronata* as a codominant (Kassas and Zahran 1967).

The tidal mud of the mangrove vegetation is usually grey or black in colour, and is often foul smelling. A notable difference between the tidal mangrove mud of *A. marina* and that of *R. mucronata* is the low content of total carbonate in the former as compared to the calcareous mud in the latter.

Reed swamps
The habitat of the reed-swamp vegetation of the Egyptian Red Sea coastal plain is provided by channels and creeks at the mouths of big wadis, and areas which represent the combined influences of the brackish-water springs, such as the Ain Sokhna (Kassas and Zahran 1962).

The reed swamps are dominated by *Phragmites australis* and *Typha domingensis*. The latter usually inhabits areas where the soil is relatively less saline and the water not too shallow, such as estuaries of wadis that collect the occasional drainage. *P. australis* grows in swamps close to the dry land, often with higher soil salt content (Zahran 1966).

Salt marshes
The vegetation of the salt-marsh ecosystems is, more or less, organized into zones following the shoreline. Within any locality only a few zones are represented, each including a mosaic of several communities depending on local topography and soil conditions (Kassas and Zahran 1967). Twelve such communities are recognized in the Red Sea coastal plains of Egypt. These are *Halocnemum strobilaceum*, *Arthrocnemum glaucum*, *Halopeplis perfoliata*, *Limonium pruinosum*, *Limonium axillare*, *Aeluropus* spp., *Sporobolus spicatus*, *Halopyrum mucronatum*, *Zygophyllum album*, *Nitraria retusa*, *Suaeda monoica*, and *Tamarix mannifera*.

Limestone plateau The plateau is represented by the mountain area of the Gebel Shayib group, the northernmost of the coastal mountains facing the Red Sea. The rich plant life of this mountain area is in contrast with the almost lifeless coastal plain. The vegetation comprises a number of community types; the following is an example demonstrating the variety of species at Bir Um Dalfa (Kassas and Zahran 1971): *Aerva persica*, *Acacia raddiana*, *Artemisia herba-alba*, *A. judaica*, *Capparis cartilaginea*, *Chrozophora oblongifolia*, *Colocynthis vulgaris*, *Cleome droserifolia*, *Fagonia tristis* var. *boveana*, *Hyoscyamus boveanus*, *Launaea spinosa*, *Lavandula stricta*, *Lindenbergia abyssinica*, *L. sinaica*, *Moringa peregrina*, *Periploca aphylla*, *Pulicaria crispa*, *P. undulata*, *Solenostemma argel*, *Teucrium leucocladum*, *Zilla spinosa*, and *Zygophyllum coccineum*.

Basement complex formations The highlands of the basement complex formations are represented by the Gebel Nugrus, Gebel Samiuki and Gebel Elba groups. Communities of *Moringa peregrina* characterize the mountain slopes of the Gebel Nugrus group. In the Gebel Samiuki area in general, the flora is much richer in species composition and in plant cover, and *M. peregrina* reaches higher altitudes. The flora of the Gebel Elba group is much richer than that of the Gebel Samiuki group. Three altitudinal zones of vegetation are recognized on the north and north-eastern slopes of Gebel Elba: a lower zone of *Euphorbia cuneata*; a middle zone of *E. nubica*; and a higher zone of *Acacia etbaica*, *Dodonaea viscosa*, *Dracaena ombet*, *Euclea schimperi*, *Ficus salicifolia*, *Pistacia khinjuk*, and *Rhus abyssinica*. Within these higher altitudes ferns, mosses, and liverworts abound. The southern slopes are notably drier; plant growth is mostly confined to the runnels of the drainage system. Communities in these runnels are dominated by *Commiphora apobalsamum*.

Red Sea wadis The wadis of the limestone country north of Qena may be represented by the wadi draining Gebel Qattar (Gebel Shayib group), Wadi Qena, and Wadi El Asyuti. The communities in the wadi draining Gebel Qattar are mostly dominated by *Moringa peregrina*. In wadi Qena, the communities dominated by *Tamarix aphylla* and *T. mannifera* occupy sandy terraces. The fringes of the main channel are occupied by a community of *Acacia ehrenbergiana*. The principal channel of Wadi El Asyuti is occupied by communities dominated by *Cornulaca monacantha*, *Calligonum comosum*, and *Tamarix aphylla*.

The communities of wadi ecosystems of the basement complex country are distinguished by Kassas and Girgis (1969) into four types: ephemeral, suffrutescent woody, suffrutescent succulent, and scrubland types. The ephemeral community type is that of *Morettia philaeana*, with *Fagonia indica* as a consistent associate, in the small runnels of Wadi Allaqi and other wadis. The suffrutescent woody types are represented by the community of *Aerva persica* with *Cassia senna* as a consistent associate, and the community of *Indigofera argentea* with *Aerva persica* and *Fagonia indica* as common associates. Both of these communities are common in small affluents of Wadi Allaqi. The suffrutescent succulent type is represented by the *Salsola baryosma* community, with *Aerva persica* as the most common associate, in the main channel of Wadi Allaqi. Each of the scrubland types is dominated by a shrub or a tree and thus comprises at least one vegetation layer higher than 150 cm. The main community types are *Acacia ehrenbergiana*, *Acacia tortilis*, *Acacia raddiana*, *Letadenia pyrotechnica*, *Tamarix mannifera*, *Salvadora persica*, and *Balanites aegyptiaca*.

Sinai wadis One of the most common communities on the elevated banks and slopes of the large wadis of Sinai is that dominated by *Ephedra alte*. A *Hyoscyamus muticus* community is common on silty and sandy beds, while gravelly terraces are commonly occupied by a community of *Achillea fragrantissima*. Around springs and in inundated depressions, the most common community is that of *Nitraria retusa*. A community of *Artemisia judaica* becomes common in wadi beds of compact sand derived from gravel; at higher altitudes this community becomes codominated by *Zilla spinosa* (Zohary 1944).

Sinai Mountains On the Sinai Mountains, vegetation composition changes with elevation from a community dominated by *Artemisia herba-alba* between 1630 and 1700 m, to a community dominated by *Phlomis aurea* and *Pyrethrum santolinoides* between 1700 and 2000 m, and further to a community dominated by *P. santolinoides* and *A. herba-alba* in the highest zone. On the northern slope of Gebel Musa, there are several tree species growing in crevices, such as *Cupressus sempervirens*, *Ephedra alata*, *Ficus carica*, *F. carica* var. *rupestris*, and *F. pseudosycomorus* (Migahid et al. 1959).

Hyperarid Western Desert Region

The following types of communities are common to all oases of the hyperarid Western Desert of Egypt (Migahid et al. 1960):

Sand dunes The mobile barkhan dunes are sterile. More stabilized dunes are dominated by *Alhagi maurorum*. At a later stage of succession, dunes are dominated by *Tamarix* spp.

Sand plains Community composition in the sand plains depends on local topographic variations, salinity, stability of sand, and depth of soil. In dry elevated locations, the community is dominated by *Alhagi maurorum*. A *Desmostachya bipinnata* community is found in localities with deep loose sand. Where the soil is more saline the community is dominated by *Sporobolus spicatus*.

Salt marshes The communities of salt-marsh ecosytems occupy locations with varying degrees of waterlogging. *Cyperus laevigatus* dominates a community in areas which are rich in clay and organic matter, and are inundated during winter. In drier locations *Juncus rigidus* dominates the community, while in areas where the water table is still deeper and the dark clayey soil is covered by sand, *Salicornia fruiticosa* dominates the community.

Arid Mediterranean Desert Region

Sinai Plains The most common community of the gravel desert (*hamada*) in the Sinai Plains is the *Anabasis articulata* community, especially in depressions and shallow water runnels. Where the runnels are covered with coarse sand the *A. articulata* community becomes co-dominated by *Zilla spinosa*. Other common communities are those of *Anabasis articulata*, *Haloxylon salicornicum*, and *Panicum turgidum* in localities where *hamada* is covered with shallow layers of sand, and *Aristida plumosa* occurs in depressions filled with coarse sand.

Gulf of Suez Coastal Plain The ecosystems of the Gulf of Suez Coastal Plain may be differentiated into two groups: littoral salt-marsh ecosystems; and desert plain ecosystems, which occupy the midland belt between the littoral salt marsh and the range of hills and mountains.

The most abundant salt-marsh communities are those dominated by *Halocnemum strobilaceum*, *Nitraria retusa*, and *Zygophyllum album*, occupying successive zones parallel to the shoreline.

The vegetation of the desert plain exhibits a mosaic pattern and not a zonal one as noted in the salt-marsh ecosystems. The following community types are described by Kassas and Zahran (1965):

(1) Desert grassland types including the communities dominated by *Hyparrhenia hirta*, *Lasiurus hirsutus*, *Panicum turgidum*, and *Pennisetum dichotomum*;

(2) Suffrutescent woody types including communities of *Artemisia judaica* and *Iphiona mucronata*;

(3) Suffrutescent succulent types including communities of *Zygophyllum coccineum* and *Haloxylon salicornicum*.

Wadis In the district of Gebel Ataqa and the Galalas, Kassas and Zahran (1962, 1965) recognized thirteen different communities. These are *Cleome droserifolia, Zilla spinosa, Haloxylon salicornicum, Retama raetam, Launaea spinosa, Iphiona mucronata, Haloxylon salicornicum, Leptadenia pyrotechnica* and *Tamarix aphylla, Acacia raddiana, Juncus arabicus, Imperata cylindrica,* and *Salvadora persica.*

In the wadis of the limestone plateau to the south and east of Cairo, Kassas and Imam (1954) described the successional trend of communities in the ecosystems of Wadi Digla, Wadi Liblab, Wadi Hof, Wadi Rashid, and Wadi Garawi. These communities are dominated by *Stachys aegyptiaca, Anabasis setifera, Zygophyllum coccineum, Zilla spinosa, Pennisetum dichotomum, Desmostachya bipinnata, Panicum turgidum, Atriplex halimus, Lycium arabicum, Nitraria tridentata,* or *Tamarix* spp.

Gravel Desert The desert country between the Muqattam Hills at Cairo and the Ataqa Mountains near Suez has two classes of biotype: (i) land forms representing different stages of the cycle of arid erosion including limestone plateaux, runnels, erosion pavements, and *hamadas*; and (ii) fluviatile deposits of sand and gravel (Kassas and Imam 1959). Wind and water remove the softer material of these deposits, leaving an accumulation of coarser material at the surface, which eventually becomes littered with flint gravel. On the gravel hillocks the community is dominated by *Centaurea aegyptiaca.*

In the runnels dissecting the gravel hills, there are communities dominated by *Artemisia monosperma, Haloxylon salicornicum, Lasiurus hirsutus, Panicum turgidum,* and *Zilla spinosa.*

Western Desert The transition between the arid attenuated and the arid accentuated provinces of the Mediterranean Western Desert of Egypt is characterized by communities dominated by *Anabasis articulata, Salsola tetrandra,* and *Thymelaea hirsuta* near the coastal region. Further south, the communities are dominated by *Artemisia monosperma, Convolvulus lanatus,* and *Helianthemum lippii* (Ayyad and El-Ghoneimy 1976). Within the northern limits of the arid accentuated province, the communities become dominated by *Moltkiopsis ciliata* (El-Ghoneimy and Tadros 1970).

Animal resources

The fauna of the Western Desert is more or less Mediterranean in its northern part: immediately south of the coastal belt and including Siwa, the Qattara Depression, Wadi Natrun and as far as the Bahariya Oasis. Further south, the fauna is typically Saharan and is related to the fauna of the central Sahara.

Mammals

According to Wassif (1976), the mammal fauna of Egypt comprises 10 to 11 Palaearctic species and has a strong African (Ethiopian) character. This is confirmed in the case of the Chiroptera (bats) where the index of faunistic affinity is 25 between Egypt and Europe and 45 between Egypt and Africa. The Nile is largely responsible for this link.

Desert mammals on both sides of the Nile include the dorcas gazelle (*Gazella dorcas*), Cape hare (*Lepus capensis*), jackal (*Canis* spp.), Ruppel's fox (*Vulpes rueppelli*), Libyan striped weasel (*Poicilictic libyca*), common genet (*Genetta genetta*), white-tailed mongoose (*Ichneumia albicauda*), striped hyaena (*Hyaena hyaena*), serval (*Felis serval*), caracal (*Lynx serval*), and rock dassie (*Procavia capensis*).

From a strictly zoogeographical point of view, Osborn and Helmy (1980) prefer to call the desert belt of Egypt the Saharo-Sindian sub-region of the Palaearctic. Eight mammalian species (four rodents, three carnivores, and *Gazella dorcas*) are Saharo-Sindian and of wide distribution in North Africa and south-west Asia. The staple food of these carnivores is rodents. The staple food of rodents, in turn, is seeds and insects, especially their soft larvae, which are also eaten by carnivores. Desert snails are a source of food as well as water to many predators, such as *Acomys* (Osborn and Helmy 1980). The rodents are in many ways the most important group of mammals in the ecosystems under discussion. One can distinguish six major habitats in which these rodents live (Osborn and Helmy 1980): rock and rugged country; desert; salt marshes; palm groves and vegetated well-watered areas; riverine habitats; and the Mediterranean littoral with its semi-desert steppe-like vegetation. Hassan and Hegazy (1968) considered *Arvicanthis niloticus* an important rodent pest for agriculture (it used to be eaten as a delicacy in Upper Egypt), because its natural enemies, snakes and mongooses, had been killed off by man. Newly settled Nubians in Kom Ombo attributed the attacks of *A. niloticus* on maize to the disappearance of owls. It reportedly damages 30 per cent of the sugar cane in Upper Egypt. *Acomys cahirinus* is also suspected of attacking crops and food stores (Osborn and Helmy 1980). Hoogstraal (1963) remarked that farmers complained of damage to maize, barley, and vegetables by *Nesokia indica*. *Jaculus jaculus* may cause some loss to Bedouins by feeding on sprouting barley and grain. One specimen was found from a groundnut field (Osborn and Helmy 1980), although Hoogstraal (1963) had observed that it never invaded established crops.

A survey on rodent pests in Egypt (Ali 1977) revealed that, apart from *Rattus* and *Mus*, *Acomys*, and *Nesokia* may be considered pests on the fringes of cultivation in Minya and Asyut. Further into the desert, *Gerbillus gerbillus*, *G. pyramidum*, *Jaculus jaculus*, and *Meriones* may be found, but their pest status was not definitely established. In laboratory experiments, *Acomys* was found to consume daily 9.6, 7.2, and 6.3 per cent of its weight of crushed maize, wheat, and barley respectively. These figures were three times higher than those for *Rattus* and *Arvicanthis* for the same seeds, but *Arvicanthis* consumed 12.2 per cent of its weight of millet. Ali (1977) estimated that rodent damage to sugar cane was 30 per cent, and that damage by *Arvicanthis* to cotton was 5–26 per cent and to wheat 1–4 per cent.

Osborn and Helmy (1978) assessed trends in population sizes of Egyptian mammals. They recognized that most rodents were not in danger of extinction. *Arvicanthis niloticus* is expanding, whereas *Allactaga tetradactyla*, *Jaculus orientalis*, *Nesokia indica*, *Psammomys obesus*, and *Spalax ehrenbergi* are threatened in areas of desert land reclamation.

Birds

Two species, the partridge *Ammoperdix heyi* and the chat *Oenanthe monacha*, do not extend west of the Nile, for they seem to need barren rocky hills which are rare in the Western Desert. The desert bird fauna is a rather specialized group with mixed origins.

Moreau (1966) noted three species that come to the Egyptian oases in summer in considerable numbers to breed: the turtle dove (*Streptopelia turtur*), the rufous warbler (*Erythropygia galactotes*), and the olivaceous warbler (*Hippolais pallida*). It is astonishing that aerial-feeding birds are absent from these oases. There are no swifts, fly catchers (although present on the coast), hirundines, or nightjars, while the accumulation of breeding doves (two or three species per oasis), whose food relations have not been worked out, is remarkable.

Opportunities for bird life are restricted to the thin line of the Nile stream, along which elements of the Upper Egyptian fauna merge with that of the Sudan (Rzoska 1976)—a situation similar to that occurring for weeds. In the Nubian Nile reaches, the encroachment of desert species on the banks is quite obvious and the riverine species experience a bottle-neck situation which reduces their numbers (Pettet *et al.* 1964). In the arid north, the desert and riverine faunas are sharply distinct from each other and each extends its range to the other's realm seasonally.

Moreau (1966) distinguished among desert birds between opportunists that can penetrate into mesic habitats and typical desert species restricted to xeric habitats. The opportunists adapt to the vegetated environment within the oases, and species such as the chat, *Oenanthe leucopygia*, which frequents houses in the Siwa and Dakhla Oases, and cemeteries in the Bahariya Oasis.

Reptiles

The Egyptian herpetofauna comprises some 93 species, with the highest affinity exhibited towards south-western Asia, western North Africa, and the Red Sea coastal regions (Eritrea, Ethiopia, Somalia) in descending order. The proportion of Egyptian fauna occurring in these regions is 68,

40, and 37 per cent, respectively (Marx 1968). Next in order of affinity is East Africa with 27 per cent of Egyptian species occurring there. Four species are endemic to Egypt. Three of them are in Sinai: *Coluber sinai, Telescopus hoogstraali*, and *Uromastix ornatus*. The fourth is from Gebel Elba: *Ophisops elbaensis*, a lizard. The number of species decreases from north to south and from east to west.

Invertebrates

Dragonflies Although dragonflies are aquatic or hydrochorous insects, they are a conspicuous and an ecologically important component of the fauna of water bodies within deserts. Few dragonflies actually breed in the Nile itself, but rather in its marginal standing or near-standing waters. Egypt has no endemic species.

Termites *Psammotermes hybostoma*, the sand termite, inhabits the fringes of the Sahara and Nubian Deserts and their oases, whenever the sandy soil supports some vegetation. It feeds on wood, vegetable debris, and dung, and on living plants, including the poisonous shrub *Calotropis*. Cloudsley-Thompson and Idris (1964) found them under stones and refuse, and in sand tunnels on *Panicum turgidum*, or mostly under animal droppings, presumably benefitting from greater moisture and food availability. The range of *Psammotermes hybostoma* extends from Faiyum to Minya in Egypt, including the Farafra, Dakhla, and Kharga Oases (Hussein 1980). *P. assuanensis* is always confined to houses, while *P. fuscofemoralis* attacks dead parts of trees and stumps, and *P. hybostoma* attacks cut palm fronds and their stumps in orchards.

Another species, of more Mediterranean affinities, is *Anacanthotermes ochraceus*; these harvester termites, associated with clayey soils and some vegetation, have subterranean nests sometimes at considerable depth. Workers carry dry grass and vegetable debris back to the nests. Soft timbers (employed in rural buildings) are eaten. Mudbrick buildings collapse as a result of extensive tunnelling to seek out the straw used in the clay mix, while grass and palm thatches are also eaten. This species occurs in the Siwa Oasis, the eastern Delta, and the Suez Canal Zone, and it extends southward to Minya and Faiyum where it co-exists with *Psammotermes hybostoma* (Hussein 1980). The other termite of importance occurring in Egypt is *Amitermes desertorum* in the southern oases (Hussein 1980).

Oligochaetes Members of this group, earthworms and their aquatic allies, are typically excluded from deserts because of their high water requirements and high rates of evaporative water loss (Ghabbour 1975). The present distribution of oligochaetes in the oases of the deserts of Egypt (Khalaf El Duweini and Ghabbour 1968*a, b*) shows that their species have originated from invasions or introductions of extra-Saharan provenance, either from the

Maghreb or from the Ethiopian region. *Gordiodrilus siwaensis*, an endemic, occurs in the Siwa Oasis. The closely related *Nannodrilus staudei* occurs in the Bahariya and Baris Oasis and in the Kharga Oasis Depression.

Human resources and impact

Man himself is the most important resource for desert development. It is therefore essential to give attention to the development and conservation of the human resource. Man is the farmer, the labourer, the technician. He drills the well, he plans, he designs, and he is responsible for implementing all these activities. In addition, man has been responsible for transferring his cultural experiences and his means of interaction with nature (Kassas 1989).

Unfortunately, we tend to neglect this valuable information. For example, in arid lands, man has discovered, developed, and adopted a number of social institutions and administrative means by which he can survive under the desert environment, which changes from fat years to lean years or from one wet season to one dry season.

It is indeed a great loss not to make use of these experiences and traditions. For example, if we observe the areas between Alexandria and Salum on Egypt's north-west coast, we find that the native inhabitants of this region have an excellent 'land use policy', and 'land use planning'. You will note that they choose the lowest parts of the hills, that is, areas where rain water collects for plant cultivation. Each person has his own borders, on his own map, between his grazing area and his cultivating area in case of a rainy season. He differentiates between locations for perennial crops, like olives and grapes, and annual crops (barley) on the one hand, and his grazing area on the other. He constructs his home on the land which cannot be used for grazing, or planting field crops, or horticulture. In other words, he gives high priority to productive operations and low priority to the location of his home; for example, on rocky hills or areas which are not useful for grazing or agricultural activities. However, the location of his home is chosen so that he can observe his sheep grazing and observe his cultivated lands.

By contrast, note that in the new land reclamation areas, the Baheeg Canal was constructed and farmers from the Delta were settled on the reclaimed land; these farmers put their homes on the agricultural land and brought with them their culture, experiences, and traditions of the old land without benefitting from the experience of the original inhabitants of these desert lands.

Man on these desert lands has a social system based on the fact that rain falls in a certain area one year and may be in another area the following year. The tribes of these areas have developed an unwritten code, where the Bedouin whose land suffers from drought can move with his sheep and agricultural activities to the land of his cousin where rain is available. This is a 'loan system' developed by the tribes of these areas as an insurance against drought. We should study all these experiences and try to develop them since they can provide us with a wealth of information on the management of man's relationship with nature (Kassas 1989).

Ancient human occupation of the deserts of Egypt is very well documented (e.g. McBurney 1960; Reed 1977). Recent important findings on the early domestication of wild cereals and on settlements showing the transition from gathering-hunting in the open desert to more or less sedentary or semi-nomadic life near the Nile, in the Late Palaeolithic, were presented by Wendorf *et al.* (1979). The impact of traditional cultures which developed during the last few millennia on the ecosystem is recognized to have been considerable, but certainly not as drastic as the new developments of the twentieth century, engendered by population explosion, poverty, and profiteering, and aided by modern technology (Baumer 1975; Cloudsley-Thompson 1971; De Vos 1975; Ghabbour 1972, 1974; Kassas 1967, 1970).

Desert development

Desertification and desert development are two poles of a global challenge now facing more than one hundred nations. Desertification, caused either by man's action or his neglect, is undermining the quality of land and transforming it into desert at an accelerating rate. Desert development is the expression of man's determination both to reverse that process and to make heretofore unused desert lands serve his needs for food, shelter, and settlement.

Worldwide, desert areas amount to roughly 36 per cent of the earth's land surface. For all arid land countries, desertification and desert development are forcibly put on the agenda of action and should be a part of their plans for national development including multi-disciplinary plans for research, education, and training. Efforts in this direction were stimulated in 1977, when an Egyptian scientist, Dr Mustafa Tolba, Executive Director of the United Nations Environment Program, chaired the first United Nations Conference on Desertification.

According to Tolba (1985); 'desertification is not the result of a lack of technology or scientific know-how. The technology and know-how are there in abundance. The problem is getting that knowledge marshalled and applied properly where it is most needed and at the scale it is needed. In Israel, Egypt, India, the USSR, or Arizona, there are marvellous applications of science that can make the deserts bloom and produce. But in Ethiopia, the Sudan, or Mozambique, and a number of other countries, there are barren hillsides and plains, starving people, and misery. We must find a way of connecting the fruits of science with the minds and bodies that are most in need of them. If we fail at this, the consequences are too tragic to contemplate. The famines of the 1990s will make those of the 1970s and 1980s look insignificant.'

Some nations have moved more decisively than others, but among them, few see desert development as a more pressing concern than does Egypt. Deserts account for roughly ninety-six per cent of its land surface, and the remaining arable land is inadequate to support its population, now in excess of 57 million and growing rapidly. The concern of Egypt's leadership was expressed more than a year before the United Nations Desertification Conference when the late President Anwar El-Sadat surveyed the problems of his country and issued the imperative, 'We must turn to our deserts.'

For Egypt, the question is not whether it should or should not develop its desert land resources for agricultural development. Given the severity of the effect of population growth upon existing arable land and the consequent undermining of the viability of Egypt's societal and ecological systems, the Egyptian Government sees no other alternative but to seek to do this. Appropriately, the nation's current Five Year Plan states as its top priority goals, (i) agriculture and food production, (ii) building materials and housing, and (iii) energy.

Egypt is seeking to achieve these goals by combining and utilizing its prime surpluses, population and deserts, for the creation of desert settlements. A desert settlement, however, is not an easy venture. It is a multi-faceted task in which human beings, resources, environments, and social organization must be synchronized. Vision and leadership are equally important factors in such an undertaking.

The national resettlement schemes undertaken by the Egyptian Government reflect its basic ideology and aspirations for agriculture and for farm people. They are designed to provide expansion of domestic production, promote local industries, help relieve population pressures in overcrowded areas, while at the same time raising the living standards of resettled families. In the past, Egypt's land reclamation strategy placed greater emphasis on large State-owned enterprises than on small individually owned farms, although some land was distributed to agricultural graduates or rented, in smaller parcels, to individual farmers. A variety of goals have been tried over the past thirty years, with varying degrees of success. Egyptian experience clearly indicates the complexity of such efforts, as well as their relatively high cost.

The new policy of the present government is to encourage individuals and private companies to pursue different aspects of desert development. Thus, privileges such as land ownership, low rents, fertilizers, seed, and energy subsidies, are given to small farmers to encourage their efforts in land reclamation. However, individuals and companies cannot count on permanent government subsidies to keep them viable. In the long run they will have to be self-sustaining. To do this they need to know more than is known today about how Egyptian desert lands can be developed.

Such development requires mobilizing vast material and human resources including farmers, labourers, agricultural specialists, physical scientists, social scientists, engineers, and managers, embracing in an integrated manner many fields of knowledge, resources, and responsibility. The means that make land poductive must be in harmony with, and reinforce those that support, related socially desirable and economically viable activities and which can sustain permanent human communities. All of these must confront two of the most important limiting factors for self-sustaining desert development: water and energy.

Tolba (1985) stressed the fact that we know the techniques to conserve soil from water and wind erosion, prevent salinization and waterlogging due to irrigation, replace vegetation that has been destroyed for human uses, increase food production, and efficiently use and create alternative, renewable sources of energy. We also know of a number of arid land plants that can be grown intensively to produce new, commercially viable products, further increasing the productivity of dry lands. The large number of international conferences held since 1977 are eloquent testimony to our advanced technological status. While these strategies can lessen the pressure on a particular area, they cannot stop farmers from once again stretching production beyond sustainable limits. Technical assistance in improving agricultural output is important, but more important still is the need to convince the local land users of the advantages of not taking more out of the land than can be returned to it in the annual cycle of rebirth.

Proper development of a desert area requires co-ordinated efforts in many disciplines related to the natural resources, technological aspects, and community aspects appropriate for the specific desert area. This integrated approach is the basis for different types of desert development systems (Fig. 15.3).

Natural resources include soil and water appropriate for desert agriculture, mineral resources (including oil and gas) for mining and related activities, raw materials for industrial and agro-industrial activities, and historical, religious and/or scenic resources for local and international tourism. Good quality water is a limiting factor in most of these activities.

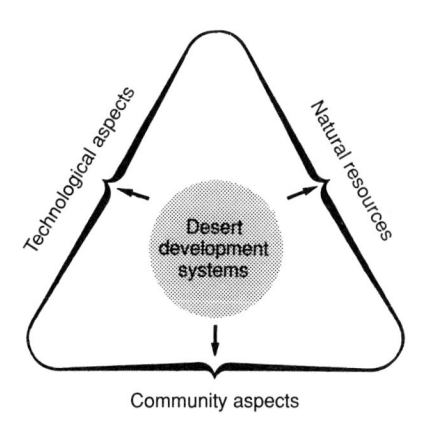

FIG 15.3. Desert development systems: the integrated approach

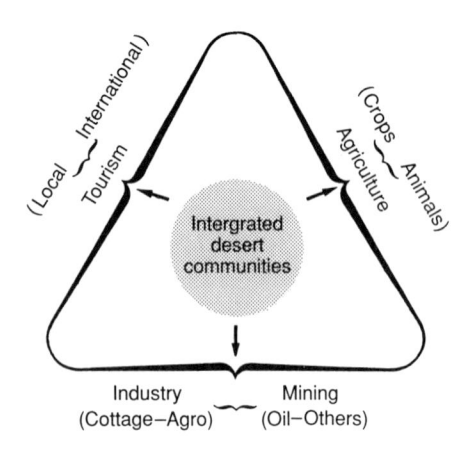

FIG 15.4. Activities contributing to the socio-economic base of desert communities

Technological aspects are based on the type and quality of natural resources, associated development activities, and level of education and culture of the original residents or new settlers in the community. Energy (conventional or renewable) is one of the major constraints in dealing with these activities.

Community aspects deal with infrastructure, socio-economic base, desert architecture, and aesthetics, all of which are very important factors in attracting some of the inhabitants of the overpopulated cities to desert communities. The socio-economic base of desert communities depends on one or more of the following activities:

(1) Agriculture, including crop and animal production;

(2) Mining, including oil, gas and other raw materials;

(3) Industry, including cottage industries, agri-industry as well as small and major industries constituting the economic base of new cities;

(4) Tourism, including religious tourism (Saint Catherine), historical tourism (East of El Minya), and environmental 'scenic' tourism (Red Sea, Sinai, Matruh);

Fig. 15.4 illustrates the above-mentioned activities, contributing to the socio-economic base of desert communities.

In this chapter the emphasis will be on the desert development systems in which agriculture constitutes the major socio-economic base. Under this system, technological aspects will be directed towards serving agricultural and domestic activities, and the community aspects will deal with settlers' needs in an agriculturally based community. Fig. 15.5 illustrates this type of desert development system.

It should be noted that the concept of desert development systems (agriculture) gives equal weight to biological, technological, and community aspects. On the other hand, the concept of desert farming systems puts emphasis on biological aspects. Furthermore, a number of desert development programmes may be based on more than one type of socio-economic activity, e.g. agriculture and tourism (Matruh and Sinai), mining and agriculture (Bahariya Oasis), industry and agriculture (Sadat City area).

History of land reclamation in Egypt (Naghmoush 1989)

The history of Egypt's land reclamation efforts can be divided into four periods:

1900–52

During this period the private sector was responsible for reclamation of about 100 000 feddans at a mean annual rate of 2000 feddans per year. Reclamation of this area of 100 000 feddans was based on the extension of flood irrigation and drainage networks within the Delta. This meant working with the same type of soil, the same agronomic practices, and the same crops (rice and cotton) as those of the Valley and Delta.

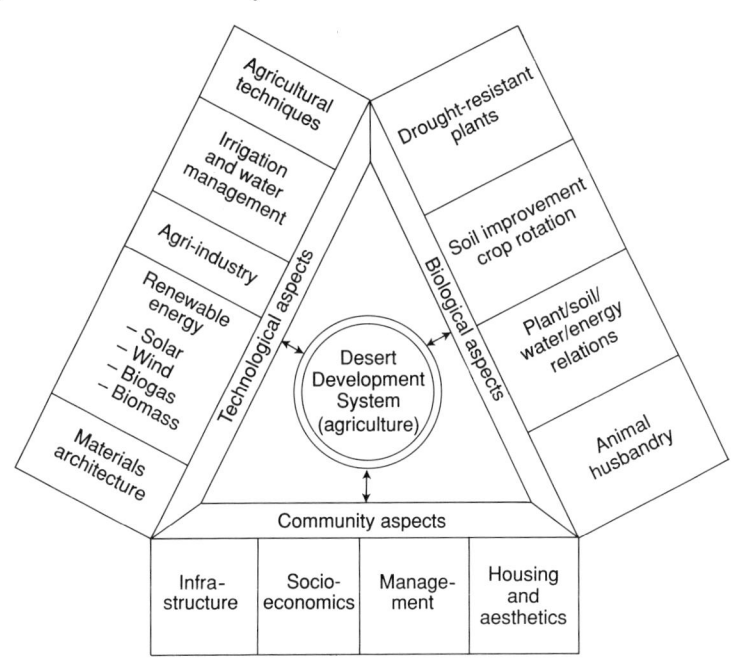

FIG 15.5. Desert development system where agriculture constitutes the major socio-economic base

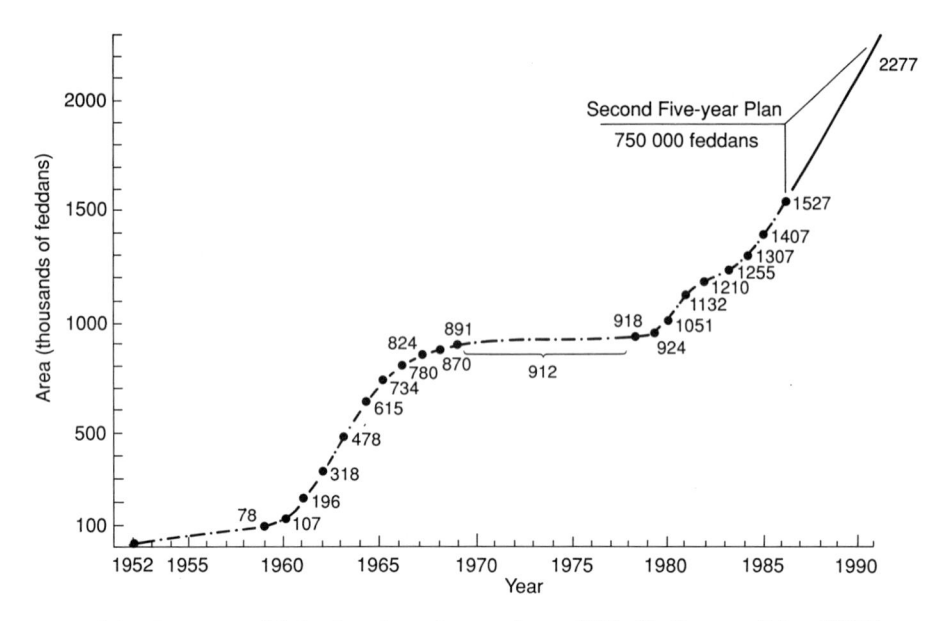

FIG 15.6. Progress with land reclamation projects, 1952–92. Source: Prior (1988)

1952–80

Reclamation and cultivation were carried out by the State and public sector (government) companies. Over the first nine years of this period (1952–60), about 107 000 feddans were reclaimed (Fig. 15.6); with an average of 12 000 feddans per year. The period from 1960 to 1970 was the golden age of land reclamation and was when the Tahrir project was initiated. The average area of land reclaimed during this period is estimated at 80 000 feddans per year and reclamation was implemented by the Egyptian Government through the General Authority for Project Reclamation established in 1954. This authority had, and continues to have, full responsibility for designing and planning different reclamation projects. It also supervizes the implementation of these projects by different companies (all public sector companies during that period). At the end of the reclamation period (with all necessary infrastructure installed and tested), the land is passed to another public sector company for cultivation.

It is now clear that agriculture can be more efficient when carried out by the private sector rather than by the government. Nowadays, even in socialist countries, many government farms are gradually being turned over to the private sector, since the private individual's incentive is usually stronger than that of a government employee. When part of the land of a public sector company was allocated to employees 'retiring' from that company, the yield of the same land was doubled. The lack of success of government farms is

TABLE 15.1. Egypt: reclaimed and productive areas by region, 1980

Location	Area reclaimed (feddans)	Productive area (feddans)
Western Delta	390 000	300 000
Middle Delta	150 000	130 000
Eastern Delta	110 000	80 000
Middle and Upper Egypt	220 000	130 000
Others	67 000	30 000
Total	937 000	670 000

Source: Naghmoush (1989).

confirmed by the fact that, until 1980, about 30 per cent of the land previously reclaimed was still below the marginal productivity level. The areas which reached the productivity level by 1980 are shown in Table 15.1. The reclaimed land and the corresponding areas reaching productivity level are shown by soil type in Table 15.2.

In the past, because of the lack of knowledge of proper irrigation and agronomic practices for the type of land, it took as long as 10–20 years for sandy soils to reach marginal productivity. Today, with modern technology and knowhow, sandy soils can reach the marginal productivity level in one or two years, if necessary funds are available. With fertilization to supply the nutrient requirements, the sandy soil is just used as a medium to anchor and support plants. This is in line with the soil-less agriculture techniques, utilizing gravel to support the roots. These techniques are used in some countries for vegetable production, and the nutrient film technique used at the Desert Development Center (DDC) Sadat City site is based on the same principle.

The sources of irrigation water for the 937 000 feddans shown in Tables 15.1 and 15.2 are shown in Table 15.3.

TABLE 15.2. Egypt: reclaimed and productive areas by soil type, 1980

Soil type	Area reclaimed (feddans)	Productive area (feddans)
Heavy deltaic soil	270 000	240 000
Calcareous soils	200 000	170 000
Sandy soils	467 000	260 000
Total	937 000	670 000

Source: Naghmoush (1989).

TABLE 15.3. Egypt: sources of irrigation water for reclaimed lands

Source of water	Area reclaimed (feddans)
The Nile	830 000
Ground water	33 000
Artesian water	48 000
Rainfall	26 000
Total	937 000

Source: Naghmoush (1989).

Because of the High Dam at Aswan, most of the water needed to irrigate reclaimed land was provided by the Nile. When choosing land for reclamation, priority was given to first and second grade soils and to low elevation areas; where the pumping water head was not more than 40 m, and in some areas it was as low as 5–10 m, the cost of reclamation was as low as £E350 per feddan.

Although appropriate irrigation and agronomic practices were not applied to the land reclaimed between 1960 and 1970, and although land reclamation and cultivation were implemented by government agencies, there was one major positive factor; namely, the Ministries of Land Reclamation and Irrigation were co-ordinated through one deputy Prime Minister. Furthermore, the availability of hard currency for reclamation did not pose many problems at that time.

For land reclamation to be effective, strong co-ordination is needed between the ministries concerned; namely, the Ministries of Land Reclamation, Agriculture, Irrigation, Electricity and Energy, Interior, Health, Education, and Social Affairs. All these organizations have to co-operate in order to develop, not simply reclaim, these areas.

1980–87

In the last decade the Egyptian Government paid more attention to land reclamation due to the increasing gap between consumed and produced food (Fig. 15.6). Thus, in 1982 an area of 85 000 feddans was reclaimed. A Five Year Plan (1982–7) was designed for the reclamation of 62 000 feddans per year. This figure was based on the Water Master Plan announced by the Ministry of Irrigation in 1981 and confirming the availability of water resources to reclaim up to 2.8 million feddans, if the second stage of the Nasr Canal was completed and drainage water was used. If the groundwater resources are taken into consideration, enough water would be available to reclaim 3.4 million feddans.

The Land Master Plan identified the regions, totalling 3.4 million feddans, which would be reclaimed within the available water resources. Priority was given to an estimated area of 1.6 million feddans where water could be lifted less than 40 m (Table 15.4).

Of the regions identified in Table 15.4, 16 per cent consists of deltaic clay soils, 6 per cent calcareous loam soils (mainly west Nubariya), 33 per cent sandy desert soils, and 45 per cent very coarse desert soils. The largest area of reclamation for present and future development consists of fine or coarse sandy soil, a factor which influenced the choice of South Tahrir for one of the research/demonstration sites for the AUC/DDC.

1987–92

During the current Five Year Plan, reclamation of 150 000 feddans per year is expected (Fig. 15.6). It is anticipated, however, that a large percentage of this reclamation plan will be implemented by the private sector. With all its other commitments and obligations, the government alone cannot undergo such an ambitious assignment.

The search for any available water resources and any available land will continue in Egypt. One example is the area of East Oweinat in the Western Desert at the Egyptian/Sudanese and Libyan borders. It is estimated that there is water available in this area to reclaim 189 000 feddans, basing land development on a solar/wind/diesel system. However, despite such advanced technology, only 200 feddans have been reclaimed to date.

In Sinai, 283 000 feddans have been identified in the Land Master Plan. The reclamation of this area will depend on bringing water through the El Salam

TABLE 15.4. Egypt: regions for reclamation identified in the Land Master Plan

Development region	Identified area (feddans)	Priority area (feddans)
East of Delta	799 000	612 000
West of Delta	685 000	264 000
Middle Delta	59 000	59 000
Middle Egypt	224 000	184 000
Upper Egypt	782 000	195 000
Sinai	283 000	212 000
High Dam lake shores	50 000	–
Total Nile irrigated area	2 882 000	1 526 000
Ground water development	546 000	82 000
Grand total	3 428 000	1 608 000

Source: Naghmoush (1989).

Canal and under the Suez Canal to Sinai. Feasibility studies in co-operation with a Japanese agency are underway.

Stages of land reclamation

The process of land reclamation and desert development has consisted of five stages:

1. Data collection and surveys of land, water, etc., i.e. area studies.

2. Planning and design of the reclamation project. This included irrigation and drainage systems, roads, power, villages, etc. and was the responsibility of the General Authority for Reclamation.

3. Implementation by specialized companies. These were mainly Egyptian companies, although some foreign companies participated during the 1960–70 period.

4. Cultivation and land improvement. This stage started when the infrastructure, water system, and other reclamation steps were concluded. It was implemented in three steps:

    (i) Studies of the physical and chemical properties of the soil, water quality, fertilizer requirements, etc. These studies should have been initiated before the implementation of reclamation since they could have affected the design of irrigation and different land treatments;

    (ii) Based on the results of soil and water analysis and other studies, the appropriate land reclamation techniques were recommended and compared with those actually implemented;

    (iii) Cultivation of the reclaimed land followed, with techniques, crop, and agronomic practices depending on the results obtained in the earlier studies. If the soil was saline, leaching was adopted, if alkaline, gypsum was needed. If the soil was both alkaline and saline, gypsum was added followed by leaching. If there was a hard crust, conditioning of the land was needed either by adding organic matter, or by cultivating the land for several years. This was preferable to growing a crop which would simply cover the cost of agriculture. Thus, in some areas rice was planted and in others, lucerne was the chosen crop. When the land was reclaimed and the soil had lost its salinity, alkalinity, or hardpan, crop rotations were initiated, using crops which were likely to give high yields.

5. The establishment of settlements is a most important stage in the land reclamation process. This process included the introduction of main services and infrastructure for the new settlers: primary schools, police stations, consumer co-operatives, and other services based on the size of the new community and the satellite system.

Cost

The cost of land reclamation has increased dramatically over the years. Costs include digging canals, installation of pumping stations, electric power, levelling or irrigation equipment, etc. Total reclamation costs in the 1960s averaged £E350 per feddan because the Egyptian Government concentrated on reclaiming low level land and depended mainly on surface irrigation. Today, most of the land being reclaimed needs water to be lifted between 20 and 40 m and in some cases more than 40 m. In 1985, the cost of reclamation was £E3000–7000 per feddan and, in the near future, it is expected to reach £E10 000 per feddan, based on today's prices.

Socio-economic aspects of desert development (Goueli 1989)

It has been mentioned earlier that, given the serious impact of population growth upon existing arable land and the consequent undermining of the viability of Egypt's societal and ecological systems, there is no alternative but to develop Egypt's deserts. By the year 2000, Egypt's population will be 70 million or more. One-third to one-quarter of this number, about 20 million, will constitute the manpower in the year 2000. At present, the manpower is estimated at 12 million, but within ten years Egypt may have to provide productive work opportunities for an additional 8 million people. This is a major challenge, considering that the State is limiting employment in the government administrative branches at present and this is a positive trend.

In 1963, in a study conducted at the Land Reclamation Institute under the supervision of the late Dr Gabali, samples of the new lands operated by individual farmers were compared with samples of land operated by public sector companies, noting that the government distributed the worst land to the individual farmers. It was observed that the farmers made a surplus in production in order to continue farming, whereas the production of government companies did not cover their operational costs, excluding wages of managers and directors, cars, etc. It was thus recommended that reclaimed lands should not be operated by the government but should be distributed among the beneficiaries and graduates of agricultural colleges. The land should be divided into small plots and there should be co-operative services. These recommendations were presented in 1968 and the Egyptian Government accepted them, but did not implement them. It was again a question of the public sector versus the private sector.

The new approach, since the 1980s is to encourage the private sector which may face many difficulties but can still achieve high economic gains. According to the 1968 study, average returns in financial terms, were above 15 per cent, excluding the appreciation of the land. The private sector makes good use of the land it chooses and depends mainly on high-cash crops in order to evade the interference of the State through pricing controls and Ministry of

Supply regulations. The private sector depends on household workers, has financial liquidity, more flexibility, and benefits from the State's subsidies. It buys diesel, electricity, and fertilizers at subsidized prices and, at the same time, can take land appreciation into consideration. In summary, according to available 1986 studies, the private sector has actually succeeded in desert development.

The second Five Year Plan (1987–1992) will give the Egyptian Government a smaller role in desert development. The government will not be responsible for land management and will limit its role to building the infrastructure and distributing the land among the private sector and graduates. The figures available indicate that it now costs up to £E10 000 to reclaim one feddan. This means that £E30 000million will be needed in order to reclaim all identified land at this rate. If it takes ten years to implement all reclamation plans, £E3000million will be required annually. With the current annual government allocation of only £E500million there would be insufficient funding to reclaim the land in 30, 40, or even 50 years, if the total burden is left to the Egyptian Government. Accordingly, the role of the private sector, with its revenues and investments, is very important in desert development. The private sector should be attracted to desert development, not alienated. While the government has one role in desert development, the private sector has another.

The size of the family plots distributed to graduates should be large enough to help the household to cross the poverty line or to be on the border of the poverty line. In fact, the size of plots actually distributed came by coincidence and was not based on economic studies. Where decisions affect every day life, laws are not based on economics but rather on political economics, because all variables and general objectives have to be taken into account. One issue may have to be at the expense of another, which is different from the economics studied at university. In view of the many contradictory and conflicting factors in every day life, the aim is to match all the factors, including social considerations involved in desert development, for the best national interest. For example, if a graduate of a university is given an opportunity to work on five feddans, it is better than if he walked the streets without a job.

Strategic factors controlling arid land utilization and development (Kassas 1989)

Desert development should be based on the integrated approach. Agricultural activities cannot proceed alone in the desert because integration will increase the efficiency of every operation and every productive activity. Furthermore, one cannot develop resources without developing services. In a new development area, one has to start with roads, power grids, schools, villages, etc. This integrated development and the building of resources and provision of services is a basic requirement for sustainable development. It is an expensive approach but is required for sustainable development.

The major problem facing desert development is the availability of sources of investment. Land reclamation operations are expensive. Planting the reclaimed land is also expensive, especially if groundwater is used or river water is raised to higher grounds. These operations need large quantities of energy for pumping and for water distribution, especially where water-saving irrigation techniques such as drip, sprinkler, and centre pivot irrigation systems are employed.

Furthermore, desert lands need very long periods of gestation before they can reach levels of productivity that are economically viable. These gestation periods are much longer than in the northern part of the Delta where the reclaimed land was cultivated for three to four years with rice, after which it was suitable for cotton and maize cultivation, reaching positive economic returns. Desert lands normally need more than five years to reach this level. Accordingly, loans from a commercial bank, at the present interest rate, are not an attractive source of capital for land reclamation. The Egyptian Government considers land reclamation ventures as part of the country's food security operations and accordingly loans for that purpose are liable to only one-third of the interest charged on normal commercial loans (i.e. 5–6 per cent).

A major policy question for newly reclaimed land is whether priority should be given to cash crops for the export market or to crops to feed the people. There is a need to define the policy of agricultural production—a very important issue to consider when planning agricultural development.

In 1984/5, people in twenty-one African countries faced starvation and the rich countries of the world, with the help of the United Nations, collected food to a value of more than $US3000million to provide relief to these countries. The paradox is that in each of these twenty-one countries, agricultural production was much higher in the 1984/5 season than in earlier years. However, in twenty of these countries, agricultural production was mainly cotton and groundnuts; in Somalia it was cotton, groundnuts, and banana. These crops were all cash crops for export. The policy issue is whether crops should be produced for food or for the market. A unit area of land planted with cucumber or cantaloupe will give a rate of return higher than if planted with barley or maize. If such cash crops are produced for export and the income from them used to buy food, then a greater number of people can be supplied with food. If, however, these cash crops are grown for the local market in cities, it will prove very difficult to obtain the means to provide the masses with wheat, maize, and staple crops.

Another issue concerning the development of arid lands, is whether to use labour-intensive or capital-intensive management procedures. Kassas (1989), felt that in Egypt, which has overpopulation and excess labour, labour-intensive techniques should be considered for desert development. In

Libya or Kuwait or a country with low population density, capital investment techniques might be more appropriate. The relationship and balance between labour-intensive and capital-intensive operations is an issue connected with the prevailing social conditions in the region.

On the other hand, according to Sherbiny (1989), it is intensive knowledge and ideas that are needed, in the current age of science and technology. Scientific knowledge on resources available in the desert must be utilized in order to build an integrated community. These include renewable sources such as solar and wind energies to obtain water. Breeds of livestock suited to the desert environment need to be developed, and advanced technology should be used for irrigation. More important, however, are the economic and social dimensions of existing knowledge. Economic feasibility studies should be made of all activities in the desert as well as anthropological studies of the desert people, particularly their survival and the conservation of their communities during the past years. With this information available, scientific knowledge could be applied, where appropriate, to the conditions of life in the desert and its available natural sources. Without economic viability, social acceptability, and environmental compatibility, large-scale activities in the technical field of desert development may not lead to aspired results. In Egypt, experience with the Water and Land Master Plans highlights the need for an economic as well as a social approach to desert development.

Desert farming systems

We will now present relevant information on the different farming systems adopted in the Egyptian deserts by indigenous nomadic and semi-nomadic communities, and by new settlers as well as entrepreneurs and agri-business ventures.

Sinai

Sinai forms 6 per cent of Egypt's land area (61 100 km^2) and contains nearly one-third of 1 per cent of her population (200 493 in 1986). Forty-one per cent of the population is rural and the average household size is 5.2 persons. The population is a mixture of Bedouin and non-Bedouin, with the former living in a dispersed settlement pattern in five sub-regions: north-west, north-east, upland, south-west, and south-east.

The north-west subregion has an overwhelming non-Bedouin population (only 35 per cent of the population are Bedouins). A high percentage of this sub-region's population is in urban settlements. Two-thirds of the Sinai population is in the north-east sub-region, despite the fact that it is the smallest in area. The settlement pattern in this sub-region depends upon cultivable land watered by rain and shallow aquifers. Over one-half of Sinai's Bedouins and 80 per cent of non-Bedouins live in this area, in addition to 8300 foreigners, mainly Palestinians.

The uplands sub-region is the largest land area, covering 42 per cent of Sinai, but it contains only 11 per cent of the population. The majority of the population in this sub-region is Bedouin, dispersed in small sub-tribal units practising livestock husbandry as their primary occupation. This ancient settlement system is based on soil and groundwater capabilities, but it is modified by a half-modern highway system, thus gaining more importance.

In the south-west sub-region, on the Gulf of Suez, there is a settlement pattern based on the extraction of hydrocarbons and minerals. Ten per cent of Sinai's population are in this sub-region, 73 per cent of whom are Bedouin. One-quarter of the population of this sub-region is concentrated in three settlements: El Tor, Abu Rudies, and Ras Sidr. The male:female ratio of the Bedouin population in this subregion is balanced, except in the settlement of El Tor. The non-Bedouin population is not balanced.

The south-east sub-region, along the Aqaba coast, covers 16 per cent of the Sinai population, of which 70 per cent are Bedouin with a balanced male:female ratio and growth. The non-Bedouin population, forming a new settlement system, has five males to one female, indicating the nature of non-Sinaian population.

The economy of Sinai (Dames and Moore 1981)

There are three key economic sectors in Sinai: (1) agriculture and fisheries, (2) industry and mining, and (3) tourism. Only agriculture and fisheries will be covered in any detail in this chapter but it should be noted that the petroleum sector is the most lucrative in Sinai. There are also numerous areas of known mineral potential, including glass sand and good quality clay. Tourism in Sinai is currently perhaps the most popular of the tourist resources in Egypt. The popular areas include St Catherine, Ras Mohamed, Taba, as well as the Mediterranean and Red Sea Coasts.

Agriculture Livestock raising is the most important branch of economic activity for Bedouin in Sinai although it is subject to the unreliability of rain and the growth of pasture. The condition of the stock is poor, in general, due to the great distances the stock has to travel and the endemic diseases affecting lambing percentage.

There is little trade beyond Sinai in the livestock products of this area, although some products like camel wool could be very valuable.

Although there is considerable potential for improvement, the sources of rangeland in this area are meagre and the commercial use of the land has led to overgrazing close to the point of eliminating some of the better forage and browse species.

Cultivation is considered as supplementary to livestock raising and there is an inverse relationship between the practice of cultivation and the size of the herd.

Rain-fed agriculture is the traditional agricultural economy which depends on the level of rainfall. One million date palm trees are supported by the sparse rainfall and barley is the most common crop in addition to watermelon and beans (*Vicia faba*). The cash value of these crops is not as high as it is for other fruit and vegetables, but they are low-risk crops and are readily absorbed by the subsistence Bedouin economy.

Irrigated agriculture in Sinai is currently mainly dependent on groundwater. An area of 12 600 feddans is planted with olives, vegetables, barley, lucerne, groundnuts, sorghum, beans (*Vicia faba*), sunflower, *bersim* (*Trifolium alexandrinum*), and fruit.

Fisheries Mediterranean catches have been falling since the mid-1960s due to the decreasing levels of enrichment from the Nile as well as to overfishing. The eastern Mediterranean is fished to the limits of its potential. Commercial fishing started in 1955 in Lake Bardawil, which has an area of 156 000 feddans but a maximum depth of only 4 m.

In the Gulf of Suez, catches are highly variable according to season and phases of the moon. Present trends in catches indicate that the maximum yield level has probably already been achieved.

The Land Master Plan The Land Master Plan identified a gross area of about 283 000 feddans in the Sinai Peninsula for development. The area to be irrigated from a mixture of water from the El Salam Canal, being constructed below the Suez Canal from the Damietta branch of the Nile to North Sinai, and drainage flows in the Bahr Hadous and Lower Serw drains. More recent studies have reduced this area for development to around 212 000 feddans. The annual water requirement for the revised area is about 1.7 milliards (about 8000 m^3 feddan^{-1}). A maximum salinity cap for the mixed Nile drain flows has been proposed at 800 p.p.m. Available data show that biochemical pollution would not be a problem in the Bahr Hadous and Lower Serw drains.

Following the Land Master Plan, a project was prepared to develop agricultural land in an area of 50 000 feddans covering the Tina Plain in north-west Sinai. It was proposed that the area would be reclaimed using surface water supplies delivered by the El Salam Canal and a methodology was outlined for developing the mixture of saline clays, loams, and alluvial clays, and sandy soils in the project area (Mahmoud 1989) (Table 15.5).

Annual water requirements for the three soil types vary with the method of water application. At the project intake on the El Salam Canal, it varies from 5586 m^3 feddan^{-1} for drip irrigation to 13 273 m^3 feddan^{-1} for surface irrigation. The mean annual requirement of the net development area is 8175 m^3 with an average cropping intensity of 200 per cent.

TABLE 15.5. Soils, irrigation, and crops in the north-west Sinai development area.

Soil type	Irrigation	Crops
Saline clay soils	Basin with drainage	Rice and berseem initially
Loams and alluvial clays	Hand movable sprinklers; gated pipes	Legumes, grapes, grains, fodder, selected vegetables
Sandy soils	Sprinklers and drip	Legumes, oil crops, fodder, vegetables, fruit trees

Source: Naghmoush (1989).

According to latest studies by the Ministry of Agriculture and Land Reclamation, the revised net area of reclamation for the Northern Sinai project is about 200 000 feddans. Accordingly, the annual water requirement for the Northern Sinai project is 1.64 milliards.

Markets for Sinai agricultural products In the markets, official wholesale prices are generally respected, indicating a balance between supply and demand. Olive oil is marketed in the area to the west of the Jordan River, while pickled olives have markets in Egypt. Vegetables from the new village of New Mit Abul Kom are exported from the region and produce from the East Bitter Lakes area has a guaranteed market to the west of the canal in addition to serving markets as far as Jordan.

The Eastern Desert

The Land Master Plan identified an area of 799 000 feddans, which was later reduced to 612 000 feddans, for reclamation in the Eastern Desert. The major water sources for irrigating this area are based on four canals from the Nile: the Isma'iliya Canal with its two branches (Suez and Port Said) which was recently widened and deepened, the Salhiya Canal, the Youth or Shabab Canal, and the El Salam Canal. These are already constucted or in the process of being completed.

Perhaps the most publicized activity in the Eastern Desert is the establishment of El Salhiya project. In 1978, Arab contractors, under the leadership of Osman Ahmed Osman, started some experimental studies by cultivating 50 feddans in the El Salhiya area: 25 feddans were irrigated by water from the Isma'iliya Canal and 25 feddans by groundwater. The experimental area was increased to 600 feddans in 1979, and to 2000 feddans in 1980. At the end of January 1981, and after the success of the large-scale experiments, the Arab contractors were assigned, by a presidential decree, an area of 50 000 feddans in Salhiya for development and cultivation. The reclamation process included the development of all the infrastructure including the installation of a large number of pivot sprinkler systems of irrigation. In 1982, the cost of reclaiming the area of 50 000 feddans was £E268million, i.e. £E5360 feddan^{-1}.

In the Salhiya/Shabab project there are 143 water pumps, 35 bulldozers, and 220 centre pivot irrigation systems. In addition to land reclamation for agricultural production which also involved animal and poultry production, El Salhiya City was constructed on 3000 feddans. It includes 10 800 housing units, four commercial centres, a central hospital, a central telephone system, and other services. The infrastructure cost for the city was about £E51million.

The Salhiya project has been a very controversial development project, heavily criticized for its costs. In 1989 the Arab contractor company was relieved from running the project and replaced by government management. It is premature to judge the implications of this change, which is contrary to the current policy of selling public sector or government companies to the private sector.

Wadi El Moulak is an old desert area in the Eastern Desert near the Salhiya/Shabab project. It has similar soil structure to South Tahrir in the Western Desert and many other desert-type sandy soils. It has the advantage of a separate canal system for itself alone, and an attractive location relative to services, labour, and markets. The potential for underground water as a back-up is, however, not very attractive due to salinity hazards which increase as pumping takes place on a continuous basis.

The South-Eastern Desert

The South-Eastern Desert inhabited by the Ababda and Busharis nomadic tribes (Fawzy 1978) is a very arid zone of about 7500 km^2, east of Aswan. It has a population of around 20 000 nomads living on the marginal pastures of this desert zone.

The area is officially divided between the Busharis in the south and the Ababdas in the north. However, following the Bedouin code, this division is elastic since it is based on production rather than area. A mountain chain forms the dividing border of water supply between the Red Sea in the east and the Nile in the west. A number of valleys provides relatively easy transportation routes between the Red Sea and the Nile and forms major areas of underground water concentration. Several water wells are scattered in this area, but these wells do not sustain large resident populations. They are used as watering stations for passing nomads during their long seasonal migrations. In winter, nomads migrate to the Red Sea coast, seeking rainwater. About April each year they return to their permanent well sites where the summer pasture is more permanent.

The Ababda tribe, numbering 17 000 individuals, is a group of camel herders whose pattern of life is determined by the nomadic search for vegetation together with the problems of personal survival and small ruminant maintenance. Camels are considered a sign of prestige and social status. They are used for transportation and milk, but rarely for meat. They are usually left to graze alone and they are watered only once a week during the period from April to July.

The small ruminants are goats and sheep. Goats outnumber all other animals since they provide a good source of milk and meat. Small ruminants, usually owned by women, need special care as they can easily get infected.

The Bedouin of this area collect wild plants indigenous to the desert. These include leaves of dom trees which the women use in craft, medicinal herbs which are packed and sold, and seeds of different kinds which form a major component of the Bedouin daily diet. They also burn dry trees to make charcoal.

The Ababdas have an inherent dislike of organized agriculture. However, they occasionally practice seed throwing in a vast wadi so as to maximize use of the little rain and humidity. They do not settle around their plots and they do not collect harmful weeds from within the crops growing in their plots. Accordingly, their yields are low and unpredictable especially when they leave camels to graze unattended in the area until they return to the plots in the following season. In spite of the abundant fisheries on the Red Sea shores, fishing is a despised activity by most of the Ababdas and may only be used for private consumption. The major source of income for this tribe comes from selling camels, charcoal, leather work, medicinal herbs, and small ruminants.

Herding of camels and small ruminants as well as the collection of wild plants are also major activities of the Bushari tribe. However, camels are not the common property of the tribe as in the case of the Ababda tribe. They are owned independently by different tribal groups, who may give them as gifts or loans to other groups in the same tribe. Camel meat is favoured by the Busharis but not by the Ababdas.

Rain is more abundant in the areas where the Busharis live. Accordingly, they practise some horticultural activities in years of exceptionally high rain. They also cultivate communally the banks of streams of silty water following torrents.

In general, socio-economic changes occurred in the South-Eastern Desert as a result of the formation of Lake Nasser which developed a more settled pattern of life and communal work, especially for women. This was accompanied by development of their cultivation activities on shore lines. The main crops of the area are maize, watermelon, cucumber, and tomatoes. However, their success in these activities is still limited due to their limited experience in agriculture.

The Western Desert

The Western (Libyan) Desert of Egypt lies on the north-eastern border of the North African Sahara (El Baz 1979). It occupies 681 000 km^2, or more than two-thirds of the land area of Egypt (Fig. 15.7). It stretches westward

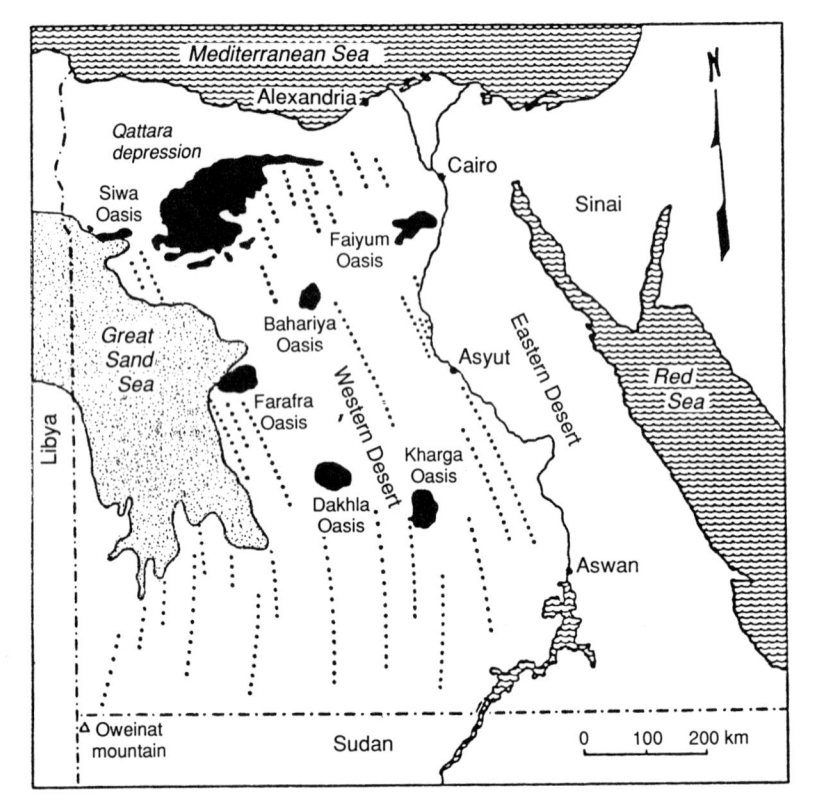

FIG 15.7. Egypt: location of seven major depressions in the Western Desert. Source: El Baz (1979) and reprinted with permission of Harwood Academic Publishers

from the Nile Valley to the Libyan border and northward from the border of Sudan to the Mediterranean Sea. It is basically a rocky plain which slopes gently to the north. The general morphology of the area was discussed earlier in this chapter; the littoral west coast is discussed in another chapter.

The most popular oases in the area called the 'New Valley' of the Western Desert are: the Bahariya, the Farafra, the Dakhla, and the Kharga.

The Bahariya depression is spindle-shaped and is elongate in a north-east direction. It is approximately 95 km long, 45 km wide, and covers an area of 1800 km^2. The Bahariya area has two essential elements for economic development. First, iron ore occurs in the area and is mined and shipped by rail to the steel mill at Helwan, south of Cairo. Second, the floor of the depression is composed of sandstone and variegated shale, many patches of which constitute arable soil. Also, groundwater exists in sandy sub-surface horizons, and wells occur along the central line of the depression and near its eastern boundary.

The Farafra depression is roughly triangular in shape and is bounded by high scarps on the east and west and a low scarp on the north. Because of the prevalence of thick sand deposits in the Farafra depression, the only areas where soil is suitable for agriculture is in the vicinity of Qasr El-Farafra, and particularly to the south near Abu Minqar. This is significant because in the latter location the groundwater reservoir is larger, less saline, and under higher pressure (up to 20 times the atmospheric pressure) than farther south-east of Dakhla and Kharga.

The Dakhla depression lies south of an escarpment that trends west – north-west from Kharga for approximately 250 km. South of the escarpment, the exposed rock belongs to the Nubian sandstone formation, which covers most of the southern half of the Western Desert. However, the erosion of the chalk and shale layers that are exposed on the scarp provides a layer of arable soil. For this reason, agricultural communities are aligned parallel with the scarp.

In addition, east of Dakhla at Abu Tartur Plateau, a phosphate layer occurs that is up to 6 m thick and 15 km long. This is one of the major phosphate deposits in North Africa.

The Kharga depression is the only one in the Western Desert that is aligned in a north – south direction. The succession of rocks from the floor upwards begins with Nubian sandstone, which is overlain by purple and variegated shales, phosphatic beds, Dakhla shale, chalk, Esna shales, Thebes limestone, and travertine on the top. The depression is open to the south and south-west. The southern opening made the Kharga Oasis easy to reach by camel caravans from the interior of Africa along the celebrated Darb El-Arbein road. The width of the depression varies from 20 km in the north to 80 km in the south. The communities in the Kharga depression are found along a line that trends in a north – south direction. This line is a major fault, which allowed the settlers to reach underground water that used the fault as a channel for upward movement. The underground water reservoir at Kharga is smaller, more saline, and is under less pressure than those further to the west. It is believed to be made of fossil water (old and not recharged) that is some 30 000 years old.

Bundles of sand dunes hamper the development of Kharga and to a lesser extent Dakhla. The sand atop the high plateau to the north accumulates in broad dunes of the longitudinal type. The prevailing sand-carrying winds are towards the south as indicated by the numerous north – south trending lines (Fig. 15.7). As these broad dunes reach the scarp, the sand creates patterns that are reminiscent of water falls. After the sand is carried south of the scarp, it forms small and relatively fast-moving barchan dunes. The speed of the Kharga barchan dunes has been measured to vary between 20 and 100 m per year.

These barchans inundate roads, telephone lines, agricultural fields, and houses along their way. In some instances, entire villages in the Kharga depression have been completely buried by the shifting sands; in 1970, inhabitants of Ginah had to abandon their village and settle elsewhere. Barchan dunes are also encroaching on the Kharga-Asyut road which is the lifeline of communications between the Kharga and Dakhla Oases and the Nile Valley.

Beyond Kharga and south-west of the other oases, the terrain of the Western Desert is flat to undulating; it is interrupted by two major high points in the south-western corner of Egypt: the Gilf Kebir Plateau and the Oweinat mountain area. The soil east of Oweinat is good for agriculture and the groundwater is sufficient to reclaim a potential total area of 189 000 feddans. Because of the remoteness of the area, the land development at East Oweinat is based on a solar/wind/diesel system. The studies made in this area were initiated by the Egyptian National Petroleum Company. There was some involvement of the Defence Ministry, and currently the area is under the auspices of the Ministry of Land Reclamation. It has been mentioned earlier that the groundwater sources in the New Valley area are estimated to give about one milliard and five hundred thousand m^3 annually, which could be used to irrigate a potential total area of about 143 000 feddans distributed among the Kharga, Dakhla, Farafra, and Bahariya Oases. In the New Valley and East Oweinat area, 35 000 feddans are to be developed during the second Five Year Plan (1987–92). The distribution of the development areas according to the plans of the General Authority for Rehabilitation Projects and Agricultural Development (GARPAD) is shown in Table 15.6.

This total area of 35 000 feddans is in addition to 43 100 feddans already cultivated in the New Valley and East Oweinat, leaving about 70 100 feddans for future cultivation in the New Valley and 183 800 feddans for future cultivation in East Oweinat.

The present population of the New Valley and East Oweinat is only 132 000. In the New Valley, the original inhabitants live on 16 100 feddans of cultivated land distributed as follows: Kharga 3500 feddans; Dakhla 12 500 feddans; Farafra 100 feddans. The new settlers live on 27 000 feddans of

TABLE 15.6. Egypt: the size of development areas according to the second Five Year Plan, 1987–1992

Location	Area (feddans)
Bahariya	5 000
Farafra, Dakhla, and Kharga	25 000
East Oweinat	5 000
Total	35 000

Source: Naghmoush (1989).

TABLE 15.7. Egypt: distribution of development areas according to the Five Year Plan, 1987–92

Location	Area (feddans)
Baris (south of the Kharga Oasis)	600
El-Zayat (halfway between Kharga and Dakhla)	1200
El-Mawhoub West (included within Dakhla)	1200
Abou Monkar (south of Farafra Oasis)	1200
El Farafra	2800
Total	7000

Source: Naghmoush (1989).

reclaimed land distributed as follows: Kharga 8000 feddans; Dakhla 16 800 feddans; Farafra 2000 feddans; East Oweinat 200 feddans.

Agricultural activities in the New Valley (Mansour and Zoghby 1988)
As part of the 1987–92 Five Year Plan, 7000 feddans are currently under reclamation/cultivation in the New Valley. The location of these areas is shown in Table 15.7.

All the New Valley settled area is cultivated with field or fodder crops, except for an area of 1500 feddans which is cultivated with fruit trees (palm, olive, citrus, and apricot). Dates are one of the most important crops in the New Valley, representing about 20 per cent of the total cultivated area.

The cultivated areas are distributed in small widely scattered plots, where water is the major constraint that determines the area of cultivation in winter or in summer. The area cultivated in summer is about 50–60 per cent of that cultivated in winter since crop water needs in summer are much higher than in winter, although available water resources are constant. Unfortunately, most of the irrigation in the New Valley is based on surface flooding since most of the soil is compacted clay with a low rate of water infiltration. When a drip system was experimented with, the problem of dripper's blockage was prevalent due to the high iron oxide content of underground water in this area. However, a lot of work is needed to select the most appropriate irrigation system and maintenance techniques under the soil, crop, water, and settler's conditions.

It is interesting to note that although we are dealing with 'desert agriculture' in an area where water is at a premium, rice is being grown on about 5000 feddans in the New Valley. Rice is a water-intensive crop, but in the New Valley they expand culture fisheries through the use of rice fields. Furthermore, rice has always been the main dish in the diet of the New Valley inhabitants, and its cultivation represents a tradition. Each feddan of rice needs about 9000 m³ of water per season. Farmers resist any move to decrease the area cultivated with rice because they are not convinced that they can find an appropriate substitute to cover their rice requirements which are far greater than the quantity needed by inhabitants in the Delta or Nile Valley.

Currently, the area cultivated with rice is 5000 feddans in the New Valley giving an average total production of 4650 t. With a population of 119 500 and an average annual consumption of 50 kg person^{-1}, 5975 t of rice are needed and this exceeds the area's production. Trials to produce improved cultivars of rice with higher yield, increased salt tolerance, and lower water requirements are in progress.

It is important to stress that there is a major difference between 'cultivable land', i.e. land that has potential for reclamation and cultivation, and land 'actually cultivated', i.e. those parts of the cultivable land for which pumping water for irrigation is economically viable. Although there are 43 000 feddans of cultivable land in the Kharga depression, only 9700 feddans are actually cultivated due to the low level of underground water which results in high pumping costs.

Dates are one of the most popular crops in the New Valley and there is an annual surplus of 3000 t that is marketed in neighbouring governorates. In addition, some 20 t (1986 data) are exported to Britain and Denmark, but there is a need to improve processing facilities before more dates can be exported.

In the New Valley an area of 11 000 feddans planted with wheat produces an average total of 11 550 t grain (77 000 ardebs). Annual wheat consumption in the local community is about 150 kg person^{-1}. Only 64 per cent of the wheat consumed in the New Valley is produced there, indicating a need to increase the area of wheat planted.

Faba beans (*Vicia faba*) planted in an area of 2500 feddans, yield 7.5 ardeb feddan^{-1} and produce an annual total of 18 750 ardebs beans (2500 t).

In the New Valley, an area of 800 feddans planted with onions produces a mean annual yield of 6400 t. Only 100 t are used by the inhabitants, leaving a surplus of 6300 t which is marketed through the El Nasr Company for Drying Food Products.

Only 20 feddans were planted with lentils at the beginning of the current Five Year Plan, 1987–92. An annual yield of 20 t (124 ardebs) is produced from this area, which is equivalent to 0.17 kg person^{-1} year^{-1}. Since the actual need is 5 kg person^{-1} year^{-1}, an area of 500 feddans of lentils is being added at an annual rate of 100 feddans until 1991/2.

About 1800 feddans are planted with maize and sorghum and these yield an average of 0.9 t feddan^{-1}, producing a total of about 1620 t for the whole area. The average monthly consumption is about 5 kg person^{-1}, equivalent to about 600 t month^{-1} or 7200 t year^{-1} for the New Valley population. The shortfall has to be secured from other sources.

Poultry production is important in the New Valley. There are some 568 000 mature hens, ducks, and turkey, 885 000 young hens, ducks and turkey, and annual egg production is about 10 million (Mansour and Zoghby 1988).

Two problems hinder poultry production in the New Valley. The first is the need to market different poultry products, the second is the lack of feed (especially yellow maize).

Livestock production is very important for supplying the milk and meat needs of the local population. There are 7500 head of livestock (including cattle, sheep, goats, and camels) owned by public sector projects in addition to 33 500 head owned by individuals. Livestock rearing provides 3000 t milk annually from the public sector projects and 14 500 t milk annually from privately owned stock.

Old – new settlements in the Western Desert: the case of South Tahrir
The South Tahrir area is a good example of old – new land reclaimed in the early 1960s and settled by university graduates and smallholders over the last thirty years. The graduates have earned university or technical degrees in agriculture and are each farming areas of approximately 20 feddans.

Graduates and smallholders as groups have widely differing backgrounds and experiences, and the constraints and opportunities that they face may be entirely different. A survey was designed by the Desert Development Center (DDC) for each group and the two surveys (graduates and smallholders) were then followed by a further three studies, of institutions and settler partnership, of marketing problems, and of water problems. The surveys and studies were supported through a grant from the Ford Foundation.

The survey of the graduate farmers, conducted in the summer of 1984, was one of the first socio-economic activities of the DDC. A comprehensive questionnaire was administered to 100 farmers in the South Tahrir area, representing about 20 per cent of the graduates in the region. The survey included questions on social and demographic characteristics, such as age, marital status, and educational level, as well as questions designed to appraise socio-economic and institutional constraints to agricultural production.

Graduate survey The first group of 114 graduates settled in the South Tahrir province in 1977. They were joined by another group of 410 graduates in 1978. The graduates are distributed over nine local agricultural co-operative societies and one umbrella co-operative has incorporated the nine societies since 1979.

The majority of the graduates (77 per cent) are in two age groups, 36–40 and 41–45 years. Ninety-five per cent are married, and 50 per cent have a university degree (B.Sc.) in agriculture. When land was allocated, priority was given to married graduates with a larger number of dependents and longer experience in agricultural work. Fifty-four per cent of graduates have thirteen to twenty years of experience working for the Egyptian Government on agricultural projects, and have experience in growing traditional crops, vegetables, and fruits. Fifty-five per cent of the graduates' wives have intermediate, above intermediate, or university education, and 17 per cent have university degrees.

The settlers marital status affects the level and variety of consumer goods desired and the educational and health needs as well as housing and recreational facilities required for the new community.

The principle applied by the Egyptian Government when selecting the first graduate settlers was that men between the ages of 25 and 45 years who are married are usually the most stable and reliable, thus enabling them to withstand and adjust to a new environment. However, this group also desires better schooling than that offered in the new land. In this case, there is a lag in the social adjustment of the family compared to that of urban families. In many cases, the graduate's wife chooses to reside in the nearest city where a satisfactory school is located.

Graduates have great aspirations for an improved standard of living, including better quality schools, hospitals, and other community services. The educational level of the wives in this group is reflected in the level of living of the family as well as in the new community; educated women can play a major role in development. In the sample interviewed, 32 out of the 55 wives with secondary or post-secondary education worked outside of the home.

The relatively high level of education among graduates suggests that they may accept new ideas and innovations more quickly than do other farmer groups. Nevertheless, lack of knowledge about appropriate desert technologies is a problem in all new land agricultural activities. Even though some parts of the deserts of Egypt have been cultivated for more than thirty years, most farmers still depend on traditional Nile Valley technologies or on a trial-and-error approach. Only 11 per cent of the farmers surveyed depend on some exogenous sources of information.

Information about land use and cropping patterns, crop efficiency measures, and livestock holdings, according to type, production goals, and feeding patterns was collected in the survey. For crop enterprises, a break-even yield for each major crop was calculated. The highest gross margin of profit in winter is provided by lupin, followed by peas, and the highest gross margin in summer results from watermelon, followed by groundnuts. Strawberries yield close to the national average, while watermelons yield higher than the national average.

The decision to grow vegetables depends upon labour availability, processing, and marketing facilities. Marketing is one of the major problems for farmers in the area.

Labour represents around one-half or more of the variable costs of most crops and only 5 per cent of the required labour comes from unpaid family labour. Thirty-eight per cent of the graduates own tractors and 23 per cent own sprayers. Graduates believe that the availability of mechanization and the provision of appropriate incentives to encourage new settlers would solve the labour supply shortage and cut costs significantly.

The most important constraint on crop production, according to graduates, is water shortage caused by frequent electricity cuts. Another major constraint is nematodes which infect groundnuts, the main crop in the area. Weeds are also a problem for fruit trees as well as crops such as groundnuts. To overcome these constraints, farmers gave highest priority to improved water supplies and irrigation facilities, and to the procurement of machinery, seeds, herbicides, and fertilizers. Most of the farmers recommendations concerning water supply and irrigation required improvements to the operations managed by the Egyptian Government. Availability of credit could make machinery and other inputs more accessible.

Where graduate farmers have cattle or buffalo they average 4.04 head per farm. The average sheep and goat holding is 3.67 head. Thirty-seven per cent of the graduate farms keep cattle for milk, and 40 per cent have buffalo for milk. The fat percentage in buffalo milk is higher than in cow's milk which gives buffalo milk a better selling price.

Livestock enterprises are constrained by marketing problems for milk and live animals, poor veterinary services, and lack of artificial insemination facilities. The major constraint—milk marketing—is due to a lack of collection centres, long distances from markets, and unfair transactions with traders.

Only 15 per cent of the graduate farmers have poultry enterprises and the size of the average poultry holding is less than 5000 chicks. Poor services, infrastructure, and marketing facilities limit expansion. Marketing is the major constraint, followed by lack of availability of inputs, either as one-day-old chicks, feed, butagas or electricity for heating, or veterinary supplies and care.

Smallholder survey The DDC conducted a second survey of smallholders, to identify their demographic, social, and economic characteristics. The population of smallholder farmers and their families residing on reclaimed lands in South Tahrir is approximately 20 000. Results show that these smallholder farmers who have made a home in the desert are committed to making the system work in spite of a life that is difficult in many ways and an economic system which suffers from a wide variety of constraints such as the irregularity of irrigation water, lack of credit, limited marketing opportunities, and low soil fertility.

The smallholder survey was conducted during the summer of 1985 in 114 households which were selected randomly from the total of 2100 families in the region. More than one-half of the households were composed of a single nuclear family with an average of four to six children. While government policy advocates no more than two children, large families are especially valued by smallholders because of their dependence on family labour to work their farms. Not only were families large, but more than one-half of the population surveyed was below the age of twenty years. This demographic structure suggests that future populations of desert communities will grow rapidly.

Sixty-one per cent of the families are originally from the adjacent governorates of Daqahliya and Minufiya, and the remainder are from eleven other governorates scattered throughout Upper and Lower Egypt. As expected, work is oriented to the farm; less than one per cent of the people are engaged in off-farm employment.

Most of the smallholders have been on their land for more than ten years. Nine per cent of them were allocated land as early as 1963 and only 5 per cent have settled since 1980. Most of them live in villages rather than on their land; these villages were built as part of the initial reclamation design.

The rate of illiteracy among the smallholders is high. Sixty-seven per cent of the men who are heads of households, and 85 per cent of their wives, have no formal education. The majority of those with some education have not gone beyond primary school.

Even though the educational level of the women is lower than that of their husbands, they have an important voice in smallholder households. Almost all smallholder wives participate in decisions concerning both farm and household operations. The women also have the major responsibility for the small-scale selling of farm products.

This survey emphasized agricultural production, but also produced some information about the community development efforts of smallholders. It was encouraging to note that some collective organizations for self-help were developing in the villages of the smallholders. For example, funds had been collected to build schools and mosques; consumer co-operatives have been formed for the purchase of basic necessities; and there had been some community efforts to maintain roads. The study also identified emerging patterns of community leadership. Certain individuals played a role either as informal political leaders or as agricultural innovators, contributing to community cohesion and agricultural development.

Alongside these positive aspects of community development, constraints to community and agricultural development were also highlighted. Data from the survey indicated that certain conditions in the new lands seriously affect the quality of life of desert settlers. It seemed important that the attention of agencies and governmental authorities should be directed towards the following issues:

1. Good accommodation and other incentives to encourage well-qualified teachers and doctors to move to the new lands.

2. Improvement of medical and educational facilities, e.g. ambulances, medical supplies, libraries, and laboratories.

3. Improvement of administrative control and supervision by the government of personnel engaged in services.

4. Provision of communication and transportation services, especially in and to remote villages and smallholder settlements.

5. Credit and insurance opportunities to enable smallholders to promote their agricultural and animal production activities.

6. Improvements to water supplies by ensuring regular electricity supplies through better maintenance, greater efforts to clean the main irrigation canals, and increasing the efficiency of pumping stations.

Settlers who have been farming the new lands for more than a decade are facing financial difficulties. Returns for fourteen crops calculated from their respective yields, total returns, production costs, and total costs revealed that many smallholders were losing money in crop production. Greater efforts are needed to bring the infrastructure and services of already settled old – new lands to a standard that leads to improved production and marketing.

Results of the smallholder survey revealed many agricultural constraints for farmers.

1. Thirty-six per cent of the smallholder farms are fragmented on more than one plot. This causes some inefficiencies (including labour) and does not allow economies of size. Survey results indicated that higher efficiency is attained with increased farm size.

2. Seventeen per cent of the farms are not owned by their operators, while 31 per cent of those who own their land rent additional land. It can be concluded, therefore, that some farmers are expanding their scale of operation while others are still having problems in obtaining their land titles.

3. Only 23 per cent of the farmers complained about soil salinity, while 81 per cent complained of low fertility, but both characterizations were based on respondents' subjective evaluations.

4. Crop yields are in general substantially lower than those for graduates in both Beheira and South Tahrir.

5. Only 23 per cent of the survey area follows a legume/cereal/legume rotation. Other rotations are: legume/legume/legume (53 per cent), cereal/cereal/cereal (15 per cent), and fallow/cereal or legume/fallow (8 per cent).

6. The most popular markets for selling animals are the governorate markets (68 per cent). The most common dairy products sold are ghee and cheese, while liquid milk is the least common. This indicates availability of cottage

industries for dairy products, although only 35 per cent of households sell these products.

7. Thirty-three per cent of the farms do not have windbreaks, but 47 per cent received their land with trees. Only 35 per cent of the farmers think positively of their trees. The rest complained that trees compete with their crops, promote diseases, and provide shelter for birds.

Smallholders were asked to identify the major constraints affecting their crop production. These problems fell into four categories: (i) agricultural inputs, (ii) irrigation, (iii) machinery, and (iv) plant diseases. The major constraint to crop production was the poor supply and low quality of available inputs including seeds, fertilizers, and chemicals. Nearly as many farmers complained about the poor functioning of pumping stations, high prices, and lack of maintenance services for machinery, the rapid spread of plant diseases and the lack of herbicides and pesticides or equipment for their application.

Economic analysis conducted by the DDC, and the numerous constraints identified by smallholders, indicate that successful desert agriculture requires the modification of many current farming practices. For example, fertilizer usage should be improved, because smallholders' crop yields are very low—although the DDC's long-term fertilizer experiment demonstrated much increased yields on equally low-fertile soils.

The majority of the smallholders complained that plant diseases have spread widely because they lack the appropriate disease control measures. Many consider the unavailability of chemicals, sprayers, and tractors as the main cause. Plant protection chemicals are required in appropriate amounts and on time to avoid smallholders having to buy at the high black market prices.

Water supply irregularities and shortages are related constraints which, in addition to the measures previously mentioned, might be alleviated if farmers were encouraged to cultivate crops with low water requirements.

The DDC has experience with sprinkler and drip irrigation, and can demonstrate a properly designed irrigation system which would be advantageous to the farmers and would save water for the whole area. At the same time, more research is required to evaluate the wisdom of investing in on-farm drainage schemes for sandy soils. At this time, 65 per cent of the surveyed area is under constant sub-surface drainage systems.

Farmers should be encouraged to employ the most productive soils for the most efficient crops (by government definition, based on value-added, maximum cereal production, import substitution, export promotion, or other factors). The practice of adding silt at increasingly high rates is uneconomic and wasteful.

The survey showed that some farmers are leaving part of their farms fallow in winter and summer, while others are expanding their operations by renting land. Public desert settlement policy should be geared towards the individual farmer's capabilities and aspirations.

Finally, smallholders depend almost exclusively upon their own technical experience. Contact with extension service agencies is either very poor or non-existent and so a national programme for training in desert agriculture is strongly recommended.

Institutions, community, and settler participation in South Tahrir One of the greatest challenges faced by the promoters of land reclamation is the problem of establishing communities towards which settlers feel a sense of belonging and participation. Without this social infrastructure, it may not be possible to induce people to move to the desert, even if technologically feasible and financially profitable systems of production have been developed. The creation of viable new communities depends largely on the development of indigenous social, political and economic organizations, as well as participation of the settlers in community affairs. In order to assess the structural and functional dimensions of existing institutions and to gain information about settlers' orientation toward these institutions, the DDC undertook a study of institutions and settler participation in South Tahrir.

The study was carried out in the summer of 1986 by a team of sociologists and anthropologists. Structured interviews focussing on: (i) personal attitudes about life in the new lands; (ii) community development; and (iii) popular participation, were conducted individually with 94 informants at three sites. The sites were selected to represent the diversity of the communities around the DDC lands: Omar Makram, one of the original villages created in the 1950s when the land reclamation programme was initiated; Baghdad, originally a 'company town' for the south Tahrir private owners; and Ma'raka, an area reclaimed only since the late 1970s and distributed to smallholders. These sites are located in the three main areas into which South Tahrir is divided, respectively Badr, El-Tahaddi, and El-Fath.

The interviews first dealt with basic demographic information. Fifty-six per cent of the sample was made up of smallholder and graduate farmers, 34 per cent were employees of the South Tahrir Company, and the remainder were involved in a diverse group of activities, mostly related to agriculture. The three sample communities differed from one another. The typical inhabitant of Omar Makram had spent 25 to 30 years in the community, having arrived around 1955–60. Most of the inhabitants of Ma'raka arrived in the 1970s, so their length of residence was around 15 years. Those interviewed in Baghdad are typically second-stage migrants, having moved to Baghdad to take advantage of land grants around 1984. Only in Omar Makram do most of the respondents live in a single nucleated settlement.

In all three communities, a double cleavage according to occupational status and geographic origin is evident. There are no large, old, or prominent families. Instead, distinctions are made between farmers and employees, employees and workers, farmers and farm labourers, the formal sector and the informal sector, smallholders and graduates. There are also differences associated with origins—whether one comes from Minufiya, Daqahliya, Beheira, or elsewhere. This is the basis for the 'folk' classification of the population of South Tahrir. Each of the three communities has its own version of stratification reflecting its history. Thus, in Baghdad there is a sharp distinction between employees according to rank; the difference at work is reflected in a visible residential difference. In recent years, a differentiation has been added between those who remain company employees and those who have opted to leave the company and have received land. In Ma'raka, the contrast is between graduates and smallholders. In Omar Makram, the main division is between farmers and employees, but there are other categories present such as those who came to round out the occupational structure after the initial period of settlement. Such a pattern of social differentiation distinguishes South Tahrir from communities in the old lands.

In general, the sample consisted of persons fairly well established in the new lands, male, mature, with a family, moderately well educated (24 per cent having completed secondary school or university), marked by an overall pattern of upward mobility, largely involved either in agriculture or in administration, typically from one of three governorates in the Nile Delta, and with a modest degree of involvement (at least one-quarter of the respondents) in various forms of community or collective activity.

The problems expressed by the respondents reiterate those of the two previous surveys. The greatest problems faced in this predominantly agricultural community resulted from an irregular electricity supply, which leads to serious problems for agriculture, all of which is dependent on irrigation. Second, the infrastructure was criticized, especially pertaining to roads which are poor or non-existent, and which affect settlers' access to many services and markets.

The survey of the institutional structure concentrated on the three villages studied, but also included information on area-wide institutions, such as the South Tahrir Company. Generally speaking, the following sets of institutions are available in the villages.

1. Some villages have a village council for local administration, but many do not.

2. Most villages (all three studied) have a local branch of the Community Development Association (CDA) that fulfills some of the functions of local government, and has an elected board. These branches are linked to the General Supervisory of Development (GSD).

3. The co-operative structure is a reflection of the history of the area and includes some smallholder co-operatives, some graduate co-operatives, some for land reclamation, and some other specialized co-operatives. The co-operatives are linked into co-operative unions which cover the entire area. Every settler who is allocated land must agree to join an agricultural co-operative. In general, these co-operatives are distributed without a

DESERT DEVELOPMENT SYSTEMS 329

great deal of regard for village boundaries; rather they reflect the layout of the land as determined by the irrigation system and the land reclamation process. The membership of the co-operatives thus does not correspond to residential units (and does not reinforce community feeling).

4. The various ministries also have a presence in health institutions, schools, and other social services.

5. In some areas, there are consumer co-operatives that play a key role in supplying the residents with certain staple goods (food and cloth). The amount of co-ordination between these various units is low.

The basic activity of the area is agriculture, and so the institutions which support agriculture are crucial. Conceived as a network of co-operatives, these institutions have many shortcomings. Their boards are often unrepresentative, and they are not always able to provide the inputs (seed, fertilizer, pesticides) that the farmers need. Some of them participate in national loan programmes for small farmers, but there are few beneficiaries. There is no extension service, so the flow of information about new practices, market conditions, and other relevant issues is practically non-existent. The position of the co-operatives in the national distribution system is weak, and, as a result, farmers must seek their inputs on the free market and also sell their product to private merchants. As small producers, they are dominated by the large merchants and market forces over which they have no control. In some areas, groundnuts had to be marketed through the co-operative until 1987, but this was a mechanism which ensured that debts to the co-operative were repaid rather than one which favoured a fair price.

In summary, the bureaucracy is very much present in South Tahrir, and many people earn their living from the State and its manifestations. At the same time, the official agencies no longer seem to be in a position to fully carry out their assigned duties. Thus the burden has been shifted to the population, which has not fully assumed it. The development of elected bodies in the area is stunted. Only South Tahrir proper, not Fath or Tahaddi, has elected local popular councils, perhaps because of the entrenched power of the two earlier bureaucratic waves (the South Tahrir Company and the GSD). One vision of development seems to have run its course, and at the moment there is no apparent substitute. The institutional structure is thus present but ineffective. More popular involvement might help, but at the moment there is little incentive to be involved.

Agricultural marketing study In both of the socio-economic surveys of the farmers and in the study of institutional development, many farmers in the South Tahrir area complained that they find it difficult to market their products. In order to gain a fuller understanding of the functioning of the marketing system, and to investigate ways in which the system might be improved, the DDC carried out a study of marketing patterns in the area. The existence of reliable marketing outlets is critical for integrated desert development; without them even a highly efficient system of production cannot be supported.

The study was conducted in two phases. In the first phase, marketing costs and factors influencing them were examined for three types of products—groundnuts, fruit and vegetables, and dairy products—based on cross-sectional data collected by the DDC in 1986 from 68 individual farmers in South Tahrir. In the second phase, relationships between farm and retail prices for the majority of individual crops grown in South Tahrir were examined by studying their changes over time (time-series data). The data were composed of farm and retail prices of commodities for the period of 1970–85.

In the survey of the individual farmers, the importance of groundnut as a cash crop was evident. About 70 per cent of production is sold commercially, while the remainder is used as seed or stored by the farmer. One aspect of groundnut marketing has recently passed through a critical juncture. In the past, farmers have been obliged to sell a minimum of 50–60 per cent of their production to the government co-operative at a price below the prevailing market level. Since 1987, however, the policy has changed. There are no more obligatory sales to the government, and groundnut prices are determined by market forces. While it is too early to make a definite evaluation, this policy change may improve the profitability of groundnut cultivation in the future, and perhaps increase incentives for its production.

It was intended that the Kaha Company should act as the primary purchaser of horticultural crops produced in the South Tahrir area. Unfortunately, Kaha has not functioned as an efficient outlet for smallholders' produce for a number of reasons. First, the company deals in larger quantities than those produced by individual farmers. In addition, farmers find it inconvenient, or lack the means, to transport their goods from their fields to the processing plant. Finally, many farmers are not comfortable with the company's purchasing procedures; they feel that they are not getting a fair deal for their goods. As a result, farmers avoid selling to the Kaha Company; instead they sell to middlemen who come to their fields to buy their products, and then transport them to wholesale markets in urban areas. Although farmers find that the production of horticultural crops is one of the most profitable activities available to them, the study indicates that the farmer's situation could be improved if they had greater bargaining power vis-à-vis the middlemen.

The marketing of dairy products also presents some difficulties, stemming largely from the decentralized nature of milk production. A large number of small farmers each own a few cows producing only small quantities of milk. Much of this is consumed by the producers, but a marketable surplus does exist. Collecting this scattered surplus for transport and wholesale trading, however, presents a problem which is yet to be solved. Furthermore, the highly perishable nature of dairy products and the need for refrigeration make marketing even more problematic.

Interestingly, the time-series data of the analysis show that farmers' share of the prices paid by consumers is actually relatively high. The highest share, which is 89 per cent, is paid for groundnuts, while the lowest share, 41 per cent, is for peas. These two values are based on the average prices for the 1970–85 period. Calculating farmers' share of the two crops using 1986 prices reveals that the individual farmer's share of the consumer's price did not change from 89 per cent for groundnuts, but for peas it rose to 64 per cent for graduates and to 74 per cent for smallholders. The same calculation could not be made for other crops such as onions and cucumbers, because the small sample sizes in the time-series analysis would have made the calculations misleading.

The commodities studied could be divided into two groups according to the behaviour of the marketing margin. For the first group, the relationship between farm-gate and retail prices is linear, while for the second group the relationship is represented by a fixed proportion. Furthermore, margin behaviour changes from one season to another which means that farm-level prices always change with changes in retail prices. For all crops except groundnuts farm-level prices do not change by more than the percentage change in retail prices. For groundnuts, a change in retail price results in an exact change in farm-level price.

The cross-sectional analysis indicated that, for all the crops studied the costs of harvesting, sorting, and packing are the major elements of the total marketing costs. The share of transportation costs to the marketing costs was much less of an obstacle than farmers or concerned observers had believed it to be. Also, a significant positive correlation between the quantity of some crops marketed, e.g. peas and cucumbers, and the production and marketing costs suggested that farmers do not know how to reduce costs by taking advantage of economies of large-scale production and marketing. Production costs did not seem to have an impact on farm-level prices for any of the crops studied. Therefore, South Tahrir settlers are pure price takers.

This conclusion implies that, in order to increase the profits of area residents, further work should be concerned with reducing production costs. At this time, South Tahrir farmers are being paid farm-gate prices which do not reflect their corresponding production costs.

A second marketing study entitled 'Estimation of the supply and demand functions for the crops grown at South Tahrir' was made in order to complete the identification of important economic relationships pertinent to crops grown in the area. In that study, the supply and demand functions for the ten major crops produced at South Tahrir were estimated at both the retail and farm levels. The own and cross-price elasticities of demand and supply functions for each crop at the retail and farm levels were also estimated. Similarly, income elasticities of demand functions were calculated. The results are summarized as follows:

1. More than one-half of the investigated crops had inelastic demand, i.e. a one per cent change in the price of the commodity resulted in a less than one per cent change in the quantities demanded. Groundnuts were the only crop possessing an elastic demand both at the farm level (represented by El-Beheira Governorate) and the retail level (represented by all of Egypt).
2. The cross-price elasticity coefficient, which measures the response of a quantity of a particular commodity to a change in the price of its closest substitute, was found to be less than one in the case of wheat – barley and more than one in the case of orange – mandarin.
3. Population size was found to be the determining factor of demand for almost all investigated crops, followed by prices and per caput expenditure.
4. All of the investigated crops possessed inelastic supply functions. This implied that an increase in the prices paid to farmers would result in an increase in the quantities of these crops by less than the increase in prices.
5. The area of land cultivated with a certain crop was the main supply determinant for the majority of crops produced at South Tahrir.
6. Production costs per feddan were not a supply determinant for any of the investigated crops.

Water – energy problems Interest is currently directed towards the pressing water – energy problems facing desert development efforts. A survey was made by the DDC in summer 1988, to determine the contribution of agricultural inputs to the production of some major crops with emphasis on water and energy inputs. DDC economists assessed the significance of water and energy to desert agriculture at the micro level against a background of both traditional and new factors in Egyptian agriculture. Traditionally, Egyptian farmers received irrigation water from the Nile free of charge, and purchased energy products at heavily subsidized prices. Distortions in the allocation of such scarce resources are thus to be expected both at the national level and at the farm level.

The new factors pertain to changes in the demand for and supply of irrigation water during the last decade, thus modifying the water balance sheet at the national level. Land reclamation efforts have accelerated throughout the Egyptian Desert, thus raising the demand for irrigation water. Meanwhile, the drought which ravaged many economies on the African continent was largely neutralized in Egypt, thanks to the Aswan High Dam. However, the water level behind the Dam diminished at an alarming rate as the severity of the drought continued and the threat to hydroelectric power generation was considerable. For the first time in recent history, national consciousness was focused on the scarcity of the Nile water, the associated hazards to production in many sectors, and to the lives of millions of Egyptians.

To estimate the impact of irrigation water on crop yields, a random sample of 260 South Tahrir farmers was selected. The sample was stratified according to farmer status (174 smallholders and 86 graduates), and method of irrigation (sprinkler and flood). Nine Cobb – Douglas production functions were estimated for the main crops (peas, clover, groundnuts, and citrus), separated by the method of irrigation used.

Analysis showed that: (i) that water and labour were the most important inputs; (ii) most sub-systems operated with increasing returns in relation to scale of production; (iii) smallholders' cultural experience outweighed graduates' acquired knowledge; and (iv) fertilizer application (organic and inorganic), although positively related to yield, was in most cases statistically insignificant.

Despite a variety of issues related to the measurement of water inputs, the statistical significance of the water coefficient in nearly all the estimated functions was a telling sign. Equally telling were the statistically significant differentials between sprinkler and flood irrigation methods. On technical grounds, the results showed that the sprinkler method was far more efficient than the flood method. On economic grounds, however, the sprinkler irrigation method was observed to be efficient in only two cases out of nine.

Agribusiness activities in the Western Desert: some case studies

The DINA Farm This farm is in effect owned and managed by Engineer Hussein Osman who was responsible for running the Salheya project in the Eastern Desert, before it was taken over by the Government.

The project is located on the Cairo – Alexandria road, about 70 km from Cairo and was carried out according to the following basic criteria:

1. In order to enjoy tax and custom exemptions provided by the Egyptian Government, a corporation with a large number of shares was formed. The structure of the corporation allowed for different sized investments, both national and foreign. This enabled the capital to be collected without the need for bank loans. Such a structure allowed for independent management, regardless of share holder individuals.

2. The project unit should not be less than 1000 feddans in order to allow for the economic establishment of all the infrastructure.

3. Animal husbandry, agricultural, industrial, and commercial activities should be integrated so that marginal production could be reached after the first year of operation.

Before this project started in 1987, there was no infrastructure in the project area. The project needed 2 MW of electricity to run all its activities. Four

500 kVA generators and one standby unit were procured, at a total cost of £E2million. Fourteen wells were drilled to a depth of 140–240 m and, in 1989 a well serving a 100-feddan area cost £E60 000.

Cultivation started a year after the start of the project, following six months of studies and six months on the infrastructure and installations. According to the DINA Farm management, annual profits were 5.4 per cent in the first year, 7.3 per cent in the second year, and 12 per cent in the third year. The major activities of the project are:

1. Plant production. This is based on five central pivot sprinkler and drip irrigation systems. The former is used for fodder crops including lucerne and clover, the latter for fruit trees including seedless grapes, pears, apples, figs, and dates. In addition, 10 plastic greenhouses have been installed to grow vegetables for the local or export market.

2. Animal production. The project started with 400 head of pure bred Holstein cows from the United States. This number will soon be increased to 800. The average daily milk production is 20 l cow^{-1} but this is expected to increase to 30 l cow^{-1} during the second year of production. In addition, the project has a flock of 2000 sheep for fattening each year, to be increased to 3000, as well as a rabbit farm, currently 360 does, but to be increased to 2000.

3. Industrial production. A concentrate factory has been established to satisfy the different animals' production requirements and green fodder is also produced on the farm. The dairy products factory is connected directly to a modern milking parlour and, for the first time in Egypt, Gouda cheese is being manufactured locally.

4. Commercial sector. The corporation has its own outlets along the Cairo – Alexandria desert road near the farm as well as in Dokki, a Cairo suburb.

It is premature to conduct an economic evaluation at this very early stage after the initiation of this corporation. However, there seems to be a high probability of success since most of the factors which have caused problems on other farms have been eliminated. Electricity and water supplies are independent of the Egyptian Government and are guaranteed, marketing is not affected by the middleman's price restrictions, and there is no capital shortage. The latter guarantees the availability of good quality inputs exactly when needed.

The Project and Investment Company (PICO) Because of the background of its Chairman, Dr Kamel Diab, and following a study of world markets, the agricultural branch of the PICO company was established in 1986 to produce fruit and vegetables for export to foreign markets.

Studies had shown the vastly increased volumes of strawberries and asparagus being imported by the European Community (EC) in the 1980s. With production conditions in Egypt as suitable or more suitable for these crops, as well as for melons, onions, tomatoes, and artichokes, for which a similar pattern of increased consumption was evident, it was concluded that these crops should be tried in Egypt.

The PICO agricultural activity started on an experimental basis in 1986. Only 30 feddans in the Um-Saber area of South Tahrir was cultivated with strawberries for export and the land was rented from the public sector South Tahrir Agricultural Company. PICO used the most advanced and appropriate technology for land preparation and cultivation, fertilizer applications based on plant requirements, harvesting, and packaging. All of these techniques resulted in maximum production rather than marginal production of strawberries. After harvest, the land was sterilized with methyl bromide before the following season. At a later stage, the company purchased 450 feddans and used the techniques already tested to produce maximum production. In addition to strawberry, they planted different cash crops for local consumption and for export. Surprisingly, banana proved to be excellent. Its production was of very high quality and was sought after. Asparagus, artichokes, grapes, Chinese garlic, potatoes, and groundnuts were also grown very successfully. For most of these crops, superior seed varieties or cuttings were imported in order to maximize yield and quality. For example, groundnuts produced 28 ardebs feddan^{-1} from the first cultivation and this crop sold for £E150 ardab^{-1}. Potatoes were exported to Saudi Arabia at a very reasonable profit.

The Pearl Farm This is a family-run farm, owned by Kamal Korra and his son Ayman, who are both mechanical engineers. The farm is one of the very first in this region of Nubariya (north-western desert). The company is now moving more aggressively toward high quality nursery operations, with apples, peaches, grapes, nectarines, along with a variety of other cash-type seasonal crops.

The Pearl Farm started to work in 1979 in co-operation with the Land Reclamation Authority. Centre pivot sprinklers and a drip irrigation system were installed with the help of an American company and the project was intended to be a pilot scheme to be expanded and replicated when it proved economically viable.

The first stage of the project encountered many difficulties. It was thought that with American knowledge and Egyptian experience, success was assured, but the centre pivot and drip irrigation systems were both, at that time, new to the Land Reclamation Authority, and the agronomic practices of the Nile Valley were still dominant. The American company providing the irrigation equipment was interested in the farm only as a demonstration site, regardless of its economic viability, and the economic returns during the early years of the project were not high enough to cover the initial costs which included most of the infrastructure and electricity. The major source of water was the Nubariya Canal from the Nile.

The initial problems faced by the Pearl Farm included the choice of an attractive product, appropriate soil management techniques, agronomic practices such as irrigation, fertilization, insect and weed control, as well as packaging and marketing procedures. All of these activities needed to be integrated successfully in order to achieve profitability. One of the first successful crops produced by the Pearl Farm was guava, but the costs of harvesting, packaging, and transport to the wholesale market were so high in relation to the market price that there was no profit. Accordingly, it was realized that for the company to survive, crops with high economic returns should be produced.

In order to reduce the high operating costs, the Pearl Farm had to decrease the cultivated area to a size which would enable them to obtain maximum productivity, and to concentrate on quality products, such as apples, peaches, and grapes, which have a high market value. The Pearl Farm also concentrated on improved management procedures leading to both horizontal and vertical expansion in the desert. Agronomic practices were improved, especially methods of fertilization, and appropriate programmes for marketing, packaging, and storing were designed and implemented.

The Nimos Farm   This operation is a prime example of the importance of the owner's interest and active involvement in a development scheme. The owners of Nimos farm, the late Niazi Mostafa and his family, were interested first in the productive potentials from their land and, second, in the appreciation of their land. They gave willingly, without great expectations of receiving, and today their farm is a model showplace for horticultural and ornamental products.

General observations on desert farming systems

Out of eight companies studied by Prior (1988) for USAID, 25 per cent were very successful; 38 per cent had ceased trading; 25 per cent had been taken over by the public sector, and 12 per cent were no longer known. The two most successful operations included in the study were both privately owned and the owners are actively involved in daily decisions.

The problems faced by most operators included electricity cuts, weeds particularly Bermuda grass (*Cynodon dactylon*), nematodes transported with manure, fertilizer supply, distribution, allocation, and quality, seed quality, and rootstock and variety constraints. It appears that basic needs are missing from the system, thus interrupting farming operations. It seems that many operators are buying the capability to ensure the regular supply of these needs at a very high cost.

The desert has some natural limitations but, overall, advances in technology have made it possible to produce nearly the same list of crops that can be produced in the old, most fertile areas. While Egypt in general has been blessed with a climate that is ideal for most crops, a geographical location that makes export potentials very attractive, and a large population that creates an attractive demand for almost anything produced, the selection of crops for desert agriculture requires some thought. In the crop selection process, a high value must be placed on those factors that make the desert different from the old, traditional lands. These include the specific nature of the wind and blowing sand, temperatures and humidity, soil chemistry, water-holding capacities of soils and infiltration rates, drainage, market distances, and labour.

If crops are evaluated in the light of these considerations, the crop mix would tend to favour the more salt-tolerant crops, the more durable and drought-resistant crops, and the soil-building crops. A rotation is always necessary to help maintain soil fertility, and prevent a build-up of certain pests, diseases, and weeds which are common with some crops. Ideally it should include a grain or forage crop, a legume, and some vegetables.

Prior (1988) has reported that crop mixes often include some of the following crops: barley on new desert soils; lucerne or *bersim*; wheat on old desert soils; groundnuts; tomatoes, potatoes, peas, beans (*Vicia faba*), onions, squash, or melons; sorghum/sordan (sorghum x Sudan grass); maize; grapes, apples, pears, peaches, or oranges (citrus).

Small landowners who plant barley, for example, on their 3–5 feddans of land will not make enough income to make a living. Some of these farmers are forced by economies of size to alter their irrigation system and plant the high-value crops such as fruit trees and vines, strawberries, asparagus, flowers, and herbs.

The future of desert development

It was established clearly in the earlier part of this chapter that land reclamation and desert development are crucial for Egypt's future. Egypt's population in 1992 is about 57 million and is expected to reach 72 million in the year 2000 and over 90 million in the year 2015.

Based on the Water Master Plan and the Land Master Plan, the Egyptian Government has decided to reclaim 150 000 feddans per year during the current Five Year Plan, 1987–92. The long-range goal is to reclaim an additional 2.3 million feddans. This goal should be achieved in about 15 years if the same rate of reclamation is maintained.

On the other hand, it is now established that the total water resources based on the river Nile, drainage, underground sources, and treated sewage will amount to 72 870 milliards by the year 2000. This will be enough to satisfy water requirements for agriculture, drinking, industrial, and reclamation in the year 2000.

No additional sources of water are expected beyond the year 2000, although plans call for additional land to be reclaimed and statistics warn against millions of additional Egyptians to be fed, housed, and clothed.

In the author's view, this alarming situation calls for drastic changes to be adopted as soon as possible:

(1) severe measures to be taken against those who use flood irrigation in desert reclaimed areas;

(2) gradual changes in irrigation techniques in the Delta and Nile Valley;

(3) remove subsidy on potable water all over Egypt;

(4) price water for irrigation;

(5) review cropping patterns with a view to decreasing water requirements of crop mixes;

(6) use drought-resistant and salt-tolerant plants whenever possible;

(7) increase desalination efforts around the Mediterranean and Red Sea (based on renewable energy techniques).

It should be noted, however, that the figure of 150 000 feddans to be reclaimed every year was based on the assumption that 60–70 per cent of this area (including its infrastructure) will be the full responsibility of the private sector. The private sector will not invest into land reclamation unless it is profitable. However, it is known that land reclamation is a long-term investment and should always be treated in this way.

There are many factors which can result in the success or failure of a desert reclamation venture:

(1) water availability at the right time;

(2) electricity dependability;

(3) availability of appropriate fertilizers (macro and micro, quantity and quality) at the right time, based on soil and plant analysis;

(4) availability of appropriate insecticides, herbicides etc. at the right time;

(5) availability of high quality seeds or rootstocks;

(6) use of appropriate desert agriculture techniques and agronomic practices.

(7) secured sources for cash or credit to enable immediate action.

In the past, although a large number of desert development efforts were financial failures, a few were successful—indicating a potential for financial and economic viability in desert lands. The reason for most financial successes or failures is illustrated in Fig. 15.8. The top of the curve (point X) is the

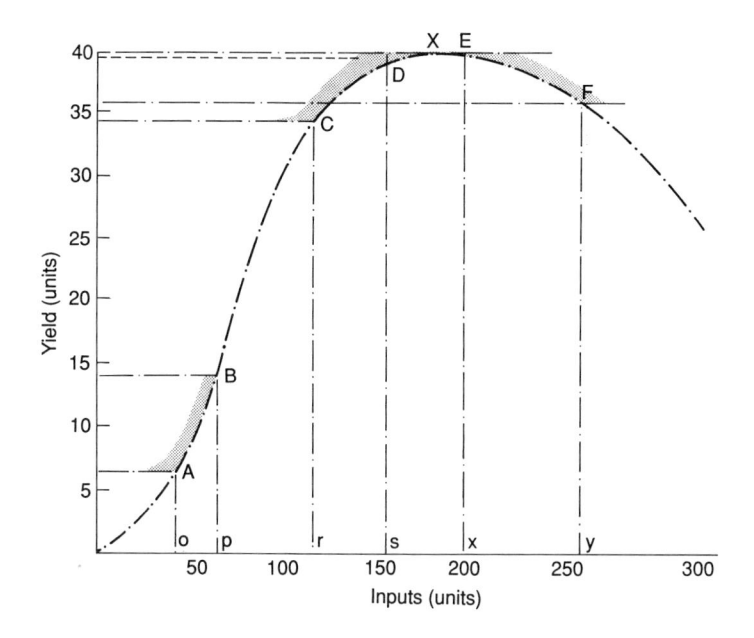

Fig 15.8. Crop production curve that defines a typical relationship between inputs and yield. Source: After Prior (1988)

optimum yield at minimum cost. European and United States operators budget for the segment of the curve marked CD. Most Egyptian desert operations are in the AB segment, while some can be found actually operating in the diminishing returns segment marked EF. Most failures can be attributed to operations that are receiving less than one-half of the inputs needed for success. As Fig. 15.8 illustrates, the operations in the AB segment of the curve are using 40–70 units of input—but losing these units. Some of the factors mentioned above are the reasons for this inability to operate in the CD segment of the curve.

Sustainable desert development and its impact on applied research

Sustainable development, or environmentally sound development, indicates that environment and development are closely linked and are mutually supportive. In other words: there will be no sustained development or meaningful growth without a clear commitment at the same time to preserve the environment and promote the rational use of resources.

Sustainable development argues that 'real' improvement cannot occur unless the strategies being formulated and implemented are ecologically sustainable over the long term, are consistent with social values and institutions, and encourage public participation in the development process.

340   THE AGRICULTURE OF EGYPT

As the primary objective is to provide lasting and secure livelihoods that minimize resource depletion, environmental degradation, cultural disruption, and social instability, this process can be viewed as an interaction among three systems: the biological and resource system, the technological system, and the socio-economic system. The general objective is to maximize the goals across all these systems through a dynamic and adaptive process of trade-offs (Hennawi 1990). In 1992, a task force sponsored by the United Nations Development Programme (UNDP) and co-ordinated by the author, was assigned to propose strategies for sustainable development for Egypt. Furthermore, the International Desert Development Commission (IDDC) is organizing its fourth International Conference based on the theme: 'sustainable desert development for our common future' to be held in Mexico City, 25–30 July 1993. Fig. 15.9 was developed by the author based on the findings of the UNDP task force.

Since its creation in 1979, the Desert Development Center (DDC) has conducted research and demonstrations based on a systems approach integrating biological, technical, and community (socio-economic) aspects of desert development. The DDC integrated approach advocates that a desert should be treated as a desert (no outside manure or silt added), and that desert development should be based on a balance of appropriate indigenous methods

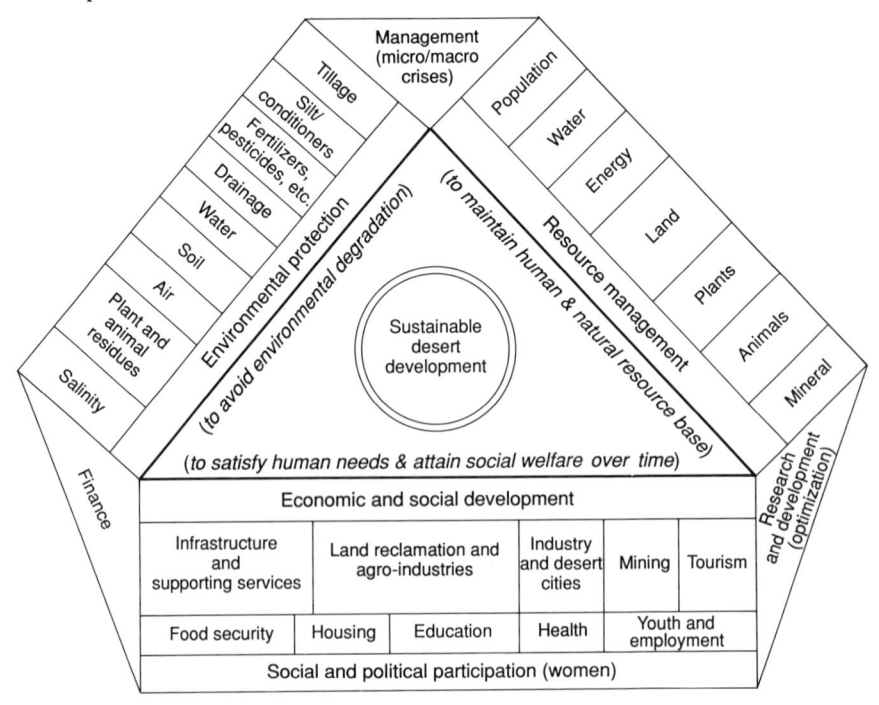

FIG 15.9.   Egypt: sustainable desert development

with modern technology. Any trials for improving productivity should be environmentally compatible with desert conditions and economically replicable under Egyptian social and technical constraints.

The DDC has two sites at Sadat City and South Tahrir (525 feddans), where activities are conducted in desert agriculture, renewable energy, desert architecture, and socio-economic aspects of desert development. The DDC research and demonstration theme has been directed since 1979 towards 'sustained desert development for Egypt's future'.

There is also the Desert Research Institute, recently renamed the Desert Research Centre which is currently under the auspices of the Ministry of Land Reclamation. Generally, this organization is well known as a water research unit for deserts of Egypt. Considerable research has been conducted in the Sinai, and this is on-going. The Desert Research Centre maintains one experimental station in Maryut area and is establishing another station in Sinai. A large number of M.Sc and Ph.D theses have been conducted at the centre in co-operation with Egyptian Universities.

The Water Research Centre, under the Ministry of Water Works (Irrigation) is the authority responsible for underground water resources in Egypt. It is a prime participant in any land reclamation plans using underground water sources.

In addition to the above centres, there is now a strong move in different national universities and agencies announcing the establishment of new 'desert development', or 'desert agriculture' centres. This is an encouraging sign, but we should ensure co-ordination and encourage complementarity between all these potential centres in order to avoid duplication of effort. The Egyptian deserts are vast enough to accommodate any number of centres. However, before re-inventing the wheel, one should confirm that existing centres are not already working on the specific proposed branch of the subject or in the same region of the country.

When the DDC was established in 1979, its founder encouraged active participation of specialists from relevant national universities, the Desert Research Institute, the Agricultural Research Centre, and the Ministries of Electricity and Energy, Land Reclamation, and New Communities. It was only through this approach that the DDC's proposed activities were guaranteed to complement rather than duplicate the work of others, and thereby satisfy some of the development needs of Egypt and the Middle East.

References

Abu Zeid, M. (1990). Conservation and management of water resources for sustainable development. *Desert Development Digest*, **3** (1), 1–3. American University in Cairo, Egypt.

Ali, A.M.H. (1977). *Final report on project of crop loss due to emergent pests.* Academic Scientific Research and Technology Center, Egypt.

Ayyad, M.A. and El-Ghoneimy, A.A. (1976). Phytosociological and environmental gradients in a sector of the Western Desert of Egypt. *Vegetatio*, **31**, 93–102.

Ayyad, M.A. and Ghabbour, S.I. (1986). Hot deserts and arid shrubrands, B. In *Ecosystems of the world, 12B*, (ed. M.E.I. Noy-Meir and D.W. Goodall). Elsevier, Amsterdam.

Baumer, M. (1975). Pastoralisme au Kordofan (Republique de Sudan): I. Les hommes et les paturages. *Bulletin de la Société Géographie d'Égypte*, **48**, 1–59.

Cloudsley-Thompson, J.L. (1971). Recent expansion of the Sahara. *International Journal of Environmental Studies*, **2**, 35–9.

Cloudsley-Thompson, J.L. and Idris, B.E.M. (1964). Some aspects of the district around Kassala, Sudan and the region south of the 13th parallel. *Entomologist's Monthly Magazine*, **99**, 65–7.

Dames and Moore (consultants) (1981). Sinai development: phase 1. In *Summary of findings on population in Sinai*. No. 7. USAID, Cairo.

De Vos, A. (1975). *Africa, the devastated continent? Man's impact on the ecology of Africa*. Junk, The Hague.

El Abd, S. (1971). Human settlements. In *Human settlements in the newly proclaimed areas of UAR: their design and development*. pp.19–30. Proceedings of the Workshop on Human Settlements, Cairo.

El Baz, F. (1979). The Western Desert of Egypt: its problems and its potentials. In *Advances in desert and arid land technology and development* (ed. A. Bishay and W. McGinnies) Vol. 1, pp.67–84. Harwood Academic Publishers, London and New York.

El-Ghoneimy, A.A. and Tadros, T.M. (1970). Socio-ecological studies of the natural plant communities along a transect, 200 km long between Alexandria and Cairo. *Bulletin of the Faculty of Science, University of Alexandria*, **10**, 392–407.

Fawzy, S.A. and Mokhtar, A. (1978). An oriented study. In *Lake Nasser and the ecological and socio-economic dislocations among the nomads of the South Eastern Desert*. USAID, Cairo.

Ghabbour, S.I. (1972). Some aspects of conservation in the Sudan. *Biological Conservation*, **4**, 228–9.

Ghabbour, S.I. (1974). Towards a classification of ecosystem manipulation. *Biological Conservation*, **6**, 153–4.

Ghabbour, S.I. (1975). Ecology of water relation in Oligochaeta I: survival in various relative humidities. *Bulletin of the Zoological Society of Egypt*, **27**, 1–10.

Goueli, A. (1989). Some socio-economic aspects of desert development. *Desert Development Digest*, **2**, 12–16.

Hassan, M.S. and Hegazy, A. (1968). Rats injurious to agriculture and their control. *Technical Bulletin of the Ministry of Agriculture, Egypt*, **28**, 31.

Hennawi, E. (1990). Conservation and management of energy resources for sustainable desert development. *Desert Development Digest*, **3**, 7–9.

Hoogstraal, H. (1963). A brief review of the contemporary land mammals of Egypt (including Sinai). *Journal of the Egyptian Public Health Association*, **38**, 135.

Hussein, M.H. (1980). Taxonomy, reconnaissance and geographical distribution of termites. In *Termites as related to food and agriculture in Egypt*, (ed. A.M. Ali), pp.12–90. Termite Project, University of Assyuit, Egypt.

Kassas, M. (1957). The ecology of the Red Sea coastal land (Sudan). *Journal of Ecology*, **45**, 187–203.

Kassas, M. (1967). *Science and the future of agriculture in the Sudan*. University of Khartoum, Sudan.

Kassas, M. (1970). Desertification versus potential for recovery in circum-Saharan territories. In *Arid lands in transition*, (ed. H. Dregne) pp.123–142. American Association for the Advancement of Science, Washington, DC.

Kassas, M. (1989). Natural resources and desert environment. In *Desert Development Digest*, **2**, 6–9.

Kassas, M. and Girgis, W.A. (1969). Plant life in the Nubian Desert, east of the Nile. *Bulletin de l'Institut du Désert d'Égypte*, **31**, 47–71.

Kassas, M. and Imam, M. (1954). Habitat and plant communities in the Egyptian Desert. III. The wadi bed ecosystem. *Journal of Ecology*, **42**, 424–41.

Kassas, M. and Imam, M. (1959). Habitat and plant communities in the Egyptian desert. IV. The gravel desert. *Journal of Ecology*, **47**, 287–310.

Kassas, M. and Zahran, M.A. (1962). Studies on the ecology of the Red Sea coastal land. I. The district of Gebel Ataqa and El Galala El-Bahriya. *Bulletin de la Société Géographie d'Égypte*, **35**, 129–75.

Kassas, M. and Zahran, M.A. (1965). Studies on the ecology of the Red Sea coastal land. II. The district from El Galala El-Qibliya to Hurghada. *Bulletin de la Société Géographie d'Égypte*, **38**, 155–73.

Kassas, M. and Zahran, M.A. (1967). The ecology of the Red Sea littoral salt marsh. *Ecological Monographs*, **37**, 297–316.

Kassas, M. and Zahran, M.A. (1971). Plant life on the coastal mountains of the Red Sea. *Journal of the Indian Botanical Society*, **50A**, 571–89.

Khalaf El-Duweini, A. and Ghabbour, S.I. (1968*a*) The zoogeography of Oligochaetes in north east Africa. *Zoologische Jahrbucher, Abteilung fur Systematik Okologie und Geographie de Tiere*, **95**, 189–212.

Khalaf El-Duweini, A. and Ghabbour, S.I. (1968*b*). The geographical speciation of north east African Oligochaeta. *Pedobiologia*, **7**, 371–4.

Kishk, M. A. and Bailey, C. R. *Technical and human aspects in desert agricultural development in Egypt: constraints and prospects*. Soil Science Department, Faculty of Agriculture, Menia University and Ford Foundation, Khartoum.

Mahmoud, K. (1989). *Egypt, Northern Sinai agricultural project, water resources assessment*. World Bank, Washington, DC.

Mansour, N. and Zoghby, S. (1988). Field studies of farming systems in the New Valley. In *Proceedings of the New Valley Conference on Land Reclamation, sponsored by IDRC*, pp.63–103. Governorate of the New Valley, El Kharga, Egypt.

Marx, H. (1968). *Checklist of the reptiles and amphibians of Egypt*. Special Publication, US Naval Medical Research Unit No. 3, NAMRU No. 3. p.91. Cairo, Egypt.

McBurney, C.B.M. (1960). *The stone age of Northern Africa*. Penguin, Harmondsworth, UK.

Migahid, A.M., El Shafei, M.A., Abd El Rahman, A.A., and Hammouda, M.A. (1959). Ecological observations in western and southern Sinai. *Bulletin de la Société Géographie d'Égypte*, **32**, 165–205.

Migahid, A.M., El-Shafei, M.A., Abd El Rahman, A.A., and Hammouda, M.A. (1960). An ecological study of Kharga and Dakhla Oases. *Bulletin de la Société Géographie d'Égypte*, **33**, 277–307.

Moreau, R.E. (1966). *The bird faunas of Africa and its islands*. Academic Press, London.

Naghmoush, S. (1989). History of land reclamation in Egypt. *Desert Development Digest*, **2**, 11–13.

Naguib, A.I., Elwan, S.H., and Rabie, M.R. (1971*a*). Studies on potentially pathogenic bacteria in desert soils of UAR. I. Description of localities and sample characteristics. *United Arab Republic Journal of Botany*, **14**, 75–89.

Naguib, A.I., Elwan, S.H., and Rabie, M.R. (1971b). Studies on potentially pathogenic bacteria in desert soils of UAR. II. Total viable bacteria. *United Arab Republic Journal of Botany*, **14**, 91–8.

Naguib, A.I., Elwan, S.H., and Rabie, M.R. (1971c) Studies on potentially pathogenic bacteria in desert soils of UAR. III. Proteolytic bacteria. *United Arab Republic Journal of Botany*, **14**, 173–87.

Osborn, D.J. and Helmy, I. (1978). *Habitat, distribution and status of the recent land mammals of Egypt*. Egyptian National Commission, Cairo, Egypt.

Osborn, D.J. and Helmy, I. (1980). The contemporary land mammals of Egypt. *Fieldiana: Zoology, N.S.*, **5**, 59.

Pettet, A., Pettet, S., Cloudsley-Thompson, J.L., and Idris, B.F.M. (1964). Some aspects of the fauna and flora of the district around Wadi Halfa. *University of Khartoum Natural History Museum Bulletin*, **2**, 28.

Prior, R. (1988). *Desert and reclamation study*. USAID, Egypt.

Radi, E. (1989). *Opening address by the Minister of Irrigation, His Excellency Essam Radi, on the occasion of the tenth anniversary of the Desert Development Center*. The American University in Cairo, Cairo.

Reed, C.A. (1977). Origins of agriculture: a discussion and some conclusions. In *Origins of agriculture*, (ed. C.A. Reed), pp.879–953. Mouton, The Hague.

Rzoska, J. (1976). *The Nile, biology of an ancient river*. Junk, The Hague.

Said, R. (1962). *The geology of Egypt*. Elsevier, Amsterdam.

Said, R. (1979). Mineral resources and desert development: a rational approach. In *Advances in desert and arid land technology and development*, (ed. A. Bishay and W. McGinnies), Vol. 1, pp.263–9. Harwood Academic Publishers, London and New York.

Sherbiny, N. (1989). Economic viability of desert development. *Desert Development Digest*, **2**, 16–18.

Tolba, M. (1985). Heads in the sand: a new appraisal of arid lands management. In *Arid lands today and tomorrow*, (ed. E. E. Whitehead, C. F. Hutchinson, B. N. Timmermann, and R. Givarady), pp.31–2. Westview Press and Belhaven Press, Boulder, Colorado and London.

Wassif, K. (1976). Mammals. In *The Nile, biology of an ancient river*. (ed. J. Rzoska), pp.95–97. Junk, The Hague.

Wendorf, F., Schild, R. El-Hadidi, M.N., Close, A.E., Kobusiewicz, M., Wieckowska, H., Issawi, B., and Haas, H. (1979). Use of barley in the Egyptian late Paleolithic. *Science*, **205**, 1341–7.

Zahran, M.A. (1966). Ecological study of Wadi Dunkul. *Bulletin de l'Institut du Désert d'Égypte*, **16**, 127–43.

Zahran, M.A. (1977). Wet formations of the African Red Sea coast. In *Wet coastal ecosystems. Ecosystems of the world*, (ed. V.J. Chapman), pp.215–31. Elsevier, Amsterdam.

Zohary, M. (1944). Vegetational transect through the desert of Sinai. *Palestine Journal of Botany*, **3**, 57–78.

16

Irrigation development programmes

M. D. SKOLD AND E. V. RICHARDSON

Introduction

Twentieth century events have brought an increasing population pressure to the cultivated lands of Egypt. In 1900 there was 0.29 ha of cropped area per caput, now only about one-third of that amount exists for each person (Table 16.1). The rapid population growth of the past 40 years is exerting a two-way pressure on the land base. Egypt is essentially a narrow valley stretching from Aswan to Cairo along the Nile River before fanning out into the Delta. Not only must the land supply a population increasing over the past three decades at a rate in excess of 2.4 per cent annually, but the agricultural land base itself is threatened by urban expansion (Adams 1985; Gardner and Parker 1985; Ikran 1980). To replace lands lost to the expanding space requirements of cities, reclamation of inherently less productive lands in the coastal areas and on the deserts has enabled the maintenance of a slightly increasing amount of cultivated land.

Irrigation water and land are of equal importance to agricultural production in Egypt; production is not possible without either resource. Egypt has responded to the increased pressure on land with a variety of measures to improve and expand its irrigation system. Over the past century a series of steps has been taken which has resulted in a shift from the ancient system of basin irrigation of one crop per year from the annual flood to perennial irrigation. Year-round cropping has been achieved by the construction of a large number of control structures, the most important and most recent of which is the Aswan High Dam. Cropping intensity, the ratio of cropped area to cultivated area (Table 16.1), increased from 1.36 in 1897 to 1.98 in 1983.

Egyptian agriculture must share its water with other users, domestic and foreign. The 1959 Nile Water Agreement with the Sudan allocates 55.5 milliards (where 1 milliard = 1000 million m^3) of Nile River water to Egypt; 18.5 milliards is allocated to the Sudan and an estimated 10 milliards is lost annually to evaporation. The recent water balance for approximately that amount of water is shown in Table 16.2. Egypt must plan for accommodating

TABLE 16.1. Growth of population and of cultivated and cropped areas, 1897–1975

Year	Population	Cultivated area		Cropped area	
	('000)	('000 ha)	ha caput^{-1}	('000 ha)	ha caput^{-1}
1897	9715	2077	0.21	2825	0.29
1907	11 190	2258	0.20	3191	0.29
1917	12 715	2231	0.18	3247	0.26
1927	14 178	2329	0.16	3581	0.25
1937	15 921	2232	0.14	3488	0.22
1947	18 967	2421	0.13	4286	0.23
1960	26 085	2479	0.10	4370	0.17
1970	33 299	2521	0.08	4580	0.14
1980	42 289	2401	0.06	4864	0.12
1983[a]	45 915	2391	0.05	4734	0.10

[a] Preliminary.
Source: Gardner and Parker (1985); Ikran (1980); Adams (1985).

TABLE 16.2. Nile water balance, estimates of use by sector, 1982

Sector	Water use (milliards)
Productive use	
Municipal	1.84
Industry	0.34
Old lands consumptive use	27.01
New lands consumptive use	0.03
Power and navigation	2.60
Total productive use	31.82
Losses	
Terminal lakes and sea	24.77
Evaporation and seepage	2.11
Total losses	26.88
Total productive use and losses	58.70

Source: Richardson (1986).

the resource needs of its rapidly growing population on a relatively fixed amount of land and water resources.

Moving towards perennial irrigation

Improving the efficiency and productivity of land and water use involves a number of resource development and reclamation measures. Since the construction of the Delta Barrage in 1863, six additional barrages were added by 1951. The small, Aswan Low Dam capable of storing one milliard of water was completed in 1902. With additions, storage capacity was increased to 5.5 milliards by 1933. Dams were also constructed in the Sudan and Uganda. The ultimate control and storage came with completion of the High Dam in 1964. Its storage capacity of 84 milliards amounts to 1.6-times the average annual flow of the Nile (El-Tobgy 1976). These constructions were necessary to provide control of the resources of the Nile to enable more effective distribution of irrigation water throughout the year, between years, and over land.

In addition to the control measures provided by the dams and barrages, the Egyptian Government has introduced a number of other programmes to enhance the availability of water, increase production, and enable expansion of its cultivated land base (Richardson 1986). The efficiency of the irrigation system on old lands, lands cultivated for hundreds, even thousands of years, has been targeted through rehabilitation and improvement of the major distribution system and by improving the management and efficiency of water use on farms. As the capacity of the irrigation system increased with greater control of the annual flow of the Nile and with increased efficiency of the distribution and on-farm application systems, potential expansion of the cultivated land base offered hope for easing the pressure on the old lands to supply food for the growing population (Ikran 1980). Consequently, land reclamation efforts were advanced on (a) swamps and marshes in the coastal zones, (b) on the beds of former lakes and around oases in the deserts, and (c) on other lands in the desert for which soils are deemed suitable for cultivation. Further, the improved management and efficiency of irrigation water on the old lands included improving the drainage system for better on-farm management and enhancing the operating efficiency of the distribution system. A substantial portion of the water applied in excess of the consumptive requirements of crops returns to the system through drains, facilitating the use of that water on other lands.

Efforts to improve the efficiency of the water delivery and removal system will be reviewed briefly in this chapter. The more important of recent efforts to reclaim new lands and to improve the management and operating efficiency of the on-farm irrigation system will then be discussed. The explanations will include descriptions of the irrigation systems as they are used under each resource situation, an identification of the important problems associated with

348   THE AGRICULTURE OF EGYPT

each system, and some evaluation of the progress and results of projects and programmes aimed at overcoming these problems.

Water delivery and removal system

The irrigation delivery system with its single source of supply (the Nile River) running the full 1300 km length of the country appears deceptively simple (Fig. 16.1). From Lake Nasser above the High Dam, the river courses 800 km before separating near Cairo into the Damietta and Rosetta branches. Between these branches the vast alluvial Nile Delta was formed. There are no tributaries entering the Nile below the Aswan Dam. The supply of water is almost entirely defined by releases from the dam plus stored groundwater from previous irrigations.

Seven barrages have been established along the natural river system; 30 000 km of public canals and 80 000 km of private canals channel the diverted water over the irrigated lands. The system also involves some 560 pumping stations and over 17 000 km of private drains. Operation of this system in conjunction with a vast number of control structures and on-farm water lifting devices makes the irrigation system very complex.

Major canals leading from the barrages are designed to serve the crop areas adequately within their reaches. The Ministry of Public Works and Water Resources (MPWWR) is responsible for all aspects of the irrigation and drainage system. The capacity of most canals is not sufficient to provide water simultaneously to all of the area served; rather, canals are designed to serve the land within their reach on a rotation basis. Depending on the predominant crops and season, canal rotations are generally four days on, eight days off, or a 5-day to 10-day rotation. All canals are closed for a period of between 10 days and two weeks in January for cleaning and repair.

The water in the Nile River is very low in salt content and the quality of water in the canals is good throughout the year. The quality of drain water and groundwater varies by site, and in the case of drains, by season (Scott and El-Falaky 1983).

Managing the system

The shift to perennial irrigation and cropping made possible by the Aswan High Dam and other control structures has presented new challenges to the operation and management of the irrigation system. When only one crop per year was taken, the system had time to drain itself and only minimal attention had to be given to surface drains and groundwater levels. However, with perennial cropping and large canals constantly carrying water to secondary and tertiary canals which are continually distributing water onto cultivated

IRRIGATION DEVELOPMENT PROGRAMMES 349

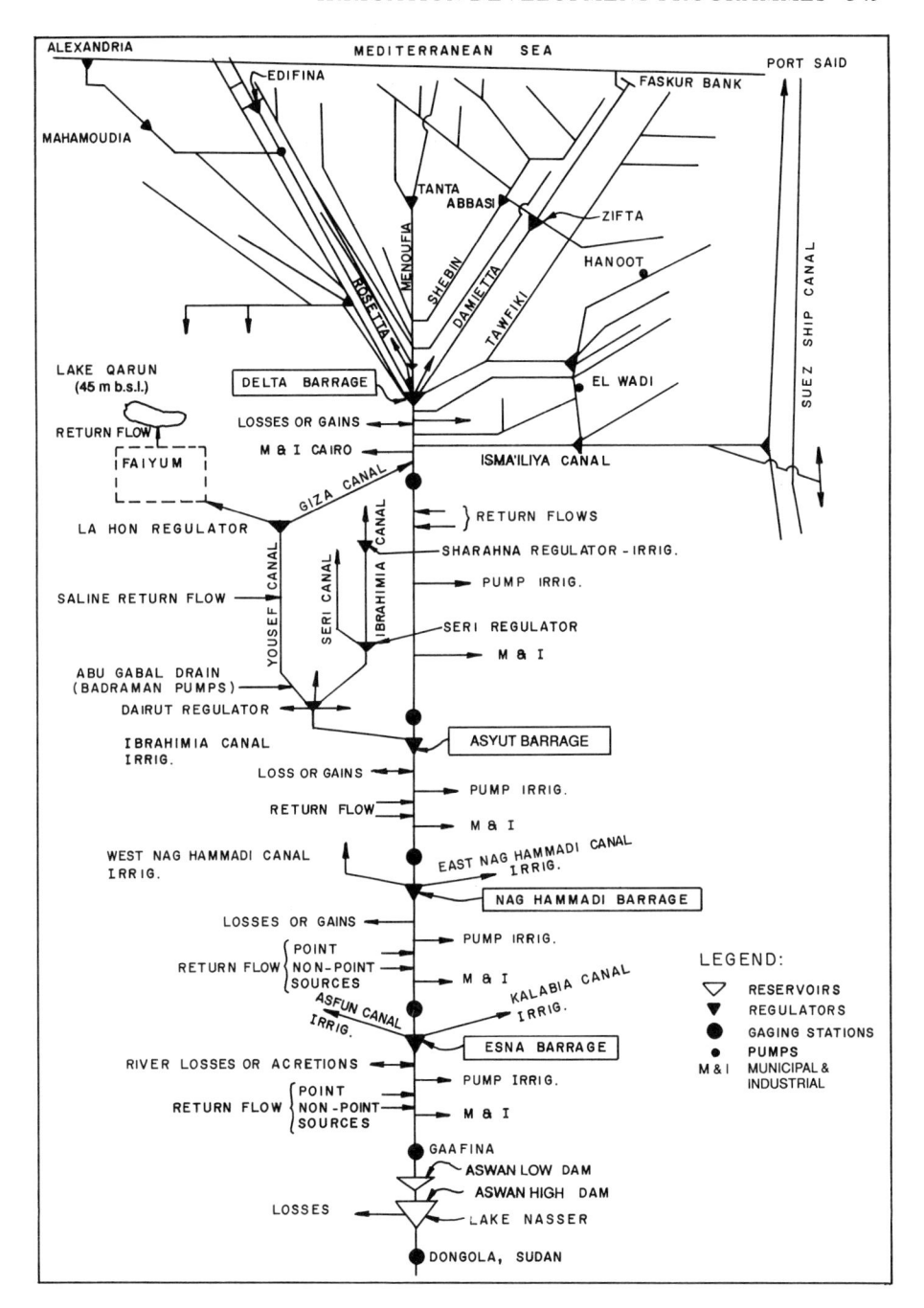

Fig 16.1. Egypt: irrigation delivery system of the River Nile

lands, a new set of problems emerge. The application of more water to a relatively fixed amount of land has caused water-logging, accumulation of salts, and excessively high water tables. The MPWWR has therefore initiated a number of changes in operations, rehabilitation, and construction programmes to reduce and/or minimize the impacts of perennial irrigation to the resource base.

Beginning in the 1970s the Ministry of Agriculture supplied the MPWWR each year with estimates of the crop area, by crop, for the area served by each canal. The MPWRR then estimated the consumptive use requirements for each crop, and water deliveries were scheduled in proportion to the water use requirements of the crops. Substantial water savings occurred and this also resulted in reduced soil deterioration, groundwater levels, and water levels in public drains. In addition, public drains were cleared to improve their efficiency. Clearing involved reshaping the drains as well as removing the weeds and debris. Public canal cross-sections were also improved, gates and weirs were supplied to branch canals, and programmes to improve their maintenance were initiated. Tile drains were constructed on agricultural land resulting, in some cases, in an immediate 30 per cent increase in agricultural production. The construction of tile drains on the old lands is an on going activity.

These improvements resulted in reduced soil degradation and significant savings in water required to irrigate the old lands of Egypt. The water savings added to the potential for horizontal expansion of the irrigation system to new lands along the fringes of the Nile and on the desert.

Reclamation of land (Ministry of Irrigation/UNDP/IBRD 1984)

Measures to provide greater control and increase the efficiency of the irrigation system made horizontal expansion of the irrigation system physically possible. Increased dependency on food imports and a commitment to food security by the Egyptian Government have stimulated several important land reclamation projects. As a result of reclamation efforts, the total cultivated area has increased by approximately 500 000 ha over the past 30 years, in spite of losing over 500 000 ha to urban expansion. Because land reclamation efforts have involved differences in climate, soil types, irrigation methods and settlement patterns, generalizations are impossible. However, a review of four of the major projects with varied manifestations of the physical and cultural differences will provide insight into the problems and potentials of reclamation efforts.

The Maryut project

The Maryut project (Fig. 16.2) is located in the Western Desert some 40 km south west of Alexandria, and reclamation began in 1963. The climate is Mediterranean; the soils are deep but are characterized by a high content of calcium carbonate throughout. The predominant type of soil is loam, and sand content increases with movement towards the sea. Drainage water from the El-Omum Drain is mixed in a 1:6 ratio with water from the Nubariya Canal. The canal, with a capacity of one million m^3 day^{-1}, serves both the Maryut project and the El-Nahda project, and the water irrigates about 18 000 ha.

A secondary canal, the El-Horreya Canal, departs from the Nubariya Canal to serve the Maryut project. Water is lifted up to 35 m by a series of pumping stations to provide gravity-flow irrigation water to farmers. Irrigation efficiencies have been low, ranging from 14.5 to 60.1 per cent depending on the month. Because of on-farm application inefficiencies and seepage from unlined canals, some drainage problems developed quickly in areas with inherently high water tables. While the major open drains seem to serve the area well, field drains are poorly maintained so that an adequate drainage system is the major constraint facing the project.

Another problem is that of energy supplies to the pumping station. Although heavily subsidized, electrical interruptions are frequent and result in water shortages during critical periods of irrigation. These technical problems associated with operation of the irrigation and drainage system have resulted in disappointingly low crop yields which range from 15 to 50 per cent of national average yields.

The land tenure system is mixed. Some 13 000 ha of land is owned and operated by the Maryut Government Company and a portion of this land is leased to smallholders. Other land (2800 ha) was sold to private agri-business investors and about 2400 ha were distributed to smallholders; distribution schemes were also devised for giving about 1800 ha of land to graduates of universities and intermediate schools, and some land is leased to private sector companies. In general, the smallholders have achieved greater production levels than that found on the larger holdings. In spite of the low yields and physical obstacles, a positive economic margin above the costs of operation has been claimed for the project.

El-Nahda project

The El-Nahda Project (Fig. 16.2) lies adjacent to the Maryut project in a north-easterly direction. It contains mostly clay soils, and a combination of smallholders, graduates, and a government company are served irrigation water through a surface irrigation system. The clay soils are more suitable for traditional field and vegetable crop production than the calcareous soils of the Maryut area. Of the 9000 ha reclaimed, 15 per cent is held by the company,

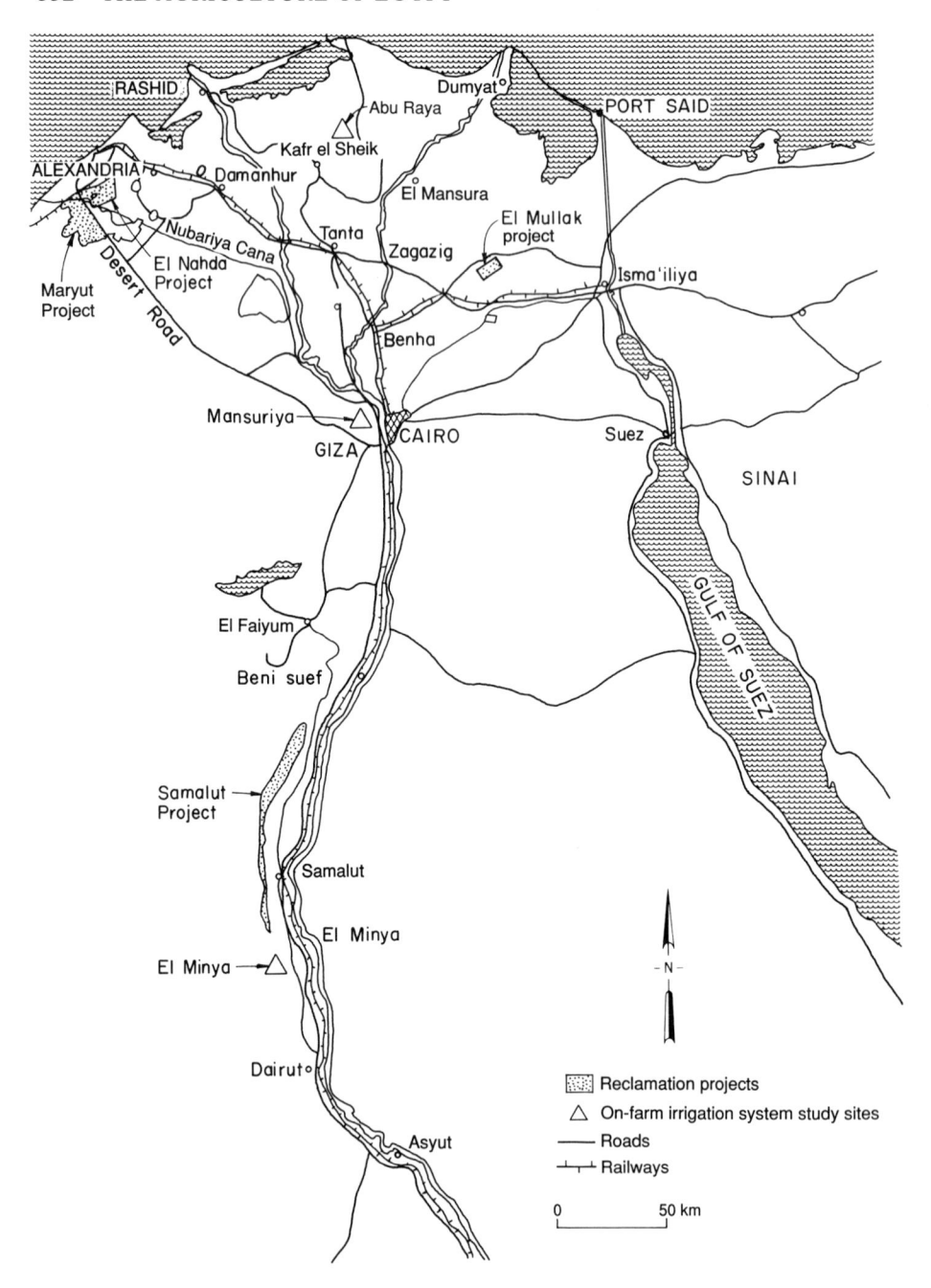

FIG 16.2. Location of reclamation and irrigation system improvement project sites

32 per cent was distributed to graduates, 49 per cent was distributed to small farmers, and about 4 per cent was sold to the private sector.

A gravity-flow surface irrigation system is present. Water from the Nubariya Canal is lifted to irrigate various portions of the area and lifts range from 12 to 18 m. Leaching of salts from the heavy clay soils proved difficult and a very intensive system of open field drains was established. These field drains discharge water into larger drains which, in turn, discharge by gravity into the Tahrir Drain. Maintenance of the drains continues to be one of the major problems for the reclamation project.

Farms cultivated by smallholders and graduates have achieved yields very close to the national average yields for corresponding crops. However, yields achieved on the company farm are much less. Economic evaluations have revealed the project to be providing a positive margin of returns above current operating, maintenance, and input costs.

The El-Mullak project

The El-Mullak project is on the edge of the eastern desert about mid-way between Cairo and Isma'iliya (Fig. 16.2). The soils are gravelly sand and the climate is like that of the south Delta. Although it was planned initially for a small settler farming system, the area has been farmed as a State-managed farm. A specific cropping pattern designed to facilitate the reclamation of the coarse soils has been practised and over 5000 ha of the potentially cultivable 7000 ha have been brought into production.

The project receives its water from the El-Mullak Canal which is fed by gravity from the right bank of the Isma'iliya Canal. The initial plan of supplementing the excellent quality Nile River water with medium to poor quality groundwater has not been implemented. Water is lifted a total of 22.5 m from the El-Mullak Canal by three electrically powered pumping stations. A network of open ditch canals distributes water to booster pumps which feed sprinkler and drip irrigation systems. Because the soils are porous and have good natural drainage and groundwater is relatively deep, a drainage system was not necessary.

Cropping patterns were initially adopted to build the desert soils which are void of organic matter; through time, cropping patterns were planned to change to be consistent with soil types and water availability per unit of land. The plan was to have about one-third of the land under citrus crops and the balance was to be used for field crops with a heavy emphasis on fodder crops (and an associated livestock industry).

Initially, water was distributed to the land through mobile sprinkler systems. However, providing the proper maintenance and technical expertise necessary to keep these relatively complex systems operating has proved to be a major problem. Further, these field distribution systems are not necessarily the best for all the crops grown, given the high water and energy costs and the need for the most efficient irrigation systems (Richardson and Koelzer 1979). Consequently, mobile sprinklers are being replaced by drip systems for the orchard crops and fixed (solid-set) sprinklers are replacing mobile systems on much of the land used to grow fodder for livestock.

Results from the El-Mullak reclamation project have been disappointing. It must be remembered, however, that when the project was initiated, the Egyptians had little desertland reclamation experience to draw upon. The soils were entirely different from any found in the Nile Valley or along the fringes of the Valley and the settlement pattern applied (State farm with hired labourers) was also a departure from the traditional smallholder system. Much has been learned during the course of the development effort. Although yields and returns have not been sufficient to cover even the variable costs for most crops during most years, the knowledge gained should enable financial variable costs to be covered by the appropriate choice of crops, cultural practices, and irrigation practices. The high energy costs associated with lifting water and the high costs of a large number of State-farm labourers may make returns in excess of economic costs difficult to attain without adopting the most efficient irrigation distribution and on-field delivery systems.

The Samalut project

The fourth and final reclamation project to be discussed is the Samalut project (Fig. 16.2). The project is located in the Beni Suef and Minya Governorate as a narrow strip along the fringe of the Western Desert, just west of the old Bahr Youssef Canal. The climate is characteristic of Upper Egypt. Sandy, gravelly, calcareous soils are found and, consequently, both sprinkler and surface irrigation systems have been employed.

Reclamation began in 1967 and during the early years much attention was devoted to leaching salts, breaking of impervious layers of clay, and soil building through cover crops, green manures, and planting of salt-tolerant crops. Of the 25 000 ha included in the area, 18 500 ha are deemed cultivable. Almost 13 000 ha remain under the management of the Upper Egypt Company, about 4500 ha were sold to private investors and private companies, and 1000 ha were distributed to graduates. A small area of 200 ha was distributed to smallholders.

Irrigation water is supplied to the Samalut project area by the old Bahr Youssef Canal which is fed from the Ibrahimiya Canal flowing from above the Asyut Barrage. The southern portion of the project is fed by the El-Kamadeer and Tarfa Canals; the northern portion is supplied water by the Sakoula and Mazoura Canals. A series of pumping stations lifts the water from 20 to 50 m. About 75 per cent of the area is irrigated by gravity flow surface irrigation and the remaining 25 per cent is irrigated with sprinkler systems.

Crop yields achieved by both the settlers and the company approximate to the national average yields for comparable crops. Similarly cropping intensities are equivalent to national averages. Such productivities result in substantial financial and economic margins for settlers and for the company farm.

Inferences about reclamation

Experience with these four projects involving different soil types, settlement patterns, and irrigation systems are indicative of the land reclamation efforts made by the Egyptian Government. The 450 000 ha reclaimed since 1952 have come under cultivation at considerable cost. Costs per hectare have exceeded £E4000 over the period and since 1973 costs have increased considerably due to increases in oil prices and capital costs. Even with the physical and social differences, some inferences can be drawn.

1. Irrigation efficiency on new lands is unacceptably low. Low irrigation efficiencies are associated with faulty design of conveyance systems, inadequate distribution systems, poor selection of irrigation methods, inadequate drainage systems, failure to re-use drainage water, and poor maintenance of the systems. Reclamation projects have been executed and operated jointly by several Ministries and many of the inefficiencies in design and operation can be related to co-ordination and communication problems between Ministries.

2. Inadequate attention has been given to the planning which would ensure that the settlement and operation of new lands would be consistent with the suitability of the area for specific crops and irrigation systems. The larger enterprises on new lands have been relatively more successful in the production of non-traditional crops such as orchards and vineyards. Smallholdings are far more productive than large-scale enterprises in the production of traditional crops. Traditional irrigation methods are efficient among smallholdings, given appropriate soil types.

3. Sprinkle, drip, and non-traditional surface irrigation systems have been relatively more successful under large-scale management schemes. However, operation and maintenance of more sophisticated irrigation systems have been a recurring problem.

Knowledge gained through the implementation of reclamation programmes under a mix of physical and institutional circumstances has been important. In its land reclamation programmes the Egyptian Government has goals other than food security; reclamation projects can also facilitate population dispersion and employment. Furthermore, continued experimentation and analysis can result in a better selection of irrigation systems, crops, and settlement patterns prior to the completion of existing programmes or to the initiation of new ones.

Irrigation of the Old Lands

The major system (Egypt Water Use and Management Project (EWUMP) 1979)

Egypt's major system involves construction of a barrage to increase the elevation of water in the Nile River sufficient to flow into a major distribution canal. Flows from major canals go into secondary branch canals and irrigation water is distributed by the secondary canals on a rotational basis.

Under the rotation method, the area served by one major canal is divided into two or three equal regions, called a double or triple rotational system, respectively. Under each of these rotations, water is admitted to only one of the regions (during the on-period) and the intakes of all the other regions are closed (the off-period). Different allocations in space and time are applied on this system according to system design, location, climate, and cropping pattern. Examples of two-rotation systems are 4 (or 7) days on, 4 (or 7) days off. Three-rotation systems involve 4, 5, or 7 days on and twice the number of days off, respectively.

Canal cross-sections are designed to carry enough water for the crop water requirements of the land it serves. The amount of water diverted from the Nile at the barrage and carried by main canals to secondary canals varies by season, depending on the crop water requirements and evaporation. The Egyptian Government through the MPWWR has full responsibility for operating and maintaining the major system through the secondary canals.

The minor system (El-Kady *et al*. 1979)

The minor system of canals (*mesqas*), which start at the end of the government's distributory canals and extend to individual farms, is in private ownership and is operated and maintained by the farmers. The *mesqa* outlets water to private farm ditches (*marwas*), which are normally steel or concrete pipes laid through canal banks with their crests 25 cm lower than the designed canal water levels. Pipe diameters are chosen in proportion to the size of the area to be served by the private ditch.

Mesqas distribute water from the government's distributory canals along and among farmers. Farmers, or a group of farmers, take water from a *mesqa* and deliver it to their fields through a *marwa* to small bunded units approximately 10 m X 10 m, called basins. The surface of basins may be furrowed for row crops or smoothed for basin crops. Excess surface water may be drained from fields into open field drains or, in some cases, back into a *mesqa*.

Irrigation water is typically delivered in *mesqas* which are 50–75 cm below the ground surface of fields to be irrigated, so farmers must lift the water to their land. Farmers are not required to pay for water; the effort of lifting water is intended to be a water-rationing mechanism. All landowners whose land is supplied by a *mesqa* have the right to take their irrigation water equally in proportion to their land holdings. Although farmers are encouraged to schedule irrigation turns on the ditch, the sequence of use is usually decided by custom which generally favours farmers at the head of the *mesqa* (EWUMP 1984).

Traditional methods for lifting water from *mesqas* to *marwas* or on to fields include the *shaduf*, the *tambour*, and the *saqqiya* (El-Kady *et al.* 1979). The *shaduf*, the most primitive of methods, consisting of a bucket on a pole with a counter-balance and a man lifting each bucket of water, is used mainly by squatters along canals and drains. This device is suitable for only very small fields of perhaps 0.02 ha or less. The *tambour* is an Archimedean screw powered by human labour. A *tambour* can supply water to a larger area than a *shaduf*, but it also has a practical limit of about 0.1 ha. Animals power water wheels known as *saqqiyas*; bullocks, camels, and donkeys are used depending on the amount of lift and the availability of the animal. *Saqqiyas* are frequently jointly owned by a group of farmers called a *saqqiya* circle. If lifts are not excessive, *saqqiyas* can irrigate 4–6 ha effectively. Increasingly, small diesel or electric (mostly diesel) pumps are replacing human and animal powered lifting mechanisms. Analyses have shown that under energy prices currently charged to farmers, diesel pumps are a lower cost means of pumping water when the amount of area served exceeds 5 ha (Wahby and Quenemoen 1981). Farmers may use these pumps to lift water directly from *mesqas*, or may have developed small tubewells to pump from relatively shallow groundwater.

Problems with the on-farm irrigation system

A study completed in 1985 identified problems and developed solutions for the problems found on the on-farm irrigation system (Richardson *et al.* 1985). To develop a cross-section of the problems which may be encountered throughout Egypt, the study identified three sites for intensive study and demonstration.

The El-Mansuriya site (Fig. 16.2) was located along the El-Mansuriya Canal in the Giza governorate. Land within the field site is served by the Beni Magdul and El-Hammami Canals and the site was selected because it represents a vegetable-producing area serving the Cairo market. Soils served from the Beni Magdul Canal are predominantly alluvial clay while those of El-Hammami are sandy. Each canal serves about 350 ha.

The Abu Raya site (Fig. 16.2) is located on new – old lands about 35 km north-east of the city of Kafr el Sheikh in the Nile Delta. This site is representative of a rice and cotton growing region. Wheat, *bersim* (*Trifolium alexandrinum*), and vegetables are also grown and the soils are heavy alluvial clay. Research and demonstration were conducted at three locations, each of approximately 100 ha.

A third site was at Abyuha (Fig. 16.2) about 20 km south of the city of El-Minya. This 500-ha site represents the up-river areas of Egypt; wheat, maize, broad beans (*Vicia faba*), cotton, and sugar cane are important crops. Some of the land in this area was served by gravity irrigation, depending on a farm's location along the *mesqa*.

Analyses of the on-farm irrigation systems of these sites revealed a number of problems; not all problems are present or present to the same extent at each site. Investigations revealed delivery systems to be operated to meet the needs of individual crops rather than considering the entire cropping pattern (Tinsley *et al.* 1984). Furthermore, water deliveries tended to assume that irrigations would occur 24 hours per day, whereas farmers irrigate only from 05.00 h until 12.00 h (El-Din *et al.* 1984). As a consequence, irregular shortages and excesses of irrigation water occurred throughout the growing season. Farmers experiencing highly variable and uncertain water deliveries have adopted very inefficient irrigation practices which result in wastage of water as well as other inputs applied to the land. Water was also found to be poorly distributed along canals (Skold *et al.* 1984). Non-uniform water distribution resulted from the poor condition of gates and regulators, the poor condition of *mesqas*, excessive water application by some farmers, ill-prepared and non-level fields to receive irrigation water, and lack of an organization or a system to affect turns among farmers using water from the same *mesqa*.

Although the principal drains are usually clean and well-maintained, the secondary and small (farmer-maintained) drains were often over-grown with weeds and silted full. As the result, many fields were poorly drained and becoming waterlogged (Litweiler *et al.* 1984).

Farmers have taken steps to adjust to the water delivery system by establishing (illegal) additional turnouts and larger turnouts than prescribed by law (EWUMP 1980). Poorly functioning canals, drains, and irrigation practices have resulted in high and variable (often salty) water tables on many lands (El-Falaky and Scott 1983).

Improved on-farm irrigation practices

Factors contributing to low on-farm water use efficiencies are numerous, but a systematic interdisciplinary approach to the solution of those problems has been effective in the identification and reduction of problems. Improved practices which involved improved crop management, irrigation scheduling, improved levelling for basin irrigation, and the introduction of long level border and furrow irrigation systems have been demonstrated. Long (100 m) level systems can replace the small basins. Various combinations of these practices were demonstrated to increase crop yields and save water (Ley 1984; Ley *et al.* 1984), increasing application efficiencies from 40 per cent to as high as 75 per cent.

Requiring farmers to lift water is not necessarily an effective means of allocating water among farmers in the required amount for a given field (El-Kady et al. 1979; Skold et al. 1984). Properly designed and operated elevated *mesqas* and canals can provide water to farmers in an orderly manner and can overcome many problems of control associated with the poorly maintained below-grade farm irrigation system. The efficacy of lined and elevated canals is greatest on sandy soils but such systems are effective for improving irrigation efficiencies on all soil types (Gates et al. 1984). Canals are designed to provide water consistent with the cropping pattern and the on-farm application system.

Headgates to control the distribution of water among branch canals were improved but their performance remained less than satisfactory. Improved headgates reduced conveyance losses which were as high as 30–45 per cent, but there remains vast potential for improving the efficiency of the irrigation system through better designed and manufactured hydraulic structures.

Improved channelization, elevated and/or lined canals can provide greater water control for increased efficiency of the on-farm irrigation system in Egypt. Significant savings of land required for the farm irrigation system can be achieved by channelization, removal of bunds for basin irrigation, and reductions in land required for traditional water-lifting devices. Land savings, or land brought under cultivation can be increased as much as 14–20 per cent (McConnen et al. 1982). Alternative rotation schedules could significantly reduce water wastage (Haider and Skold 1984; Helal et al. 1983).

The efficiency of the farm irrigation system is affected by a number of parameters beyond the total control and responsibility of the MPWWR. Agricultural policies and programmes, in particular, have a significant effect on the use of resources, including water, on farms. For example, *bersim* represents one of the few crops which does not come under direct government regulation. As it is also one of the most profitable crops for farmers they have an incentive to allocate water and other scarce resources to *bersim* in deference to food crops.

Problems inherent in the traditional farm irrigation systems of Egypt invariably remain after physical improvements have been made to those systems. The private, farmer-owned *mesqas* and drains do not get maintained by the farmers served by the system. There is a lack of organization to operate co-operatively in order to maintain improved and/or elevated *mesqas*, pumps to lift water, and other control structures. Consequently, major attention must be given to the establishment of water-users associations and organizations in order to disseminate technical knowledge about improved irrigation practices to farmers. In situations where such organizations and associations were established, improved water management resulted in decreased irrigation time, increased crop yields and water savings (Layton et al. 1984).

The Egyptian Government is correct in its expectations of achieving significant increases in irrigation efficiencies. Efficiencies can result in (a) reductions in the amount of land claimed by the irrigation system, (b) greater amounts of agricultural output from less water, (c) reductions in soil degradation due to water-logging, salinization, and nutrient leaching, and (d) better accommodation of the potentials for perennial cropping and irrigation made possible by the Aswan High Dam and its associated control structures.

Implications

The Egyptian irrigation system is complex, procedures to gain greater control over the resources of the Nile have been evolving for over 100 years. With the construction of the High Dam, Egypt gained increased control over the flow of the Nile and a supply of water throughout the year. Perennial irrigation and cropping permits three crops per year on some land resulting in an overall cropping intensity approaching 2.0. The water delivery and removal system must be operated in conjunction with a farm irrigation system serving almost three million farmers farming 2.1 million ha of land. Perennial irrigation, which has been in operation for only about one-quarter of a century, greatly increased the irrigation potential and the problems. The Egyptian Government and its people have done well to adjust to an irrigation system substantially different from the one which had served it for over 50 centuries.

The High Dam and the control it provides over the flow of the Nile made possible the complete adoption of a perennial irrigation system. Thereafter it was land rather than water that became the most limiting resource. Through efficient management of the water delivery and removal system and of the on-farm irrigation system, there is potential for the horizontal expansion to new lands. Drawing upon information from previous research, the MPWWR is presently pursuing activities to update and improve the irrigation system. These include improving the on-farm system by establishing long, level furrows, where appropriate, and more level basins for irrigation. Farmers are being organized at the *mesqa* level for co-operative maintenance of *mesqas* and to schedule the receiving of water from the *mesqa*. The MPWWR is also forming an irrigation advisory service to promote the exchange of technical information between the MPWWR and farmers. At the same time, the total delivery system, from major canals to *mesqas* is being improved in design and structure (Consortium for International Development 1986).

Many problems remain, but much has been learned about the reclamation and cultivation of new lands. Although success, to date, has been limited, there is great potential when reclamation schemes and irrigation systems introduced take cognizance of the physical environment, the planned settlement pattern, and the biological potentials of the area. Evaluations are on-going to determine the appropriate irrigation system for varied conditions of soil types, crops, energy and labour costs (WDISRI 1984).

Horizontal expansion of the irrigated area can be enhanced by further implementation of proven measures to improve the efficiency of the irrigation system on old lands. Unfortunately, time and capital have become a constraint. Because of the pressure on food supplies associated with a rapidly increasing population, Egypt requires instant success. While knowledge to improve the irrigation system of Egypt is known, its dissemination is limited by the availability of personnel trained to design, construct, operate, manage, and maintain the improved system. The improved and better managed physical system will also require irrigation advisers to organize farmers to maintain *mesqas* and equitably schedule water among farmers. While the economic costs are high, the costs of not attending to improvement and further development of the irrigation system are totally unacceptable.

References

Adams, R. H., Jr. (1985). Development and structural change in rural Egypt, 1952 to 1982. *World Development*, **13**, 705–23.

Consortium for International Development (1986). *Egypt irrigation improvement project; technical progress report. Project No. 263–0132*. Tucson, Arizona.

El-Din, K. El-Din, E., and Litweiler, K. (1984). *Baseline data for improvement of a distributory canal system*, EWUMP, Technical Report No. 72. Colorado State University, Fort Collins.

El-Falaky, A. and Scott, V.H. (1983). *Water quality of irrigation canals, drains and groundwater in Mansuriya, Kafr El-Sheikh, and El-Minya project sites*, EWUMP, Technical Report No. 62. Colorado State University, Fort Collins.

El-Kady, M., Clyma, W., and Abu-Zeid, M. (1979). *On farm irrigation practices in the Mansuriya District, Egypt*, EWUMP, Technical Report No. 4. Colorado State University, Fort Collins, Colorado.

El-Tobgy, H. A. (1976). *Contemporary Egyptian agriculture*, (2nd edn). The Ford Foundation, Cairo.

EWUMP (Egyptian Water Use and Management Project) (1979). *Preliminary evaluation of the Mansuriya canal system Gisa Governorate, Egypt*, EWUMP, Technical Report No. 3. Colorado State University, Fort Collins, Colorado.

EWUMP (Egyptian Water Use and Management Project) (1980). *Problem identification report for the Kafr El-Sheikh study area*, EWUMP, Technical Report No. 6. Colorado State University, Fort Collins.

EWUMP (Egyptian Water Use and Management Project) (1984). *Improving Egypt's irrigation system in the old lands: findings of the Egypt Water Use and Management Project*, Water Distribution Research Institute, Water Research Center, Ministry of Irrigation, Cairo.

Gardner, G. R. and Parker, J. B. (1985). *Agricultural statistics of Egypt, 1970–84*. United States Department of Agriculture, Economic Research Service Statistical Bulletin No. 732. USDA, Washington, DC.

Gates, T.K., Ree, W.O., Helal, M., and Nasr, A. (1984). *Hydraulic design of a canal system for gravity irrigation*, EWUMP, Technical Report No. 46. Colorado State University, Fort Collins.

Haider, M.I. and Skold, M.D. (1984). Impact of agricultural policies on resource allocation and farm income: the case of Egypt. *International Journal for Development Technology*, **2**, 223–9.

Helal, M., Nasr, A., Ibralim, M., Gates, T.K., Ree, W.O., and Semaika, M. (1983). *Water budgets for irrigated regions of Egypt*, EWUMP, Technical Report No. 47. Colorado State University, Fort Collins.

Ikran, K. (1980). *Egypt, economic management in a period of transition*. World Bank Country Economic Report. Johns Hopkins University Press, Baltimore.

Layton, J. et al. (1984). *Experiences in developing water users associations*, EWUMP, Technical Report No. 65. Colorado State University, Fort Collins.

Ley, T.W. (1984). *Precision land leveling on Abu-Raya farms, Kafr El-Sheikh Governorate Egypt*, EWUMP, Technical Report No. 38. Colorado State University, Fort Collins.

Ley, T.W. et al. (1984). *The influence of farm irrigation system design and precision land leveling on irrigation efficiency and irrigation water management*, EWUMP, Technical Report No. 41. Colorado State University, Fort Collins.

Litweiler, K., Deweeb, H., and Ley, T. (1984). *Infiltration studies on Egyptian vertisols*, EWUMP, Technical Report No. 57. Colorado State University, Fort Collins.

McConnen, R.J., Farouk Abdel Al, Skold, M.D., Ayad, G., and Sorial, E. (1982). *Feasibility studies and evaluation of irrigation projects*, EWUMP, Technical Report No. 12. Colorado State University, Fort Collins.

Ministry of Irrigation/UNDP/IBRD (1984). *Detailed examination of existing land reclamation projects*, Water Master Plan, Technical Report No. 29. Ministry of Irrigation, Cairo.

Richardson, E.V. (1986). Joint UNDP/FAO/IBRD mission to review technical requirements in agriculture and food security sector in Egypt. Unpublished report, Colorado State University, Fort Collins.

Richardson, E.V. and Koelzer, V.A. (1979). Irrigation technology for desert land reclamation. In *Advances in desert and arid land technology and development*, (ed. A. Bishay and W. G. McGinnies). Harwood Academic Publishers, New York.

Richardson, E.V., Quenemoen, M.E., and Horsey, H.R. (1985). *Final administrative report, Egypt Water Use and Management Project, Contract No. AID/NE-C-1351*. Consortium for International Development, Tucson, Arizona.

Scott, V.H. and El-Falaky, A. (1983). *Conjunctive water use: the state of the art and potential for Egypt*. EWUMP, Technical Report No. 44. Colorado State University, Fort Collins.

Skold, M. D., Shinnawi, S. A., and Nasr, M. L. (1984). Irrigation water distribution along branch canals in Egypt: economic effects. *Economic Development and Cultural Change*, **32**, 547–67.

Tinsley, R.L., Ismail, A., and El-Kady, M. (1984). *A method for evaluating and revising water rotations*, EWUMP, Technical Report No. 48. Colorado State University, Fort Collins.

Wahby, H. and Quenemoen, M.E. (1981). *A procedure for evaluating the cost of lifting water for irrigation in Egypt*, EWUMP, Technical Report No. 7. Colorado State University, Fort Collins.

WDISRI (Water Distribution and Irrigation Systems Research Institute) (1984). *Technical and socio-economic evaluation of the irrigation systems in the new lands and their impact on crop production, soil properties and water requirements*. Water Research Center. Ministry of Irrigation, Cairo.

17

Fertilizers

M. M. EL-FOULY

Introduction

Egypt has a long tradition of using chemical fertilizers, its first use of Chilean nitrates dating back to 1902. For over thirty years all chemical fertilizers were imported, until the local production of phosphate fertilizers started in 1936 and later the production of nitrogen fertilizers in 1951. The great demand for food, highly productive soil, availability of good quality irrigation water, optimum climatic conditions, and the use of new technology have contributed to a continuous increase in fertilizer production and thus, in fertilizer use. The increase in fertilizer use is influenced mainly by the following factors: cropping index and rotation, use of high yielding cultivars, degree of cultivation of newly reclaimed areas, requirements for subsidized fertilizers determined by the Ministry of Agriculture each season, amount of subsidies available for fertilizers, local production and the availability of fertilizers when they are needed. Tables 17.1 and 17.2 show the quantities of fertilizers distributed to farmers by the Principal Bank for Development and Agricultural Credit (PBDAC) between 1970 and 1990, together with the official fertilizer requirement (an amount fixed by Ministerial Decree each season) for the years 1981/2 until 1988/9.

Until recently, fertilizers have been used mainly in the form of straight fertilizers (Table 17.3). The major forms of nitrogen fertilizer used are calcium nitrate, ammonium sulphate, ammonium nitrate, and urea, while phosphate fertilizers are superphosphate, improved or double superphosphate, and triple phosphate. Only one potassium fertilizer is used, namely potassium sulphate. Straight fertilizers are still the major fertilizers used in traditional agriculture. Compound soil fertilizers were once imported for trial purposes (Table 17.4). Very small amounts of imported soluble compound fertilizers (Table 17.5) are used mainly in the new agricultural areas with unconventional irrigation systems—drip, pivot, and sprinkler irrigation.

From 1970 until 1973 the use of N, and until 1974 the use of P_2O_5, was more or less stagnated; since then in general, there has been a gradual increase. The use of K has remained very low compared with that of N and P because it was believed that the high exchangeable-K content of the soil was adequate for most crops except those with a high starch content, as well as some vegetables and fruits (Serry 1980). However, as the research on potassium has increased (Faizy 1980) and, taking modern concepts of plant nutrition into consideration, it has been found that the K requirement of crops is much greater than was previously estimated (El-Fouly 1984, 1989; Abdel-Hadi 1989). Since 1975, a gradual increase in K consumption has been recorded (PBDAC 1986). This trend is leading to a more balanced ratio of fertilizers applied, instead of favouring only N; it also reflects the real needs of crops more accurately.

Table 17.2 shows an interesting comparison between official fertilizer requirements, fixed by Ministerial Decree each season (an estimation of fertilizer need), and the actual quantities of fertilizer distributed to farmers.

TABLE 17.1. Amounts of NPK fertilizers, distributed by PBDAC during the period 1970–1990 ('000 t)

Year	Fertilizer		
	N	P_2O_5	K_2O
1970	317	55	1.4
1971	308	59	1.9
1972	332	61	1.4
1973	322	57	2.4
1974	360	50	1.9
1975	403	78	2.4
1976	408	83	3.4
1977	418	68	2.9
1978	474	87	3.8
1979	483	98	6.7
1980	490	99	11.5
1981	626	134	12.0
1982	660	143	10.0
1983	746	150	17.0
1984	639	164	25.0
1985	775	184	25.0
1986	750	177	29.0
1987	791	184	29.0
1988	805	190	33.0
1989	790	180	27.0
1990	793	165	21.0

Source: data for 1970–80, MOA (1986); 1981–90, PBDAC (1991, personal communication).

TABLE 17.2. Fertilizer need estimation and distribution during the period, 1981–9

Year	Fertilizer	Fertilizer requirement[a] (need estimation) ('000 t)	Fertilizers distributed by PBDAC ('000 t)	Fertilizer distributed relative to fertilizer requirement (%)
1981/2	N	692	626	90.5
	P_2O_5	158	134	85.0
	K_2O	–	12	–
1982/3	N	701	660	94.0
	P_2O_5	165	143	86.7
	K_2O	–	10	–
1983/4	N	617	746	121.0
	P_2O_5	168	150	89.3
	K_2O	–	17	–
1984/5	N	733	639	87.2
	P_2O_5	168	164	97.6
	K_2O	–	25	–
1985/6	N	752	775	103.1
	P_2O_5	180	184	102.0
	K_2O	29	25	86.2
1986/7	N	781	750	101.3
	P_2O_5	189	177	97.4
	K_2O	27	29	107.4
1987/8	N	786	791	102.4
	P_2O_5	187	190	101.6
	K_2O	34	29	97.1
1988/9	N	801	772	98.6
	P_2O_5	192	181	93.8
	K_2O	38	27	71.1

[a] An official rate fixed by Ministerial Decree each season, which can be purchased by all farmers.

Source: data for 1970–80, MOA (1986), 1981–90, PBDAC (1991, personal communication).

TABLE 17.3. Distribution of fertilizers through PBDAC during the period, 1981/2–1989/90 ('000 t)

	1981/2	1982/3	1983/4	1984/5	1985/6	1986/7	1987/8	1988/9	1989/90
Calcium nitrate (15.5%N)	36.5	39.6	41.9	34.7	22.4	38.2	35.6	38.2	37.9
Ammonium sulphate (20.6%N)	25.5	31.9	37.3	43.7	75.3	72.3	68.9	66.8	82.3
Ammonium nitrate (33.5%N)	246.1	233.4	233.9	234.5	269.5	254.1	300.1	291.9	272.0
Urea (46%N)	315.9	372.0	455.9	339.8	314.1	440.5	385.8	396.4	348.6
Superphosphate (15%P_2O_5)	74.3	86.1	110.1	129.3	160.6	151.8	142.7	140.4	154.1
Double superphosphate (37%P_2O_5)	–	–	–	–	11.8	38.1	37.1	24.5	30.1
Triple superphosphate (45%P_2O_5)	59.6	56.8	49.5	34.8	10.9	–	–	–	–
Potassium sulphate (48%K_2O)	12.4	10.4	18.9	26.3	26.4	32.7	27.0	21.2	27.7

Source: El-Khashab (1987); PBDAC (1991, personal communication).

TABLE 17.4. Compound fertilizers (for soil application) imported during the year 1973/4

Fertilizer	t
Complex 25/5/5	4000
Complex 30/10/0	5000
Diammonium phosphate 16/48	5000

Source: El-Khashab (1987).

It shows that farmers are becoming more convinced of the need for potash. In 1985, the first official estimation of the K_2O requirement was carried out. The difference between the actual distribution of fertilizers by PBDAC and the estimated need is due mainly to the unavailability of fertilizers at certain times rather than the unwillingness of the farmers to use them.

Production and importation of fertilizers

Local production of nitrogen and superphosphate

The fertilizers industry in Egypt started in 1936 by producing superphosphate from locally available raw material. In 1951 the first locally produced N fertilizer, namely calcium nitrate, came on to the market. Since then, the capacity of existing production plants has increased and new plants have been built (Aglan 1987; NSC 1981). Details of Egyptian companies producing chemical fertilizers are given in Table 17.6. No potash fertilizers are produced in Egypt due to the lack of raw material, but it was reported recently that some local potash deposits had been found. The production of N and P fertilizers

TABLE 17.5. Compound fertilizers for drip irrigation, imported in the years 1986–9

Fertilizer	1986/7 (t)	1988/9 (t)
13/40/13 (Kristalon 66)	261	100
15/5/30 + 3 Mg (Kristalon White)	306	1606
19/6/20 + 4 Mg (Kristalon Blue)	329	1000
19/19/19 + 2 Mg (Kristalon Special)	627	200

Source: El-Khashab (1987).

TABLE 17.6. Companies producing chemical fertilizers in Egypt

Company/location	Product	Start of production
El-Malia Wael Senaiaa (EFIC)	Superphosphate (15% P_2O_5)	1936
Kafr El-Zayat branch	Superphosphate (15% P_2O_5)	1936
Asyut branch	Superphosphate (15% P_2O_5)	1969
Abu Zaabal Company (fertilizers and chemicals)	Superphosphate (15% P_2O_5)	1948
	Double superphosphate (37% P_2O_5)	1984
El Nasr Company for Fertilizers and Chemicals (SEMADCO)	Calcium nitrate (15.5% N)	1951
	Ammonium sulphate (20.6% N)	1963
Suez branch	Calcium ammonium nitrate (26% N)	1975
	Calcium ammonium nitrate (31% N)	1976
	Urea (46% N)	1980
Talkha branch	Ammonium nitrate (33.5% N)	1988
Egyptian Chemical Industries (KIMA)	Calcium ammonium nitrate (20.5% N)	1960
	Calcium ammonium nitrate (26% N)	1964/5
	Calcium ammonium nitrate (31% N)	1968/9
	Ammonium nitrate (33.5% N)	1988
El Nasr Company (coke industries and basic chemicals)	Ammonium sulphate (20.6% N)	1964
	Calcium ammonium nitrate (20.5% N)	1971
	Calcium ammonium nitrate (33.5% N)	1973
Abu-Qir Company (fertilizers)	Urea (46% N)	1980
	Ammonium nitrate (33.5% N projected)	1991

Source: NSC (1981).

has increased markedly over the last decade. In the late 1980s, production exceeded 600 000 t N and 128 000 t P compared with 118 000 t N and 59 000 t P in 1970. Detailed data on the production, consumption, and trade in N, P, and K since 1962 is given in Appendix 4 Tables 4.1, 4.2, and 4.3.

Studies on the future of fertilizer industries (Aglan 1987; NSC 1981) stress the need for expansion and describe possible plans. Wherever existing plants are producing at less than full capacity, it is recommended that bottlenecks, caused by the shortage of hard currency required for repairs and maintainence, should be overcome (CAPMAS 1982). It is recommended that the locally available natural gas should be used less as a source of energy and more as a raw material for fertilizer production. In 1991, the Abu Qir Company began producing ammonium nitrate, its annual capacity reaching 750 000 t.

Rock phosphate for production of phosphatic fertilizers is available, but sulphur is mainly imported and partially produced by the Red Sea oil industry. Another major constraint in the fertilizer industry is the subsidy policy, for fertilizers are delivered to the PBDAC at a price far below the production cost (CAPMAS 1982).

A private fertilizer company in Egypt is now providing farmers with a variety of different NPK formulations, using local sources of N and P and imported K.

Importation of fertilizers

Inspite of the increasing local production of N and P fertilizers, their imports are considerable. The value of imports has reached two maxima in excess of £E80million, the first in 1974–5, followed by a decrease, followed by a second peak in 1985/6 (Table 17.7 and 17.9) The main countries exporting fertilizers to Egypt are given in Table 17.8, and Table 17.9 shows public sector imports until 1987/8. Since 1988, the private sector has also been allowed to import fertilizers.

Distribution of fertilizers

Until recently the PBDAC has been the only outlet for straight NPK fertilizers in Egypt. All local production was delivered to the PBDAC from where it was distributed to farmers through its branches in villages all over the country. In the meantime, imported fertilizers are imported through the General Organization for Agricultural Price Stabilization (GOAPS) on behalf of the Bank which distributes them. In a meeting in April 1987, the Deputy Prime Minister and Minister of Agriculture indicated that the Ministry planned to authorize dealing in straight fertilizers outside the PBDAC. This was put into action in 1989.

TABLE 17.7. Quantity and value of imports of chemical fertilizers (1970/1–1982/3)

Year	N fertilizers (15.5% N)		P fertilizers (15% P_2O_5)		K fertilizers (48% K_2O)	
	'000 t	£E '000	'000 t	£E '000	'000 t	£E '000
1970/1	1385	1 0792	–	–	1.6	32
1971	1524	10 146	–	–	1.6	38
July/Dec. 1972	955	8 246	–	–	–	–
1973	1250	16 800	–	–	8.0	221
1974	2094	81 854	–	–	15.0	611
1975	1928	97 557	–	–	6.1	316
1976	824	13 890	–	–	5.0	291
1977	1957	32 291	20	302	10.0	615
1978	1930	41 302	39	651	10.0	644
1979	1400	29 661	41	970	20.0	1296
1980	1021	45 159	80	3912	15.5	2082
Jan./June 1981	1241	74 584	85	8432	15.5	2596
1981/2	520	31 320	145	29 102	25.0	5018
1982/3	160	10 261	160	23 248	25.0	4838

Source: MOA (1986).

TABLE 17.8. Major countries exporting fertilizers to Egypt

Fertilizer		Imported from
Type	Content	
N	Ammonium sulphate	France, Italy, Belgium, Bulgaria
	Ammonium nitrate	USSR, Romania, Czechoslovakia
	Urea	The Netherlands
P	Triple superphosphate	Tunisia, Lebanon, Belgium, France, USA
K	Potassium sulphate	Switzerland, USA

Source: MOA (1986).

Subsidized NPK fertilizers are currently distributed by the PBDAC, through its stores which are distributed all over the country (Table 17.10). These stores can accomodate up to about three million tonnes of fertilizers (El-Khashab 1989). Fertilizers are distributed to farmers according to a Ministerial Decree issued each agricultural season. This indicates fertilizer rates for each crop (Keleg *et al.* 1987). Appropriate amounts of fertilizer to meet the statutory rates of application are given to farmers all over the country with very limited differentiation. They are considered a compromise between different factors

TABLE 17.9. Imports of fertilizers by the public sector

	1982/3	1983/4	1984/5	1985/6	1986/7	1987/8
Fertilizer (t)						
Urea (46% N)	–	–	–	78 659	–	–
Ammonium nitrate (33.5% N)	–	–	52 306	183 214	70 366	119 230
Ammonium sulphate (20.6% N)	131 237	68 474	136 410	493 376	193 687	196 952
Triple superphosphate (44% P_2O_5)	160 421	71 725	49 387	–	–	–
Potassium sulphate (48% K_2O)	24 961	55 211	35 175	50 011	60 301	69 872
Nutrient (t)						
N	27 035	14 106	45 623	199 195	63 472	80 514
P_2O_5	70 585	31 559	21 730	–	–	–
K_2O	11 981	26 501	16 884	24 005	28 944	33 539

Source: El-Khashab (1989).

TABLE 17.10. Storage areas in different regions in Egypt (June 1986)

Region	Closed stores		Open stores		Free storage area	
	Number	Area ('000 m²)	Number	Area ('000 m²)	Number	Area ('000 m²)
Alexandria	31	4.26	–	–	5	41.8
Cairo	–	–	–	–	1	70.0
East Delta	59	26.90	27	54.2	123	983.9
West Delta	26	22.13	11	14.02	117	854.6
Middle Delta	19	27.00	5	22.75	53	389.3
Suez Canal	14	5.82	–	–	12	62.8
Middle Egypt	47	34.00	7	17.03	133	929.0
Upper Egypt	63	51.18	27	30.80	77	558.0

Source: El-Khashab (1987).

including real crop needs in different locations, availability, price policy, and subsidy.

Farmers are entitled to purchase specific quantities of fertilizers, by credit, according to the decree and their acreage. The PBDAC finances the whole purchase and allocates the required funds (Table 17.11). The value of fertilizers purchased by credit during the years 1983/4–1989/90 is shown

TABLE 17.11. Funds allocated by the PBDAC for the purchase of different quantities of fertilizers (1981/2–1990/91)

Year	Quantity of fertilizer ('000 t)	Value of fertilizer (£Emillion)
1981/2	2798	244
1982/3	2675	178
1983/4	2493	143
1984/5	2814	170
1985/6	3711	270
1986/7	3197	244
1987/8	3342	263
1988/9	3660	600
1989/90	3430	611
1990/1	3227	782

Source: El-Khashab (1987); PBDAC (1991, personal communication).

TABLE 17.12. Fertilizers purchased by credit

Year	Value of total distribution (£Emillion)	Value of distribution purchased by credit (£Emillion)	Percentage on credit (%)
1983/4	225.0	162.1	72
1984/5	199.5	162.9	82
1985/6	241.2	186.7	77
1986/7	204.0	–	75
1987/8	243.6	–	75
1988/9	361.3	–	75
1989/90	436.4	–	75

Source: El-Khashab (1987); PBDAC (1991, personal communication).

in Table 17.12. Until 1986, only about 20–25 per cent of the total volume of fertilizers distributed were purchased with cash. There are no figures to show the private sector's share of fertilizer distribution since 1989. Since 1990, companies producing fertilizers are also allowed to distribute their products directly to farmers or through channels other than the PBDAC.

Fertilizer subsidy

Consumer (farmer) prices and production prices of fertilizers are both fixed by the Egyptian Government. The production price does not allow the fertilizer companies to cover their production costs.

Previously, the fixed delivery price to the PBDAC was far below the fixed production price, but the difference between these prices was paid by the Fertilizer Fund through the General Organization for Agricultural Price Stabilization (GOAPS).

Production cost = Production price + Deficit (balanced by the Government)

Production price = Delivery price to the PBDAC + subsidy from the Fertilizer Fund (public finances).

For example, 31 per cent ammonium nitrate had a delivery price of £E47 and a subsidy from the Fertilizer Fund of £E48 making a production price of £E95 t^{-1}. However, the actual production cost plus a very low marginal return was £E120 t^{-1} (Aglan 1987). As the demand for fertilizers is increasing and the Egyptian Government is trying to cut the subsidy or at least to freeze it, the 'requirements' given to farmers at the subsidized price are fixed. Any other fertilizers, additional to the fixed requirement are sold at higher prices and may be distributed by the private sector. Until now, however, these additional amounts have usually been distributed through the PBDAC.

For imported fertilizers, the GOAPS imports according to world market prices but delivers to the PBDAC at the fixed prices; the difference is subsidized. The subsidy on imports reached a peak in 1975, a year of low local production and high fertilizer usage, then decreased to a minimum in 1977. Thereafter, it increased gradually to more than £E100million in 1981/2 (Table 17.13). The total subsidy in the past few years has been in excess of £E100million, not including the difference between local production cost and the fixed production price.

From the increase in fertilizer prices shown in Tables 17.14 and 17.15, it is apparent that the fertilizer subsidies are being cut each year. By 1993, or even before, it is estimated that the subsidy on at least N and P fertilizers will be cut totally. Subsidies were reduced from £E183million in 1989/90 to £E62million in 1991/2 (PBDAC 1991, personal communication).

Fertilizer requirements, distribution rates, and crop needs

In a system such as that described above, one has to distinguish between the official fertilizer requirement (need estimation), an amount fixed by Ministerial Decree each season, the amount of fertilizer actually used by

TABLE 17.13. Subsidy to local (N and P) and imported chemical fertilizers (£E'000) (1973–82/3)

Year	Local fertilizers	Imported fertilizers	Total
1973	980	495	1475
1974	3732	48 721	52 453
1975	5736	71 504	77 240
1976	3735	31 398	35 133
1977	7694	6180	13 874
1978	15 269	16 742	32 011
1979	28 937	47 265	76 202
1980	29 422	12 313	41 735
Until 30:6:1980			
80/1	66 409	35 082	101 491
81/2	83 312	101 417	184 729
82/3	83 600	35 000	118 600

Source: CAPMAS (1982); MOA (1986).

TABLE 17.14. Development of fertilizer retail prices (£E t^{-1})

	1983/4	1984/5	1985/6	1986/7	1987/8 (up to Jan. 1988)	1987/8 (up to Aug. 1988)
Subsidized prices						
Urea (46% N)	126.80	126.80	131.00	131.00	151.00	263.00
Ammonium nitrate (33.5% N)	91.20	91.20	91.20	91.20	108.00	141.00
Calcium ammonium nitrate (31% N)	84.50	84.50	88.50	88.50	98.00	98.00
Calcium nitrate (15.5% N)	47.70	47.70	47.70	47.70	47.70	84.00
Ammonium sulphate (20.6% N)	57.90	57.90	57.90	57.90	74.00	127.00
Superphosphate (15.5% P$_2$O$_5$)	30.30	30.30	30.30	30.30	42.00	71.00
Double superphoshate (37.5% P$_2$O$_5$)	–	–	–	75.00	104.00	188.00
Triple superphosphate (44% P$_2$O$_5$)	86.80	86.80	86.80	86.80	86.80	86.80
Potassium sulphate (48% K$_2$O)	57.00	57.00	57.00	57.00	57.00	57.00
Unsubsidized prices						
Urea (46% N)	213.70	226.17	226.17	226.17	246.17	293.20
Ammonium nitrate (33.5% N)	182.60	277.54	277.54	283.20	283.20	283.20
Calcium ammonium nitrate (31% N)	182.90	182.90	182.90	182.90	262.20	262.20
Calcium nitrate (15.5% N)	74.70	117.24	117.24	117.24	131.00	131.00
Ammonium sulphate (20.6% N)	121.30	221.28	221.28	221.28	221.30	221.30
Superphosphate (15.5% P$_2$O$_5$)	73.80	142.70	142.70	142.70	142.90	142.90
Double superphosphate (37.5% P$_2$O$_5$)	–	–	–	301.00	301.00	301.00
Triple superphosphate (44% P$_2$O$_5$)	217.50	189.62	189.62	189.62	321.00	321.00
Potassium sulphate (48% K$_2$O)	211.10	211.10	206.58	206.58	211.00	211.00

Source: El-Khashab (1989).

TABLE 17.15. Ex-factory selling prices, 1982/3–1988/9 (£E t^{-1})

Company	Product	1982/3–1986/7	1 July 1987	1 Jan. 1988	1 July 1989	1990	1991
SEMADCO	Calcium nitrate	26.60	35.90	116.00	126.0	139	194.21
SEMADCO	Ammonium sulphate	32.50	48.60	140.44	205.0	232	295.05
El Nasr		32.50	48.60	171.00	181.0	203	264.6
SEMADCO	Calcium ammonium nitrate	46.20	56.10	98.00	–	189	276.11
KIMA		48.00	61.50	139.00	–	235	298.20
SEMADCO	Ammonium nitrate	–	62.18	152.00	166	271	362.25
El Nasr		46.20	63.00	239.00	249	–	–
KIMA		–	71.58	209.00	249	189	276.11
SEMADCO	Urea	123.50	143.50	209.40	209	236	303.45
Abu-Qir	Urea	127.70	147.70	209.40	209	236	303.45
Abu Zaabal	Superphosphate	23.66	35.36	101.00	167	159	191.10
Asyut	Superphosphate	23.66	35.36	101.00	147	152	183.75
EFIC	Superphosphate	23.66	35.36	101.00	147	152	183.75
Abu Zaabal	Triple superphosphate	60.74	89.74	301.00	390	422	458.85

Source: El-Khashab (1989); PBDAC (1991, personal communication).

farmers, and the amount of fertilizer required by a specific crop to give the most economic yield.

The amounts of fertilizers fixed by Ministerial Decree are the quantities of fertilizer which can be purchased by all farmers at subsidized prices and on credit. Apart from a few exceptions, these quantities do not change very much from place to place because they are heavily subsidized and subsidies are divided equally between farmers. In recent years, however, additional quantities of fertilizer have also been purchased at market prices (Hamissa 1989).

It is well known that in practice neighbouring farmers use different rates of fertilizers for the same crop. This happens in spite of all the experiments carried out by the Ministry of Agriculture to determine the afore-mentioned requirements. The reason for this is that the rates given by the Ministry of Agriculture are averages and not specifically tailored according to specific crop needs in a specific area. The result is the sale of additional fertilizer to farmers, if required, at the higher prices (unsubsidized or only partially subsidized), in order to increase yields.

It is expected that by continuing the policy of cutting subsidies, the rate of fertilizer use should be optimized. In the meantime, the rapid shift to high-yielding cultivars, together with other changes in production make the use of modern fertilizer requirement estimations, carried out by soil testing and plant analysis, a necessity. There are already two central laboratories, in Cairo and at Maryut near Alexandria, belonging to the National Research Centre. These laboratories serve farmers by carrying out soil testing and plant analysis. The Ministry of Agriculture has also established some other regional laboratories which are still in their initial phase of development (El-Fouly and El-Baz 1990). Their aim is to determine crop fertilizer requirements at a local level. Once these laboratories are functional, the country's fertilizer needs, the existing estimation of fertilizer requirements until the year 2010, and also the fertilizer policy, will face a big change. Current and projected estimations of the production and demand for N and P fertilizers until the year 2010 are shown in Appendix 4, Tables 4.4 and 4.5. These laboratories will help to identify micronutrient problems which are believed to place constraints on production (USDA *et al.* 1976; El-Fouly 1984). They will also enable farmers to be provided with accurate field information on fertilizer requirements for optimizing production (MOA *et al.* 1982), and help to overcome negative environmental side effects resulting from unbalanced fertilization, especially the over use of nitrogen.

Foliar fertilizers and micronutrients

It has already been mentioned that only small quantities of compound fertilizers are imported. The Abu-Qir Company is producing urea fortified with micronutrients in experimental quantities (Massoud 1983). Zinc sulphate is used extensively as a soil fertilizer in rice production. Some is produced locally, the rest is imported from different sources.

Foliar fertilizers have been registered in Egypt since 1967. In the 1960s and early 1970s characterized by the lack of NPK soil fertilizers, foliar NPK fertilizers showed positive effects on most crops. However, this is not the case now. The trend has been towards micronutrient foliar fertilizers (El-Fouly 1987). New regulations for fertilizer registration issued in 1986 resulted in 121 foliar fertilizers being registered until the end of March 1988, and it is estimated that the number will increase with time. Of these foliar fertilizers, 11 are produced locally, but the majority are imported. No data are available for the quantities of foliar fertilizers produced locally and those utilized all over the country. These fertilizers are not subsidized and can be handled freely in the market place, which makes data collection very difficult. Most of these fertilizers are used for cotton, where no distinction is made between NPK and micronutrient foliar fertilizers. It was estimated that in 1982, 2400 t of fertilizers were used in cotton (El-Gala 1987). In 1982-3, the PBDAC distributed about 2000 t of foliar fertilizers (Table 17.16), but the amount has decreased since then. The decrease is due mainly to the stopping of NPK foliar fertilizer use in cotton. Attempts were made to rationalize the use of foliar fertilizers by categorizing them (Abdel-Hadi 1987; El-Fouly 1987) and identifying crop needs in different regions (El-Fouly 1987). An estimation of foliar fertilizer usage is given in Table 17.17, while Table 17.18 shows preliminary estimations of the need for foliar micronutrient fertilizers.

The use of micronutrient foliar fertilizers is likely to show a gradual increase in the future. In the meantime, and with the increasing availability of NPK soil fertilizers, the use of foliar NPK fertilizers will decrease.

Future outlook

Estimations of fertilizer requirements and availability until the year 2000 (Appendix 4, Tables A4.4 and A4.5) (Aglan 1987; Mazen 1987; NSC 1981) show that for N and P a negative balance will exist, provided that the production capacity does not increase during this period. Needs for potash will double between 1985 and 2000 (NSC 1981), and this is a low estimation in view of new trends in potash research (El-Fouly 1989). There are, as yet, only preliminary estimations of the need for micronutrient fertilizers, in spite of the general recognition of their value and the increasing information on the increased needs of micronutrients for different crops. Furthermore, the whole fertilizer sector will face drastic changes in the 1990s as a result

TABLE 17.16. Distribution of foliar fertilizers by the PBDAC, 1981/2–1985/6 [a]

Year	Distribution (t)
1980/1	1159
1981/2	2174
1982/3	2112
1983/4	2065
1984/5	1588
1985/6	1506
1986/7	938
1987/8	926
1988/9	749
1989/90	397

[a] Excludes quantities distributed by other organizations and agents.
Source: El-Khashab (1987).

of changing production and trade policies, e.g. the removal of the price control and subsidy, production of new types of fertilizers, the increased role of the private sector, and the use of modern methods for determining crop fertilizer requirements according to varieties, location, and farming system, using soil testing and leaf analysis. In view of these changes,

TABLE 17.17. Total imported and locally produced foliar fertilizers for the period 1981–88

	1981–5	1986	1987	1988	Total
Macronutrient compounds[a]	4784	432	442	2280	7938
Macro- and micronutrient compounds[b]	900	157	127	54	1238
Single micronutrient chelated compounds	312	35	52	47	446
Complex micronutrient compounds	962	25	160	160	1307

In addition to approximately 900 t zinc sulphate per year for rice.
[a] With less than 2% micronutrient (Mn + Fe + Zn) in chelated form.
[b] With higher than 2% micronutrient (Mn + Fe + Zn) in chelated form.
Source: El-Fouly (1989).

TABLE 17.18. Estimated crop requirements for zinc, manganese, iron, and copper, used as foliar fertilizers in chelated form.

Crop	Micronutrient requirement (g feddan^{-1})			
	Zn	Mn	Fe	Cu
Wheat	40 (80%)	30 (70%)	15 (60%)	22.5 (5%)
Maize	50 (80%)	45 (30%)	20 (60%)	22.5 (5%)
Sorghum	50 (80%)	40 (30%)	20 (10%)	20 (5%)
Cotton	35 (85%)	22 (70%)	15 (70%)	10 (5%)
Sugar-cane	60 (80%)	45 (80%)	25 (25%)	20 (5%)
Rice	50 (90%)	40 (60%)	20 (30%)	20 (5%)
Vegetables	75 (100%)	60 (100%)	50 (50%)	10 (5%)
Orchards	150 (100%)	100 (100%)	100 (100%)	20 (15%)
Legumes	50 (100%)	40 (100%)	40 (100%)	20 (15%)

Numbers in parentheses are percentage of total crop area requiring micronutrients.
Source: El-Fouly (1989).

research institutions and agricultural extension services will also have to change their policies, viewpoint, and methodology, in order to give the best advice to farmers, and to inform the industry about farmers' needs. The economics of fertilizer use in different crops will also be changed, and the environmental aspects of using fertilizers will play a greater part than at present. In view of these changes, two major projects are in progress to co-ordinate industrial and agricultural activities. One project is the work of the Egyptian Fertilizer Development Centre (UNIDO/UNDP) in collaboration with IFDC, USA (Klada 1989). The second is the NRC/GTZ project on micronutrient and other plant nutrition problems in Egypt (El-Fouly 1989).

Acknowledgements

The author wishes to acknowledge with great appreciation the assistance given by the PBDAC, especially Engineer E. El-Khashab, in the preparation of the chapter.

References

Abdel-Hadi, A. H. (1987). Effect of Zn, Mn, Fe-chelates and some fertilizers on the yield of different field and horticultural crops in Egypt. In *Proceedings of a Symposium on Special Fertilizers, Alexandria, Egypt, 21–23 February, 1986*, (ed. M. M. El-Fouly, R. Eibner, and G. Hahr), pp.49–72. Scherig A.G., Federal Republic of Germany and National Research Centre (Micronutrients Project), Cairo. (in Arabic).

Abdel-Hadi, A. H. (1989). *Potassium and its effect on crop production in Egyptian soils*. MOA/ARC Soil and Water Research Institute, Cairo. (in Arabic).

Aglan, S. (1987). Fertilizer industry in Egypt. In *Proceedings of a Special Symposium (First Fertilizers Conference: Needs and Availability), Cairo, 13–16 April, 1987*. (ed. M. M. El-Fouly), pp.3–18. SWRI, ARC-MOA, Cairo and NRC (Micronutrients Project), Cairo. (in Arabic).

CAPMAS (Central Agency for Public Mobilization and Statistics) (1982). *Fertilizer industry in Egypt*. CAPMAS, Cairo. (in Arabic).

Dawoud, O. (1989). *Egyptian fertilizer industry*. Position Paper presented at a special workshop, IFDC, Muscle Schoals, Alabama.

El-Fouly, M. M. (1984). Increasing production of food crops in Egypt through balanced nutrition: role of micronutrients. In *Fertilizer use and food production in the Middle East and North Africa*. Proceedings of an IFA/AFCFP Regional Seminar, Bahrain, November, 1984. pp.1–12. International Fertilizer Association, Paris.

El-Fouly, M. M. (1987). Use of micronutrients under practical conditions in Egypt. In *Proceedings of a Symposium on Special Fertilizers, Alexandria, Egypt, 21–23 February, 1986*. (ed. M. M. El-Fouly, R. Eibner, and G. Hahr), pp.71–86. Schering A.G. Federal Republic of Germany and National Research Centre (Micronutrients Project), Cairo. (in Arabic).

El-Fouly, M. M. (1989). *Micronutrient and potassium needs in Egypt*. Unpublished position paper presented at a special workshop, International Fertilizer Development Centre, Muscle Schoals, Alabama.

El-Fouly, M. M. and El-Baz, F. K. (1990). Experiences and future of soil testing and plant analysis services in Egypt. *Communications on Soil Science and Plant Analysis*, 21, 1745–65.

El-Gala, A. (1987). Foliar fertilization in Egypt. In *Proceedings of a Special Symposium (First Fertilizers Conference: Needs and Availability), Cairo, 13–16 April, 1987*. (ed. M. M. El-Fouly), pp.57–68. SWRI, ARC-MOA, and NRC (Micronutrients Project), Cairo. (in Arabic).

El-Khashab, E. (1987). Handling and marketing of fertilizers and manures in Egypt. In *Proceedings of a Special Symposium (First Fertilizers Conference: Needs and Availability), Cairo, 13–16 April, 1987*. (ed. M.M. El-Fouly, R. Eibner, and G. Hahr), pp.45–56. SWRI, ARC-MOA, and NRC (Micronutrients Project), Cairo. (in Arabic).

El-Khashab, E. (1989). *The role of the Principal Bank for Development and Agricultural Credit in marketing and distribution of fertilizers in Egypt*. Unpublished position paper presented at a special workshop held at International Fertilizer Development Centre, Muscle Schoals, Alabama.

Faizy, S. (1980). Consumption and use of fertilizers in Egypt, and its effect on yield of different crops. In *Proceedings of the International Workshop. Role of potassium in crop production, 20–22 November, 1979, Cairo*. (ed. A. Saurat and M. M. El-Fouly), pp.159–70. NRC (Micronutrients Project), Cairo.

Hamissa, M. R. (1989). *Fertilizers and fertilizer use in Egypt*. Unpublished position paper presented at a special workshop, International Fertilizer Development Centre, Muscle Schoals, Alabama.

Keleg, A., Khedr, M.S., Hamam, A., and Abdel Aziz, I. (1987). *Fertilizer requirements for different crops in Egypt*. Soils and Water Research Institute, Information Leaflet No. 1. ARC-MOA, Cairo. (in Arabic).

Klada, R. T. (1989). *The Egyptian Fertilizer Development Centre*. Unpublished position paper presented at a special workshop, International Fertilizer Development Centre, Muscle Schoals, Alabama.

Massoud, A. M. (1983). *Egyptian fertilizers, research and practice*. Abu-Qir Company for Fertilizers and Chemical Industries. (in Arabic).

Mazen, M. (1987). Fertilizer availability and needs in Egypt. In *Proceedings of a Special Symposium (First Fertilizers Conference: Needs and Availability), Cairo, 13–16 April, 1987*. (ed. M. M. El-Fouly), pp.19–44. SWRI, ARC-MOA, NRC (Micronutrients Project), Cairo. (in Arabic).

MOA (Ministry of Agriculture) (1986). *A study on fertilizers in Egypt and the world*. Agricultural Foreign Relations International Economical Studies, MOA, Cairo. (in Arabic).

MOA, ARE, USAID, IADS, and USDA (1982). *Egypt: strategies for accelerating agricultural developments*. MOA, Cairo.

NSC (National Specialized Councils) (1981). *Egypt year 2000—chemical fertilizers*. Arab Centre for Research and Publishing, Cairo. (in Arabic).

PBDAC (Principal Bank for Development and Agricultural Credit) (1986). *Consumption of sulphate of potash*. PBDAC, Cairo. (in Arabic).

Serry, A. (1980). Fertilizing practice in Egypt and concepts for future development. In *Proceedings of the International Workshop. Role of potassium in crop production, 20–22 November, 1979*. (ed. A. Saurat and M. M. El-Fouly), pp.3–6. NRC (Micronutrients Project), Cairo.

USDA, USAID, and MOA (1976). *Major constraints to increase agricultural productivity*. Foreign Agricultural Economic Report No. 120. USDA, Washington, DC.

18

Feedstuffs

T. J. BARKER

Introduction

In Egypt feeds and feedstuffs are still in short supply particularly in summer when a high proportion of the winter fodder lands are replaced by the cotton crop. In spite of this livestock continues to form a major part of the agricultural economy. It is estimated that livestock production, including animal power and manure, account for as much as one-third of Egypt's agricultural output.

Historically, animals have been kept mainly for draught power—water lifting, land preparation, transport—as well as for meat and milk, but with the advent and increasing use of tractors and diesel water pumps the need for work animals has declined. According to the 1982 census, actual numbers of cattle and water buffalo have increased at a greater rate than the annual government estimates suggested. This emphasizes the growing importance of meat and milk in the economy and therefore of feed and fodder on the farm.

Egypt's domestic livestock is almost entirely confined to the intensely farmed irrigated cropping lands that are also relatively densely populated. The main exceptions to this are the herds of camels, sheep, and goats that live partly on natural grazing along the northern coastal belt. Up to the end of the last decade it was estimated that the bulk of the cattle (96 per cent) was owned by the traditional small farm sector (Winrock International 1980). In this decade it is clear from observation that this pattern is changing: government policy has encouraged the development of the large-scale farms and also stimulated the development of the desert areas, where a significant amount of livestock production is emerging. This has led to an adjustment in the ownership pattern whereby the small farmers now own between 80 and 90 per cent of livestock. Eighty-two per cent of milk production still comes from the small farmers (Soliman and Abd El-Azim 1983), and they also supply feeder stock to the larger enterprises for growing and fattening.

Although these changing patterns affect feed and feeding systems, the feeding of farm animals nevertheless still centres around the intensive use of leys, mainly *bersim* (*Trifolium alexandrinum*), crop residues, by-products, and grains. Some of these are combined into concentrate mixtures primarily in government feedmills but there is now also a growing private-sector milling industry, mainly attached to or associated with large-scale feedlots or poultry production units.

In the poultry sector, government policies have spawned an almost entirely modern broiler and layer industry in both the public and private sectors, which has largely removed egg and white meat production from the traditional farmers.

Fodder and the cropping pattern

As stated above, *bersim* is the main fodder crop—in fact it is by far the most important livestock feed in Egypt. Two types are grown: 'catch crop' *bersim*, and 'long season' *bersim*. Together these take up about 50 per cent of the winter rotation (Egypt operates two distinct rotations annually, based on winter and summer seasons). Fig. 18.1 shows the traditional cropping pattern of the Delta, but the relative proportions of the crops are changing. The desert has much greater areas of year-round grazing.

Bersim is planted after the harvest, from September to November, each year. The catch crop variety is grown with a view to planting cotton after it the following spring, usually in March. In the intervening period two or three cuts are taken. The long season variety is cut four or five times and is grown through to May or June the following year. Irrigation of *bersim* is stopped by law on 15 May each year, because it acts as a haven for cotton pests. Generally *bersim* is fed by cutting and carting the crop back to the village where the cattle are housed, although in some instances animals are tethered in the fields as a means of rationing their grazing. Small quantities of excess production are made into sun-cured hay which is usually sold to feedmills and larger farms.

In the new lands of the reclaimed desert areas, *bersim* is replaced by lucerne (*Medicago sativa*) which is grown as a perennial crop and gives a more continuous source of green fodder all the year round. It is also cut and carted for feeding. In the relatively unproductive soils of the desert, the lucerne crop is usually a sparser crop; moreover, it does not grow as well in winter as *bersim*.

Other sources of forage are maize, forage sorghum, and elephant grass (*Pennisetum purpureum*). The latter, although actively encouraged by the Egyptian Government in the past (Makky 1976), has failed to become established to any large extent. It is a perennial that grows in summer but there are some management problems with the tussocked, woody bottom growth that develops with time. In contrast, sorghum, and particularly maize, are increasing in acreage in the summer rotation, and therefore shortages of green fodder at this time of year are less significant than they were ten years

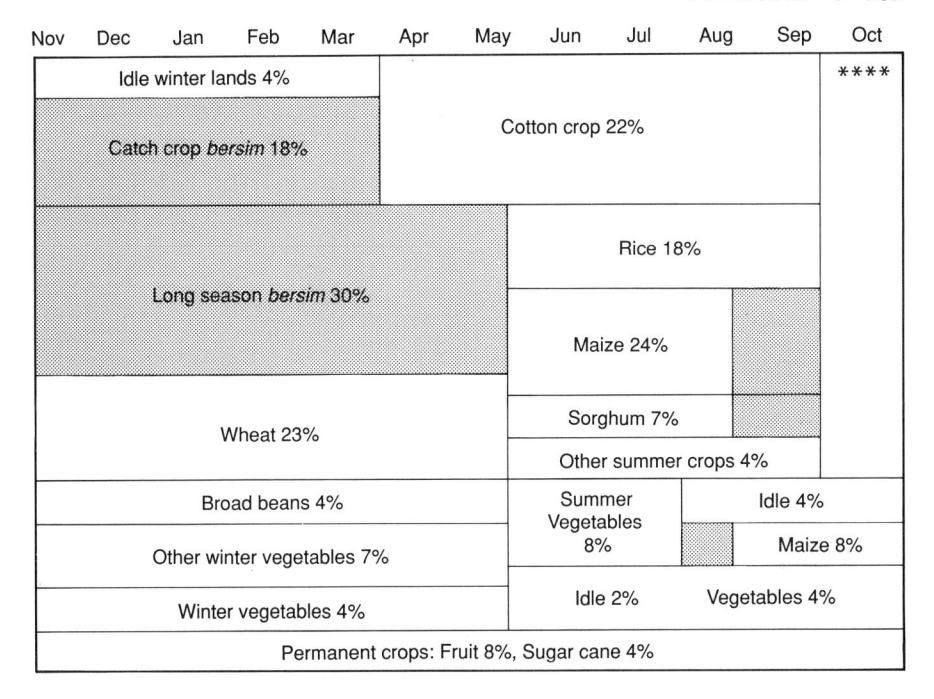

	Nov	Dec	Jan	Feb	Mar	Apr	May	Jun	Jul	Aug	Sep	Oct
	Idle winter lands 4%											****
	Catch crop *bersim* 18%					Cotton crop 22%						
	Long season *bersim* 30%							Rice 18%				
								Maize 24%				
	Wheat 23%							Sorghum 7%				
								Other summer crops 4%				
	Broad beans 4%						Summer Vegetables 8%			Idle 4%		
	Other winter vegetables 7%									Maize 8%		
	Winter vegetables 4%						Idle 2%			Vegetables 4%		
	Permanent crops: Fruit 8%, Sugar cane 4%											

Fig 18.1. The traditional cropping pattern of the Delta, 1977–9. Percentages are proportions of crop acreages in the rotation (winter and summer). The shaded areas indicate the production of ruminant fodder crops. ****, Land which is temporarily idle between summer crops (cotton, rice), and winter crops (*bersim*). Source: Habashi and Fitch (1980)

ago. The greater quantity of forage crops in summer, together with improved availability of manufactured and imported feed, has reduced, if not removed, the practice of leaf stripping whereby the farmers remove individual leaves from crops, including rice and maize, to feed their animals. This practice was clearly detrimental to the yields of those crops.

Each district has its own feeds of local importance. For instance in Damietta governorate a local grass, called *amshut* (*Echinochloa stagnina*), is fed extensively by farmers in summer. It is encouraged to grow on all unused bits of land such as canal and road margins, headlands, waste areas, and commons.

Concentrates

In Egypt the name 'concentrates' more often than not refers to the government-manufactured pelleted feed mix usually known as 'unified feed'. Private mills were first started in the 1950s, and the Egyptian Government became involved in the manufacture of unified feed in the 1960s (Soliman and Abd El-Azim 1983). Since 1966, the Egyptian Government has controlled feed mixtures and has monopolized cottonseed cake. Unified feed has become a much sought after product because it is a highly subsidized source of protein and energy for the ruminant diet. It is available at a very low cost, but has always been in short supply. More recently, modern private-sector feedmilling enterprises have been re-established, although they have no access to cottonseed cake.

Unified feed has always been based on cottonseed cake as the main source of protein. As much as 84 per cent of the annual cottonseed production is used to make unified feed, usually in its undecorticated form, and in the early days of the mix it formed as much as 65 per cent of the mixture. Over time this percentage has been steadily reduced, mainly in order to try and increase the volume of unified feed supplies. Table 18.1 shows how the production of unified feed has increased from about 200 000 t in 1961 to nearly 2 000 000 t in 1989. Table 18.2 shows how the unified feed composition has changed over the years; cottonseed cake content is now as low as 23 per cent.

In recent years, in order to increase the amount of unified feed on the market without loss of quality, products such as urea, soyabean meal, and

TABLE 18.1. Egypt: production and use of cottonseed cake and the production of unified feed

Year	Production ('000 t)	Direct feeding (%)	To feed plants (%)	Export (%)	Unified feed production ('000 t)
1961	578	83.2	16.8	–	194
1963	571	82.7	17.3	–	198
1965	618	86.3	13.7	–	169
1967	547	24.1	31.4	44.5	264
1969	509	31.9	41.4	26.7	324
1971	611	66.8	33.2	–	372
1974	605	35.5	36.0	28.5	454
1976	436	40.4	59.6	–	660
1979	474	15.8	84.2	–	1052
1980	540	–	100.0	–	1270
1983	–	–	–	–	1360
1985	–	–	–	–	1251
1986	–	–	–	–	1840
1989	–	–	–	–	–

Source: 1961–1983, Soliman and Abd El-Azim (1983); 1985–1989, Ministry of Agriculture and Land Reclamation (1989, personal communication).

TABLE 18.2. Egypt: trends in unified feed composition (%)

	1970	1973	1976	1979	1980	1983	1988[a]	1989
Cottonseed cake	65	65	45	42	40	35	25	23
Wheat bran	9	18	20	25	30	25	40	46
Rice bran	23	11	10	5	4	12	5	4
Molasses	–	3	2.5	3	3	3	3	4
Limestone/minerals	2	2	1.5	2	2	2	3	3
Salt	1	1	1	1	1	1	1	1
Yellow maize	–	–	20	22	20	22	20	19

[a] Figures rounded.
Source: 1970–1980, Soliman and Abd El-Azim (1983); 1983–1988, Capper *et al.* (1988); 1989, MOALR (1989, personal communication).

linseed cake have been added to a small proportion of the mix as a substitute for the protein supplied by cottonseed cake. This, together with the steady removal of the government subsidies on unified feed, has in turn led to the development of formulated feeds for specific purposes, for example, dairy mix and beef fattening mix.

The allocation of government-subsidized concentrates was, and still is, decided on a quota basis. Quotas are controlled through an insurance scheme and only those farmers with more than five head of cattle are given a quota; this amounts to 3 kg head^{-1} day^{-1} in winter and 5 kg head^{-1} day^{-1} in summer. The individual farmers have to collect these concentrate quotas from the feedmills, or more usually from the stores of the Principal Bank for Development and Agricultural Credit (PBDAC). Thus, in the past decade 96 per cent of the national herd (reducing to perhaps 80 per cent) was excluded from purchasing this feed unless they sold their milk supplies to government-controlled dairies, which give concentrates in part payment for milk.

Private-sector feedmills are a relatively new phenomenon in recent times and, although they sell pelleted feeds and mashes to farmers, the successful ones are usually attached to large-scale feedlot or poultry enterprises. This ensures that these enterprises have constant feed supplies and feed quality, which are particularly important in egg and milk production. Their mixtures of cattle feeds are known as 'non-traditional' feeds because government rules state that they have to include 30 per cent of roughage in the mix. This is generally something like milled *bersim* hay or even rice straw, although the latter is difficult to grind successfully. A list of currently operating and planned feedmills is given in Tables 18.3*a* and 18.3*b*.

The products from the non-traditional feedmills are sold at free market prices and are subject to market forces, unlike the subsidized unified feed.

However, the Egyptian Government has reduced this subsidy considerably in recent years.

Crop by-products

The most important commodity under this heading is cottonseed cake, and this forms the basis of unified feed (see above). Cottonseed cake is closely followed in importance by wheat bran, of which there is a very large supply from the flour milling and baking industries. Wheat bran is sold direct to farmers as well as to the feed industry, particularly the government mills.

Molasses, mainly from cane but some from beet, are of increasing importance in ruminant feeding. About 350 000 t of molasses are available annually, of which two-thirds are either exported or manufactured into alcohols and other industrial chemicals. About 100 000 t are left for animal

TABLE 18.3a. Livestock feed manufacturing plants: government unified feed plants — Feed Industry Corporation (FIC)

Company	Location	Capacity (t year^{-1})
Main FIC plants		
Alexandria Company for Oil and Soap	Kafr el Zaiyat	235 000
Egyptian Salt and Soda Company	Kafr el Zaiyat	365 000
Extraction Oil and Oil Products	Kabbary/Alexandria	125 000
Tanta Oil and Soap Company	Benha/Zifta Tanta	150 000
Misr Oil and Soap Company	Kafr Sa'ad/Biqas/MitGhamr	190 000
Cairo Oil and Soap Company	Badrshein/El Aiyat	180 000
El Nil Oil and Soap Company	Sakha/Beni Mazar/Ktam Said/Minya	285 000
Extensions to existing FIC plant		
Alexandria Company for Oil and Soap	Kafr el Zaiyat	120 000
FIC hired plants		
Alexandria Company for Oil and Soap	Kafr el Zaiyat	120 000
Egyptian Salt and Soda Company	Kafr el Zaiyat (2 units)	240 000
Extraction Oil and Oil Products	Damanhur	120 000
Tanta Oil and Soap Company	Mahalla	120 000
Misr Oil and Soap Company	Zagazig	120 000
El Nil Oil and Soap Company	Sohag	120 000
	Total	2 490 000

Source: Adapted from FIC figures (FIC 1990, personal communication).

TABLE 18.3b. Livestock feed manufacturing plants: non-traditional feed mills and non-FIC plants

Company	Location	Capacity (t year^{-1})
Ministry of Supply—existing and extensions		
Zagazig Plant	Zagazig	90 000
Shirbin/Damietta	Shirbin	160 000
	Rashid/Beheira	120 000
	Delengat/Damanhur	120 000
Private and other organizations		
El Eqtisadia Company (private)	Noubariya	9 000
3 smaller plants	Faiyum/Qena	125 000
	El Marg/Qalyubiya	25 000
Misr Meat and Dairy Company	3 plants	36 000
	Total	685 000

Source: Adapted from FIC figures (FIC 1990, personal communication).

feeding in Egypt and of this 55 000–60 000 t are used in the manufacture of unified feed (Tables 18.1 and 18.2). About 50 000 t are sold to the private sector or to projects and government farms and attempts are also being made to manufacture supplements based on both liquid and solid molasses

A wide variety of other by-products are available in varying quantities, some of which are imported. These include such protein-rich commodities as soya beans, linseed cake, and groundnut meal, which are now in greater demand by the private farmers because of the much higher protein requirements of the Friesian and Holstein cattle, breeds that are now popular in Egypt. A number of other by-products are used as feeds, some of which have only local importance. These include: citrus and tomato wastes from canning factories, date kernels, brewer's grains, sunflower meal, and rapeseed meal.

Crop residues

Crop residues have always had a place as an animal feed within the farm system. More recently, however, trading in roughages and straws has become commonplace in order to overcome the scarcity of concentrates and roughages in the summer hunger gap. The most commonly traded is rice straw but there is a large quantity and wide variety of straws and stalks that can be tapped

TABLE 18.4. Availability of various crop residues in 1987

Residue	t (millions)
Wheat straw	2.100
Rice straw	1.150
Other straws	0.750
Maize stalks	4.700
Maize cobs	0.600
Rice hulls	0.500
Bagasse	2.100
Cane molasses	0.300
Total	12.200

Source: Ministry of Agriculture and Land Reclamation (1989, personal communication).

for ruminant feeding. In 1989, there was an estimated 12.2 million t of usable residue (Table 18.4).

The small farmers who produce rice will retain some rice straw, which they store on the roof of their houses to use as a feed for their own animals; the rest is sold loose to baling contractors. These contractors use stationary balers which make high-density bales fastened with wire, which they deliver to their customers, usually medium or large farmers or even co-operative societies. The wire fastenings are a risk and it is not unknown for cattle to die from ingested wire. As a result there has been some movement towards the use of plastic string in the last few years.

Wheat straw is also widely traded, but as a loose, chopped material which is transported in large net bags. It is chopped automatically in the threshing process and some grain passes through into the straw as well, giving it a higher nutritional value than straw alone. Nevertheless, the price of wheat straw has for many years been much more expensive than its nutritional value justifies; this was an adjustment to the heavily subsidized price and scarcity of the unified feed concentrate. It is usually too expensive for the cattle industry and so is sold to poultry enterprises.

Other crop residues are available and some, including maize and sorghum stalks, bean straw, cotton stems, and vegetable crop wastes, are used as feeds. However, there are other needs that compete for these materials; maize stalks are used in the home for cooking and heating fires, while rice straw is used in the paper-making industry.

In order to further improve nutrition and to reduce the summer hunger gap, the improvement of low quality straws through chemical treatment has been developed and promoted during the last ten years (Creek et al. 1984; Yacout et al. 1985). At the present time, treatment of baled straws with anhydrous ammonia, under a plastic sheet, seems to be the most appropriate method. Treatment with caustic soda, urea and various other chemicals have also been investigated.

Minerals and supplements

Although there is some local manufacture of mineral supplements and lick-blocks, the formulas are not sufficiently comprehensive. Even in the unified feed mix, the mineral supplementation is generally inadequate and, as a result, most balanced mineral supplements are imported. The main purpose behind the attempts to produce molasses-based supplements is to supply this vital need from local resources as far as possible. Both liquid and solid molasses supplements are being produced (Sansoucy 1986), and the Animal Production Research Institute, for example, provides not only minerals (in particular phosphorus and sulphur) but also protein sources such as urea.

Trends

Radical changes are taking place in Egypt as a result of government policies. First there were the 'open door' policies of the early 1980s which made it possible to import commodities that had been banned previously. This was followed by the gradual reduction of government subsidies on commodity prices. For example, in 1980, unified feed was sold to the farmer ex-factory at £E38 t^{-1}, but in 1989 the price was £E260 t^{-1}. This has had knock-on effects: in 1980 wheat straw cost about £E145 t^{-1} on the open market and in 1989 it was £E65. Care has had to be taken with these kinds of readjustments. If cattle feed became too expensive relative to bread, then farmers would purchase the latter for their livestock at the expense of human consumption. A phasing operation has therefore been undertaken which is not yet complete.

A faster than expected increase in cattle numbers owing to the favourable economic climate, including the rise in demand for meat and milk, has led to rising returns for the farmer and increased acreages devoted to cattle production. However, the lack of land available for cultivation prevents the farmer from keeping abreast of demand; moreover a conflict develops between the requirement for human food and the farmers' economic preference to produce cattle feed. The Egyptian Government is currently trying to persuade or coerce farmers into growing winter wheat at the expense of the *bersim* crop and, in consequence, imports of meat, milk products, and feeds are all rising while importation of maize and wheat flour remain relatively stable (Table 18.5). The importation of feed concentrates is likely to continue to rise because of the increase in livestock numbers and because of the static local poduction levels of cottonseed cake and molasses (Table 18.6).

TABLE 18.5. Trends in the value (£Emillions) of imported milk, meat, feed and related products, 1982–7

Product and trend	Year					
	1982	1983	1984	1985	1986	1987
Meat (+23.7%)	157	97	193	176	217	349
Livestock (+22.5%)	117	159	166	171	159	301
Milk Products (+5.2%)	386	405	540	493	476	794
Maize (+0.8%)	210	142	174	145	136	232
Wheat (+4.1%)	492	348	376	343	464	572
Wheat flour (+0.8%)	192	206	272	213	181	229

Source: Animal Production Research Institute, Ministry of Agriculture and Land Reclamation (1989, personal communication).

TABLE 18.6. Changes in the level of production of cottonseed cake and molasses, 1953–87 ('000)

	1953	1982	1983	1984	1985	1986	1987	Trend (%)
Cottonseed cake	410	548	539	507	480	420	440	−4.82
Molasses	100	295	336	294	324	348	362	+3.88

Source: Animal Production Research Institute, Ministry of Agriculture and Land Reclamation (1989, personal communication).

Acknowledgements

With acknowledgements to the Director and members of the Animal Production Research Institute, Ministry of Agriculture, Cairo; to Dr Moustafa Hathout, Mr Hussein El Nouby, and Mr Helmy Yacout for their valuable assistance; and to members of the EEC Delegation in Cairo.

References

Capper, B.S., Coulter, J.P., Garforth, C.K., and Levieux, G.P. (1988). *Innovations in ruminant feeding in Egypt*. Overseas Development Natural Resources Institute (ODNRI) Consultancy Report No. C0807. ODNRI, Chatham.

Creek, M.J., Barker, T.J., and Hargus, W.A. (1984). The development of a new technology in an ancient land. *World Animal Review*, **51**, 12–20.

Habashi, N. and Fitch, J. (1980). *Egypt's agricultural cropping pattern*. Micro-Economic Study Unit, Ministry of Agriculture, Cairo.

Makky, A. M. (1976). *Introducing Elephant grass in Egyptian agriculture to solve problems of animal production*. El Megalla, El-Zerayia.

Sansoucy, R. (1986). The Sahel: manufacture of molasses-urea blocks. *World Animal Review*, **57**, 40–8.

Soliman, I. and Abd El-Azim, M. (1983). *An appraisal of livestock concentrate feed policy in Egypt*. Agricultural Development Systems Project, Economics Working Paper No. 38, Ministry of Agriculture, Cairo.

Winrock International (1980). *Potential for on-farm production and utilisation by the Egyptian small farm sector*. Study prepared for the Catholic Relief Services and USAID, Cairo.

Yacout, M.H., Barker, T.J., Hathout, M.K., and El-Nouby, H.M. (1985). On-farm trials with ammoniation of straw in Egypt. In *Proceedings of a workshop, 25–28 March, Aleppo, Syria*. Report No. IDRC-242e, pp. 173–89. International Development Research Center, Ottawa.

Further reading

Abd El-Aziz, A.E. (1986). The utilization of agricultural by-products in ruminant feeding. Unpublished Ph.D. thesis. University of Ain-Shams, Egypt.

Abou-Raya, A.K., Abou-Hussein, E.R.M., Abou-El-Hassan, A. and El-Shinnawy, M.M. (1975). The use of some by-products in feed mixtures for animals. In *Proceedings of a conference on animal feeds of tropical and subtropical origin*. Tropical Products Institute, London.

Borhami, B.E.A. and Johansen, F. (1981). Digestion and duodenal flow of ammonia treated straw and sodium hydroxide treated straw supplemented with urea, soybean meal or fish viscera silage. *Acta Agriculturae Scandinavica* **31**, 245.

Chesson, A. and Orskov, E.R. (1984). Microbial degradation in the digestive tract. In *Straw and other fibrous by-products as feed*. (ed. F. Sundstøl and E. Owen), pp. 305–39. Elsevier Science Publishers, Amsterdam.

Han, Y.W. (1978). Microbial utilization of straw (a review). In *Advances in Applied Microbiology*, (ed. D. Perlman), pp. 119–53. Academic Press, New York.

Hathout, M.K. (1984). *Feed resources for livestock in Egypt*. Animal Production Research Institute, Ministry of Agriculture, Egypt.

Hathout, M.K., Barker, T.J., Hargus, W.A. and Creek, M.J. (1983). The on-farm ammoniation of straw and distribution of liquid supplement in Behera Governorate. In *Proceedings of the 2nd workshop on utilization of low quality roughages with special reference to developing countries, 14–17 March*. University of Alexandria, Egypt.

Klopfensten, T.J. (1978). Chemical treatment of crop residues. *Journal of Animal Science*, **46**, 841–8.

Morgan-Rees, A.M., Williams, T.E., Smith, A.J., and Capper, B.S. (1977). *Report on ODM mission to Egypt to undertake a pre feasibility study of forage production and animal feeds*. Tropical Products Institute No. R635. Tropical Products Institute, London.

Naga, M.A. and El-Shazly, K., (1983). Inclusion of low quality roughages in the diets of fattened ruminants. In *Proceedings of the 2nd workshop on utilization of low quality roughages with special references to developing countries, 14–17 March*, p. 148. Alexandria, Egypt.

Shahla Shapouri; Kelley White, T. and Hassan Khedr (1985). *Egyptian feedlot practices, costs and returns*. ERS Staff Report No. AGES851101. International Economics Division, Economic Research Service, United States Department of Agriculture, Washington, DC.

Sundstøl, F. and Coxworth, E. (1984). Ammonia treatment. In *Straw and other fibrous by-products as feed*. (ed. F. Sundstøl and E. Owen), pp. 196–247. Elsevier Science Publishers, Amsterdam.

19

Agricultural mechanization and farm equipment supply

A. U. KHAN

Introduction

The cultivated land area in Egypt is about six million feddans which is about 3 per cent of the total land area. This is concentrated primarily in the Nile Valley and in the Nile Delta. The entire crop area is irrigated except for some rain-fed areas on the Mediterranean coast. The availability of additional water, due to the construction of the Aswan High Dam, led to increased cropping intensities on the old lands and the reclamation of over one million feddans of new desert lands. However, year-round irrigation has magnified drainage and soil salinity problems and this has affected the supply and quality of land.

The major agricultural crops, in order of the area planted, are *bersim* (Egyptian clover), maize, wheat, cotton, and rice. The other important crops are sorghum, broad beans, soya beans, and barley. With long periods of high temperature and abundant solar radiation, crop yields are generally quite high compared with average world yields.

The average size of farm holdings is only 1.6 feddans, with 95 per cent of all holdings averaging only 0.9 feddans. Table 19.1 gives the distribution of land ownership in Egypt. Since most farm land is subdivided further into smaller plots that are usually planted in rotation, fragmentation of the cropped area is even greater. Almost all farmers, however, are members of government-sponsored co-operatives which often permit consolidated land preparation.

There are a number of large state-owned farms occupying approximately one million feddans. These farms are well mechanized, utilizing a wide range of modern equipment imported from the industrially advanced countries. Except for the poor repair and maintenance of equipment, inadequate availability of spare parts, and lack of adequate training, mechanization on such farms does not present any serious technological problems. The mechanization of small privately owned farms presents more challenging problems and hence emphasis is placed in this chapter on the small farm sector.

TABLE 19.1. Egypt: distribution of landownership, 1981

Farm size (feddans)	No. of land owners[a] ('000)	Percentage of all land owners (%)	Area owned ('000 feddans)	Percentage of all land owned (%)
<5	3 479	95.6	2 916	53.0
5–9.9	87	2.4	577	10.5
10–19.9	43	1.2	553	10.1
20–49.9	23	0.6	620	11.3
50–99.9	6	0.2	418	7.6
More than 100[b]	2	0.0	413	7.5
Total	3 640	100.0	5 497	100.0

[a] State farm lands and newly reclaimed desert land not included.
[b] Includes organizations, companies, and individuals.
Source: CAPMAS (1989).

Most farm inputs are available to farmers at fairly subsidized prices but, gradually, input prices are being increased by the government. Fertilizers are sold at 25–90 per cent of the cost, insurance, and freight (CIF) value. Similarly diesel fuel is heavily subsidized and sells for about 30 per cent of world prices. Electricity is priced at about 30 per cent of international prices. Credit for tractors and machinery is available to qualified buyers at 16 per cent per annum (Winrock International 1986).

Farm-gate prices of agricultural crops are generally below international prices. Part of the cotton and all of the sugar cane production, must be sold to the government at prices which are considerably lower than the international prices. Farmers generally encounter seasonal shortages of labour, mainly during the months of May to July and September to November. This coincides with the harvesting and threshing of the winter crops, and the land preparation and planting of the summer crops. The labour shortage has provided considerable impetus for the mechanization of agriculture in Egypt.

According to a study by Louis Berger International (1985) the critical determinant of machinery use was not the magnitude of the general labour supply but the availability of family labour and the high wage rates during the peak periods. When the supply of family labour was insufficient, there was a forced entry into the hired labour markets resulting in a greater tendency to mechanize. The study also found that villages with inadequate infrastructure (markets, roads, repair facilities, etc.) showed the lowest incidence of mechanization, but land fragmentation appeared to have no significant effect on the incidence of mechanization. Farmers with a greater number of plots showed a higher use of machinery than those with a smaller number of plots.

Current level of mechanization

Although agricultural mechanization in Egypt has had a long history, substantial progress has been made only during the last thirty years. A stage has now been reached where almost 90 per cent of farmers utilize machines to carry out at least some of their operations.

The pattern of agricultural mechanization in Egypt has been based more on farming operations than on specific crops. Thus, certain farming operations are quite heavily mechanized for all relevant crops whereas other operations have very little or no mechanization. Most farmers utilize a combination of mechanized and non-mechanized practices for growing their crops.

Table 19.2 gives the degree of mechanization for different agricultural operations in the country. Land preparation, threshing, and water pumping are most heavily mechanized whereas, planting and harvesting are carried out primarily by manual methods. It is important to note that the use of draft animals for agricultural operations has been drastically reduced because of the widespread use of tractors, particularly for land preparation.

Mechanization has progressed more in some regions than others. A survey in 1981 (Hopkins et al. 1982) covering 1000 farms in the governorates of Beheira, Gharbiya, and Qalyubia in the Delta and Minya along the Nile Valley, indicated a higher level of mechanization in the Delta (Table 19.3). This survey also indicated that wheat and rice are the most mechanized crops, maize and clover are the least mechanized, and cotton and vegetable production fall somewhere in between.

TABLE 19.2. Egypt: degree of mechanization for various agricultural operations

Operation	Tractor mechanization (%)	Animal mechanization (%)	Animal and man (%)
Ploughing	90	4	6
Levelling	60	10	30
Ridging	56	10	34
Planting[a]	–	–	100
Irrigation	62	3	35
Harvesting	(rare)	–	100
Agricultural transportation	15	10	75
Threshing	80	5	15
Winnowing	30	10	60

[a]Planting includes seeding operations, planting on ridges, and transplanting.
Source: Hopkins et al. (1982).

TABLE 19.3. Farmers reporting mechanized operations for all relevant crops in selected areas

Farming operation	Delta (%)	Minya (%)
Ploughing	91	98
Levelling	86	18
Furrowing	69	27
Pest control	46	7
Irrigation	74	34
Harvesting	1	0
Transport	22	1
Threshing	85	86
Winnowing	78	3

Source: Louis Berger International Inc. (1985).

Except for irrigation, most of the mechanized operations are tractor-dependent. The predominant size of tractors in the country is 50–65 horsepower (hp). The principal tractor-operated implements and machines are tyne cultivators which are known locally as chisel ploughs, wheat threshers, trailers, and land-forming equipment. Table 19.4 lists the number of tractors and agricultural machines, the available irrigation power, and the total cultivated area of the different governorates.

Prior to the Second World War there were 3800 tractors working in Egypt. After the war, the tractor population more than doubled to 8302 units during the years 1945–55. By 1973, Egypt adopted an open-door policy which freed tractor and equipment imports, resulting in a five fold increase in the annual tractor imports. By 1981 the tractor population had reached almost 39 000 units, most of which were of the 50–65 hp size. This population was considered adequate to cover 92 per cent of the total ploughed area in the country (El-Hossary 1987).

According to rough predictions, the total tractor population is estimated to reach 65 000 by the year 2000 (Winrock International 1986), which amounts to a net annual increase of 2.9 per cent compounded. Until only recently, about 14 per cent of farm tractors were used exclusively for non-agricultural purposes while the remainder were used about 30 per cent of the time for non-agricultural operations. The use of tractors for non-agricultural operations, however, has declined substantially due to the development of alternative means of transport and a ban on the manufacture of red bricks.

The predicted total numbers of agricultural machines by the year 2000 and the compound annual growth rates between the years 1982 and 2000

TABLE 19.4. Cultivated area and number of agricultural machines in use by governorates, 1981–2

Governorate	Total area[a] cultivated ('000 feddans)	Wheel[b] tractors 50–65 hp	Ploughs	Pesticide spraying machines	Threshing machines	Trailers	Total available power for irrigation (hp)
Alexandria	74	469	135	36	15	28	5447
Beheira	621	4397	3511	998	599	673	82 787
Bharbia	388	3650	4339	1708	813	2476	44 335
Kafr El-Sheikh	467	2755	3007	1607	514	491	53 871
Daqahliya	592	4893	6314	2676	711	2871	113 178
Damietta	98	813	535	286	47	182	38 327
Sharqiya	611	4180	4286	1949	730	2958	45 060
Isma'iliya	63	612	562	90	45	410	11 630
Suez	6	76	71	9	9	22	2275
Port Said	–	4	5	7	–	6	592
Minufiya	287	1811	1686	482	671	499	40 127
Qalyubiya	146	1070	1257	392	343	874	18 600
Cairo	7	31	27	3	–	10	979
Giza	162	1250	259	96	247	107	35 629
Beni Suef	245	1093	305	101	143	17	23 620
Faiyum	296	2073	1485	322	9	685	3 990

continued overleaf

TABLE 19.4. contd.

Governorate	Total area[a] cultivated ('000 feddans)	Wheel[b] tractors 50–65 hp	Ploughs	Pesticide spraying machines	Threshing machines	Trailers	Total available power for irrigation (hp)
Minya	382	2616	1625	1445	662	831	62 292
Asyut	296	2271	3198	43	686	2756	91 622
Sohag	285	2196	1941	250	872	1162	75 985
Qena	167	1773	1276	67	511	725	101 872
Aswan	69	296	140	5	2	35	2178
Matruh	–	118	–	3	71	–	2934
New Vally	31	192	32	–	12	–	4312
North Sinai	–	–	1	8	–	1	–
Total	5293	38 639	35 997	12 583	7712	17 819	861 642

[a] Except areas under sugar cane and horticultural crops.
[b] Tractors except those delivered before 1970.
Source: El-Sahrigi and El-Haddad (1982).

are: 60 000 ploughs (2.8 per cent), 35 000 threshers (3.8 per cent), 35 000 mower reapers (17.5 per cent), and 3 000 000 water pumps (4.9 per cent) (Winrock International 1986). Fig. 19.1, which shows the rate of adoption of the main agricultural implements and machines in selected villages in the Delta and the Nile Valley, indicates an accelerated rate of adoption during the last two decades.

Until twenty years ago, threshing in Egypt was done almost entirely by driving a team of bullocks over the crop. Threshing of sorghum was done manually by beating sorghum heads with sticks. In the 1960s, the number of tractors increased and some farmers started to thresh by driving tractors over the crop. Others adopted tractor-powered threshing drums, popularly known as *baladi* threshers, for wheat threshing and straw fodder making. The *baladi* threshers did not winnow, and hence winnowing was done manually or with simple winnowers.

Since 1982 onwards, beater-type threshers (Fig 19.2) imported from Turkey started to gain popularity. These machines can make the wheat straw fodder, popularly known as *tibin*, and can also winnow the crop. The threshers are now manufactured locally and are used by about 90 per cent of the farmers for threshing winter crops such as wheat, barley, sorghum, and beans. Paddy, which is the second most important cereal crop and is grown during the summer season, cannot be threshed with beater-type machines and hence continues to be threshed by treading under tractors.

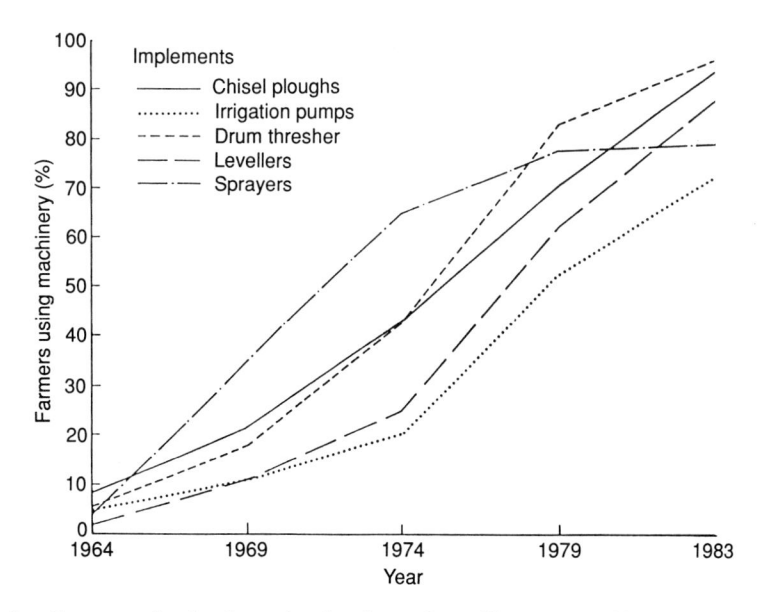

FIG 19.1. Farm mechanization adoption in project villages. —, chisel ploughs; ....., irrigation pumps; ---, drum threshers; – –, levellers; –·–, sprayers. Source: Louis Berger International (1985)

FIG 19.2. Turkish-type threshers popularly used in Egypt for threshing wheat and for fodder making

Western-type combine harvesters are used primarily on the large state farms. Of late, small Japanese-type combine harvesters have started to gain some acceptance for harvesting paddy. Tests with Western-style combine harvesters have indicated that the time required for wheat and paddy harvesting can be reduced from 102 and 116 h feddan^{-1}, respectively, to 1.16 and 1.2 h feddan^{-1}. With combine harvesters, grain losses were found to be very high (12.5 per cent) and straw losses were about 16 per cent (Louis Berger International 1985). Since wheat straw is used as animal fodder, this magnitude of straw losses was found to be unacceptable to farmers. Substantial investment requirements and the large size and complexity of combine harvesters are factors which have hindered their widespread use on small farms.

The use of mechanical water-lifting devices has increased significantly over the past decade, as these have reduced the cost of irrigation by 25 per cent compared with animal-driven water wheels (El-Sahrigi and El-Haddad 1982). Table 19.5 gives comparative costs of irrigation with animal-driven water wheels and mechanical pumps. About 65 per cent of farmers are currently using mechanical pumps for irrigation. About 25 per cent of the pump owners rent their pumps to other farmers. Small, portable, 5 hp pumps driven by a diesel engine (Fig. 19.3) are preferred for irrigation, since farmers often cultivate many small plots at scattered locations. Larger-capacity jointly owned pumps have not succeeded because farmers prefer to maintain independent

TABLE 19.5. Comparison of pump and water-wheel irrigation costs for several crops

Crop	Water needs (m^3 feddan^{-1})	Cost of irrigating 1 feddan (£E)		Cost reduction through mechanical pumping (%)
		Animal driven water wheel	Mechanical pump	
Clover	2 200	24.30	6.00	75
Wheat	1 310	14.46	3.57	75
Cotton	3 640	31.38	9.93	68
Rice	7 740	68.86	21.14	69
Sugar cane	10 000	96.27	27.26	72

Source: Ministry of Agriculture (1982).

control on their irrigation water supply during the limited periods of water availability (Louis Berger International 1985).

The use of tractor-powered equipment for land levelling and shaping has gained much popularity. More recently precision land levelling, using laser-guided equipment, has been gaining popularity for improving irrigation efficiency and crop yields. Pressurized irrigation techniques, using sprinklers and drip irrigation have also gained much popularity, especially in the newly

FIG 19.3. Locally manufactured 5 hp Kirloskar-type vertical diesel engine pump sets popular in Egypt

reclaimed desert lands. Many state farms now use large-centre pivot or side-roll sprinkler irrigation equipment and with portable or permanent underground water distribution systems for field crops. Drip and microjet irrigation techniques are also used for irrigating orchards and vegetable crops. Similarly the use of plastic greenhouses has gained much popularity for vegetable production during winter months.

Pattern of machine use

The Egyptian Government recognized at an early stage that the majority of small farms in Egypt could not be mechanized individually. Hence, in the 1960s, it initiated a major effort on mechanization through co-operative-based custom-hire services. These co-operatives were general purpose co-operatives providing a wide variety of services to village societies. During the 1960s, with government encouragement, mechanization spread rapidly to reach most small farms. Basically, the co-operatives helped to establish a viable market for custom-hire services in the country. As this market developed, private custom-hire operators entered the field in rapidly increasing numbers.

From 1963 to 1968 a mechanization network covering one million feddans was also established by the government in the newly reclaimed lands. This included about 80 farm-machinery renting stations, 20 middle-level workshops, four specialized workshops, and a machinery testing centre. (In 1981 all these operations were consolidated under the West Nubariya Company for Agricultural Engineering and Mechanization). In 1973, Egypt adopted an 'open-door policy' for tractor imports which encouraged further development of the private tractor and equipment hire services with a concurrent decline of the general purpose co-operative-based mechanization services.

In 1984, a national apex organization for mechanized farming co-operative societies was established with headquarters at Cairo and about fifty offices at village level and eight at the governorate level. This co-operative society is a government-funded institution which rents machinery through general agricultural co-operatives and is planning a substantial increase in its rental activities. It also pools the machinery purchase requests of co-operatives and individual farms and issues tenders for their purchase. The society is exempt from custom duties and taxes and collects a handling fee of one per cent on the value of items imported.

The pattern of custom hire operations is now well established and will probably continue to be the basis for the progressive mechanization of small farm holdings in the country. A survey of farm machinery owners by the USAID supported Agricultural Mechanization Project in 1982 (Hopkins *et al*. 1982) found that 54 per cent of tractor owners and 25 per cent of pump owners provide custom services. Whereas 90 per cent of farmers used tractors for ploughing, fewer than 2 per cent of farmers owned tractors, indicating a widespread reliance on custom hiring.

From 1983 onwards, the Ministry of Agriculture established 72 machinery rental stations with substantial investments and assistance from the World Bank. Like most government-run machinery rental stations worldwide, these stations in Egypt have not been very successful. Although the original plan was to establish a total of one hundred and fifty government-owned machinery rental stations, it seems that these plans have now been shelved. More recently, these machinery rental stations have been transferred to a government company, the Agricultural Mechanization Service Station Company, in the hope of improving their management and services.

The unfavourable experience of the government-owned machinery rental stations has helped to provide greater encouragement to the private sector services. Until recently, however, the government has continued to set up new machinery rental facilities such as the conversion to rental operations of the twenty-seven mechanization demonstration units under the Agricultural Mechanization Research Institute. Quasi-government joint-venture development companies, formed to increase crop production in specific geographical regions such as Nubariya, Aswan, and Qena, are also providing machinery rental services to farmers. Because of these developments, it is difficult to discern any clear-cut government strategy on the future role of government machinery-rental stations in the country.

Private rental constitutes the main means of renting the Turkish beater-type threshers which are popular for the simultaneous threshing of winter crops and the production of fodder from wheat straw. For every thresher owner, there are up to one hundred thresher users, for the seasonal threshing capacity of a Turkish thresher is about 75 feddans (Kerr 1988).

Local manufacturing

Tractors

Table 19.6 gives some data on the local production of tractors and agricultural machinery together with an estimated market demand (El-Hossary 1987). Since 1983, imports of the 20–85 hp tractors have been banned and tractor manufacturing has been reserved for the nationalized sector. No private company is permitted to assemble or manufacture tractors. At present El Nasr and Company (NASCO) a public sector company, is the major assembler of Nasr tractors (Fig. 19.4) in the country. This company has licensing arrangements with the manufacturers of Rakovica tractors in Yugoslavia and Universal tractors in Romania. It assembles about 3000–4000 tractors annually in the 60–65 hp range. Currently, the basic Nasr tractors sells for about £E22 000.

Since there are no mandatory conditions for the progressive deletion of imported components, tractor manufacturing in Egypt has continued primarily as an assembly operation with rather low levels of local components. Monopoly tractor production, coupled with a ban on imports has guaranteed

TABLE 19.6. Local production of tractors, implements, threshers, and pumps with estimated market demand in the mid–1980s

Machine	Local production (units year^{-1})	Estimated market demand (units year^{-1})
Threshers (350–450 kg h^{-1})[a]	900	2500
Mounted chisel plough (7–9 tines)	2000	2600
Trailed chisel plough (9–11 tines)	600	600
Subsoiler, mounted	400	400
Ditcher, trailed	600	1000
Leveller, trailed	1400	2000
Water pump	18 000	20 000
Diesel engines (6–16 hp)	13 000	20 000
Field sprayer (600 l capacity)	750	1500
Tractors (60–65 hp)	3000	6000
Trailers (4 wheel, 4 t)	1600	2000
Back-hoe trailed	180	600

[a] Thresher production has increased substantially and is meeting the current (1990) market demand.
Source: El-Hossary (1987).

an assured market and has provided few incentives for setting up local production, or improvement, of after-sales services.

Very recently a second public sector company, the Helwan Diesel Engines Company has introduced a model IMT 35 tractor (Fig. 19.5). It uses a locally assembled two-cylinder air-cooled 35 hp Deutz diesel engine and an imported IMT trans-axle from Yugoslavia. This basic tractor is being sold for £E18 000, but it has only recently been offered on the market and hence it is difficult to comment on its acceptability.

Diesel engines

The Helwan Diesel Engines Company, a public sector enterprise, is the only company in Egypt which is authorized to produce diesel engines. At its Shubra Diesel Engine factory the company produces 11–16 hp slow-speed horizontal water-cooled diesel engines of a fairly old design. The company also assembles

FIG 19.4. Nasr tractor (60–65 hp) assembled locally by NASCO, a public sector company

FIG 19.5. The IMT 35 tractor (35 hp), assembled locally by Helwan Diesel Company, a public sector company

air-cooled vertical diesel engines of approximately 3.0, 5.5, 7.5, 11.0, and 15.5 hp under licence from the Deutz Company of West Germany. Private imports of all types of diesel engines below 125 hp have been banned in order to protect production and imports of diesel engines by the public-sector factories. The import ban has created many problems of engine availability, since the type and size of locally assembled diesel engines are limited and prices are generally high. Lack of availability of appropriate diesel engines has also hampered the development of machines which could be used individually on smallholdings.

Prior to the diesel engine import ban, single cylinder 5 hp water-cooled vertical diesel engines, popularly known as the 'Kirloskar type', were imported from India for use with irrigation pumps. The Kirloskar type engines are of a very simple and sturdy design and have been widely credited for bringing about an engine revolution in Indian agriculture. These engines were well accepted by Egyptian farmers and are still in demand in the country. Since the implementation of the ban on imports of diesel engines, many small workshops have started local assembly of the Kirloskar type engines (Fig. 19.3) with partly imported and some locally fabricated components.

Since production of diesel engines is reserved exclusively for the public sector, assembly of such diesel engines by small private workshops is against the government policy. However, farmers like the Kirloskar-type engines as these sell for between one-fifth and one-third of the price of locally assembled Deutz engines of comparable size. The Egyptian Government has, on many occasions, tried to stop the production of such engines but production continues on account of their attractive low price, simple design, ease of repair and maintenance, and widespread acceptance by farmers.

Lightweight, imported air-cooled gasoline – kerosene engines are used widely in the country for sprayers and pumps. The Egyptian Government is now planning to encourage local assembly of such engines.

Agricultural machinery

A wide range of machines are manufactured locally by the public and private sector manufacturers in Egypt. These include chisel ploughs, furrowers, ridgers, scrapers, threshers, trailers, stationary balers, corn shellers, irrigation pumps, and tractor front loaders. In addition, many small firms assemble irrigation pump sets and field sprayers using imported engines, pumps, and locally produced chassis.

The main public sector company which produces various agricultural implements and machines is Beheira & Company in Alexandria. This company produces light and heavy ploughs and light earth-moving equipment. It is planning to produce a wider range of implements under licence from a West German company. In addition the Helwan Machine Tools Company, a military factory, is producing a Turkish-type thresher, and limited quantities of tractor-powered mowers and rotary tillers.

Several well established private companies manufacture or assemble machines that are essentially identical to that produced by the public-sector companies. The leading private-sector company, Tanta Motors, has a long history of machinery production. Recently, it has established a modern factory (Fig. 19.6) under its subsidiary EMAGRO for the production of Turkish-type threshers and tractor power-take-off powered hoes and excavators.

The major centres for the production of agricultural machines in the country are at Tanta and Mansura in Gharbiya and Daqahliya governorates, respectively. Most smaller towns and villages have workshops which repair and make simple agricultural equipment including chisel ploughs, water wheels (*saqqiya*), threshers (Fig. 19.7), and tractor power-take-off powered irrigation pumps. These shops cater for the low-cost markets within their geographical areas and generally make products of lower quality than the larger well established firms.

Most equipment produced by the private-sector manufacturers is on a made-to-order basis. Because of this, machines such as threshers, which have high seasonal demand, are always in short supply during the season.

The following are the major constraints that manufacturers are facing in developing local production.

1. Most small manufacturers have weak marketing arrangements as they sell mostly to walk-in customers. This system does not create a large enough demand and forces manufacturers to continue with inefficient low-volume production methods. The Principal Bank for Development and Agricultural Credit (PBDAC) has a national system for marketing machines to farmers on credit through its branch offices. Because of the cumbersome procedures, however, most small manufacturers cannot use this facility.

2. Machines are produced on an individual basis or in very small batches with

FIG 19.6. A modern private sector thresher factory established by EMAGRO, a division of the Tanta Motors Company

Fig 19.7. Small-scale production of Turkish-type threshers in Egypt

simple production tools. Lack of proper jigs, fixtures, tools, and modern production equipment does not permit economies of scale in production and creates many problems in the manufacture of high-quality machines with interchangeable parts and standardized designs.

3. Very few manufacturers produce under licence or technical agreements with established manufacturers abroad and consequently do not follow rigid international quality standards in production. Most manufacturers have yet to establish their own brand names in the market. Lack of brand name recognition also results in insufficient emphasis on product quality.

4. Many restrictions and problems in importing complex machine components for use in local production hamper long-range planning and continuous production.

5. Fully assembled agricultural machines can be imported at fairly low customs duties (about 3–10 per cent), whereas machine components and production materials bear rather high customs duties (10–35 per cent). Special industrial tariff rules, which permit lower customs duties (10 per cent) for local producers, are mostly available to the larger organized manufacturers who import components directly.

6. There are no clearly defined areas of machinery production reserved for the heavily subsidized and generally favoured public sector and the small

private manufacturers. Often the same machines are produced by small rural workshops as well as by the large public-sector or military companies. Some of the military companies produce agricultural machines primarily to utilize their excess production capacity rather than as profitable and economic operations. The public sector factories often initiate production of the same machines for which a market has been developed by private manufacturers. This creates unhealthy competition and often hampers the development of small-scale private-sector manufacturing in the country.

Although the Egyptian Government generally favours local production over imports, there is no government organization responsible for assisting and encouraging local production of agricultural machines. Most local manufacturers do not have sufficient political power, or access to authorities or to information, to take full advantage of the national and international assistance programme. Several government institutions, such as the Agricultural Mechanization Research Institute (AMRI), the Industrial Design and Development Centre (IDDC), the General Organization for Industrialization (GOFI), and the Industrial Development Bank (IDB) can assist local manufacturers, but communication between such organizations and manufacturers is either poor or virtually non-existent.

Imports of tractors and equipment

In 1973 the Egyptian Government liberalized the imports of tractors and other farm equipment. Between 1974 and 1981, the total annual value of agricultural machinery imports, in constant 1975 terms, grew from £E1million to £E25million (World Bank 1985), the number of tractors imported grew annually from fewer than 1000 to nearly 12 000, and the number of other imported agricultural machines grew from less than 300 units in 1974 to 15 000 units in 1981.

In spite of the liberalized import regulations, there are a number of restrictions on imports of tractors and machines which have at times hampered mechanization. Imports of 20–85 hp tractors have been banned since 1983 in order to protect local assembly of tractors. Similarly, imports are banned of diesel engines up to 125 hp, threshers, mowers, post hole diggers, rotary tillers, and sprayers with engines. Most tractor and machinery imports from the Eastern Block countries are channelled through two public sector companies, the General Company for Engineering Equipment and the General Chemical and Trading Company.

There are several government authorities involved with regulating imports of farm machinery. These include the Under Secretariat of Engineering Affairs of the Ministry of Agriculture (MOA), the different military factories of the Ministry of Military Production, the Central Bank, and the Tarsheed

Committee, and the import rationalizing committee in the Ministry of Economy.

Currently, private importers must obtain permission from the respective military factories to import farm machines. About five years ago, this regulation was put into effect because the Ministry of Military Production wanted to utilize its excess production capacity and had indicated to the government its desire to manufacture most of the agricultural machines that were being imported into the country. Although the real purpose of this regulation was to encourage and protect local production by the military factories, its effect has been the opposite.

The military factories currently produce only a few items of agricultural equipment, namely, a limited range of diesel engines, Turkish-type threshers, a limited range of rotary tillers, mowers, and other implements. So far the military factories have not announced specific plans for the production of any other agricultural equipment, and yet they exercise almost absolute control on all imports of agricultural equipment into the country. Since the implementation of the regulation, all imports of agricultural machines must be approved by the military factories. Private importers have to pay up to 15 per cent of cost, as an assembly charge, for obtaining a certification of no objection from the military factories for importing equipment.

Most importers, however, claim that machines continue to be imported into the country in a fully assembled condition. It seems that the net effect of this regulation has been a substantial increase in machine prices and a continuation of imports rather than the development of local manufacturing in the country. Many developing countries such as India, Pakistan, and Turkey have made rapid progress with the development of local production because they have linked market protection with mandatory schedules for the progressive deletion of imported components. Unfortunately, this has not been the case in Egypt.

Credit availability

The Principal Bank for Development and Agricultural Credit (PBDAC), which at present encompasses 17 governorate level banks and an extensive network of about 5000 village level banks and agencies, is the main source of agricultural credit in the country. The PBDAC also handles the supply and marketing of all agricultural inputs. The PBDAC system enjoys a virtual monopoly on the supply of credit for tractors and agricultural machinery, since machinery dealers and manufacturers rarely extend credit to individual farmers.

The PBDAC also acts as a selling agent for tractors and machinery and charges a 5 per cent mark-up as a 'marketing service charge'. The PBDAC has a major role in marketing tractors in the country. For example, from 1981 to 1984, 43 per cent of all the tractors sold in the country were sold through the PBDAC system. Only registered manufacturers and dealers can sell their machines through the PBDAC system. Although in principle it is easy to register with the PBDAC, in practice a number of stringent requirements make it difficult for small private manufacturers to register.

In general farmers apply for a loan from a village bank, a branch of the PBDAC. The bank requires a 25 per cent down payment with the balance payable over 4 years at an interest rate of 16 per cent. The bank displays some of the machines at its branch offices but it does not provide any repair or after-sales services. Since the bank has a monopoly on the financing of tractors and machinery, it competes advantageously with private machinery dealers in the market.

Until recently, farmers had to own 5 feddans of land to qualify for loan collateral for tractors. These collateral requirements are less rigid now, but they are still based on property ownership, which excludes the vast majority of farmers. Owners of machinery contract hire services can borrow from the PBDAC but the collateral requirements are equally stringent.

Most locally made machines, except Turkish threshers, are purchased by farmers on a cash basis. About three years ago, The Helwan Machine Tool Company, a military factory, was successful in limiting the PBDAC thresher financing exclusively to its own production machines. This seriously affected the sales of threshers produced by the other manufacturers in the country. After considerable efforts, three private manufacturers, Tanta Motors (EMAGRO), Diamond and Company, and Gabr and Company were successful in obtaining the approval of the PBDAC for thresher financing. However, the majority of small manufacturers are still not able to avail themselves of the PBDAC financing for selling their threshers.

Although the PBDAC marketing system is difficult to harness by small manufacturers, a widespread marketing structure with a network of village banks reaching the smallest villages, is already in place in the country. With some effort, this network could be developed as an effective mechanism for selling machines produced by the numerous small manufacturers in Egypt.

Repair, maintenance, and training

Major tractor and farm machinery repair and maintenance facilities are located in the larger towns and the governorate capitals. Since tractors have been imported from many countries, there is a an assortment of tractor makes in Egypt (Table 19.7). This has complicated the problem of parts availability and has hampered the development of specialized repair and maintenance services in the rural areas. The ban on tractor imports and the local assembly of tractors has, however, improved the situation, especially in relation to the repair and service of Nasr tractors.

In a sample survey of 27 workshops (Hopkins *et al.* 1982), 63 per cent had been operating for over 10 years, whereas 85 per cent were at least 5 years old, indicating a certain level of stability in the repair and maintenance services. Seventy per cent of these shops were limited to repairs of tractors and irrigation pumps. The most common repair work was a general overhaul (92.6 per cent), gear box repair (81.5 per cent), hydraulic system (70.4 per cent), body work (37 per cent), and petrol pump (29.6 per cent).

Most mechanics receive their training through apprenticeship. Less than 50 per cent of tractor and machinery operators have been estimated to have had some operation and maintenance training. Inadequate training of mechanics and operators leads to excessive misuse and mismanagement of equipment. Even though machinery importers, by law, must import spare parts equal to 20 per cent of the value of the machines, the survey indicated that repair shops were hampered more by lack of equipment and spare parts, than by lack of training.

The Ministry of Agriculture (MOA) operates a major training centre, the Farm Machinery Training Center at Ma'mura near Alexandria. This is a unique and highly successful centre which has been developed with assistance from the West German Government. This centre provides excellent training

TABLE 19.7. Tractors used in Egypt with country of origin, 1982

Model	Size (hp)	Country of origin
Nasr (all models)	60–65	Assembled in Egypt
Ford (all models)	45–65	USA and UK
Universal 65 hp, and Zetor (all models)	45–65	Romania
Ballarus	60–80	Russia
David Brown	45–70	UK
Massey-Ferguson	35–75	Canada, USA
Deutz	65	West Germany
Hanomag	55–70	West Germany
Fortschrit ZT–300	80–100	East Germany
Ebro	75–80	Spain
IMT	65	Yugoslavia
Rakovica	65	Yugoslavia
John Deere	45–100	USA
Caterpillar	80–125	USA
Fiat	45–100	Italy
Others	20–45	Japan

Source: El-Sahrigi and El-Haddad (1982).

for tractor and machinery mechanics and operators. A number of other centres also provides mechanic and operator training, but their facilities and programmes are relatively less developed.

Research and testing

Eleven universities currently offer agricultural engineering/mechanization education at undergraduate, and in some cases, graduate levels. Most universities operate under severe budgetary limits, so the research conducted by graduate students on agricultural mechanization topics at these institutions is limited. Such research is generally not sufficiently applied and rarely reaches the farm level.

The Agricultural Mechanization Research Institute (AMRI) of the MOA is the primary institution charged with conducting applied research on agricultural mechanization in the country. This institute was organized in 1983, by bringing together under its umbrella the Tractor and Agricultural Machinery Testing Center at Sabahiya and the Rice Mechanization Center at Meet el-Diba. Since AMRI is a relatively new organization and government salaries are not as attractive, it has only been successful in attracting a limited number of professional staff to its headquarters in Cairo. Consequently its research programmes have yet to be fully developed.

The Institute had a USAID supported Agricultural Machinery Project during 1982–5 under which a range of mechanization extension and research activities were undertaken. This project was successful in introducing laser land-levelling techniques for precision levelling in the country. This practice is now slowly gaining popularity for saving irrigation water and improving crop yields. More recently, under the USAID-supported National Agricultural Research Project (NARP), AMRI has been focussing on the development of critically needed farm machines. The major focus of this activity is on the development of cereal crop harvesters and threshers which could be manufactured in Egypt.

In view of the special requirements for fodder production from wheat straw, Egyptian farmers use beater-type threshers for wheat, their main winter crop. Rice, the second major cereal crop, which is grown during the summer season, continues to be threshed by treading under animals or tractors. Recently, the NARP project has been successful in developing a unique all-crop thresher which for the first time enables Egyptian farmers to thresh both winter and summer crops and make fodder with a single threshing machine. This machine is now being produced by a number of local manufacturers.

The Tractor and Machinery Testing Center, which is now a part of the AMRI, was established in 1963 at Sabahiya to test both imported and locally produced machines. This centre also conducts limited mechanization research.

The centre was expected to conduct regular machinery tests and issue periodic reports on commercially available tractors and machinery. These would help to educate farmers and end-users on machine suitability and performance. However, very few reports have been issued by this centre.

A team of experts (Winrock International 1986) producing a report on the development of policy guidelines for agricultural machanization in Egypt provided the following viewpoint about testing of tractors and machines in Egypt.

'Contrary to the claimed benefits of machinery testing, we found little evidence that the private sector needs government assistance in identifying good machines. In fact the main examples of successful introduction of new machines (domestically manufactured scrapers, Turkish threshers, Indian water pumps), appear to have been private-sector achievements. We believe that the claimed potential contributions of public testing and evaluation have been overly optimistic.'

The Rice Mechanization Centre was established at Meet el-Diba, in 1983, for conducting research on rice mechanization with assistance from the Japanese Government. Research at this centre has been focussed primarily on Japanese rice production equipment and more specifically on combine harvesters and rice trasplanters. Because of the small size, high costs, and relative complexity of Japanese farm equipment, farmer acceptance of such equipment has been slow in Egypt.

Extension

The Directorate of Agricultural Extension is in the central administration of the Ministry of Agriculture under the National Production Programmes. There is currently no group within the extension service specializing in agricultural mechanization (Louis Berger International 1985). According to a survey of farmers in the Nile Delta (Hopkins *et al*. 1982) almost 50 per cent of farmers felt that the Egyptian Government was not helping them enough in mechanization. About 25 per cent of the farmers felt that mechanization extension services should be given high priority.

To a large extent, the various government tractor and machinery rental stations have helped to popularize new machines and mechanization techniques to farmers, even though this was not their primary function. Some activities on introducing new mechanization technologies are currently being carried out by the PBDAC under a USAID-supported Agricultural Production and Credit Project. This project is evaluating and popularizing different machines and mechanized crop production techniques in selected areas of the country.

Mechanization policies

In the past, the Egyptian Government has promoted mechanization through a 'saturation approach' in the belief that it would help to increase crop yields and cropping intensities, and reduce production costs. Since 1973 a host of incentives have propelled mechanization to an extent that more than 90 per cent of farmers utilize machines for at least some of their farming operations. While progress in mechanization has been significant, it has been achieved at a rather high cost and with mixed results. Many conflicting policies are in effect, and these seem to indicate a lack of co-ordination among various government branches and between the public and private sectors in Egypt.

The mechanization of agriculture and the Egyptian farmers have suffered considerably due to the somewhat antagonistic relationship that exists between the heavily subsidized public sector and the private sector. Since the roles of the two sectors are not clearly defined, the two sectors often compete rather than complement each other. The private sector is active in machinery importation, manufacture, sales, services, and rental but so is the Egyptian Government. Generally, government institutions compete in highly regulated and protected markets at subsidized prices with a heavily controlled private sector, which is forced to remain relatively small and expensive. Thus, excessive government controls have constrained the balanced growth of agricultural mechanization and the development of local machine production. Some examples of the conflicting mechanization policies are given below.

The Egyptian Government is committed to bringing mechanization technology to farmers at the lowest possible costs, yet many regulations put into effect have resulted in substantially higher prices to farmers. For example, officially mandated charges, mark-up, and tariff on imported machines sold on credit to farmers are:

Customs duty on machines	3–10%
Customs duty on spare parts	25–35%
25-Military factory mark-up on imported farm equipment	15%
PBDAC marketing services charges	8%
Interest rate on machinery loans (p.a.)	16%

According to government regulations, most private sector imports of farm machines must now be approved and channelled through the military factories, which add a 15 per cent mark-up, as a local assembly charge. This mark-up is eventually paid by farmers in the form of higher machine prices. Importers claim that the military factories are not providing any assembly services and that most machines are being imported in fully assembled condition.

The Egyptian Government has recognized that efficient machinery contract hire services can be best provided to farmers through private machinery rental contractors and by neighbourhood farmers. Yet the government continues to operate public sector machinery rental stations in most parts of the country.

The stated goal of the Egyptian mechanization policy is a free competitive machinery market. Nevertheless one finds that the Egyptian Government has in effect created an absolute or near absolute monopoly for the manufacture of tractors and diesel engines, the import of a wide range of farm machines, and the marketing and financing of tractors and equipment. Lack of competition has resulted in higher prices, inefficient production and poor after sales services, all with detrimental consequences for Egyptian farmers.

Conclusion

During the last thirty years, considerable progress has been achieved towards the mechanization of agriculture at the farm level. However, developments in agricultural mechanization have been somewhat isolated. For example, the operations of land preparation and wheat threshing are almost fully mechanized, whereas harvesting of wheat and paddy and threshing of paddy still remains almost completely non-mechanized. Similarly, progress towards the local production of tractors and agricultural machines has not been commensurate with the mechanization developments at the farm level. Undoubtedly, better co-ordination of the government policies, reduction of various restrictions, clear demarcation of public and private sector roles in mechanization, and greater emphasis on local manufacturing can help in achieving a broad-based mechanization of agriculture in Egypt.

Note on recent policy changes effecting agricultural mechanization in Egypt

In the interest of economic liberalization, the Egyptian government has recently instituted a number of policy changes, which have an important bearing on the future of agricultural mechanization in the country. These policy changes are summarized below.

a) Farm input subsidies on fertilizer, diesel fuel, electricity are now being gradually removed.
b) Farm gate prices of all crops, except cotton and sugar cane, are no longer controlled. Prices of cotton and sugar cane have also been increased substantially.
c) The Principal Bank for Development and Agricultural Credit (PBDAC) has no longer a monopoly for financing tractors and agricultural machinery. PBDAC is now competing with the commercial banks in this area.
d) The PBDAC is being asked to limit its activity to financing of agricultural inputs and to divest itself from its business involvement in marketing farm inputs.
e) Tractor manufacturing is no longer limited exclusively to NASCO, the public sector tractor manufacturing firm. It is now open to any interested parties in both the public and the private sectors.
f) Most agricultural machines have been included in the open general import list. The

requirement of a no-objection certificate from the military factories for imports of machines has also been removed.

g) Egyptian manufacturers, dealers, and importers of agricultural machinery have organized themselves in an association which has successfully drawn the attention of the government to their problems resulting in a number of favorable policy changes.

Acknowledgement

Opinions expressed in this chapter are solely those of the author and do not in any way reflect the views and policies of the institutions in Egypt with which he is affiliated.

References

CAPMAS (Central Agency for Public Mobilization and Statistics) (1989). *Statistical yearbook of the Arab Republic of Egypt, 1982–3*. CAPMAS, Cairo.
El-Hossary, A. M. (1984). *Farm mechanization development in the Arab Republic of Egypt*. Ministry of Agriculture, Dokki, Cairo.
El-Hossary, A. M. (1987). *Agricultural mechanization of Egypt*. Ministry of Agriculture, Dokki, Cairo.
El-Sahrigi, A. and El-Haddad, Z. (1982). *Egyptian agricultural mechanization five years development plan 1982/83–1986/87*. Agricultural Mechanization Project, Ministry of Agriculture, Cairo.
Hopkins, N. S., Mehanna S., and Maksoud, B. M. A. (1982). *The state of agricultural mechanization in Egypt, results of a survey 1982*. Agricultural Mechanization Project, Ministry of Agriculture, Cairo.
Kerr, J. (1988). *The Egyptian agricultural machinery market. Report of a case study of the Turkish thresher*. Food Research Institute, Stanford University, Stanford, California.
Louis Berger International Inc. (1985). *Agricultural mechanization project. Final report*. Louis Berger International, Inc., East Orange, New Jersey.
Ministry of Agriculture (1982). *A comparative study of water raising costs in Egypt*. Agricultural Mechanization Project, Agricultural Mechanization Research Institute, Giza, Egypt.
Winrock International (1986). *Policy guideline for agricultural mechanization in Egypt*. Winrock International, Morillton, USA.
World Bank (1985). *Staff appraisal report, Arab Republic of Egypt*. Second Agricultural Development Project. Ministry of Agriculture, Cairo.

20

Marketing channels and price determination for agricultural commodities

J. ROWNTREE

Introduction

Egypt was essentially food self-sufficient 30 years ago, whereas today it imports more than one-half of its food requirements. This declining food self-sufficiency can be attributed to several factors, notably: the rapidly growing population, increasing demands for food associated with rising per caput income and an extensive food subsidy system, slow growth in agricultural productivity associated in part with government production and marketing policies which have repressed the agricultural sector, and extensive food wastes in a 'backward and inefficient' marketing system (Wally *et al.* 1983). The inadequacies of the agricultural marketing and distribution system have been repeatedly indentified as being a significant constraint on increasing agricultural production and food self-sufficiency (USDA 1976; USAID 1982; IADS 1984; NCBA 1986).

With the complex interaction of a traditional agricultural farming system and an interventionist government, up until the 1990s the marketing channels for agricultural commodities in Egypt fell into two general categories: the regulated and unregulated marketing channels. The regulated marketing channels were grounded in the cropping pattern of Egypt which is mainly based on an agricultural rotation that emphasizes exportable, and largely government-controlled non-perishable crops, including cotton, rice, wheat, maize, broad beans, lentils, and sugar cane. Modern storing, grading, and standardizing operations have been restricted to these non-perishable crops, particularly cotton, where marketing is based on well-defined grades. Prices for the non-perishable crops were determined, or substantially influenced, by government policy. At one extreme, for example, the prices of cotton and sugar cane were set by the Egyptian Government which required that farmers deliver all of their output to government marketing agents. For rice, broad beans, lentils, sesame, and groundnuts, on the other hand, the Government set prices for a production quota that farmers were obliged to deliver to pooling centres; production in excess of the required deliveries entered the unregulated market. While there were no government-required deliveries for wheat and maize in the late 1980s, their prices, being formally determined in the free market, were significantly influenced by government control over the supply of imported cereals.

The Egyptian Government has also been involved in a wide range of food-processing activities and supported an extensive consumer food subsidy and rationing system that was, until recently, a principal marketing channel for food commodities.

The unregulated markets, on the other hand, were grounded in the traditional integrated crop – livestock farming system on which the major cereals and fibre crop rotations were imposed. Perishable crops, such as vegetables, fruits, and livestock products, are produced primarily to meet domestic consumption requirements. Since these are produced for the local market, there has been little incentive to promote the efficiency of marketing operations, for example in storing, grading, standardizing, packing, or transporting. The prices of the vegetables, fruits, red and white meat, dairy products, and *bersim* (Egyptian clover) in unregulated markets are determined substantially by the forces of supply and demand.

The agricultural marketing system in Egypt has thus been a mosaic of regulated and unregulated markets and of traditional and modern practices. The discussion that follows emphasizes the elucidation of the institutional framework and pricing mechanisms in these regulated and unregulated markets as they existed in the late 1980s, providing concrete examples of the details of post-harvest practices and produce handling.

Faced with stagnating agricultural production, declining food self-sufficiency, growing budget deficits, and continued pressure from the International Monetary Fund, the World Bank, and United States Agency for International Development, and other international donors to dismantle the command-economy system imposed on the agricultural sector and to move toward freer markets, the Egyptian Government recently embarked on a revolutionary restructuring of the marketing channels and determination of prices of agricultural commodities. In 1990, the Egyptian Ministry of Agriculture announced 'comprehensive plans to liberalize agricultural input and output markets over the next three years.' (USDA 1991) By the end of 1991, only cotton and sugar cane remained under government production control, and the food subsidy and rationing system for consumers is being dismantled. Since neither the pace of implementation nor the impact of this liberalization programme is known, the discussion which follows focuses on the agricultural marketing system that existed through to the end of the 1980s, indicating some of the known changes and recent developments.

Following an overview of the development of Egypt's agricultural marketing channels, the government food subsidy and rationing system will be discussed. This will be followed by detailed examinations of the regulated marketing channels, particularly for cotton and the major cereals, and the unregulated marketing channels, particularly for meat, milk, vegetables, and fruits. Finally, some of the unintended consequences of the food subsidy system, wholesale market concentration, and post-harvest losses and food wastes will be discussed.[1]

Development of contemporary agricultural marketing channels

Egyptian agriculture has a long history of producing an abundance of products which have provided more than the subsistence needs of the traditional farmer. For centuries this surplus has been processed and shipped through a marketing system which involved an array of administered and free market prices.

The contemporary Egyptian agricultural marketing and pricing system bears the imprint of the highly centralized agricultural planning system that was put into place during the years of President Nasser's Government. Following the Agrarian Reform Act No. 178 of 1952, the Egyptian Government increasingly implemented control mechanisms on the agricultural production, marketing, and pricing system, with the primary purpose of transferring the surplus from the agricultural sector to other sectors of the economy. The policy emphasized industrialization and was aimed at providing cheap food to the urban population, at the expense of the agricultural sector. This development policy was based on the notion that the rural sector should subsidize the urban sector, despite the fact that the average rural income was only about one-half that of the average urban income. The government monopolized the marketing of major cash crops, controlled the pricing and distribution of non-farm inputs, imposed firm controls on the organization and marketing of agricultural production, and implemented a food-rationing system to support a low-wage policy designed to promote industrialization.

During the early 1960s, an array of new institutions were created to implement this policy. Agricultural co-operatives were established in each village to link the farmers to the governmental institutions. Each year the Egyptian Government allocated areas for the production of major crops, especially cotton, wheat, rice, onions, groundnuts, and sesame. A State monopoly was created for marketing cotton and sugar cane. A system with the misnomer 'co-operative marketing' compelled farmers to deliver a given quota of rice, wheat, broad beans, groundnuts, and sesame to pooling centres at fixed prices. The government extracted the surplus from agriculture by requiring farmers to sell their crops to the government at substantially less than their international or free market values.[2]

Several other policies also contributed to the repression of the agricultural sector. For example, an overvalued exchange rate discouraged agricultural exports and encouraged low-priced imports to compete with domestic producers. The agricultural sector also received little public investment. For over two decades during which agriculture produced about 23–30 per cent of national income, generated 25–75 per cent of the value of exports, and provided 37–45 per cent of total employment, public investment in the agricultural sector accounted for only 7–9 per cent of the total. The combined effects of these repressive policies on the agricultural sector has generated very modest growth rates in agricultural productivity. In the 1970s, the average annual growth rate of agricultural production did not exceed 2.0 per cent, while population was growing at 2.7 per cent, and the industrial sector was growing at 4–7 per cent each year (Dethier 1989).

Another aspect of the government's development policies was the low-wage, cheap-food policy which led to the implementation of an extensive food-rationing and subsidy system. The food subsidy system, a cornerstone of the contemporary equity policies of the Egyptian Government, was designed to guarantee a minimum level of nutrition to the Egyptian people, to protect consumers from wide fluctuations in the world prices or extreme shortages of particular items, and to promote a more efficient and equitable regional distribution of food in Egypt. Furthermore, inexpensive and available basic food has been considered essential to maintaining Egyptian social and political stability.

The food-subsidy programme has been a part of a broader welfare system in which the government subsidizes energy, housing, transportation, clothing, and several other items. This food subsidy system, in combination with a vast food-rationing system, turned the government institutions into a crucial marketing channel for food and agricultural products in Egypt.

While food subsidies are not new in Egypt, the scope of the contemporary food-subsidy system represented a dramatic break with the past. Up to 1973, it was a modest programme, with the costs accounting for less than one per cent of national income. The real growth in the food-subsidy system occurred under President Sadat in the mid-1970s, as shown by the fact that between 1970–1 and 1980–1 the food subsidies rose as a share of total public expenditures from 0.3 to 33.3 per cent and rose as a share of gross domestic product (GDP) at market prices from 0.1 to 12.4 per cent (Alderman et al. 1982; Scobie 1981, 1983). Thus, under President Sadat, the government institutions in Egypt became major marketing channels for food commodities.

The low agricultural prices paid to farmers, discouraging production, and low food prices paid by consumers, encouraging consumption, contributed to Egypt's rapidly growing food gap. In 1961, Egypt imported only 6 per cent of its food requirements; by 1970, Egypt imported 14 per cent of its food; since 1980, Egypt has had to import more than 50 per cent of its food requirements. This declining food self-sufficiency became a major concern to policy makers in the 1980s (USAID 1982; Wally et al. 1983).

Government involvement in agricultural markets under Presidents Nasser and Sadat was one of the world's great market intervention experiments, one that is only now being dismantled. Although the efforts in the late 1980s to reduce government intervention in agricultural markets were being retarded by the rapidly growing supply requirements of the food-subsidy and rationing system, government involvement in agricultural marketing did decline. The agricultural sector is still viewed as the generator of the surplus required to promote economic development. Policy makers treated the negative aspects of controls on agricultural markets as a necessary and justifiable cost of maintaining a certain degree of social equity in food and income distribution. However, by 1990, the array of problems generated by the government control of production, marketing, and price of agricultural commodities forced the government to begin a drastic restructuring of its role in the agricultural sector. The 1990s are the beginning of a new era in Egyptian agricultural marketing and price determination. The structure of the command system imposed on the agricultural sector described below suggests the enormity of the adjustments which will be required during the 1990s as Egypt turns to free markets.

The government's food-subsidy and distribution system

A major agricultural commodity marketing channel is the government's food-subsidy and distribution system, through which more than one-half of the food consumed by Egyptians is channelled (Abdou 1982; Alderman *et al.* 1982; Von Braun and De Haen 1983). The scope of the system is illustrated by the following: the government's food subsidy expenditures amount to about 15 per cent of GDP; and about three-quarters of the rice, two-thirds of the edible oil, and one-third of the sugar are distributed via the rationing system.[3].

The Ministry of Supply and Home Trade is responsible for administering the system, including the procurement of supplies, the allocation of regional quotas of commodities and the determination of consumer prices and quotas for subsidized and rationed goods. Under the Ministry of Supply there are many agencies, including the General Authority for Supply Commodities (GASC); twelve wholesale food companies and co-operatives, meat and fish procurement companies, and processing and packaging companies; nine rice milling and marketing companies; ten mill, silo, and bakery companies; and eight clothing, shoe, and other non-food commodity distribution companies.

The GASC is responsible for importing the quantities of wheat, flour, maize, beans, lentils, meat, and poultry that are necessary to fill the gaps between domestic requirements and production. The commodities are then distributed to storage or wholesale companies at subsidized prices, so that most of the cost of the food-subsidy system shows up in the budget of the GASC. The Principal Bank for Development and Agricultural Credit

(PBDAC), the other principal purchaser of subsidized commodities, is responsible for obtaining domestic production to enter the subsidy and rationing system. The PBDAC, also the supplier of agricultural credit to farmers, acts as the middleman between the farmer and the marketing channels for subsidized foods. The PBDAC purchases domestic wheat at pre-announced fixed prices and accepts quota deliveries as a percentage of the rice, lentil, and bean crops at fixed prices from Egypt's farmers. It then channels rice and wheat to the milling companies and co-operates with the GASC in channelling beans and lentils to wholesale companies and imported maize to private users. The wholesale and milling companies then channel the subsidized commodities to bakeries and the retail level. The General Wholesale Company and the Egyptian Wholesale Company together have about 2000 branch warehouses to service the subsidized food-marketing channels.

At the consumer level the government food-distribution system consists of outlets for the three principal types of subsidized goods: rationed goods, semi-rationed goods, and unrationed goods. The outlets are a combination of some 32 000 private grocers registered with the government as rationed goods supply shops, the *tamween*, and some 4500 government food stores and co-operatives, the *gamayya*. In the city of Cairo there are about 2000 *tamween* and about 800 *gamayya* out of a total of about 40 000 retail food outlets (Alderman *et al.* 1982; Abdou 1982; Gardner and Abdou 1982).

As Egypt enters the transition to privatization and market liberalization in the 1990s, there is little understanding of the roles that the GASC, PBDAC, the *tamween*, and the *gamayya* will play in the future, especially as the prices of rationed goods are raised to free market levels.

Subsidized and rationed commodity marketing channels

Cooking oils, sugar, rice, and tea are both subsidized and rationed, with quantities per individual guaranteed. The principal outlets for rationed commodities include both private grocers and government food stores. The *tamween*, the licensed supply shops, receive rations from wholesale companies and distribute them to holders of green ration cards. The green ration card issued to a household is registered with a grocer who is required to pick up the monthly volume of rations from the wholesaler at fixed dates and to hold the rations of a registered household for the month. Grocers make little profit from the rationed commodities and thus have little incentive to serve as *tamween* except that it ensures a regular patronage to whom the grocer may sell non-subsidized items. About 95 per cent of the resident Egyptian population is served by this food ration system.

In 1985–6, the rationed items were distributed on a monthly basis as follows: sugar, cooking oil, and tea were rationed at 750, 450, and 40 g per person, respectively, at prices which were 20–40 per cent of their international values.

Additional quantities could be bought at government and private food stores at significantly higher prices. Low quality rice was rationed at 3 kg for 1–2 person households at about one-third of the non-rationed price. Imported frozen red meat was rationed at 1 kg for each two family members at £E1.50 per kg, which was only about 20 per cent of the unregulated market price for domestic red meat. Imported poultry and flour are not rationed but are distributed by the *tamween*.

Subsidized and semi-rationed commodity marketing channels

Beans, lentils, imported frozen beef, poultry, and wheat flour are both subsidized and loosely rationed, whereby quantities per family are guaranteed at the regulated prices except in times of supply shortages.

The principal outlets for the semi-rationed food are government food stores and consumer co-operatives. Co-operative members receive a red ration card identifying the number of persons in the household, but quantities depend on availability. Al-Ahram, El-Nile, and the Alexandria Company for Consumer Goods, popularly know as co-operatives (*gamayya*), are public sector companies with about 1800 associated retail outlets, operating primarily in the urban areas. These government food stores and co-operatives also offer, at unregulated prices, unrationed items such as jams, cheeses, meat, chocolate, local and imported canned goods and juices, eggs, and fresh produce, along with some non-food items such as matches, batteries, etc. The unsubsidized items are generally sold at a 10 per cent profit margin, and prices here are usually lower than those for comparable items at private stores because of economies of scale and because the Al-Ahram co-operatives can bypass the wholesaler, passing on to the consumer some of the government-authorized four per cent margin for wholesalers (Alderman *et al.* 1982; Gardner and Abdou 1982).

These public-sector companies receive quotas of commodities from the governorates in which they function, and deliveries by company truck to outlets are distributed principally according to population density and outlet size. These stores record purchases of sugar, lentils, rice, and beans in the consumer's ration book. Purchases of imported frozen meat, limited to 1 kg per month for each two family members, are recorded in the outlet's register.

While the public-sector retail outlet chains serve most urban areas, the other main distribution channel for the subsidized and semi-rationed commodities is through consumer co-operatives established by governorates, neighbourhoods, and government and factory workplaces. These co-operatives are true 'co-operatives in which members obtain membership by paying £E1 in dues and receive one vote in co-operative management and dividends according to their purchases, as recorded in their red ration books. Governorate co-operatives vary in size, but have up to 10 000 members.

Subsidized but unrationed commodity marketing channels

Bread is the principal example of a subsidized but unrationed commodity, where quantities bought are unrestricted. The Egyptian Government, recognizing that *baladi* (country or native) bread is the most important single element in the Egyptian diet, has made it readily available at a very inexpensive price. The huge direct subsidy to bread has amounted to approximately £E1000million per year in recent years.

Most bakeries are small family-operated private enterprises, employing simple technology and using stone ovens heated with kerosene to bake the bread. The Ministry of Supply provides these bakeries with subsidized flour and regulates bread weights, moisture content, and prices. For many years the three principal bread types have been *baladi* bread made of 93 per cent extraction rate flour which sold for one piastre per 135 g loaf, *shami* bread made of 72 per cent extraction rate flour which sold for 2 piastres per 148 g loaf, and *fino* bread, a small, long loaf, made of 72 per cent extraction rate flour which sold for one-half piastre per 62 g loaf. Extraction rates for flour have varied from time to time according to governmental regulations, and a larger variety of loaf sizes than the main ones mentioned above are often available (Rizk 1983). Facing growing fiscal and trade deficits, the Egyptian Government has been raising the average price of bread in Egypt in recent years by selectively reducing the availability of the cheaper *baladi* breads. By late 1991, the 1 and 2 piastre loaves of *baladi* bread had been phased out, being replaced by a 5 piastre loaf with a price still only about 40–50 per cent of the cost of production.

Approximately 5000 commercial bread bakeries registered with the Ministry of Supply obtain about 60 per cent of the imported flour at heavily subsidized prices and make inexpensive bread readily available in unlimited quantities in the urban areas. Another 25 per cent of the imported flour is sold to about 100 registered pasta factories, with the remaining 15 per cent of imported flour being sold to individual consumers or used for other baked goods (Alderman *et al.* 1982).

Regulated agricultural markets

The Egyptian Government plays a dominant role in the marketing and price determination for many major agricultural products, particularly cotton, the major cereals, sugar cane, and pulses (Nassar *et al.* 1981).

The cotton economy and marketing channels

Cotton is the most important fibre and oil crop in Egypt. Cotton acreage currently accounts for about 20 per cent of the cultivable land area and about

10 per cent of the cropped area of Egypt. It has been and remains Egypt's most significant export crop, as well as a major source of government revenue. Government control of cotton is so extensive that cotton could be called the 'government crop'.

The history of cotton in Egypt is also the history of Egypt's involvement in the international economy. Egypt is the home of extra-long staple cotton which, following ruler Muhammad Ali's construction of the modern irrigation system of barrages in the early part of the nineteenth century, opened the way for Egypt's integration into the world economy. President Nasser's Government substituted the State for the landlords in taking the economic surplus from farmers, and changed the use of the surplus from supporting the luxuries of the landlords to promoting economic development for Egypt. Cotton, continuing to play its pivotal role in the formulation of Egyptian economic policy, was the major export crop around which the state's interventionist policies in agriculture were begun.

Cotton production and marketing are highly modernized through a command-economy system. Cotton planting is in large blocks, mandated by the Government. All farmers with farm plots within these blocks are required to plant them with cotton. There is a nationwide pesticide spraying programme, and inputs are supplied through government institutions. The cotton crop is marketed through the government alone, and the government reaps declining but significant revenues from its controls by the international marketing of cotton at prices substantially above farm-gate prices.

This command system in cotton production and marketing is not, however, without problems. Low prices would normally lead to reductions in the quantities supplied, but the Egyptian Government has attempted to keep production, and its revenues, high by mandating requirements for farmers on cotton acreage planting. With government control over cotton planting and pricing, the political forum was the only avenue available to farmers for adapting cotton production to expected returns. The mandated cotton planting requirements have been gradually reduced, as Egyptian farmers 'politically' effected a normal upward-sloping supply curve for cotton. Cotton planting is currently less than 1.2 million feddans, down by more than one-third compared with plantings of twenty years ago.

The cotton produced must be delivered after harvest to cotton collection centres of the State Cotton Organization. The cotton is weighed, graded, and then taken to gins for processing. About 60 per cent of the cotton supply is processed domestically, the rest being exported. Most domestic consumption is of long staple cotton rather than of the extra-long staple which is grown principally for export (Hafez *et al.* 1982)

Cotton prices are set by a multi-ministerial high council on the basis of concerns from the interested ministries. Farm prices vary by grade and variety, taking into account the farmers' costs of production; transport, milling and other processing costs; the selling price to domestic mills; the expected export price, and the State's revenue requirements. Prices and production targets, or acreage requirements, are jointly determined. Prices tend to be set at the average cost of production of a supply quantity sufficient to meet domestic cotton requirements and export earnings targets. For many years farmers received only one-third of the international price for their cotton. However, in view of the reluctance of farmers to produce the government crop, this price has been raised in recent years, currently reaching about 45–50 per cent of the international price. The government plans to raise the cotton procurement price to about 66 per cent of the international price in 1992.

The cereals economy and marketing channels

Major cereals—wheat, rice, maize, and sorghum—provided almost two-thirds (64.9 per cent) of the daily caloric intake of Egyptians during the period 1979–81. Wheat alone provided one-third of the daily caloric intake. Despite the importance of major cereals in the diets of Egyptians, self-sufficiency in major cereals has been declining. While population has been growing at 2.5–2.7 per cent annually, the annual growth in total Egyptian grain production averaged 2.26 per cent during the period 1956–65, 2.50 per cent during the period 1966–75, and only 0.59 per cent during the period 1976–85. Maize is the only major cereal crop for which the annual percentage growth rates in output have exceeded the rate of population growth. In 1985–6, Egypt's major cereal production was 7.8 million tonnes (t), while net imports amounted to about 9.0 million t. Total usage of major cereals amounted to 12.9 million t for human consumption, and 3.7 million t for animal feed. Domestic production of major cereals was only 47 per cent of total usage (USDA 1986).

Self-sufficiency in major cereals production has been declining, partly due to the slow growth in production and partly due to the rapid increase in consumption. While self-sufficiency in wheat was about 70 per cent in 1960, it dropped to 42 per cent by 1974, falling to 25–33 per cent in the early 1980s. Human consumption of wheat and its derivatives per caput increased from 80 kg per year in 1960 to 123 kg by 1974 and then to about 177 kg by 1980. This increased per caput consumption of wheat was largely due to the government subsidy system discussed above. In addition the low prices of wheat and wheat products relative to other commodities have encouraged producers to use wheat, flour, and bread to feed livestock and poultry. In the early 1980s the price of 1 kg flour was cheaper than 1 kg straw or bran (Wally *et al.* 1983). The introduction of new high-yielding varieties of wheat, combined with price liberalization, led to a roughly 50 per cent increase in domestic wheat production in the late 1980s. This has provided encouragement to continue and to extend the market liberalization programme.

Prior to being channelled to the bakeries and consumer outlets, about 90 per cent of the imported wheat is milled into flour by public mills under the Ministry of Supply, the remainder being milled primarily by licensed private-sector mills. In addition there are some private village mills outside the control of the Ministry of Supply which process domestic production for local use.

Rice has also been a highly controlled major cereal crop in Egypt. While forced deliveries of wheat at fixed prices were eliminated in 1976, such deliveries for rice continued, with about two-thirds of the crop being delivered to the government in 1985. The PBDAC accepted the quota deliveries from the farmers and then channelled the rice to the rationed-food distribution channels. About one-half of the domestic rice crop was distributed via outlets controlled by the GASC, with the remainder of the domestic crop being distributed through village markets for home or local use.

During the period 1978–82, forced deliveries of rice accounted for about 50 per cent of production at a price less than one-half of the world market price. During the same period, the consumers' subsidized price was only 22 per cent of the world market price, and the 'unregulated' market price for the rice quantities above the quota was about 62 per cent of the world market price (Carter et al. 1983). Due to the forced deliveries of rice, producers indirectly paid more than 90 per cent while the government budget absorbed the remaining ten per cent of the consumer subsidy for rice (De Janvry et al. 1981). Not surprisingly, Egypt's rice exports, which amounted to about 700 000 t at the end of the 1960s, had virtually vanished by the early 1980s (Wally et al. 1983). Rice became a free market commodity under the new liberalization programme as forced deliveries, price controls, and consumer subsidies on rice were all eliminated during 1990–1.

Maize is Egypt's largest cereal crop, about three-quarters of which is used as livestock or chicken feed. Domestic production goes via village markets to village mills. While there are no acreage allotments for farmers and no forced deliveries to the Egyptian Government, the government programme of subsidizing quota deliveries of imported maize to livestock producers significantly affects the quantities and prices of domestic production.

Other regulated marketing channels

Although farmers are free to make planting decisions about sugar cane, they must dispose of their sugar cane through government marketing channels, which are similar to marketing channels for cotton. The marketing and processing of sugar cane and the marketing and distribution of sugar are controlled solely by government authorities. Beans and lentils, on the other hand, have been controlled in a similar way to rice, with required deliveries to the PBDAC in recent years amounting to about one-third of the beans and two-thirds of the lentils at approximately 60 per cent of the world market prices at the farm gate. The crops delivered to the government agencies have then been channelled into the government's food-rationing and distribution system. Bean and lentil production in excess of the quota deliveries were channelled into local unregulated markets. Quota deliveries for beans and lentils have recently been eliminated, leaving sugar cane and cotton as the only commodities for which deliveries to government authorities are mandated.

Unregulated agricultural markets

While the Egyptian Government has superimposed a command system on the production and distribution of the major fibre, grain, pulse, and sugar crops, Egyptian agriculture is still fundamentally a competitive industry. Composed of 11–12 million farms, 95 per cent of which are five feddans or less in size, the traditional farming system remains the driving force behind agricultural production, marketing, and pricing. The markets for vegetables, fruits, and livestock exemplify the operations of the unregulated markets in Egypt.

The livestock economy and traditional agriculture

Egyptian agriculture is characterized by the traditional integrated crop–livestock production system. Playing a crucial role in the agricultural production, marketing, and distribution system in Egypt, livestock has remained one of the freest agricultural sectors. In the traditional integrated crop–livestock production system, animals play a key role in supplying milk, meat, energy for agricultural work, and dung for organic fertilizers and for cooking fuel, while requiring a feed supply system which dictates that a substantial portion of the cultivatable area be planted for fodder.

Livestock production, including the value of animal power and manure, accounted for about one-third of the value of Egypt's total agricultural output in 1981. However, when assessed in terms of value added at the farm level, which includes home milk-processing and livestock work for transportation not included in the national statistics, it was found that a surprisingly high 43 per cent of agricultural output was derived from livestock, with live animals accounting for 18 per cent of the total value of livestock products, milk and dairy products accounting for 35 per cent, and eggs for 5 per cent, while animal work and manure accounted for 27 per cent and 13 per cent, respectively. In addition, animal products provide 10 per cent of the calories and 15 per cent of the protein in Egyptian diets (Fitch and Soliman 1982, 1983; Shapouri and Soliman 1985).

The crop–livestock linkages are significant. On the one hand, livestock contributes about 40 per cent of the total value of farm livestock production as direct inputs in the form of animal work and manure to crop production. On the other hand, crop production contributes about 22 per cent of crop products as direct inputs in the form of *bersim* to livestock production. A trend towards increasing livestock production, with the concomitant increases in *bersim* areas, is due in part to government pricing policies. The combination of extensive government feed subsidies and the absence of effective government controls of livestock price contribute to establishing a set of prices for domestic livestock and livestock products that are substantially above their equivalent world prices.[4]

About 90 per cent of Egypt's approximately 4 million cattle and buffalo are held in the traditional farming sector. About 84 per cent of all farm animals are held by farmers with 5 feddans of land or less, and 65 per cent of all animals are held on farms of 3 feddans or less. Furthermore, the small farmers are more efficient in livestock production than the larger ones; one survey found that farms with 3 feddans or less had net returns of £E190 per animal unit, compared to £E111 for farms of 3 to 5 feddans and £E35 for farms of 5 to 10 feddans (Soliman *et al.* 1982).

The small farm also devotes a large portion of its labour to livestock production. On average, Egyptian farms devote almost one-half of their labour to livestock production, but on farms of one feddan or less, almost three-quarters of the labour is for livestock. Women play a significant role in livestock care. On farms of 3 feddans or less, women provide 40 per cent of livestock-related labour (Fitch and Soliman 1982).

Milk marketing channels and price determination

The principal role of large animals in the Egyptian farming system is for milk production, with milk and milk products accounting for 35 per cent of the value of livestock products at the farm level. Almost 80 per cent of this milk is consumed or processed on the farm. About 40 per cent of Egypt's total milk consumption in liquid milk equivalents is farm milk and milk products consumed on farms, and another 38.5 per cent is farm milk and farm-processed milk products consumed off farms. Another 10.8 per cent is milk or milk products of the government's Egypt Milk Company plus other privately owned modern dairy plants, and 10.9 per cent is imported cheeses and butter, plus imported powdered milk sold directly to consumers (Soliman 1985).

The dairy buffalo is the principal milk animal in Egypt, with buffalo providing about two-thirds and native cattle providing about one-third of Egypt's total milk production. Imported foreign cattle and cross-breeds produce less than two per cent of the total milk output, despite efforts by the Egyptian Government to improve the genetic composition of the Egyptian herd by encouraging importation of Friesian and other foreign breeds (Soliman *et al.* 1983).

Milk is produced under two different systems. The principal milk producing system is the small farmer in the traditional integrated agricultural system. These farms typically have one to three native cows or buffaloes and produce about four-fifths of the total milk production. The other main milk production system is the commercial dairy herd. These are known as 'flying herds' and consist mainly of buffalo, located on the outskirts of the major cities. Producing about one-sixth of the total milk production, these herds are composed of lactating buffalo cows which are purchased from outlying rural villages and sold for slaughter when they have completed one lactation. There are also a few large-scale State farm and joint venture dairy herds which currently account for only about one per cent of the milk.

Milk yields for the buffalo and native cattle are low by international standards, reaching only about one-third and one-quarter, respectively, of the typical yield of milk cows in developed countries. However, costs are also lower, so Egypt is close to having a comparative advantage in domestic milk production, with the buffalo milk produced under the traditional integrated farming system being the lowest cost Egyptian milk (Fitch and Soliman 1983).

Milk is handled and processed in the farm household much as it has been for centuries. Women do most of the home processing of milk on the small farms where, due to the lack of refrigeration, more than 70 per cent of the milk is processed into products such as cheese, butter, and ghee for preservation. Few villages have improved facilities for cream separation or butter churning. The cream is churned into butter, most of which is then processed into ghee, the clarified butter oil which is used for cooking and can be stored for a long time without becoming rancid. A clabbering agent is added to skimmed buffalo milk to process it into *gibna beyda*, a fatless white cheese which is one of the most popular Egyptian foodstuffs. Large quantities of salt are usually added to the *gibna beyda* for preservation, so the whey is usually wasted. Only a small amount of the milk is processed into a cheese that can be aged without spoiling.

The government-owned Egyptian Milk Company is a major processor of milk and milk products for sale in urban areas. It has nine processing plants located throughout the country. Currently about one-half of the company's output is based on reconstituted imported milk, much of which is received on concessional terms from the European Community (EC). The Egyptian Milk Company has difficulty in procuring sufficient quantities of locally produced milk because it cannot afford to pay a competitive price. It is required to sell its products at prices which are 50–66 per cent of the prices for competitive products supplied by the traditional sector (Carter *et al.* 1983; Soliman *et al.* 1983).

Meat marketing channels and price determination

Red meat production, accounting for only 18 per cent of the value of total livestock products, is essentially a by-product of the traditional dairy industry. A large portion of the buffalo calves are slaughtered each year for veal without any finishing, and the small farmer is the only source of feeder calves, the main type of red meat in Egypt. Feedlots are usually very small, handling only 10–50 animals at a time.

Livestock markets in Egypt are still very primitive. There are no regulations concerning animal features, such as age, weight, quality, etc. Often there are no scales to weigh animals. Only about one-half of the slaughtering takes place in slaughter houses due to their inadequate number. There is inadequate cold storage and virtually no packaging or meat grading system.

The feedlot operators, after taking about a 20 per cent profit margin over costs, sell their stock to wholesalers and brokers. The wholesale market is highly centralized, with about 15 traders controlling the livestock market in Cairo, where about 50 per cent of the animals slaughtered in Egypt are marketed. The wholesaler also sells the edible by-products and the hide. Wholesaler profit margins are typically about 10 per cent of the net cost. Receiving the full carcass, the butcher divides it into parts of different quality, which he sells at different prices. The butcher typically obtains about 15 per cent of the retail price as net profit (Fitch and Soliman 1982, 1983).

Meat prices have been essentially free market prices. Meat price controls are generally ineffective. The Egyptian Government has attempted to impose price controls on some livestock products, particularly red meat prices. The controls are applied to retail (butcher) outlets, to wholesale carcass prices, and to fed live animals, but not to feeder calves. These controls have been difficult to enforce, however, since a feedlot operator would have to operate at a loss if he purchased his calf at free market prices and had to sell at the controlled prices.[5] Furthermore, most livestock sales yards are without weighing scales, so it is difficult to enforce price ceilings on a per kg basis. A 1989 ban on the importation of red meat, still in effect in late 1991, has recently led to dramatic increases in domestic livestock and meat prices.

Poultry and egg marketing channels and price determination

Compared with milk or beef production, poultry production enterprises are much larger. Until two decades ago the dominant poultry system in Egypt was the traditional village flock, and most of the chicks were produced in the traditional village hatcheries. In recent years, however, the small, private commercial broiler operations have become the dominant poultry producers. Now more than one-half of the broilers are produced in modern, confined units, and almost one-half of the chicks produced each year come from modern hatcheries. However, only about two per cent of the broiler producers have an annual capacity of greater than 100 000 broilers per year; about 75 per cent have an annual capacity of only 25 000 broilers per year. Although the industry is dominated by private-sector broiler operations, about one-quarter of all broilers are produced by large-scale government companies. The small broiler producers usually sell an entire cohort of about 5000 birds to a wholesaler who then distributes them live to retail outlets, while the government companies distribute their broilers through the *gamayya* (Soliman and Ibrahim 1983).

Government subsidies on imported poultry feed have encouraged this modernization. A family with a broiler farm receives a net farm income about ten times the average family income in Egypt, of which about 80 per cent comes from the government subsidies to feeds, baby chicks, and credit (Fitch and Soliman 1983).

The inadequacy of broiler production and marketing institutions is highlighted by reference to the modern sector. The supply of chicks, concentrated rations, and vaccine is restricted to one government institution, the General Company for Poultry, and a limited number of large-scale private producers and importers. Broiler farms are operating at only 60 per cent of their maximum capacity. The small number of wholesalers who control marketing operations appear to prolong the time intervals between successive rotations of birds in order to restrict the supply and maintain prices. There are no abattoirs or cold storage facilities available to private broiler producers (Wally et al. 1983).

Egg production, like poultry production itself, is a mixture of the traditional and the modern. Most eggs for home and rural village consumption are produced by the village flocks. The private layer industry is primarily composed of small producers; about two-thirds of the private layer industry have a capacity of 300–3000 layers, while the remaining third has a capacity of 3000–10 000 layers. Several governorates also operate a number of large-scale automated egg factories. About one-half of the chicks are still produced by the approximately 800 traditional village hatcheries (Carter et al. 1983).

Free markets in vegetables and fruits

Egyptian farmers produce a wide range of vegetable and fruit crops that include tomato, cucumber, pepper, squash, broad bean, lentil, onion, potato, garlic, eggplant, cabbage, cauliflower, orange, banana, mango, strawberry, and many others. By weight, fruits and vegetables make up almost 25 per cent of all food consumed in Egypt (NCBA 1986). Tomatoes are the most important vegetable crop in Egypt, accounting for about one-third of the total planted area of vegetables and up to 40 per cent of the total value of produced vegetables (Ragab and Simmons 1983).

The private sector is dominant in the marketing and pricing of vegetables and fruits. The two government companies which deal with these commodities, El-Naql and El-Ahram, handled only 1 per cent of the tomatoes, less than 5 per cent of the potatoes, and only about 3 per cent of the courgettes in 1981–2. The private sector is responsible for about 95 per cent of the tomato exports, 75 per cent of the potato exports, and 60 per cent of the grape exports. Of all the vegetable and fruit-marketing channels, the public sector is dominant only in the exportation of citrus, onions, and garlic.

Vegetables and fruits are collected at the edges of the fields and are usually transported to the nearest roads on the backs of animals or by donkey cart. There are no cooling facilities, and most shipping for tomatoes or other delicate products is done in the traditional domestic palm crate. Brokers or middlemen, who usually buy a farmer's crop for themselves or for others in the wholesale markets, handle the shipping of the crop (Ragab and Simmons 1983).

There are two main wholesale markets in Cairo, Rod El-Faraq and Athr El-Nabi, the latter specializing primarily in garlic and onions. A similar wholesale market, El-Nozha, serves Alexandria. These wholesale markets handle only about one-quarter of the vegetables and fruits entering Cairo. The remainder enters through marketing channels that link the farmers to a chain of private brokers, wholesalers, jobbers, and retailers. The wholesalers, and often the jobbers who buy huge amounts for distribution to secondary city markets or directly to retailers, extend market information and credit services to retailers as well as providing them with inventories. Transportation is primitive and small-scale. One study found that donkey carts provided 21 per cent of the transportation of vegetables and fruits to retailers in Cairo from the Rod El-Faraq, with another 68 per cent of the transportation provided by 1.0 and 1.5 t pickup trucks (Carter *et al.* 1983; Ragab and Simmons 1983).

The Egyptian Government sets formal 'fixed prices' for vegetables and fruits, but these are established without regard to quality and are generally ineffective. In practice, prices for these perishable commodities are determined by competitive auction at levels that clear the markets daily.

A 1982 survey indicated the variety of marketing channels that exists for various horticultural crops, and also indicated the share of the retail price that goes to the farmer, the wholesaler, and the retailer. For example, in that survey 47.7 per cent of tomatoes were distributed at the farm level, 1.3 per cent in governorate-level markets, 50 per cent in the wholesale market, and one per cent in the export market. The shares of the consumer price for these tomatoes were distributed as follows: 47.4 per cent to the farmer, 16.1 per cent to the wholesaler, and 36.5 per cent to the retailer. For potatoes, on the other hand, with 10.1 per cent distributed at the farm level, 1.9 per cent at the governorate level, 32.5 per cent in the wholesale market, and 55.5 per cent in the export market, the shares of the consumer price were distributed as follows: 42.1 per cent to the farmer, 17.0 per cent to the wholesaler, and 40.9 per cent to the retailer. In the case of grapes, where 73.2 per cent of the crop was distributed at the farm level, 8.8 per cent in the governorate market, and 18 per cent in the wholesale market, the farmer received 48 per cent, the wholesaler 17.4 per cent, and the retailer 34.6 per cent of the consumer price, respectively. The market shares of the retail price going to the farmer, the wholesaler, and retailer appeared to be unrelated to marketing channel, to perishability, or to the value of the crop (Habashy 1982).

There are two private export companies dealing with non-traditional exported crops, particularly fresh vegetables and fruits, the El Nile and the El Wadi companies. These companies buy produce from brokers who themselves usually buy all of a grower's harvested crop and take responsibility for grading, packing, and transportation to exportation points. However, quality control is generally poor, packing and shipping to collecting centres is often done in the palm crate, and packing and grading is often done at the farm in the shade of a tree or a shed. Inadequate quality control and the inability to guarantee regular and timely deliveries are the principal constraints on significant expansion to the export of Egyptian winter vegetables to European markets (Simmons 1982).

Food processing in Egypt is limited by inadequacy of marketing channels, bureaucratic controls on the companies, and the availability of competitive, inexpensive imported items. There are two large public-sector food-processing companies, Kaha and Edfina, and some relatively small-scale private food processors. Tomato sauce accounts for about one-sixth of the total production of the public food-processing companies, while private-sector companies have not processed tomatoes over the last decade. The administered price for tomato sauce restricts the prices which the public-sector firms are willing to pay for fresh tomatoes to less than one-half of the average wholesale price. This limits their purchases to the periods of the most extreme surpluses of raw produce. During the period 1979–83, Kaha and Edfina had losses in the production of tomato sauce amounting to one-third of production costs. Not surprisingly, the companies operate at only 50 per cent of capacity, which results in high average fixed costs per unit processed. These companies limit production to a level at which the government subsidy will cover their losses when their products are sold at the low administered prices.

Problems and inadequacies of the agricultural marketing channels

Handling, storing, transporting, processing, and marketing agricultural output are recognized as potentially serious constraints on the development of the agricultural sector. There are problems and inadequacies in both the regulated and unregulated marketing channels for agricultural commodities in Egypt. In the regulated marketing channels the controlled prices are unresponsive to market signals of surplus and shortage, and this bureaucratic rigidity leads to inefficiencies in the distribution system. In the unregulated marketing channels, inadequate marketing infrastructure and traditional practices are the principal sources of marketing inefficiencies.

Unintended consequences of the food-subsidy and rationing system

Some unintended consequences of the elaborate food subsidy and distribution system include the following: interpersonal and intersectoral resource misallocation, rapid growth in consumption of basic subsidized food items, the channelling of some subsidized food commodities into unintended uses, and declining food self-sufficiency with increased dependency on food imports (Abdou 1982; Gardner and Abdou 1982).

Maintaining low consumer food prices adds to the government's fiscal and trade deficit burdens which necessitates continuing the government enforcement of low procurement prices on agricultural commodities and the general repression of the agricultural sector (Scobie 1983; Von Braun and De Haen 1983). The consequences of the food-subsidy and rationing system, however, extend to the interpersonal level as well. For example, the distribution of the semi-rationed and unrationed food items depends upon the availability of supply. Long queues are almost always seen outside the government food stores and co-operatives. Availability of the consumer's time rather than price or rationed quota actually determines the distribution of many of the semi-rationed and unrationed goods.

Low food prices have also contributed to increased per caput consumption. Per caput consumption of basic subsidized food items, including bread, sugar, rice, tea, and cooking oil, increased by 2–6 per cent each year between 1974 and 1984. Domestic requirements for subsidized food items also increased due to the diversion of subsidized foodstuffs to unintended uses, such as the feeding of livestock. Many subsidized items are purchased by consumers who intend to resell them at higher prices than they paid, thus channelling subsidized commodities to unintended uses. One survey yielded estimates of the percentage diversions to unintended uses as follows: wheat, 7.2 per cent; flour, 5.1 per cent; *baladi* bread, 6.1 per cent; *fino* bread, 3.1 per cent; rice, 7.2 per cent; and broad beans, 24.4 per cent (Gardner and Abdou 1982).

This increased domestic food utilization, when combined with the slow growth in agricultural production due to the low procurement prices paid to farmers, has led to significant reductions in agricultural exports and increases in food imports. There was an absolute drop in agricultural exports by nearly one-half during the 1970s, with agriculture's share of all goods exported falling from 80 per cent to 16 per cent during the decade. During the same period, food imports increased from 21 per cent to 34 per cent as a share of total imports.[6] By 1980, less than one-third of the food import bill was paid by agricultural exports, compared with the late 1960s and early 1970s when the value of agricultural exports was more than twice the value of food imports (Von Braun and De Haen 1983).

With the advent of the 1980s, Egypt entered a period in which food self-sufficiency fell below 50 per cent (USAID 1982). In 1982, aggregate food self-sufficiency in cereals, pulses, sugar, oil, and meats was about 70 per cent. However, the self-sufficiency in calories in these commodities was less than 50 per cent because wheat, with a self-sufficiency ratio of 32 per cent, and oil, with a self-sufficiency ratio of 25 per cent, account for high shares of total imports and have high caloric values relative to their monetary values.

Inadequate competition in the unregulated wholesale markets

The competitive markets not under government control are generally characterized by few wholesalers. Economic studies are unable, however, to confirm any significant monopoly power. A survey of tomato handlers in the Rod-El Faraq wholesale market found that the operations range in size from very small, handling less than 2 t per year, to quite large, handling about 12 000 t per year, with the nine largest of the approximately 100 tomato wholesalers handling more than 50 per cent of the tomato volume. For potatoes, almost one-half of the wholesalers distributed only 2 per cent of the total, while 9 per cent of them dealt with 72 per cent of the total distributed. Despite this concentration, however, market prices were generally inversely responsive to quantities available and there appeared to be a good correspondence of tomato prices among all the major governorates during the various seasons (Ragab and Simmons 1983).

The large number and limited financial capacity of small livestock, fruit, and vegetable producers and middlemen restrict their ability to establish or to use efficient marketing. There are few general-purpose food-store outlets. Of Cairo's 40 000 retail food outlets, only about 200 are 'supermarkets', but these are generally small-scale operations which are called supermarkets only because they have refrigerated and frozen storage and display cases and usually carry some imported food items. Similarly, there are no full-line wholesalers to serve retailers. Wholesalers, like retailers, are generally specialists in one or several commodities (NCBA 1986). Traditional elements in the marketing channels inhibit the development of improved marketing efficiency, with the appearance of some monopoly by wholesalers and brokers being more strongly linked with the lack of specialization of marketing functions than with the small number of marketing agents. These agents supply packing, transportation, information, credit, and other services to the farmers and the retailers, with the prices for each of these services being included in a single marketing margin.

Post-harvest losses and food waste

One of the greatest inadequacies in the agricultural marketing channels is revealed in the widespread food waste. Packing and storage facilities are inadequate, expensive, and often inappropriately located in relation to the producing and consuming regions. The share of packaged food items accounted only for 2.1 per cent of the total of approximately 14.2 million t of vegetables and fruits, sugar, cooking oil, meats, and grocery items available for consumption in Egypt in 1985 (ASRT 1985). The bulk handling of food, especially when combined with the lack of refrigeration for meat and dairy products, contributes to sanitation and health problems. Grading and quality control are limited almost solely to the export crops. Food waste in the marketing and handling of vegetables and fruit is almost one-half of the output for some crops. Post-harvest handling, marketing, and distribution system losses are estimated at about 30 per cent for all fruits and vegetables. These compare with distribution system losses estimated at 35–50 per cent for fish, 10–15 per cent for beef, and about 10 per cent for frozen poultry (NCBA 1986).

Avoiding this waste is the problem of agricultural development. At present, many of the changes which would reduce losses are not economically feasible. One survey in 1982, focussing on potatoes, tomatoes, and grapes, highlighted the problems. The percentage of post-harvest losses at the farm, wholesale, and retail levels for potatoes were found to be 11.8 per cent, 1.5 per cent, and 4.2 per cent, respectively, for a total loss of 17.5 per cent. For tomatoes, the losses at each level were 9.0 per cent, 17.9 per cent, and 16.3 per cent, respectively, for a total loss of 43.2 per cent. For grapes the losses at each level were 15.1 per cent, 6.9 per cent, and 6.0 per cent, respectively, for a total loss of 28 per cent. Solutions for correcting these losses are available, but they do not appear to be cost-effective.

Storage of potato seeds in refrigerators, currently done for only about 30 per cent of the total stored, rather than in the traditional cooling house, called the *nowlaut*, would reduce losses from about 12 per cent to less than 2 per cent. Using refrigerators, however, would increase costs by tenfold, making the switch too expensive. Storing and shipping potatoes in 75 kg containers rather than 25 kg containers would save about 5 per cent of the national crop, but at a net cost to the purchaser because the cost of the larger containers exceeds the extra value created. Using carton boxes rather than palm crates for the packing of grapes would reduce the national losses by one-half, but the cost of the carton boxes exceeds the extra value received by the individual broker or wholesaler (Habashy 1982). These examples illustrate that the inadequacies of the unregulated marketing channels are primarily those of the low level of economic development.

Conclusions

The command-economy system which has been imposed on the agricultural sector and has repressed Egyptian agricultural production, was the product of the Nasser Government. The food subsidy and rationing system, on the other hand, which has dramatically increased domestic food requirements, was the product of the Sadat years. Together these policies have left Egypt with a legacy of dependency on world markets for feeding its population. In the late 1980s, the marketing system in Egypt combined the inefficiencies of government price and quantity controls on production and consumption with the inadequacies of an underdeveloped economy's marketing infrastructure.

The wide and growing food gap has alerted Egyptian policy-makers to the severity of the problems to which the government's production and marketing policies have contributed. The Egyptian Government has recognized that the repression of the agricultural sector has been too severe, that the food-subsidy system needs to be better targeted to the poor, and that the Government needs to create a market-driven environment for the development of its traditional farming system and marketing channels. The command system imposed on the agricultural sector is being dismantled as Egypt enters the 1990s. The transition to free agricultural and food markets will undoubtedly be a difficult one, requiring new roles for the government institutions in promoting agricultural development and new approaches to achieving social equity.

Notes

[1] This chapter emphasizes the institutional marketing channels and mechanisms of price determination for agricultural commodities in Egypt, with limited discussion of post-harvest handling practices. The marketing of fish is not discussed but the reader is referred to Reid *et al.* (1984). Two valuable sources in English of Egyptian agricultural marketing data are by the Arab Organization for Agricultural Development (1985) and Gardner and Parker (1985).
[2] While the agricultural sector has received substantial input subsidies which offset the Egyptian Government's implicit taxation on the sector, Cuddihy (1980) estimated that the net impact of these effects was that agriculture faced a net tax rate about 20 per cent higher than the average for the economy as a whole.
[3] Alderman *et al.* (1982) in a study in a rural governorate found that farm households, in addition to home-produced consumption, spent about 13 per cent of their total household food expenditures on subsidized and rationed commodities, which accounted for about one-third of their calorie requirements.
[4] Cuddihy (1980) calculated the effective protection coefficients (the ratio of value added expressed in domestic prices to value added expressed in border prices, indicating the net effect of taxes and subsidies on outputs and inputs) for 1976-7 as follows: 0.68 for cotton; 0.53 for rice; and 2.80 for meat. Thus, while cotton farmers received only 68 per cent of the equivalent world values for cotton, livestock producers received 280 per cent of the equivalent world values for meat. Fitch and Soliman (1982) suggest that the pricing policies are not solely responsible for the trends towards increasing livestock and *bersim* production. The increasing number of small farms and the declining average farm size, which strengthen the crop – livestock linkages in the traditional agricultural setting, appear to be a significant contributor to the shift towards more livestock production.
[5] The 'fixed prices' for meat at slaughter and at the consumer level are inconsistent. While government prices for red meat in 1986 were set at £E4.50 per kg, the retail prices were actually £E7–10 per kg. If animals were sold by farmers at the fixed price of £E2.55 per kg liveweight, then, with boned meat accounting for about 40 per cent of the liveweight, the butcher would have to sell meat at about £E6.40 per kg just to cover the costs of the live animal (NCBA 1986).
[6] Von Braun and De Haen (1983, p. 17) point out that 'If factor and non-factor services are included, agriculture's share of all goods exported was only about 8 per cent in 1980 and food was 26 per cent of all goods imported.'

References

Abdou, D.K. (1982). *Government distribution and price policy for major subsidized food commodities in Egypt: an overview.* Agricultural Development Systems Project Economics Working Paper, No. 23. Ministry of Agriculture, Cairo, Egypt.

ASRT (Academy of Scientific Research and Technology) (1985). *A general economic description for upgrading packaging for five goods*. Industrial Research Council, Academy of Scientific Research and Technology, Cairo, Egypt.

Alderman, H., Von Braun, J., and Sakr, S. A. (1982). *Egypt's food subsidy and rationing system: a description*. International Food Policy Research Institute, Research Report, No. 34. International Food Policy Research Institute, Washington, DC.

Arab Organization for Agricultural Development (1985). *Yearbook of agricultural statistics*, (Vol. 5). League of Arab States, Arab Organization for Agricultural Development, Khartoum.

Carter, H., Goueli, A., and Rowntree, J. (1983). The economics sub-project. In *The accomplishments of a California-Egypt research collaboration*, (ed. R. D. Blond), pp.99–111. University of California, Davis.

Cuddihy, W. (1980). *Agricultural price management in Egypt*. World Bank Staff Working Paper, No. 388. World Bank, Washington, DC.

De Janvry, A., Siam, G., and Gad, O (1981). *Forced deliveries: their impact on the marketed surplus and the distribution of income in Egyptian agriculture*. Agricultural Development Systems Project Economics Working Paper, No. 38. Ministry of Agriculture, Cairo, Egypt.

Dethier, J. (1989). *Trade, exchange rate, and agricultural pricing policies in Egypt*, Vol. I, *The country study*. World Bank Comparative Studies. World Bank, Washington, DC.

Fitch, J. B. and Soliman, I. (1982). *Livestock and crop production linkages: implications for agricultural policy*. Agricultural Development Systems Project Economics Working Paper, No. 92. Ministry of Agriculture, Cairo, Egypt.

Fitch, J. B. and Soliman, I. (1983). *An overview of livestock in the Egyptian economy*. Agricultural Development Systems Project Economics Working Paper, No. 142. Ministry of Agriculture, Cairo, Egypt.

Gardner, B. D. and Abdou, D. K. (1982). *Food consumption and distribution: an overview*. Agricultural Development Systems Project Economics Working Paper, No. 89. Ministry of Agriculture, Cairo, Egypt.

Gardner, G. R. and Parker, J. B. (1985). *Agricultural statistics of Egypt 1970–84*. Economics Research Service, Statistical Bulletin, No. 732. International Economics Division, United States Department of Agriculture, Washington, DC.

Habashy, N. T. (1982). *Economic aspects of the estimation of post harvest losses in some horticulture crops*. Agricultural Development Systems Project Economics Working Paper, No. 64. Ministry of Agriculture, Cairo, Egypt.

Hafez, A., El-Amir, M. R., and Nassar, S. (1982). *Some remarks on the institutional framework of cotton in Egypt*. Agricultural Development Systems Project Economics Working Paper, No. 66. Ministry of Agriculture, Cairo, Egypt.

IADS (International Agricultural Development Service) (1984). *Increasing Egyptian agricultural production through strengthened research and extension programs*. International Agricultural Development Service, Arlington, Virginia.

Nassar, S., El-Amir, M. R., and Moustafa, A. E. (1981). *Determinants of agricultural price policy in Egypt*. Agricultural Development Systems Project Economics Working Paper, No. 50. Ministry of Agriculture, Cairo, Egypt.

NCBA (National Cooperative Business Association) (1986). *Egyptian food distribution systems: an assessment and recommended plan of action*. National Cooperative Business Association, Washington, DC.

Ragab, M. E. (1983). *Economic study for the supply of tomatoes in Egypt*. Agricultural

Development Systems Project Economics Working Paper, No. 154. Ministry of Agriculture, Cairo, Egypt.
Ragab, M. E. and Simmons, R. L. (1983). *Tomato supply, demand and market structure in Egypt.* Agricultural Development Systems Project Economics Working Paper, No. 139. Ministry of Agriculture, Cairo, Egypt.
Reid, T. R., Rowntree, J. T., El-Kholei, O., and Abou-Auf, A. (1984). Fisheries management in the northern Nile delta lakes of Egypt. In *Management of coastal lagoon fisheries.* General Fisheries Council for the Mediterranean, Studies and Reviews, No. 61, Vol. 2. FAO, Rome.
Rizk, F. H. (1983). *Economic impact of ordinances of the ministry of supply regulating the processes of production and distribution of balady bread.* Agricultural Development Systems Project Economics Working Paper, No. 158. Ministry of Agriculture, Cairo, Egypt.
Scobie, G. M. (1981). *Government policy and food imports: the case of wheat in Egypt.* International Food Policy Research Institute, Research Report, No. 29. International Food Policy Research Institute, Washington, DC.
Scobie, G. M. (1983). *Food subsidies in Egypt: their impact on foreign exchange and trade.* International Food Policy Research Institute, Research Report, No. 40. International Food Policy Research Institute, Washington, DC.
Shapouri, S. and Soliman, I. (1985). *Egyptian meat market: policy issues in trade, prices, and expected market performance.* Economics Research Service Staff Report, No. AGES841217. International Economics Division, United States Department of Agriculture, Washington, DC.
Simmons, R. L. (1982). *Feasibility of exporting winter tomatoes to the EEC under changing market conditions.* Agricultural Development Systems Project Economics Working Paper, No. 67. Ministry of Agriculture, Cairo, Egypt.
Soliman, I. (1985). Milk marketed surplus of the Egyptian farm. In *The proceedings of the 20th annual conference in statistics, computer science, and information and operation research,* Vol. 1. *Applied statistics and econometrics*, pp.22–36. Institute of Statistical Studies and Research, Cairo University, Cairo, Egypt.
Soliman, I. and Ibrahim, A. (1983). *The productive efficiency of the broiler industry in Egypt.* Agricultural Development Systems Project Economics Working Paper, No. 122. Ministry of Agriculture, Cairo, Egypt.
Soliman, I., Fitch, J. B., and El-Aziz, N. A. (1982). *The role of livestock production on the Egyptian farm.* Agricultural Development Systems Project Economics Working Paper, No. 85. Ministry of Agriculture, Cairo, Egypt.
Soliman, I., El-Zaher, T. A., and Fitch, J. B. (1983). *Milk production systems in Egypt and the impact of government policies.* Agricultural Development Systems Project Economics Working Paper, No. 121. Ministry of Agriculture, Cairo, Egypt.
USDA (United States Department of Agriculture) (1976). *Egypt: major constraints to increasing agricultural productivity.* Foreign Agricultural Economic Report, No. 120. United States Department of Agriculture, Washington, DC.
USDA (United States Department of Agriculture) (1986). *World food needs and availabilities, 1986/87.* Economics Research Service, United States Department of Agriculture, Washington, DC.
USDA (United States Department of Agriculture) (1991). *Agricultural situation annual.* Foreign Agricultural Service Report, No. EG1024. Foreign Agricultural Service, United States Department of Agriculture, American Embassy, Cairo.
Von Braun, J. and De Haen, H. (1983). *Effects of food price and subsidy policies*

on Egyptian agriculture. International Food Policy Research Institute, Research Report, No. 42. International Food Policy Research Institute, Washington, DC.

Wally, Y., El-Kholei, O., Abbas, M., and Heady, E. O. (1983). *Strategies for agricultural development in the eighties for the Arab Republic of Egypt.* International Development Series, DSR-9. Center for Agricultural and Rural Development, Iowa State University, Ames.

USAID (United States Agency for International Development) (1982). *Strategies for accelerating agricultural development: a report of the presidential mission on agricultural development in Egypt.* United States Agency for International Development, Cairo, Egypt.

21

Agricultural education, research, and extension

M.E.I. MANSOUR AND S.A.M. ISMAIL

Agricultural education

An overview

Contemporary cultural trends in Egypt necessitated the speeding up of the growth of education (Boktar 1963). Agricultural education has not been an exception. Rather, it has been treated as a matter of priority in the educational expansion programmes. Table 21.1 illustrates the substantial growth in agricultural education during the period 1952–89.

Such a development in the size of agricultural education in Egypt is related to several factors. First, education, of which agricultural education is an integral part, is regarded in Egypt as an important means of social regeneration and integration. Second, attempts are being made to achieve economic and social development through the expansion of cultivated land, the establishment of new factories, and the shift from a traditional to a modern economy. In consequence, agricultural education is considered indispensable for providing the appropriate manpower required in different parts of the agricultural sector. Third, the need for more agricultural education has been felt strongly in the 1970s and 1980s due to the expansion of the agricultural sector in other Arab countries, especially Iraq and the Gulf States.

The broad lines of the educational policy (Soroor 1989) have been as follows:

(1) generalization of primary education (now basic education);

(2) equal opportunity at all stages of education;

(3) planning post-primary education in the light of the country's requirements and potentials;

(4) extending girls' education;

(5) extending vocational education, expanding its services, and distributing them on a proper geographical basis;

TABLE 21.1. Egypt: development of agricultural education, 1952–89

Education level	Number of institutions		Female students		Total students	
	1952	1989	1952	1989	1952	1989
Agricultural secondary level	11	77	–	22 255	4201	111 226
Agricultural higher level	3	16	78	8784	1991	26 759

Source: CAPMAS (various years); Ministry of Education (1960c); Department of Statistics and Census (1962).

(6) developing higher and university education, giving particular attention to scientific colleges and higher technological institutes;

(7) encouraging scientific research and keeping pace with world development in scientific theory and practice;

(8) co-operating with all friendly states and strengthening the bonds of cultural unity with the Arab States.

The agricultural educational system: the educational ladder

In 1956, public education in Egypt was reorganized on a three-part system of six years, three years, and three years (Ministry of Education 1960a). In 1981 it was changed to a two-part system of nine years basic and three years secondary education. Later, in 1989 it became eight years (Table 21.2). Basic education was made a unified stage, free and compulsory for all children aged between 6 and 15 years. Plans were designed to generalize this education and make it, *de facto*, available to every child. Between 1952–3 and 1988–9 the number of children enrolled in basic education rose from 1 550 516 to 9 880 224. In this basic education, 15–20 per cent of the total school hours are to be allocated to practical studies (agriculture, commerce, home economics, etc.).

Agricultural secondary education, which is found in 77 schools, and is more limited in enrolment than basic education, is free, but not compulsory. Agricultural secondary schools are located in all the main centres of population in the Delta, the Northern Littoral Region, Middle and Upper Egypt, and the New Valley.

Up to July 1961 (Ministry of Education undated; 1960a) higher education was mainly in the hands of two agencies, the Ministry of Education, and the four State universities. After that date, however, it came exclusively under the new Ministry of Higher Education. Although universities are thus officially

TABLE 21.2. Egypt: the agricultural education system in relation to labour status

Age (years)	Duration (years)	Level	Educational establishment	Labour status
22	2–6	Higher	University postgraduate courses	Researcher and staff member
18–20	4	Higher	University and higher institute undergraduate courses	Agronomist, agricultural engineer, other agricultural specialists
15	3	Secondary	Agricultural secondary schools	Skilled labourer and technicians
15	5	Secondary	Technical agricultural schools	Skilled labourer and technician
6	9	Basic	Schools	Semi-skilled labourers

under the Ministry of Higher Education, they are considered independent bodies with independent councils, policies, budget, etc. It is only in very important matters that their decisions should be subject to the authorization of the Ministry.

In Egypt, there are now fourteen faculties of agriculture, four faculties of veterinary science, and two agricultural higher institutes (Table 21.3). Fig. 21.1 shows their location and Tables 21.4 and 21.5 give details of the various departments in the faculties of agriculture and veterinary science. Since 1962, agricultural higher education has been free for all those who have the ability and aptitude to continue their schooling beyond the secondary level.

Administration of agricultural education

Agricultural education in the Arab Republic of Egypt comes mainly under the direct control of the Ministry of Education and the Ministry of Higher Education (Ministry of Education 1960b). The former is responsible for basic

TABLE 21.3. Egypt: establishments for higher education

Faculties of Agriculture
1. Faculty of Agriculture at Cairo University
2. Faculty of Agriculture at Cairo University at Faiyum
3. Faculty of Agriculture at Alexandria University
4. Faculty of Agriculture at Ain Shams University
5. Faculty of Agriculture at Asyut University
6. Faculty of Agriculture at Tanta University
7. Faculty of Agriculture at El Mansura University
8. Faculty of Agriculture at Zagazig University
9. Faculty of Agriculture at Zagazig University at Benha
10. Faculty of Agriculture at El Minya University
11. Faculty of Agriculture at Minufiya University
12. Faculty of Agriculture at Suez Canal University (Isma'iliya)
13. Faculty of Agriculture at Suez Canal University (El Arish)
14. Faculty of Agriculture at El Azhar University

Faculties of Veterinary Science
1. Faculty of Veterinary Science at Cairo
2. Faculty of Veterinary Science at Asyut
3. Faculty of Veterinary Science at Zagazig
4. Faculty of Veterinary Science at Edfina

Institutes
1. The Higher Institute for Agricultural Co-operation at Ain Shams University
2. The Higher Institute for Agricultural Extension in Asyut governorate

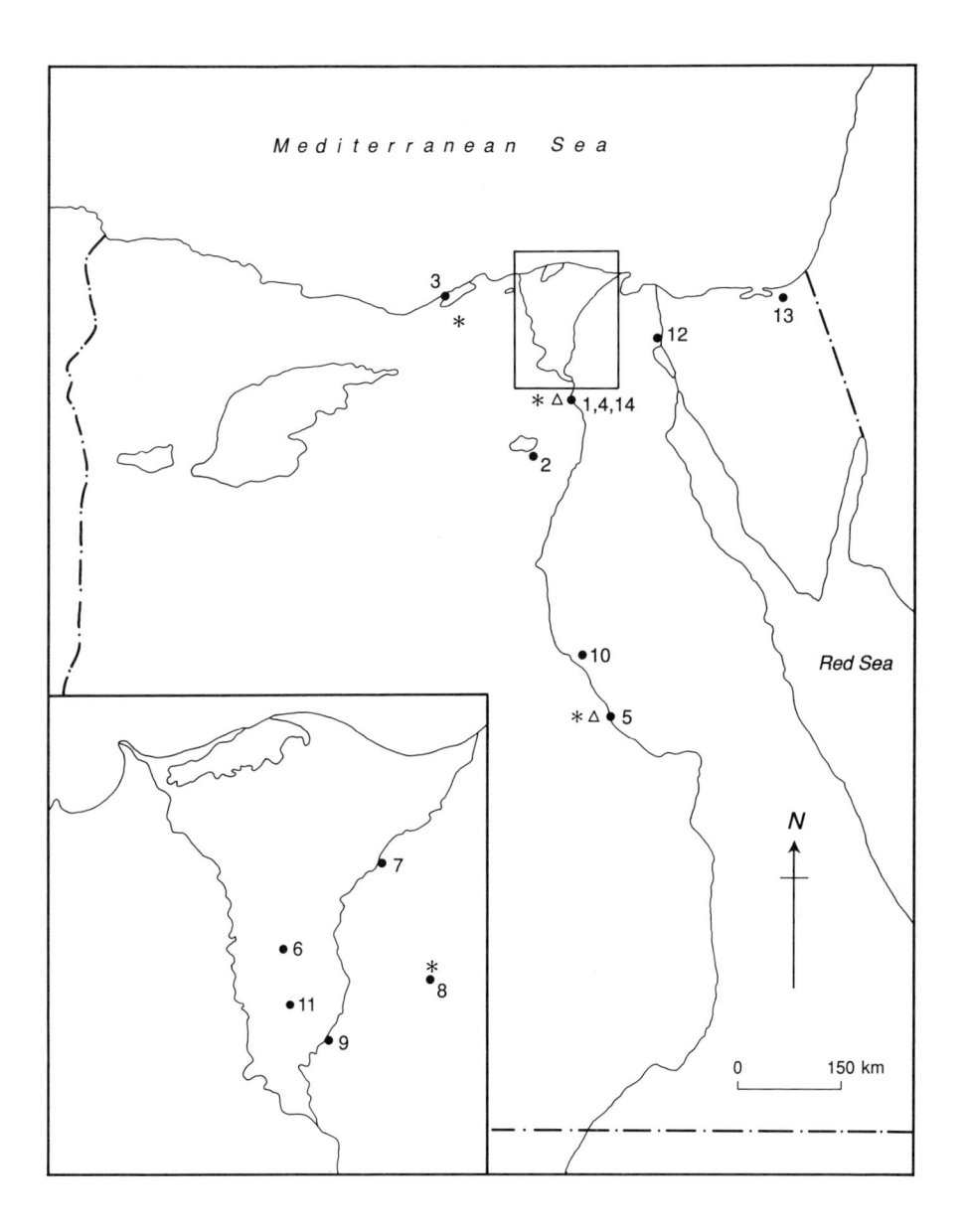

FIG 21.1. Egypt: location of faculties of agriculture (●), faculties of veterinary science (*), and agricultural higher institutes (△)

TABLE 21.4. Egypt: faculties of agriculture, location, year of establishment, and departments

University	Location	Year of establishment	Departments
Cairo	Giza	High school 1908 Faculty 1935	Plant Production and Plant Pathology; Agricultural Economics; Soil Science; Agricultural Engineering; Animal Production; Genetics; Food Industries; Dairy Science; Agricultural Microbiology; Animal Husbandry; Agricultural Biochemistry; Entomology and Insecticides; Horticulture; Agronomy
Cairo	Faiyum	Faculty 1976	Agricultural Economics; Soil Sciences and Chemistry; Plant Protection; Food Industries; Animal Husbandry; Plant Production
Alexandria	Alexandria	Faculty 1942	Soil Sciences; Agricultural Extension; Plant Production; Agricultural Economics; Plant Pathology; Animal Husbandry; Food Industries; Horticulture, Agricultural Engineering; Genetics; Plant Protection; Rural Sociology
Ain Shams	Shubra el Kima	Faculty 1950	Agricultural Economics; Soil Sciences; Horticulture; Animal Husbandry; Food Science; Genetics; Plant Pathology; Microbiology; Plant Protection; Agronomy
Asyut	Asyut	Faculty 1959	Horticulture; Genetics; Agronomy; Agricultural Economics; Rural Sociology; Soils; Plant Pathology; Plant Protection; Animal Husbandry; Food Industries
Tanta	Kafr el Sheik	Institute 1957 Faculty 1969	Plant Protection; Dairy Science; Animal Husbandry; Agricultural Botany; Genetics; Soil Sciences; Agricultural Mechanization; Agricultural Economics; Agronomy; Horticulture; Food Industries
Mansura	Mansura	Institute 1967 Faculty 1963	Agronomy; Agricultural Botany; Chemistry; Soils; Animal Husbandry; Horticulture; Food Science; Plant Protection; Agricultural Economics; Agricultural Mechanization

451

TABLE 21.4. contd.

University	Location	Year of establishment	Departments
Zagazig	Zagazig	Institute 1957; Faculty 1969	Horticulture; Agronomy; Food Science; Agricultural Economics; Soils; Agricultural Botany; Plant Protection; Animal Husbandry
Zagazig	Moshtoher (Benha)	Institute 1957; Faculty 1969	Agricultural Economics; Soils and Chemistry; Food Science; Horticulture; Plant Protection; Agronomy; Agricultural Engineering; Animal Husbandry; Plant Pathology
Helwan[a]	Alexandria	Institute 1959; Faculty 1986	Arid Agriculture; Cotton Technology; Fisheries Technology; Planning and Development
El Minya	El Minya	Institute 1957; Faculty 1969	Chemistry and Soil Science; Microbiology; Animal Husbandry; Agronomy; Horticulture; Agricultural Economics; Dairy and Food Industries; Plant Protection; Botany and Plant Pathology; Genetics
Minufiya	Shibin el Kom	Institute 1957; Faculty 1969	Animal Husbandry; Agricultural Economics; Food Science; Soil Sciences; Agricultural Botany; Plant Protection; Horticulture; Agronomy; Genetics
Suez Canal	Isma'iliya	Faculty 1977	Agricultural Botany; Plant Protection; Agronomy; Horticulture; Agricultural Economics; Soil and Water; Food Industries; Animal Husbandry
El Azhar	Cairo	Faculty 1961	Agricultural Extension; Plant Protection; Agronomy; Horticulture; Agricultural Economics; Soils; Agricultural Engineering; Agricultural Botany; Dairy Science; Food Industries and Technology; Animal Husbandry

[a] The Faculty of Agriculture at Helwan University was merged with the Faculty of Agriculture at Alexandria University in 1991 with the same departments.

TABLE 21.5. Egypt: faculties of veterinary science, location, years of establishment, and departments

Location	Year of establishment	Departments
Cairo	1935	Animal Medicine; Surgery and Obstetrics; Anatomy and Histology; Pathology and Parasitology; Health and Food Control; Pharmacology and Forensic Medicine; Physiology and Biochemistry; Microbiology
Asyut	1961	Pathology; Anatomy and Histology; Obstetrics; Surgery; Animal and Poultry Medicine; Health and Food Control
Zagazig	1969	Pathology; Surgery; Physiology; Health; Anatomy; Medicine
Edfina	1975	Anatomy and Histology; Physiology and Pharmacology; Animal Care; Health and Food Control; Microbiology and Animal Health; Pathology and Parasitology; Animal Medicine and Forensic Medicine; Birds and Fish Medicine; Surgery and Obstetrics

and agricultural secondary schools, the latter is responsible for universities, higher agricultural institutes, and vocational centres.

The Ministry of Higher Education is responsible for co-ordinating the activities of its institutes with those of the colleges of the universities. Both the Ministry of Education and the Ministry of Higher Education take over the function of financing, planning, organizing, supervising, and following up agricultural education, as well as other types of education which are under their jurisdiction. Established bodies within these Ministries undertake these functions. Since the current trend is towards decentralization, more authority and freedom of action are given to educational zones, school units, and individual higher institutes.

Organization of agricultural education

In Egypt, there are two stages of agricultural education: secondary education and higher education. These are the two successive stages of education above the basic education level. Agricultural secondary education is regulated by Law No. 139 of 1981, and agricultural higher education is regulated by Law No. 40 of 1963.

Agricultural secondary schools

These schools are for children between the ages of 15 and 20 years. They aim to prepare students to manage farms, engage in private agricultural enterprises, hold technical jobs related to agriculture in government departments, work as agricultural advisers in rural areas, and teach agricultural subjects in schools. Pupils are selected for admission to these schools from those who have obtained the general basic education, providing they meet additional requirements to be specified by the Minister of Education. Candidates for admission to these schools are further required to pass a vocational test to indicate their ability to follow this type of education. Priority for admission is given to children of farmers and to those who are strongly bound to agriculture.

The course of study for this type of school is for 3–5 years, the successful completion of which leads to the agricultural secondary school diploma. This diploma qualifies the successful students for the above-mentioned jobs and also qualifies those who distinguish themselves in their studies for admission to higher institutes and colleges of agriculture where they can acquire greater specialization and technical skill.

To ensure that students acquire sufficient practical experience, a farm of 31 ha is attached to the agricultural secondary school. The students are required to perform as many activities as possible concerned with agricultural production.

Agricultural higher education

The general philosophy of university education in agriculture is based on a system of teaching which emphasizes the value of basic scientific information, and lays particular stress on principles rather than the amount of instruction in specific techniques.

The faculties of agriculture in Egyptian universities have the right to participate in the following duties:

(1) provide the country with the required number of specialists in the agricultural field and keep abreast of agricultural progress;

(2) train personnel capable of contributing to wide-ranging scientific activities;

(3) put forward and study planned scientific agricultural research for the determination of fundamental facts and their evaluation under actual operational conditions.

(4) disseminate proven and practical agricultural knowledge through advisory and educational channels to those engaged in the field of production.

In most faculties of agriculture in Egypt, all students follow a general course of basic science for two years. In the third and fourth year students choose a specific course in one of the agricultural subjects and at the end of the fourth year the student is awarded a B.Sc. degree in agriculture with major knowledge in his line of specialization. In the third year students opt for one of eleven options: agricultural biochemistry, soil science, agricultural economics, food sciences, agricultural production, plant protection, animal production, genetics, horticulture, agronomy, and agricultural microbiology. In the fourth year and in certain departments, the student can even opt for one of two or three different lines of specialization under one option. For example, in food sciences, the options are food industries or dairy science in the fourth year.

The universities in Egypt offer advanced post-graduate studies at diploma, master and doctorate levels for undergraduates who achieve the highest academic success in their academic careers. Parallel to the university faculties of agriculture are two higher institutes. These concern themselves primarily with the training of technicians and specialists in agricultural co-operation and extension.

The course of study at the higher institutes lasts for four years and, like the universities, these institutes recruit their students from the graduates of the academic secondary schools.

Agricultural research and extension

Introduction

Blessed with unusually favourable water, soil, climate, and experienced farmers, Egyptian agriculture has, over several thousand years, developed local technologies, plant and animal varieties, and management procedures, and it has achieved the highest average yields in the world for many agricultural commodities.

By the middle of the nineteenth century, scientific investigations began to have a bearing on agriculture. From that time, Egypt realized that improvements in agricultural production would depend mainly on scientific research in agriculture. Experimental agriculture, which emphasized the role of scientific research for improving agricultural production, began in 1898 with the Royal Society for Agriculture. This later became the present Agricultural Research Centre. Field experiments on fertilization of cotton and wheat began on research farms in 1900. The first pathology laboratory for the production of vaccines for farm animals was established in 1904 and in 1910 the research group was placed with the agricultural department under the Ministry of Public Works. The School of Agriculture and the Royal Society for Agriculture started to undertake joint experiments and research especially on the use of fertilizers.

In November 1913 the first Ministry of Agriculture was founded and it was responsible for agricultural research. Over the years the various sections of the Ministry have brought about enormous improvements in the agriculture of the country, notably improved cultivars of crops, improved quality of seed supplied to cultivars, and pest control.

The agricultural research system in Egypt

At present the agricultural research system in Egypt comprises about 50 centres and institutes belonging to various ministries and organizations (IADS 1984). A significant level of research relating to food and agricultural production is being conducted throughout the country and financed from many sources. Similarly, there is a variety of extension efforts aimed at enabling farmers to use improved technology.

It should be noted that all agricultural research and extension activities are not centred in the Ministry of Agriculture and Land Reclamation (MALR) (or bodies relating to the MALR). Although MALR-related activities constitute a substantial part of the total agricultural research and extension effort, other organizations and institutions, both within and outside the Egyptian Government are also involved, and must be considered a vital part of the national research and extension effort in agriculture.

The recently reorganized Agricultural Research Centre (ARC), which is under the administrative responsibility of the Minister of Agriculture, has the primary responsibility for agricultural research and extension within the Egyptian Government. Other ministries, such as the Ministry of Public Works and Water Resources, because of its concern with agriculture, also has important research responsibilities. Other Egyptian institutions, such as the National Research Centre, the Academy for Scientific Research and Technology, and the Universities, are also involved in research and, to a certain extent, extension. In addition, other organizations, such as the international agricultural research centres, are conducting research in Egypt, either independently or in co-operation with government agencies. A description of some of these organizations and their activities follows.

The Agricultural Research Centre

The Agricultural Research Centre (ARC) is the focus for research activities in the agricultural sector (ARC 1988*a*). It is the largest research establishment in Egypt and is the largest agricultural research authority in a developing country, with the exception of India.

In 1971 the ARC was established initially as a semi-autonomous organization governed by an Agricultural Research Board (ARB) under the chairmanship of the Minister of Agriculture. The ARC is headed by a director general who has three deputies, one for research, one for the agricultural research stations and State farms, and a third for the extension service. The board of directors of the ARC consists of the Minister of Agriculture (chairman of the board), the director general of the ARC, who is responsible for the day-to-day operations, the three deputy directors, the directors of the research institutes, a representative of the Academy of Scientific Research and Technology, and five independent external consultants, versed in agriculture, and selected by the Minister of Agriculture.

The main functions of the ARB are to organize, co-ordinate and promote agricultural research, to plan and review agricultural research, to utilize the results of research, to identify problems and develop closely co-ordinated research programmes, to plan and develop extension programmes, to administer agricultural research stations, to publish and disseminate research results and agricultural information by other methods, to release the results of research in the form of recommendations to be carried out by extension service workers and to foster relations with national and foreign scientific institutions. The ARB may establish subject matter committees to review research projects and report on matters that may be referred to.

The ARC includes 14 research institutes, three central laboratories, 36 agricultural research stations, 33 production farms (12 134 ha), 18 nurseries (243 ha), and 15 animal production units. The research institutes cover the following areas: Soils and water; cotton; field crops; horticultural crops; plant protection; plant pathology; animal production; animal health; animal reproduction; veterinary serum and vaccines; agricultural mechanization; sugar crops; agricultural economics; agricultural extension. The three central laboratories are the Statistical Central Laboratory, the Pesticides Central Laboratory, and the Food Analysis Central Laboratory.

Research institutes and central laboratories have research departments and support sections for administration and technical services. Both the research institutes and departments are headed by a chief of research or research professor. In 1989 the ARC employed 29 749 people. These included 1434 Ph.D. research workers, 1125 research assistants (M.Sc. and B.Sc.), 7736 technical assistants (B.Sc., etc.), 6118 employees for administration and 13 336 labourers. Numbers of staff change continuously because staff with M.Sc. and B.Sc. degrees have the right to continue to the Ph.D. level. Scientific staff members (2559) and supporting staff (27 190) are listed in Table 21.6.

Of the research activities of the institutes, the dominant ones are breeding programmes, agronomic research, plant and animal protection, and the economic evaluation of major operations.

Egypt is fortunate to have established within the respective research institutes of the ARC, a competent, professional staff of plant breeders, who have supplied Egypt and other countries with a consistent and steady flow of new improved, disease-resistant, and high-yielding cultivars of the major crops.

Agronomic research covers the determination of cultural practices recommended for the major crops in different parts of the country. In particular, it covers water and nutritional requirements of these crops as well as rates, form, and time of application of chemical fertilizers. Soil research includes soil survey and classification, soil fertility, soil micro-organisms, recycling of organic wastes, biofertilizers, bioenergy, single cell protein production, and soil reclamation and management, especially of problematic saline, alkaline, sandy, calcareous, and gypsiferous soils.

TABLE 21.6. Egypt: numbers of staff in the Agricultural Research Centre, 1989

Type of staff	Number	Total
Research staff (Ph.D)		
Chief of research (Research Professor)	321	
Senior researcher (Assistant Professor)	284	
Researcher (Lecturer)	829	
Total research staff		1434
Junior research assistants		
Assistant researchers (M.Sc.)	793	
Research assistants (B.Sc.)	332	
Total junior researchers		1125
Supporting staff		
Technical assistants (B.Sc. etc.)	7736	
Administration	6118	
Labourers	13 336	
Total supporting staff		27 190
Total working force		29 749

Source: ARC, personal communiction (1989, 1992).

Plant protection research covers the identification of all the species of insects, mites, nematodes, fungal, bacterial, and viral diseases as well as biological and ecological studies of these organisms and effective methods for their control by agronomic, biological, and chemical methods.

Animal production research covers studies of cattle, sheep, and poultry breeding and husbandry. The field of animal nutrition is particularly important in view of the country's limited resources of forage and other feedstuffs. Animal health research is responsible for the identification of all animal diseases and their control, including the production of the necessary serums and vaccines. Animal reproduction research deals with sterility problems, artificial insemination, semen production, and embryo techniques.

Agricultural economics research includes financial and economic assessment of agricultural projects and technical research results, analysis of demand and supply of commodities, the economic classification of agricultural land, the economics of production and consumption of major agricultural commodities, agricultural credit and financing, consolidation of land use through grouping of crop rotations at the village level, conducting studies on sampling methods, and agricultural statistics.

The main headquarters of agricultural research is located at Giza, where work at the central national station is divided between two locations:

(1) Dokki—agricultural economics, plant protection (entomology, and pesticides), animal production, animal health, agricultural mechanization, and vegetable crops

(2) Orman—soil and water, cotton, field crops, fruit crops, plant pathology, sugar crops, and the agricultural statistics central laboratory.

The central station is now assisted by a well-distributed network of some 35 further agricultural research stations. These are listed in Table 21.7 and their distribution is shown in Fig. 21.2. The first research station at Giza was founded in 1900, the second at Bahtim in 1909. Of the total research and supporting staff in the ARC, 102 Ph.D. staff members and 4206 assistants and labourers are working at the research stations.

Foundation seed (i.e. seed produced by plant breeders) and registered seeds and seedlings of selected fruit trees are produced in 33 production farms (12 134 ha) and 18 nurseries (243 ha) distributed all over the country. More than 4300 agricultural engineers, technicians, and labourers work in these farms and nurseries.

Training facilities including laboratories, glasshouses, farms, and also accommodation, are available in most of the stations for both local and foreign trainees.

Contribution of the ARC to agricultural development in Egypt Faced with an exploding population, 1.2–1.4 million annually, (ARC 1988*b*) and limited

TABLE 21.7. Egypt: Agricultural Research Centre research stations

	Governorate
Field crop stations	
1. Agricultural Research Station, Isma'iliya	Isma'iliya
2. Agricultural Research Station, Sers el Layan	Minufiya
3. Agricultural Research Station, Gemmeiza	Gharbiya
4. Agricultural Research Station, Sakha	Kafr el-Sheik
5. Agricultural Research Station, Ityai el Barud	Beheira
6. Agricultural Research Station, El Serw	Damietta
7. Agricultural Research Station, Nubariya	Beheira
8. Agricultural Research Station, Sabahiya	Alexandria
9. Agricultural Research Station, Sids	Beni Suef
10. Agricultural Research Station, Mallawi	Minya
11. Agricultural Research Station, Sandaweel	Sohag
12. Agricultural Research Station, Mataena	Qena
13. Agricultural Research Station, Kom Ombo	Aswan
14. Agricultural Research Station, Alexandria	Alexandria
15. Agricultural Research Station, Marsa Matruh	Marsa Matruh
16. Agricultural Research Station, New Valley	New Valley
Horticultural stations	
1. Horticultural Research Station, Kanater	Qalyubiya
2. Horticultural Research Station, Tahrir	Beheira
3. Horticultural Research Station, Nubariya	Beheira
4. Horticultural Research Station, Sabahiya	Alexandria
5. Horticultural Research Station, Kassaeen	Isma'iliya
6. Horticultural Research Station, Sids	Beni Suef
Animal production stations	
1. Animal Production Research Station, Sakha	Kafr el-Sheik
2. Animal Production Research Station, Mahalet Moussa	Kafr el-Sheik
3. Animal Production Research Station, El Serw	Damietta
4. Animal Production Research Station, Gemmeiza	Gharbiya
5. Animal Production Research Station, Sids	Beni Suef
6. Animal Production Research Station, Burg el Arab	Marsa Matruh
7. Animal Production Research Station, Anshass	Sharqiya
8. Animal Production Research Station, Faiyum	Faiyum
9. Animal Production Research Station, Mallawi	Minya
10. Animal Production Research Station, Gaziret el Shaeer	Qalyubiya
11. Animal Production Research Station, Sabahiya	Alexandria
12. Animal Production Research Station, Karda	Kafr el-Sheik

water and land resources, Egyptian yields, although quite high by world standards, must continue to increase. As the main institution for agricultural research in Egypt, the ARC contributes to agricultural production by strengthening agricultural research, by continuing the present successful biological and chemical technologies of breeding and releasing high-yielding cultivars of field and horticultural crops and of livestock, by improving soil fertility and increasing the utilization of fertilizers, by developing effective methods for pest control, and by improving animal health.

The Academy of Scientific Research and Technology (ASRT)

The Academy of Scientific Research and Technology undertakes the following activities in its efforts to strengthen programmes relating to science and technology throughout Egypt (ASRT 1979):

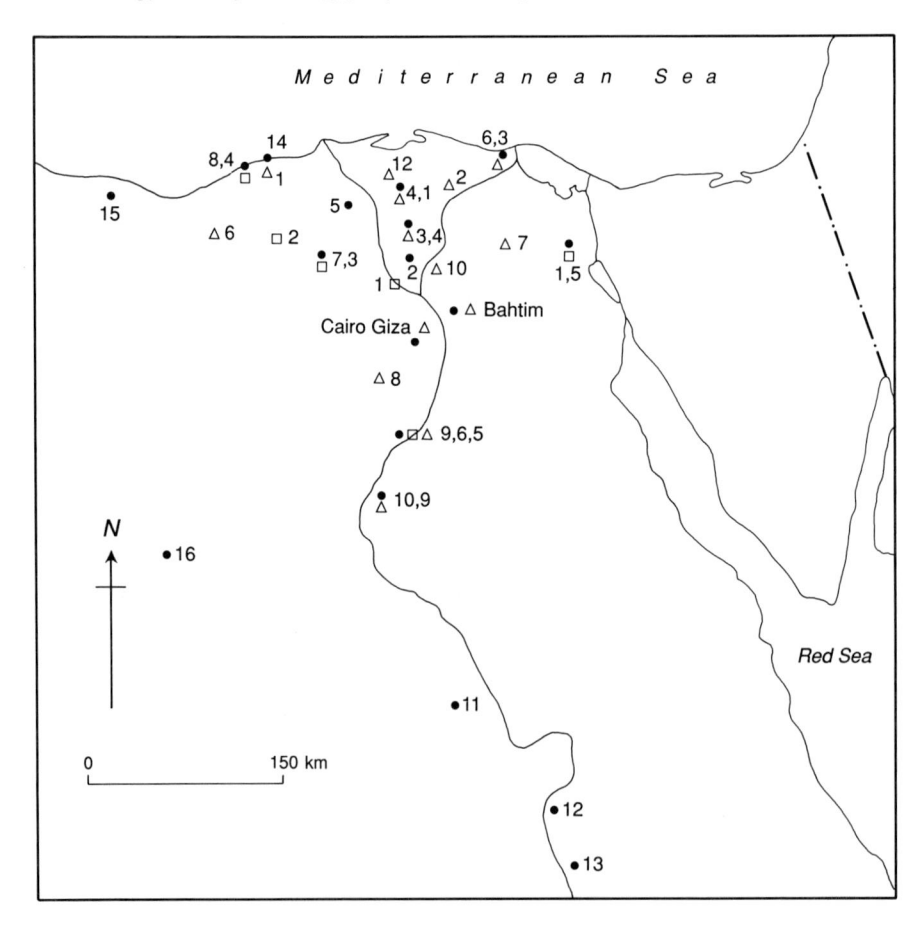

FIG 21.2. Egypt: location of Agricultural Research Centre research stations. (●) Field crop stations; (□), horticulture stations; (△), animal production stations

(1) supporting research directed towards solving problems of national interest;

(2) encouraging the application of modern technology;

(3) formulating policies that strengthen linkages between scientific and technological organizations;

(4) defining priorities for scientific and technological research in the major development areas;

(5) encouraging basic research and supporting research schools;

(6) participating with universities in manpower development for training of research specialists;

(7) organizing state awards in branches of science;

(8) organizing scientific publishing and the popularization of science;

(9) supporting scientific societies;

(10) developing international relations in science and technology.

There are 11 somewhat autonomous research organizations included under the ASRT 'umbrella'. These are the National Research Centre, the Theodor Bilharz Research Institute, the Central Metallurgic Research and Development Institute, the Institute of Oceanography and Fisheries, the Institute of Astronomy and Geophysics, the Egyptian Petroleum Research Institute, the Technology Transfer Organization, the Opthalmology Research Institute, the Electronics Research Institute, the National Institute of Standards, and the Remote Sensing Centre.

Many of the ASRT's activities and programmes are formulated and carried out by 11 specialized councils: (1) Management and economics; (2) Food and agriculture; (3) Transport and communication; (4) Industry; (5) Construction and housing; (6) Petroleum energy and mining; (7) New settlements; (8) Health and medicine; (9) Social science and demography; (10) Environment; (11) Basic sciences. These councils plan and organize research which is then conducted under their supervision. The research is organized either through direct contact with a specific agency or by advertising research projects in the newspapers and providing an opportunity for many agencies to submit proposals. The councils make a selection on the basis of technical superiority and financial viability.

The National Research Centre

The National Research Centre (NRC) was organized in 1956 as an Egyptian Government programme to 'conduct basic and applied research in natural sciences' which would 'contribute to the national welfare'. It now functions as an individual unit under the ASRT 'umbrella'.

Since 1975 research and development activities have been organized under the following five programmes: technology transfer (with particular emphasis on the industrial sector), food and agriculture, health and environment, energy, and natural resources. The centre is also involved in scientific and technical consultancy, technical advice and training, and human resource development.

The NRC is under the management of a director and it receives guidance from a governing board. The basic administrative unit is the laboratory and, currently, there are 52 laboratories grouped into 13 divisions, including food industries, agricultural and biological, environmental, and basic science research.

The divisions dealing with food industries, agricultural and biological research, and basic science are involved in an array of research activities relating to food and agriculture. While most of the programmes involve research, some are concerned with extension and these include work to increase the production of certain crops.

The NRC employs 4000 people and one-quarter of the research staff have a Ph.D. The agricultural research division is the largest, with some 600 employees, including 200 with Ph.D. degrees.

In the agricultural research division, the NRC develops programmes to improve the output of certain crops. Among these programmes is the *More and better food project* initiated in 1978 as part of the applied science and technology programme funded by USAID. This project has confirmed the value of simple technology packages which, when fully developed, can be applied for technology transfer on a broader scale.

The Water Research Centre

Research in the Ministry of Public Works and Water Resources was consolidated into the Water Research Centre (WRC) by Presidential Decree in 1975. The WRC now consists of eleven research institutes, a project for the *Water Master Plan*, a Department for Technical Training, and a Department for Research Services. Three of the institutes (the Water Distribution and Irrigation Systems Research Institute, the Drainage Research Institute, and the Ground Water Research Institute) and the project for the *Water Master Plan* are conducting studies that are closely related to the research being done at the Soil and Water Research Institute and the Agricultural Mechanization Research Institute of the ARC.

The Desert Research Institute

The Desert Research Institute, recently renamed the Desert Research Centre, was established in 1934 and was inaugurated officially in 1951. Its main objectives are to conduct studies useful for the development of the desert.

It places emphasis on research into water and soil resources and on plant and animal production.

The institute has four divisions: water resources, soil resources, plant production, and animal production. The staff consists of 61 researchers, 40 assistant researchers and 94 technicians. Of these 53 have Ph.D. degrees and 35 are studying for a Ph.D. Fifty-five staff members have M.S. degrees and 87 are studying for an M.S. The Desert Research Institute library has about 5000 books and periodicals and publishes *The Desert Institute Bulletin* biannually. It has a 50 ha experimental station at Maryut in the Mediterranean littoral zone where studies on sheep production, range management, horticulture, reclamation of saline soils, irrigation techniques, and the use of saline water for crop production have been made. Another field station is under development in Sinai within the arid belt of the Gulf of Suez.

Research centres in the Ministry of Electricity

The National Centre for Radiation Research and Technology and the Nuclear Research Centre are two research centres with agricultural interests within the Ministry of Electricity. The Nuclear Research Centre has a soil and water research division and an animal production research division.

The Ministry of Land Reclamation

Before merging with the Ministry of Agriculture in 1987/8, the Ministry of Land Reclamation had little research capability. A few project-related studies had been conducted by the staff of the International Centre for Rural Development (within the Ministry of Agriculture and Land Reclamation) to guide the management and settlement of some of the new lands. Research on well drilling, groundwater and geological behaviour, corrosion and pump performance, and soil surveys and land classification have been conducted by the General Company for Research and Ground Water (REGWA) which is within the Ministry of Construction and New Settlements.

The National Aquaculture Research Centre

Several steps have been taken in recent years to expand the availability of high quality protein by developing the fish-farming industry and its supporting institutions. Among these developments was the establishment of a National Aquaculture Centre in Abbasa under the Administration of the Authority for Fisheries Resources Development. The centre is constructed on a 620 ha site and includes 160 ponds and a research, training, and production building. The project is conducting research in four areas: breeding, disease, nutrition, and hatching for inland fisheries. Extension services and training will be provided to homestead fish farmers, along with credit facilities through the Principal Bank for Development and Agricultural Credit (PBDAC). In addition to the Aquaculture Centre at Abbasa (IADS 1984), there are five main stations at different locations, with one of them on Lake Nasser at

Aswan. Moreover, many universities are conducting research on development of fisheries.

Universities

Egypt has fifteen universities and fourteen of the universities contain faculties of agriculture. The universities are associated with and funded by the Ministry of Education and Scientific Research, but many decisions concerning universities are made by the Supreme Council of Universities, which consists of the Minister of Education as chairman and the presidents of the universities.

Each university's agricultural programme is administered by a dean and includes a wide range of commodity-oriented and discipline-oriented departments. Human nutrition and more broadly based home economics programmes are included in the agricultural faculties of Alexandria, Cairo, Ain Shams, and Zagazig.

Although the universities do not have an organized research programme, they do conduct considerable agricultural and related research through their graduate students, many of whom are staff members of the ARC pursuing advanced degrees and who often use ARC as well as university facilities. There is also general recognition that academic faculties traditionally conduct research, and there is considerable independent research being conducted by the faculty despite meagre support from the Ministry of Education and Scientific Research.

The university faculties have some excellent scientists who could make a significant contribution to agricultural and related research in addition to the teaching programme in which they are engaged. The universities also have considerable amounts of functional laboratory space for research, although such desirable features as hot water, compressed air, gas, and air-conditioning are usually lacking.

There are some well-equipped laboratories, and a considerable amount of excellent equipment is scattered among various departments of agriculture. Faculties of agriculture can also utilize the equipment of other faculties in the university. For example, electron microscopes in the faculties of science at the Suez Canal University and Asyut University are available to members of the faculty of agriculture.

Most, but not all, universities have at least small farm areas for teaching and research. None compare to the large farms of the ARC, but some are 42–84 ha and the Suez Canal University has over 420 ha being developed into a farm. Al Azhar University, on the other hand, has no farm, only a small amount of land for greenhouses, and a nursery. Cairo University has a small farm.

University libraries are in general better than the ARC Central Library, but they are still deficient. The best libraries are at Alexandria, Cairo, and Ain Shams Universities.

Massive funding would be required to bring laboratories and land up to a satisfactory standard for research at all departments and in all faculties of agriculture. However, there is an excellent base of faculty laboratory space and instrumentation that could, with well-directed support, make a substantial contribution to a national agricultural research effort. Such research would enhance the quality of the graduate programmes in agriculture, which provide a major source of the staff for the ARC.

Private research units

Parallel to the above-mentioned governmental research units in the different ministries and research centres, there are some other research units in private companies and industries. Their main interests lie in the development of new cultivars of cereals such as wheat and the breeding of poultry, fish, and other animals.

Detailed information on the research interests of private organizations is difficult to obtain and would be beyond the scope of a narrative report.

Agricultural extension

The use of agricultural chemicals, especially fertilizers and insecticides, is well established in Egypt. Farmers are rapidly accepting improved seed varieties, especially for vegetables. However, research and extension have a major role to play in developing technologies suited to Egyptian agriculture and in ensuring that the farmer can use them effectively and safely.

The Agricultural Extension Agency is the link through which research findings, recommendations, and new technologies are transferred to farmers for adoption. The Extension Agency in Egypt is a part of the MALR and is organized at governorate, district, and village level. In each district there is one agricultural extension unit with a demonstration field attached to it. In addition, there are the 'individual farmer's extension fields' where the recommended practices are implemented in demonstration plots to teach farmers by seeing and doing. The 'grouped farmer extension fields' are utilized in rice areas by making a demonstration plot of 5–10 feddans in area, which may include 6–10 farms. These extension fields work in the same way as the individual fields. The 'extension villages and regions' have been used in the past but are no longer of use.

Bottlenecks between agricultural education, research, and extension

A number of factors lead to significant problems in the area of agricultural education, research, and extension.

The lack of accurate data about the labour requirements of the production and service sectors of agriculture result in the misuse of agricultural graduates in the labour market. In addition, agricultural graduates tend to be of a low scientific standing due to their low academic rating at the time of admission to the universities. The shortage of the agricultural education facilities, particularly of farms and laboratories, also has a detrimental effect on the practical performance of graduates and it is regrettable that most organizations, factories, and production units refuse training for agricultural students and graduates.

One of the major constraints on the Extension Agency in Egypt is the very small amount of feedback on agricultural practices and improved inputs coming from the farm into the extension and the research systems.

The dissemination of the results of agricultural research in the form of simple production recommendations suitable for widespread adoption by farmers is the corner-stone of agricultural development in any country. In Egypt in recent years there have been considerable research results but recommendations derived from these results have not spread into farmers' fields at a satisfactory rate.

Increases in agricultural production have been much smaller than the potential indicated by on-farm verification trials and a major contributing factor was the lack of co-ordination between research and extension. Other factors, such as the lack of trained extension staff, the acute need for a specific work plan, and the absence of a clear educational philosophy, have also contributed to the slow rates of information dissemination among the farming community.

Little research has been directed towards solving production constraints. Meantime, applied research has not been expanded enough to reflect the major development objectives of vertical and horizontal agricultural expansion. Furthermore, the interrelationships among specialized institutes have been almost absent in research programmes.

The distribution of investment in the agricultural sector has not been optimal and the relatively small amount of investment in agricultural research and extension programmes has prevented improved technology being developed and made available to the farmer.

Although Egypt is short on research resources, no prioritization of research areas or commodities has been adopted. The country is in pressing need of better defined research priorities in agriculture. Narrowing the food gap must become the major concern and ultimate target of any agricultural research activity.

In addition to the above constraints, there is an acute shortage of transportation facilities, extension equipment, and aid as a result of the low budget. Incentives are inadequate to activate the extension system in order that relevant information be passed from researchers to farmers, and information relating to farmers' problems be relayed back to research specialists. This issue would, however, be a major concern of a recently started National Agricultural Research Project (NARP).

The goal of NARP is to increase agricultural productivity by improving the quality of technologies available to farmers. The purpose of the project is to develop the capability of the agricultural research community to provide a continuous flow of improved appropriate technology.

NARP's management, consisting of MOA and ARC administrators and programme advisory groups, plays a key role in providing overall guidance and direction necessary to meet the MOA's agricultural targets. A research support system (RSS) was designed to provide standardized procedures to assist ARC researchers in the development of disciplinary, multidisciplinary, and interdisciplinary research plans to manage the research activities effectively.

Conclusions

Contemporary cultural trends in Egypt necessitated an acceleration of the growth of education, and agricultural education has not been an exception.

Agricultural education includes two stages: agricultural secondary stage is found in 77 schools (111 226 students) and agricultural higher stage is found in 14 faculties and two institutes (26 759 students).

Agricultural education is not in general compulsory and is free of charge. It comes under the direct control of the Ministry of Education and the Higher Council for Universities.

At present, the agricultural research system in Egypt comprises about 50 centres and institutes belonging to various ministries and organizations. A significant level of research relating to food and agricultural production is being conducted throughout the country and it is financed from many sources.

ARC has the primary responsibility for agricultural research and extension within the Egyptian Government. Other institutes, such as the National Research Centre and the Academy for Scientific Research and Technology, are also involved in the research. In addition, international agricultural research centres are conducting research in Egypt.

The Agricultural Extension Agency is the link through which research findings, recommendations and new technologies are transferred to farmers for adoption. In Egypt, there remain several factors which lead to significant problems in the areas of agricultural education, research, and extension.

References

ARC (Agricultural Research Center) (1988*a*). *NARP News*, **2** (11), 3–5. Agricultural Center Press, Cairo.
ARC (Agricultural Research Center) (1988*b*). *NARP News*, **2** (12), 2–5. Agricultural Center Press, Cairo.

ASRT (Academy of Scientific Research and Technology) (1979). *Directory of research centers and organizations.* ASRT, Cairo.
Boktar, A. (1963). *The development and expansion of education in the United Arab Republic.* The American University in Cairo Press, Cairo.
CAPMAS (Central Agency for Public Mobilization and Statistics) (various years). *Statistical yearbook.* CAPMAS, Cairo.
Department of Statistics and Census (1962). *Ten years of revolution.* Statistical Atlas, Department of Statistics and Census, Cairo.
IADS (International Agricultural Development Service) (1984). *Increasing Egyptian agricultural production through strengthened research and extension programs.* Report of a United States team of consultants to the Ministry of Agriculture of the Arab Republic of Egypt. IADS, Arlington, USA.
Ministry of Education (undated). *Report on education progress in 1959–60.* Ministry of Education, Cairo.
Ministry of Education (1960a), *Education in eight years, Southern Region, Egypt.* Ministry of Education, Cairo.
Ministry of Education (1960b). *General guide to educational systems in the Southern Region (Egypt).* Ministry of Education, Cairo.
Ministry of Education (1960c). Unpublished statistical tables, General Directorate for Statistics and Computers. Ministry of Education, Cairo.
Soroor, A.F. (1989). *Developing education in Egypt.* Ministry of Education, Cairo.

Further reading

Goering, T.J. (1981). *Agricultural research sector policy paper.* World Bank, Washington, DC.
Ismail, S.A.M. (1984). A mathematical model for studying the relationship between education and development in the agricultural sector in Egypt. Unpublished Ph.D. thesis. Agricultural Economics Department, Zagazig University, Benha. (in Arabic).
Ismail, S.A.M. (1989). Higher studies and agricultural scientific research. In *Egypt in the year 2000, eighth conference of the Friends of Egyptian Scholars Abroad, 28–30 December*, pp.1–15. Friends of the Egyptian Scholars Abroad, Cairo (in Arabic).
Ismail, S.A.M. (1989). Agricultural secondary education and rural development in Egypt. In *Egypt in the year 2000, eighth conference of the Friends of Egyptian Scholars Abroad, 28–30 December*, pp.16–31. Friends of the Egyptian Scholars Abroad, Cairo (in Arabic).
Mansour M.E.I. (1988). *Need for evaluation of agricultural research in Egypt*, Working Paper, No. 88–13. University of California, Davis.
Ministry of Agriculture (1983). Utilization, efficiency and distribution of human and material resources of the agricultural scientific research sector in Egypt. Unpublished Report, Ministry of Agriculture, Cairo (in Arabic).
Sanyal, B. *et al.* (1984). *University education and the labour market in the Arab Republic of Egypt (ARE).* UNESCO, International Institute for Educational Planning, Pergamon Press, Oxford.

Glossary

afrangi	a loaf similar to French bread available in urban areas
al-adl w'al-kifayah	justice and efficiency
al-badu	desert dwellers
amshut	*Echinochloa stagnina*
badwiyyin	desert dwellers
baladi	from *balad*, a village, often used to describe the common local variety of crop or animal; *baladi* bread is a flat loaf of bread available from licensed government bakeries and vendors
bersim	Egyptian clover or berseem (*Trifolium alexandrinum*)
bir	well
bettai	large wafer-like bread consumed in middle Egypt and parts of the Delta
bitello	veal
chetwi	winter crops
corvee	labour
falafel	fried bean cakes
fateer	a bread rolled with shortening and eaten as a breakfast food
fas	short-handled hoe with a large blade
feddan	an area of land approximately equal to an acre; 1 feddan = 1.038 acre = 0.420 ha
fino	a loaf similar to French bread available in urban areas
foul madames	boiled faba beans (*Vicia sativa*)
gammaya	government food stores and co-operatives
gellabiya	loose tunic
gibna beyda	fatless white cheese made from buffalo milk
hamada	a level to gently sloping surface covered mainly by stone and gravel; gravel desert
hayy	urban district, quarter, or neighbourhood, introduced in 1975
hosha	aquaculture where fish are confined to an enclosed section of water
howash	north Delta lakes
hurr	free tribes
infitah	openness of the economy
izbah	hamlet system

kirat	one twenty-fourth of a feddan
karkadé	*Hibiscus sabdariffa*
kankur	gorge
koshari	a mixture of lentils, rice, pasta, and onions, an inexpensive meal sold from pushcarts in urban areas
madina	city or town
mahbas	water storage basin for irrigation
markaz	county, or a central or large town with a number of satellite or related villages, unit introduced in 1975
marwa	private farm ditch, usually steel or concrete pipes which take water from *mesqas* and deliver it to small bunded units called basins
mesqa	minor system of canals that distribute water from the government distributory canals
molokhiya	leaves and tender shoots of young jute (*Corchorus olitorius* plants
morabiteen	literally 'the bound' or 'the tied'; groups integrated into an Awlad Ali tribe
muhafaza	governorate or province
nili	'during the flood' season; traditionally the autumn-sown crops following the Nile flood
nowlaut	traditional cooling house or potato seed
omdas	clan leaders
pasha	class of wealthy landowners
qameradeen	apricot leather, the traditional food for the nightly breaking of the fast during Ramadan. Apricot leather is made by drying apricot pulp in sheets about one-eighth of an inch thick. These have a leathery texture
qamhi	light brown, wheaten colour of Egyptian skin
qanatir	barrage
qariya	village
sabkha	saline lagoon; shallow depression for draining off excess irrigation water
samna	ghee
saqqiya	the Persian water-wheel; an animal-operated device for lifting water
sebakh koufri	residues of excreta from old inhabited sites, used as a fertilizer at the end of the nineteenth century
sefi	summer crops, irrigated from February to August
shaduf	a hand-operated water lifting device using a counter-weighted bucket
shami	a flat loaf similar to *baladi* but made of more refined flour and available in urban areas
sun	leavened bread consumed in the south of Egypt

tamaliyya	resident, permanent labourers
tambour	Archimedean screw powered by human labour
tamiya	fried bean cakes
tamween	private grocers registered with the government as rationed goods supply shops
tarabeles	mounds occupied by *Acacia ehrenbergiana*
tarahil	migrant labourers
tibin	wheat straw fodder
waqf	land set aside ostensibly for charitable purposes and not taxed (Islamic Law)
zabaleen	garbage collectors
zweigwirtschaft	'dwarf agriculture', small farms.

Appendix 1

Price weights and tonnages of agricultural products, value of agricultural imports and exports, orange and potato exports by destination

The tables and graphs which follow were supplied by J.B. Parker and are based on unpublished data collected and collated by the Trade and Analysis Division of the USDA Economic Research Service from varying sources, including the Ministry of Agriculture and Land Reclamation (MOALR), Cairo, the Central Agency for Public Mobilization and Statistics (CAPMAS), Cairo, FAO, and the Agricultural Counsellor, Cairo.

FIG A1.1.   Egypt: value of total agricultural exports, 1961–91

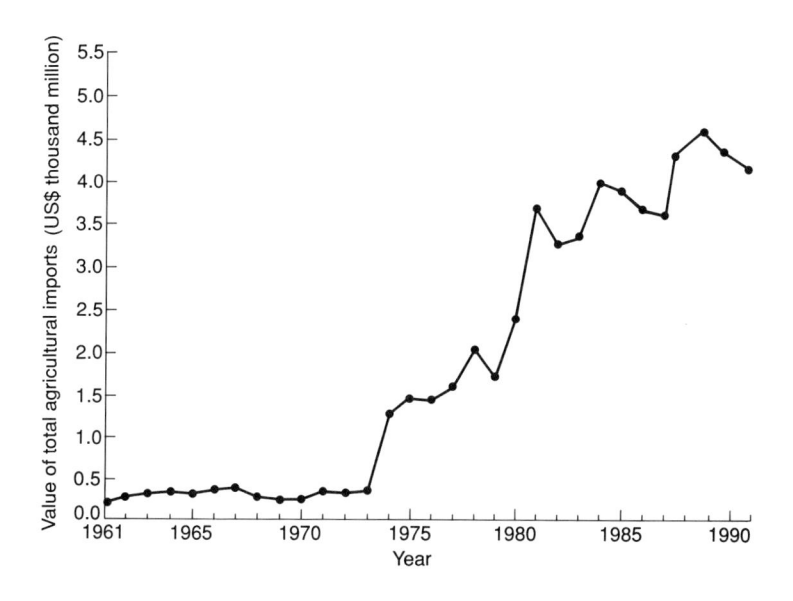

Fig A1.2. Egypt: value of total agricultural imports, 1961–91

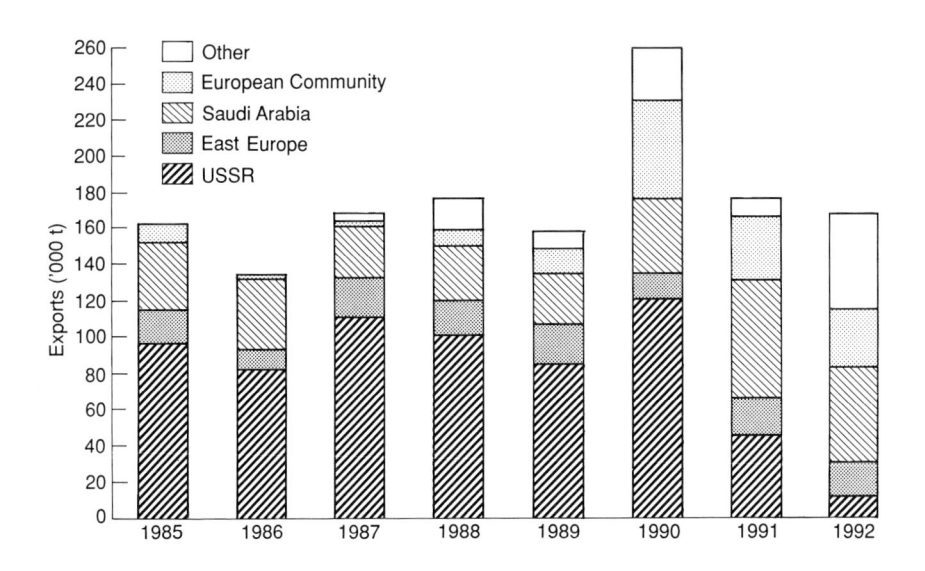

Fig A1.3. Egypt: orange exports to specified markets, 1985–92

TABLE A1.1. Egypt: price weights and tonnages of agricultural products, 1983–91

Commodity	Price weight ($US) (1988–90 average)	Production ('000 t)								
		1983	1984	1985	1986	1987	1988	1989	1990[a]	1991[a]
Wheat	185	1996	1815	1874	1929	2722	2839	3183	4268	4483
Rice, paddy	167	2442	2336	2312	2445	2279	2132	2679	3168	3447
Barley	151	120	144	145	152	137	109	138	142	140
Sorghum	155	622	540	551	606	552	578	599	628	568
Maize	192	3507	3698	3699	3608	3619	4088	4529	4798	4400
Broad beans	292	295	305	302	448	499	362	460	451	283
Lentils	429	6	10	13	15	18	17	19	12	12
Soya beans	393	162	143	140	133	134	129	91	106	120
Cotton lint	1374	419	399	435	401	351	306	296	310	294
Cotton seed	137	674	680	698	647	561	487	440	494	485
Groundnuts	561	21	23	23	21	23	32	34	31	30
Sunflowers	370	8	9	10	15	11	22	32	26	60
Tomatoes	172	2862	2993	3576	4456	4921	4212	3997	4123	3806
Sweet potato	80	115	112	130	132	111	63	67	70	73
Potatoes	183	1095	1250	1478	1400	1673	1725	1657	1638	1655
Onions	66	619	643	653	877	897	1009	1040	903	990
Cabbage	57	384	374	392	414	431	452	450	465	480
Vegetables	195	3567	3669	3790	4103	4398	4370	4350	4300	4500
Lemons	268	115	117	120	122	121	122	125	127	130
Oranges	172	1243	1182	1168	1234	1387	1400	1397	1400	1780

TABLE A1.1. cont.

Commodity	Price weight ($US) (1988–90 average)	Production ('000 t)								
		1983	1984	1985	1986	1987	1988	1989	1990[a]	1991[a]
Sugar, raw	216	780	740	767	887	909	955	949	957	1025
Grapes	228	344	357	395	435	510	540	621	615	527
Tangerines	92	120	104	106	108	109	110	112	120	126
Dates	199	470	474	509	528	491	542	560	570	600
Other fruit	214	627	626	632	687	715	729	740	750	880
Meats	2660	521	623	639	666	704	614	680	771	805
Milk	327	2080	2235	2341	2346	2403	2433	2474	2534	2635
Eggs	980	111	112	118	146	152	142	143	145	155
Wool	2287	5	5	5	5	5	5	5	5	5

[a] Estimate.
Source: Parker (1992, personal communication based on unpublished data, Economic Research Service, USDA).

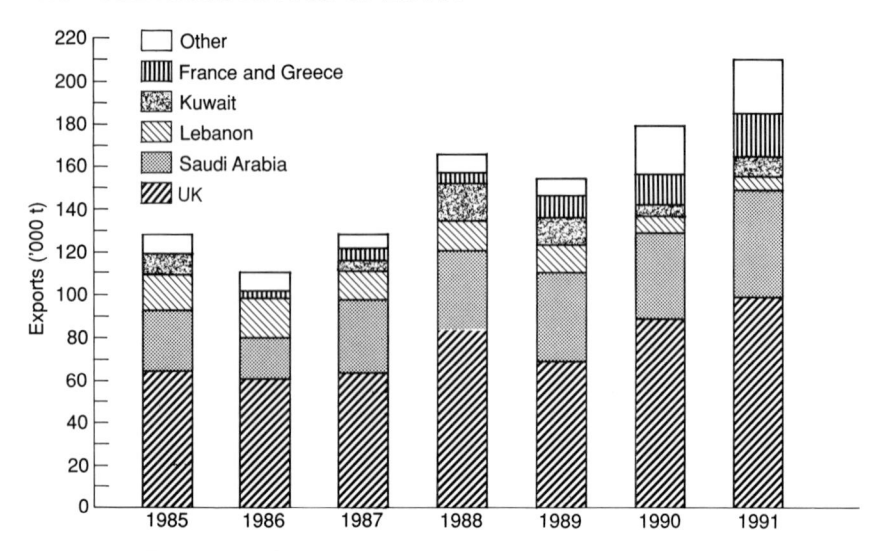

Fig A1.4. Egypt: potato exports to specified markets, 1985–91

Appendix 2

Cotton production and trade

J. B. PARKER

In 1990, the area planted to cotton in Egypt was 417 000 ha, but yields remained below expectations and production was only 300 000 t[a]. Competition from other crops, especially wheat and rice, caused the area planted to reach the lowest level in three decades in 1991, falling to 360 000 ha. Use of improved technology helped yields to rise from 720 to 820 kg ha^{-1}, and total production was down 2 per cent in 1991 to 291 000 t.

Cotton production increased in Egypt in the early 1980s because of higher yields, but declined in the late 1980s as other crops became more profitable and attractive to farmers. Cotton production declined 6 per cent in 1989 to 288 000 t, following a decline of 13 per cent in 1988. While problems with ineffective insecticides and unusual rainfall adversely affected yields in 1989, the major problem related to pricing policy. Higher prices were expected to contribute to a rebound in 1991, but farmers shifted heavily to other crops. The reasons why the area planted in cotton has declined include the following:

1. Profits from other crops have increased, while increased production costs have caused profits from cotton to remain less attractive.

2. Cotton takes about seven months to produce, while other crops occupy the land for a shorter time, thus giving a higher profit per month for use of the land. In Egypt, the land can be used for twelve months of the year, and the time a crop occupies the land is important.

3. Hand labour for picking cotton has become much more expensive, and the time of harvest in September tends to interfere with school attendance in some areas. The rising interest in jobs beyond the farm or the village has caused farmers to give a much higher priority to the education of their children, rather than to their work as cotton pickers.

4. The marketing system lacks flexibility, and farmers are not paid for the

---

[a] Production and yield of cotton are for lint. Cotton farmers sell seed cotton at the gin for a fixed price and it is estimated that 37 per cent of the weight of the seed cotton is lint before ginning.

cotton seed separately. Required government procurement has ended for other crops, except for sugar cane, and farmers are upset at a marketing system which does not allow them to share in the higher prices an open market system may bring.

Government control of marketing has been viewed by some as a problem, but if prices were more rewarding it would be less of a hindrance to obtaining a rebound in production. Egypt has a comparative advantage in producing long staple cotton. The addition to Egypt's gross national product (GNP) from the production of a feddan of long staple cotton is usually greater than that of any other traditional field crop, yet wheat production is being promoted and the government also aims to avoid rice imports. In addition to the exports of raw cotton[b], which were usually over $US400million annually in the 1980s, a surprisingly large textile export trade has emerged. While some people may question the exchange rates used for barter arrangements for exporting textiles to East Europe, the value registered by the Central Agency for Public Mobilization and Statistics (CAPMAS), the Egyptian Government official statistical agency, was over $US700million in 1987, and again in 1988 and 1989. This has caused Egypt to become a net importer of cotton.

Egypt's cotton exports declined from 148 340 t in 1986 for $US442.6million to 81 683 t in 1988. Higher prices kept the export value at $US457million in 1988, and to 39 438 t for $US186million by 1990. Exports of long staple cotton reportedly fell to 20 000 t in 1991. Most of the cotton exports went to the Soviet Union, East Europe, the European Community (EC), and Japan in the late 1980s. The lack of Egyptian cotton deliveries in 1990 contributed to the larger United States cotton exports to the European Community (EC) and East Europe. In addition to the Egyptian shortfall, East Europe is also finding it much more difficult to obtain deliveries of cotton from China and Pakistan and the quantity from the Soviet Union is also down.

United States cotton exports to Egypt increased from 59 768 t in 1990 to 61 633 t in 1991, and the value rose from $US115.6million to $US118.1million in 1991, and the average price showed little change, falling from $US1933 to $US1917 per t. Egypt needs large imports of raw cotton to help meet export contracts for textiles, and to allow some exports of the more valuable long staple cotton to continue. United States cotton exports to Egypt increased by 66 per cent in 1989 when the upward trend began. The average export price for Egyptian long staple cotton was over $US3000 per t in 1991.

Government agencies dominate cotton marketing in Egypt. Price increases to farmers tend to be partly offset each year by inflation and the declining value of the Egyptian pound. While prices have increased to a relatively respectable level, farmers are still not happy. The average price of about

Raw cotton is the lint which has not been processed further into yarn or textiles.

$US543 per t is still less than one-half the export price. The Cotton Organization is hesitant about a great price increase for cotton because it may affect the profitability of the textile industry. Most of the textile factories are public-sector companies but, by 1988, three private firms were operating. Some compromise is needed. A bonus dividend for exports of raw cotton and textile exports may help. If Egyptian cotton farmers received a series of bonus payments for cotton delivered to government gins, their incentive to grow cotton would increase.

Currently, no dividends are paid to farmers related to world prices or textile exports. Since multiple cropping allows over two crops per year in Egypt's climate, the time cotton occupies the land is a concern. Farmers in the Delta already have onions or some other vegetable growing in the field when rows of cotton are planted between the onion rows. As the onions are lifted in late April and May, the harvesting process serves to prepare the soil for the new cotton plants.

Problems have arisen in recent years related to the second picking. Most farmers prefer to have their children in school rather than working for the second picking of cotton. Furthermore, with the large number of tractors available and also custom hire available for small farmers without tractors, a new system has recently emerged. A day or two after the first picking, tractors can quickly disc in the cotton stalks and prepare the fields for profitable vegetable crops. For example, snap beans will provide a crop in 45 to 52 days, and the profit may be nearly as great as the cotton which was on the land for about six months.

Deregulation of rice farming recently contributed to a sharp rise in the area planted in this crop, and in the area where efforts to increase cotton cultivation were planned. Also, the high priority on producing more wheat and maize has increased competition for land which might have been used for either cotton or vegetables; the area of both cotton and vegetables has declined significantly since 1987. Vegetables are not regulated. In addition to the increased competition from cereals, vegetables have lost land to urbanization in the major vegetable area north of Cairo.

Farmers can sell vegetables at attractive open-market prices. However, the area planted to vegetables was hampered in 1990 and 1991 because of the great push to increase maize and rice production. Maize production rose to a record 4.6 million t in 1990, but the area planted to cotton was less than it should have been. It appears that increasing the area planted to cotton back up to 1 million feddans or 420 000 ha, and ideally to a higher level, may require a willingness to allow more maize imports and some willingness to shift land from maize to cotton. Egypt can get about $US1800 per ha for long staple cotton, compared with about $US1200 per ha for maize.

It would be better to get an extra $US1000million from rebounded cotton and textile exports than to save about $US200million on imported maize. Imports of animal feed to expand output of livestock products rapidly, combined with striking gains for cotton, could give Egypt an aggregate agricultural production index increase of 5 per cent in 1993, compared with the estimate of about 8 per cent in 1990, and only 3 per cent in 1991, and possibly 2 per cent in 1992.

Trade statistics prepared by CAPMAS, for the United Nations (UN) indicate strong growth in the export of textiles and clothing, while the decline in the quantity of raw cotton exports has been partly offset by a rise in prices.

Textile exports at the official exchange rate and values used in trade agreements increased from only $US290.7million in 1985 to $US1340million in 1988, with shipments to the USSR rising from $US89million to $US426million, and to the EC from $US95million to $US426million, and to the US tripling to $US54million.

By exporting textiles, Egypt has sought to make strong gains in cash markets, while at the same time expanding sales of cotton to traditional customers through trade agreements and also promoting textiles in these markets.

In addition to strong gains in the EC and United States in the 1980s, Egypt increased textile exports to various wealthy markets. Exports to Switzerland rose from $US412 000 in 1985 to $US23million in 1988, partly because of a duty-free opportunity. Another boom market was Sweden, with a rise from $US2million in 1985 to $US12million in 1988. Exports to Australia rose from $US88 000 in 1985 to $US3.6million in 1988. Exports to Japan zoomed from $US10 000 in 1985 to almost $US38million in 1988. The duty-free trade with Sudan allowed Egyptian textile exports to this market to reach $US50million in 1989, up from only $US2million in 1985. Exports to a number of Arab markets soared. Exports to Morocco rose from $US948 000 in 1985 to $US7.2million in 1988, while exports to Iraq tripled and reached $US52.8million in 1988. Exports to Jordan, tripled and reached $US11.9million in 1988. Trade agreements helped boost exports to many markets, but Saudi Arabia was a cash market where sales quadrupled, reaching $US4.95million in 1988.

Exports to Bangladesh rose from $US2.4million in 1985 to $US15million in 1988. Efforts to penetrate markets in India, Burma, and Mauritius for Egyptian fabrics which could be used for their expanding clothing industries did not work well in the 1980s. Problems in selling fabrics to some Asian markets increased Egypt's interest in returning to familiar customers in East Europe and the USSR.

Trade with East Europe was greater for raw cotton in the 1960s and 1970s. To offset declining raw cotton deliveries to East Europe after the shift to cash markets in Japan and Europe, textile exports to these markets increased, partly to provide sufficient value for Egypt to obtain the machinery and forest products which East Europe provided at reasonable prices.

Egyptian textile exports to Czechoslovakia increased from $US5million in 1985 to $US36million in 1988, and exports to East Germany rose from

$US25million in 1985 to $US43million in 1988. Albania appeared as a market for $US3million worth of Egyptian textiles in 1988, when Albania became Egypt's top tobacco supplier for $US14million, in exchange for a range of commodities. Bulgaria was another new market for Egyptian textiles, rising from only $US59 000 in 1986 to $US7million in 1988. Textile exports to Poland fluctuated from $US5.4million in 1985 to a peak of $US35.8million in 1987, and back down to $US5million in 1988 as Polish raw cotton imports from the USSR and China expanded, and reduced demand for textile imports. Also, Poland changed from trade agreements to more reliance on market forces. Romania purchased raw cotton and avoided purchases of Egyptian textiles beyond token arrangements related to trade shows.

Egyptian clothing exports soared from only $US21million in 1985 to $US67million in 1988 and $US134million in 1990. The United States became the top export market for Egypt's clothing exports in 1989 with a value of $US42.8million up from $US14million in 1987 and less than $US1million in 1985. The other rapid growth market was the EC rising from $US5.5million in 1986 and $US17.3million in 1988 to $US47.5million in 1989 and over $US50million in 1990. The third major market was the USSR at $US8.8million in 1988, and triple that value in 1987, with further gains in 1989 and 1990. East Germany was the fourth major market in 1988 at $US5million, and triple that value in 1985, and fifth was Czechoslovakia at $US4.2million up from only $US160 in 1985. Iraq was the sixth major clothing customer for Egypt at $US2.7million in 1988, and triple that value in 1987.

Egyptian all-cotton flannel shirts and T-shirts are popular. Underwear for both men and women from Egypt has a very receptive foreign market. Greater use of trade shows and advertizing could probably bring further gains, although quotas have hampered growth in some markets. As Egyptian clothing exports become more diversified, the quota problem may be less. The shortage of raw cotton, private marketing flexibility, and hesitation at making imports of cotton or other fibres when needed, are major constraints for further gains in export activity by the cotton and textile industries of Egypt.

Cotton area harvested, production, yield, imports, and exports

The graphs included in this appendix were supplied by J.B. Parker and are based on unpublished data collected and collated by the Agriculture and Trade Analysis Division of the USDA Economic Research Service from varying sources including the Ministry of Agriculture and Land Reclamation (MOALR), Cairo, the Central Agency for Public Mobilization and Statistics (CAPMAS), Cairo, FAO, and the Agricultural Counsellor, Cairo.

482 THE AGRICULTURE OF EGYPT

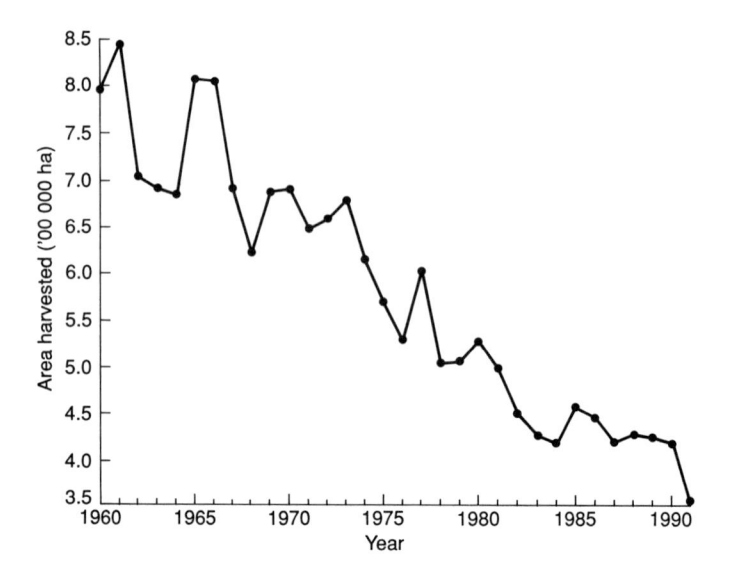

Fig A2.1. Egypt: cotton area harvested, 1960–91

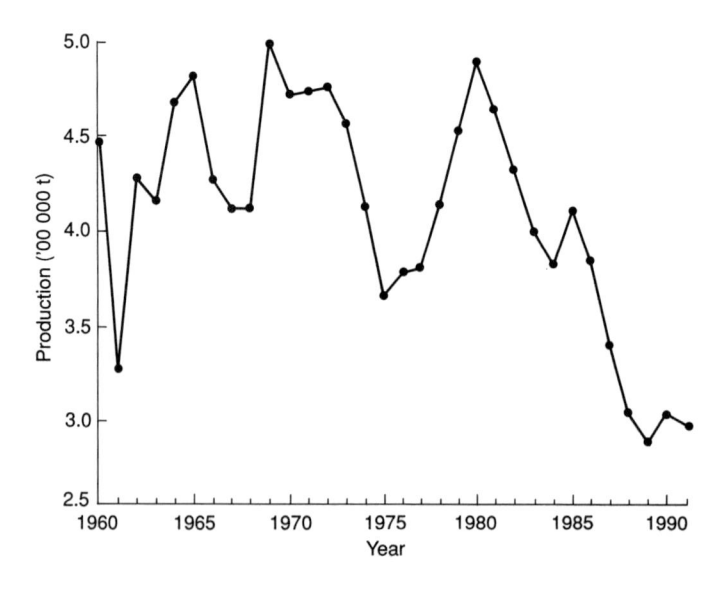

Fig A2.2. Egypt: cotton lint production, 1960–91

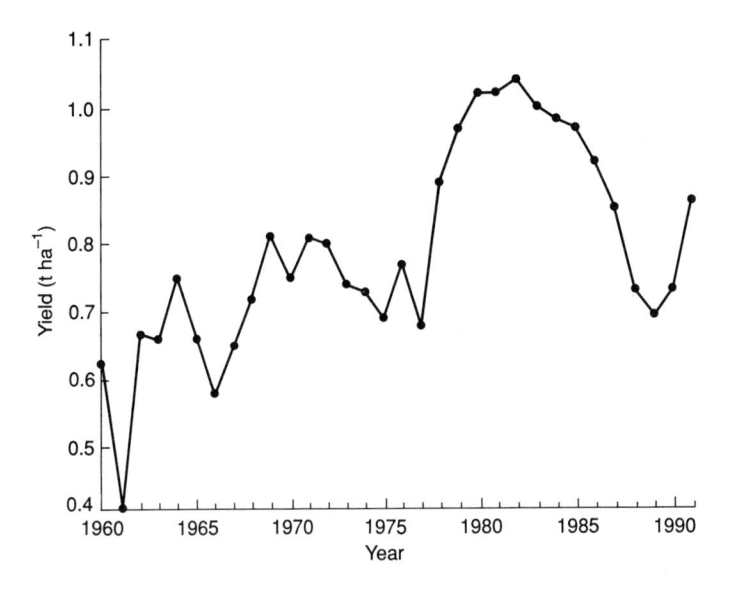

Fig A2.3. Egypt: cotton lint yield, 1960–91

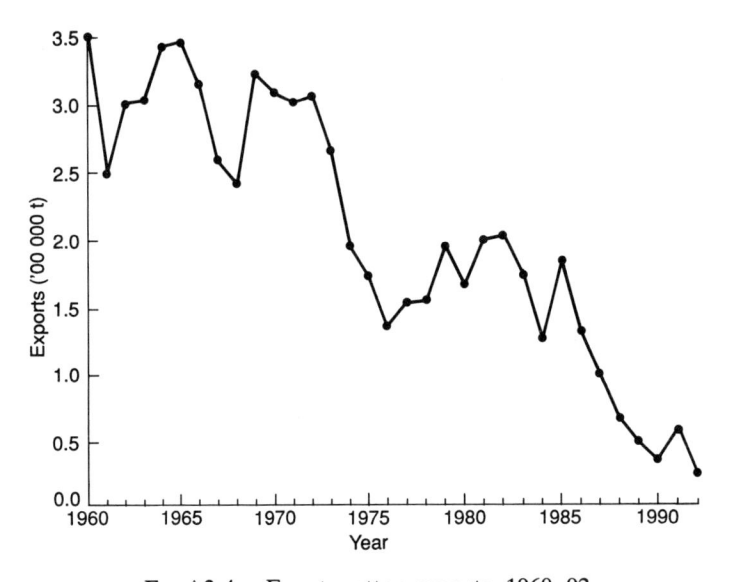

Fig A2.4. Egypt: cotton exports, 1960–92

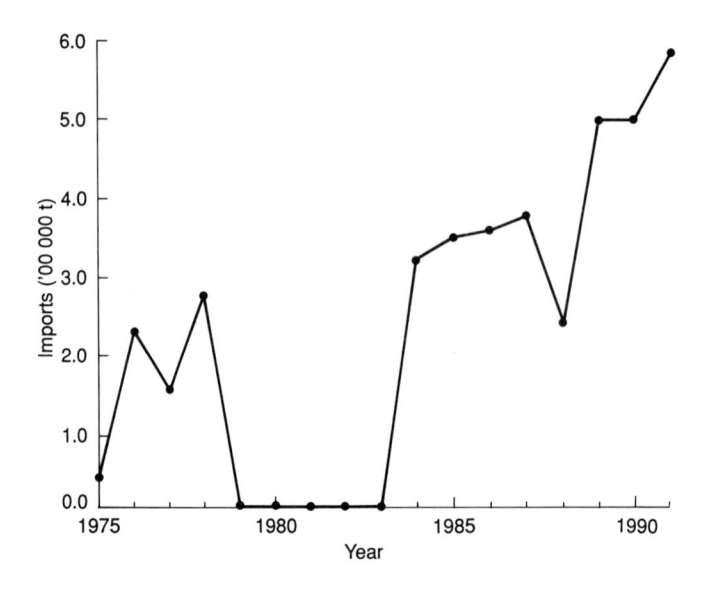

FIG A2.5. Egypt: cotton imports, 1975–91

Appendix 3

Wheat area harvested, production, yield, purchase prices, wheat and flour imports and distribution flow chart, maize production and imports, rice consumption

---

The graphs and charts which follow were supplied by J.B. Parker and are based on unpublished data collected and collated by the Agriculture and Trade Analysis Division of the USDA Economic Research Service from varying sources including the Ministry of Agriculture and Land Reclamation (MOALR), Cairo, the Central Agency for Public Mobilization and Statistics (CAPMAS), FAO, and the Agricultural Counsellor, Cairo.

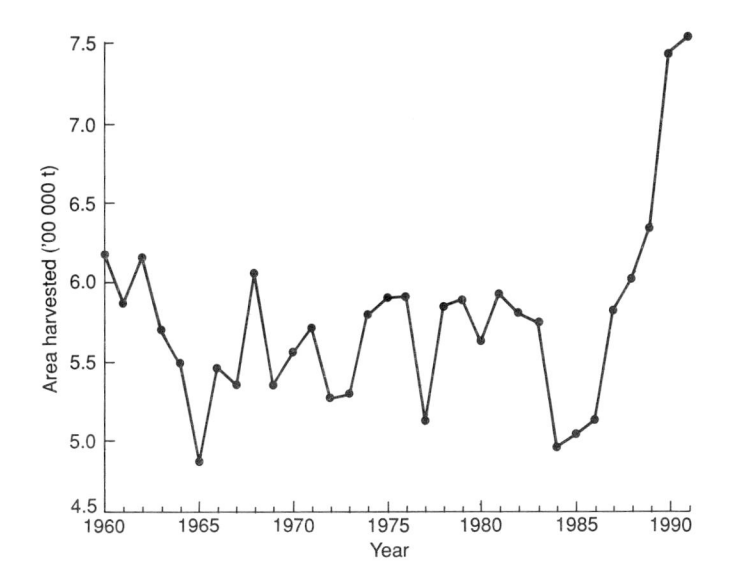

FIG A3.1.   Egypt: wheat area harvested, 1960–91

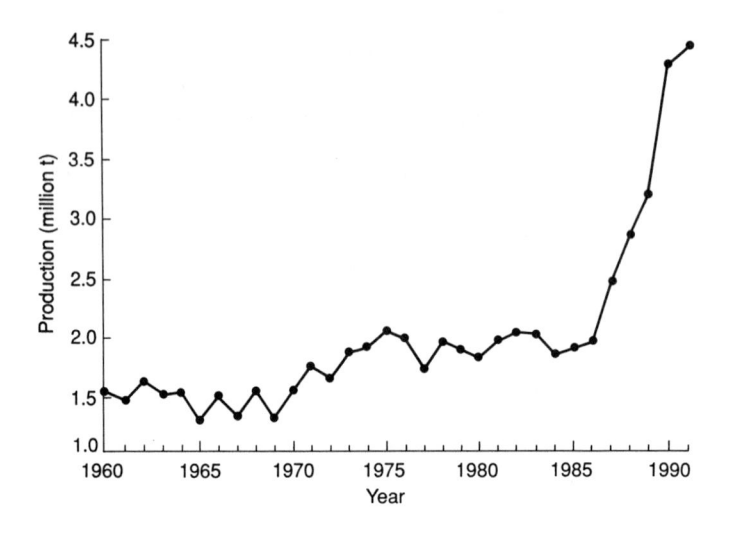

FIG A3.2. Egypt: wheat production, 1960–91

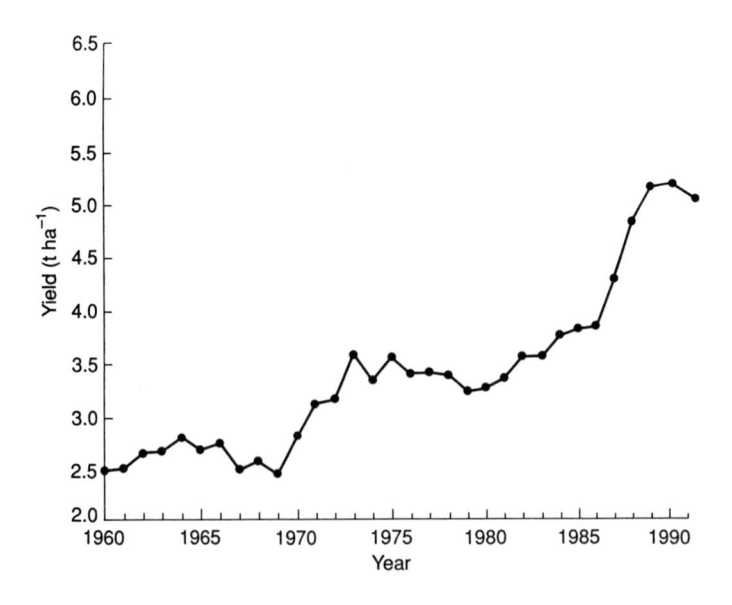

FIG A3.3. Egypt: wheat yield, 1960–91

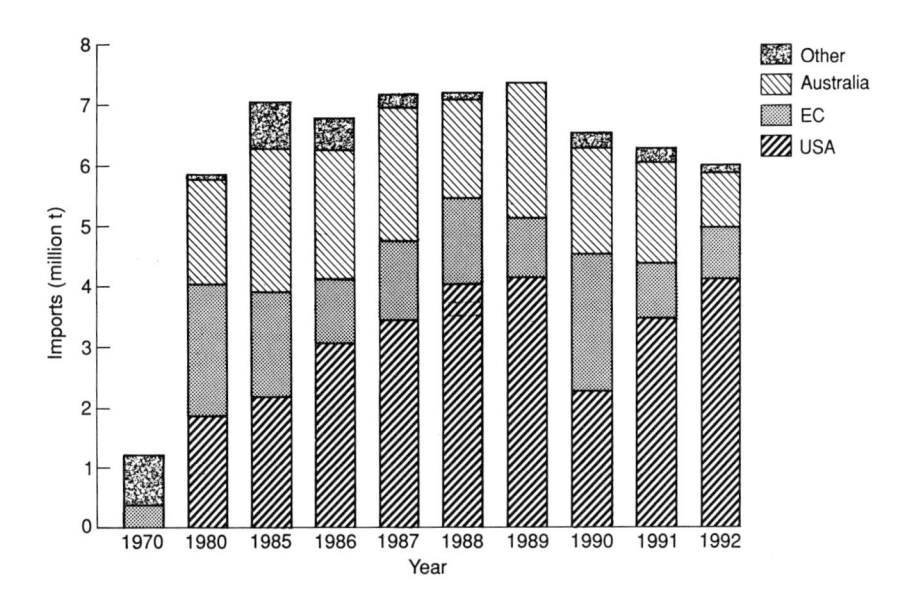

FIG A3.4. Egypt: wheat and flour imports from specified suppliers, 1970–92

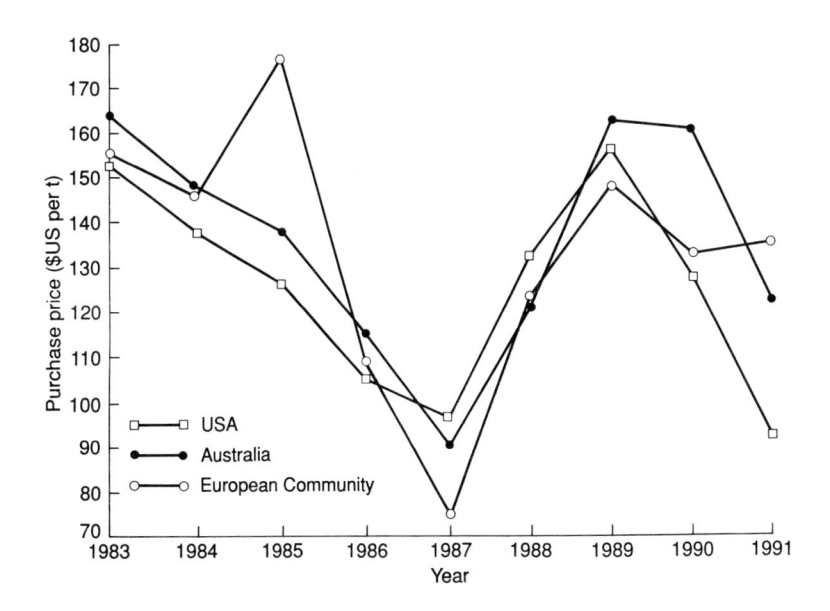

FIG A3.5. Egypt: purchase prices of wheat imports from the United States, Australia and the European Community (EC), 1983–91

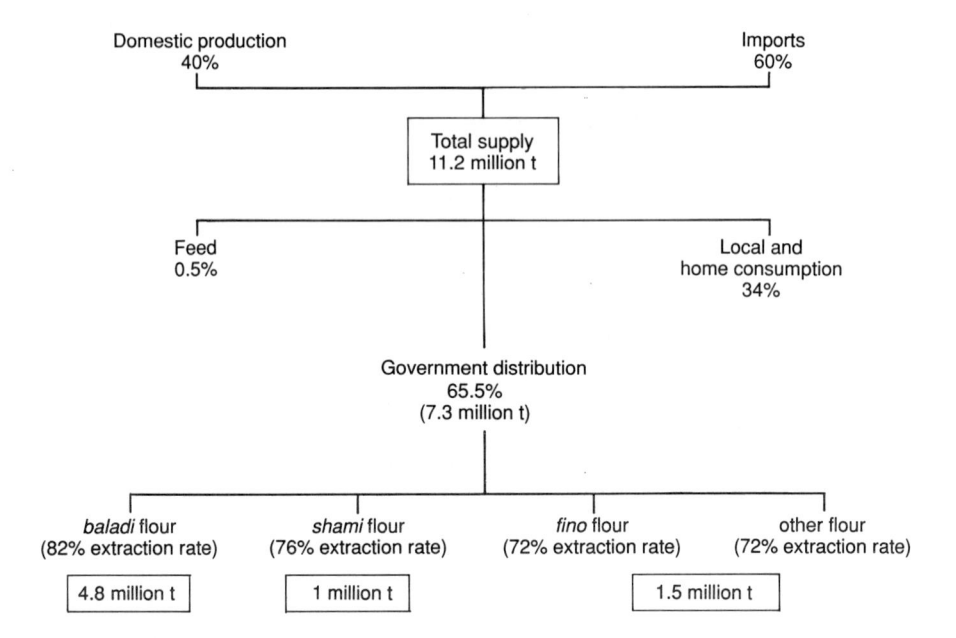

FIG A3.6. Egypt: wheat and flour distribution flow chart, 1991

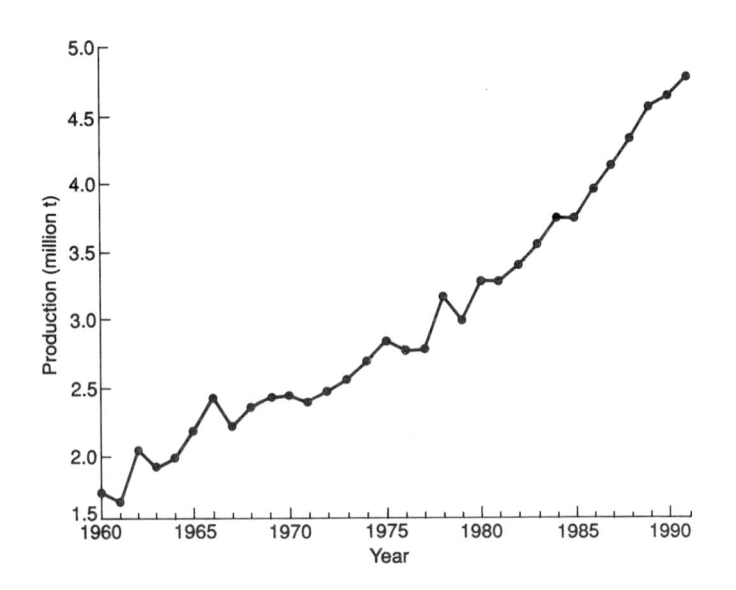

FIG A3.7. Egypt: maize production, 1960–91

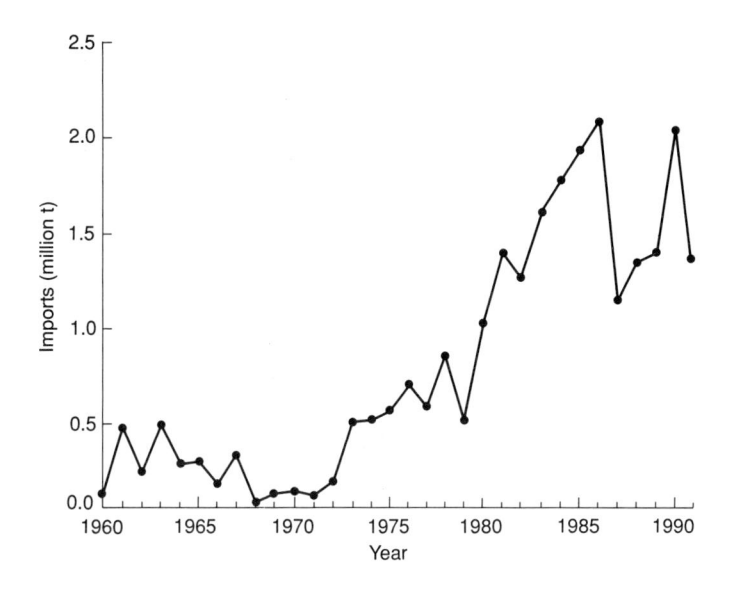

FIG A3.8. Egypt: maize imports, 1960–91

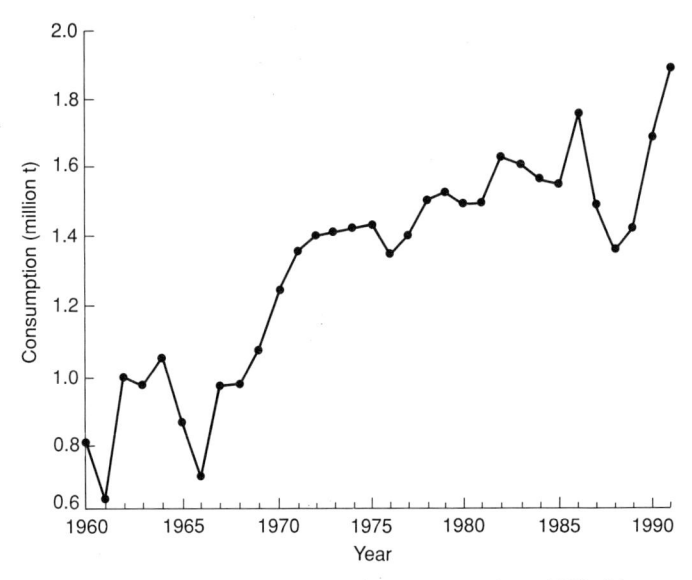

FIG A3.9. Egypt: rice consumption, 1960–91

Appendix 4

Production, imports, exports, and consumption of fertilizers, and current situation and expectations until the year 2010

The following tables accompany Chapter 17 on Fertilizers by M. M. El-Fouly.

TABLE A4.1. Production, imports, exports, and consumption of nitrogen ('000 t N)

Data source/Year	Production	Imports	Exports	Consumption
FAO[a]				
1962	106.5	85.4	0	191.9
1963	111.2	85.0	0	196.1
1964	114.2	58.8	2.2	227.1
1965	128.1	120.3	2.6	260.6
1966	149.1	136.5	0	284.8
1967	163.3	84.0	0	243.8
1968	146.1	145.0	0	244.1
1969	139.8	134.0	0	281.4
1970	117.8	158.0	0	310.1
1971	118.3	148.1	0	330.8
1972	120.0	238.1	0.3	324.5
1973	151.8	290.0	0	360.9
1974	150.7	260.0	0	358.2
1975	100.2	263.5	0	360.0
1976	150.5	227.4	0	415.0
1977	169.9	227.3	0	427.7
1978	195.2	258.2	0	459.5
1979	216.5	333.1	0	490.5
1980	263.9	208.1	0	500.0
1981	400.5	164.4	1.0	554.0
1982	482.0	162.6	1.6	585.0
1983	622.8	36.0	0.4	667.8
1984	639.2	26.0	82.0	722.2
1985	625.9	143.5	138.0	649.2
1986	575.5	146.9	9.2	640.3
1987	601.8	53.6	0	655.4
PBDAC[a]				
1984	572.0	14.1	82.0	743.3
1985	701.0	45.6	138.0	638.0
1986	690.0	199.2	9.2	775.8
1987	662.3	63.5	0	776.0
1988	663.6	80.5	0	791.7

[a] Note the difference in figures between the different sources, FAO, Rome and the PBDAC (Principal Bank for Development and Agricultural Credit), Cairo.
Source: El-Khashab (1989).

TABLE A4.2. Production, imports, exports, and consumption of phosphorus ('000 t P_2O_5)

Data source/Year	Production	Imports	Exports	Consumption
FAO[a]				
1962	25.6	19.7	1.0	48.4
1963	29.4	20.0	5.0	41.6
1964	31.6	20.0	4.6	47.9
1965	46.4	19.3	3.6	43.1
1966	42.5	9.8	4.4	52.2
1967	46.3	4.4	5.4	43.4
1968	52.2	3.7	4.3	35.5
1969	56.8	0	7.2	38.6
1970	59.4	0	9.0	36.2
1971	74.3	0	12.1	40.2
1972	73.8	0	11.8	46.3
1973	115.5	0	18.2	55.7
1974	81.0	3.1	9.2	44.4
1975	95.0	2.0	5.1	65.0
1976	77.0	1.5	4.3	83.0
1977	73.7	4.8	4.7	66.4
1978	88.4	9.0	10.1	80.8
1979	97.8	1.0	15.3	86.9
1980	93.0	11.6	18.2	97.5
1981	106.0	10.5	13.2	102.0
1982	116.6	32.0	23.2	110.0
1983	145.8	32.0	30.5	149.6
1984	122.8	64.4	11.9	159.6
1985	112.0	69.0	9.7	181.0
1986	149.1	59.8	10.7	198.2
1987	128.1	0	6.5	121.6
PBDAC[a]				
1984	139.9	31.6	0	158.5
1985	150.1	21.7	0	163.3
1986	184.1	0	0	183.2
1987	186.6	0	0	185.4
1988	189.4	0	0	190.8

[a] Note the difference in figures between the different sources, FAO, Rome and the PBDAC (Principal Bank for Development and Agricultural Credit), Cairo.
Source: El-Khashab (1989).

TABLE A4.3. Production, imports, exports and consumption of potash ('000 t K_2O)

Data source/Year	Production	Imports	Exports	Consumption
FAO[a]				
1962	0	2.00	0	2.00
1963	0	1.25	0	1.25
1964	0	1.02	0	1.03
1965	0	0.92	0	0.92
1966	0	0.59	0	0.59
1967	0	0.69	0	0.69
1968	0	1.20	0	1.20
1969	0	1.96	0	1.10
1970	0	1.74	0	1.52
1971	0	3.09	0	1.91
1972	0	3.72	0	1.64
1973	0	4.00	0	2.12
1974	0	0.75	0	1.92
1975	0	5.30	0	3.60
1976	0	5.33	0	3.19
1977	0	2.50	0	2.84
1978	0	0	0	2.94
1979	0	5.00	0	3.80
1980	0	5.00	0	6.70
1981	0	7.50	0	7.50
1982	0	12.91	0	12.91
1983	0	8.32	0	9.20
1984	0	16.90	0	17.50
1985	0	34.40	0	21.00
1986	0	14.27	0	25.00
1987	0	37.20	0	30.10
PBDAC[a]				
1984	0	26.5	0	17.5
1985	0	16.9	0	24.2
1986	0	24.0	0	24.4
1987	0	28.9	0	28.9
1988	0	33.5	0	29.5
1989	0	38.4	0	31.2

[a] Note the difference in figures between the different sources, FAO, Rome and the PBDAC (Principal Bank for Development and Agricultural Credit), Cairo.
Source: El-Khashab (1989).

TABLE A4.4. Nitrogeneous fertilizers: current situation and expectations until the year 2010 ('000 t N)

Year	Production of existing plants[a]	Demand[b]	Deficit in production
1991/2	660	874.0	214
1992/3	653	893.0	240
1993/4	647	911.0	262
1994/5	640	930.0	290
1995/6	634	949.0	315
1996/7	627	967.0	340
1997/8	621	985.0	364
1998/9	615	1004.0	389
1999/2000	609	1023.0	414
2000/1	603	1042.0	439
2001/2	597	1060.0	463
2002/3	591	1078.0	487
2003/4	585	1097.0	512
2004/5	579	1116.0	537
2005/6	573	1135.0	562
2006/7	567	1153.0	586
2007/8	562	1172.0	610
2008/9	556	1190.0	634
2009/10	551	1209.0	658

[a] Based on the actual production in 1987/8, then decreased by a rate of 1% yearly because of the very old factories.
[b] Based on the estimates of the Ministry of the Agriculture.
Source: Dawoud (1989).

TABLE A4.5. Phosphorous fertilizers: current situation and expectations until the year 2010 ('000 t P_2O_5)

Year	Production of existing plants[a]	Demand[b]	Deficit in production
1991/2	181	203	22
1992/3	179	208	29
1993/4	177	212	35
1994/5	175	217	42
1995/6	173	221	48
1996/7	171	226	55
1997/8	169	230	61
1998/9	167	235	68
1999/2000	165	239	67
2000/1	163	244	81
2001/2	161	248	87
2002/3	159	253	94
2003/4	157	257	100
2004/5	155	262	107
2005/6	153	266	114
2006/7	151	271	120
2007/8	149	275	126
2008/9	147	280	133
2009/10	145	284	139

[a] Based on the actual production in 1987/8, then decreased by a rate of 1 per cent yearly because of the very old factories.
[b] Based on the estimates of the Ministry of the Agriculture.
Source: Dawoud (1989).

Index

Ababda tribe 314–15
Academy of Scientific Research and Technology 460–1
Acomys cahirinus 293–4
afrangi bread 115
Agency for the Reconstruction and Development of the Egyptian Village (ARDEV) 140–1
  programmes 141–3
Agrarian Land Reform Law 148–50
  see also land tenure
Agricultural Act, Law No. 53 (1966) 152–3
Agricultural Extension Agency 465, 467
Agricultural Mechanization Research Institute (AMRI) 415
Agricultural Research Centre 455, 456–60
  activities 457–8
  contributions to agricultural production 458, 460
  functions 456
  history 455
  organization 456
  staff 457
  stations 458, 459, 460
aid
  for drainage materials 134
  for fig orchards 220
  see also United States Agency for International Development (USAID)
alfalfa, for reclaimed desert 60
almonds, in Northern Littoral Region 220
Amitermes desertorum 295
ammonia treatment of straw 391
Ammoperdix heyi 294
amshut 385
Anacanthotermes ochraceus 295
anaemia in women/children 124
Anglo–Egyptian Agreement (1929) 30
animals, *see* birds; fauna, desert; livestock
aphids 75
apricots 252
  dried 117

aquaculture 226–7
  National Aquaculture Research Centre 463–4
  potential in canals 223
aquifers
  Nile Valley/Delta 35
  Western Desert 35–7, 283, 286
Arabian Desert, *see* Eastern Desert
Arab settlement, pre-Islamic, ethnic influence 82
artificial insemination 66, 257, 262
Arvicanthis niloticus 293, 294
Aswan High Dam 30–3, 133, 190, 345, 347
  and drought 332–3
  and Nile Basin vegetation 59
  problems 31, 33
  and sea fishing decline 223
Aswan Low Dam 30, 133, 171, 190, 347
Atbara Nile tributary 26
Awlad Ali 213

bacteria in desert soils 287
Bahariya Oasis 54, 316
Baheeg Canal 296
Bahra el Burullus 6
Bahra el Manzala 6
Bahr Youssef Canal, serving Samalut project 354
bakeries 119, 157, 159, 427
  see also bread; retailing of food, regulated
baling of straw 390
bananas
  Nile Delta 252
  Western Desert 335
banks
  village, for subsidized inputs 236
  World Bank loans, for drainage 134
  see also Principal Bank for Development and Agricultural Credit (PBDAC)
barchan dunes, Kharga depression 317–18
Bardawil, Lake, fishing 312
Barki goat 69

Barki sheep 68, 70
  Northern Littoral Region 214–15
barley
  Nile Delta 247
  production system, Northern Littoral
      Region 216–18
  Sinai 312
  wind damage 216
barrages 347, 348
  Nile Delta 133
bats 293
beans
  broad beans
    Nile Delta 249
    Northern Littoral Region 218
  marketing control 430–1
  Sinai 312
  in street food 118
  Western Desert New Valley area 320
  yields, and location 269, 270
bedouin
  in Northern Littoral Region 213
    settlement 213–14
  population 83
  settlement 213–14
    and fig orchard aid 220
  Sinai 48, 310–11
  social system 296–7
  South-Eastern desert
    Ababda tribe 314–15
    Bushari tribe 314, 315
Beheri buffalo 65
Beja ethnic group 83
Berber population 83
bersim 198, 201, 384, 385
  Nile Delta 245
bettai bread 115
beverage consumption 118
birds
  desert 294
  grain pests 76
Bir Kisseiba 53
Bir Safsaf 53
black market for food 159
Blue Nile 26
  flow 27
bollworms 75
bread 115
  price 125
  subsidized 118–19, 427
  see also bakeries; flour
broad beans
  Nile Delta 249
  Northern Littoral Region 218
broilers 384, 434–5
brucellosis, slaughter policy 261
buffalo 64–6
  animal characteristics 253
  feeding 256–7
  fertility 257
  large-scale units 259–60
  management 254–6
  meat production 259
  milk production 258–9, 432
  role 253–4
  productive characteristics 65–6
  types 65
  veal calves 65, 433
  see also milk
Bushari tribe 314, 315
butchers 434

Cairo, population growth 276
calcisols 11
calcrete 11
camels 71–2
  Ababda tribe herding 314–15
  Nile Delta 260
  Northern Littoral Region
    north-west zone 215–16
    population 214
canals 33–5, 348
  Baheeg 296
  Bahr Youssef 354
  and desert reclamation 283–4
  early development 133
  Eastern Desert 313
  El Bustan 284
  El Mullak 353
  El Salam 284, 305–6, 312, 313
  fishing 222–3
  Jonglei 33, 283
  for Maryut project 351
  Nasr 283–4
  Nubariya 351, 353
  for old-lands 356–7
  Salhiya 284, 313
  Toshki 138
  Youth 284, 313
carnivores, desert 293
cattle 66–8
  animal characteristics 253
  diseases 261
  feeding 256–7
  fertility 257
  large-scale units 259–60
  management 254–6
  meat production 259
  milk production 258–9
  population increase 391
  role 253–4
  see also milk
censuses, 20th century 84, 85
Central Agency for Public Mobilization and
    Statistics (CAPMAS) 129

INDEX 499

Central Auditing Agency (CAA) 129
cereals
  marketing system 429–30
  self-sufficiency decline 429
  see also flour
  see also by particular cereal
chat (bird) 294
cheese 433
  consumption 117
chickens 72–3
  broiler/layer industry 384
  broiler production 434–5
  consumption 117
  Nile Delta 260–1
  see also poultry
children
  cotton pest control 198, 252
  girls' role in agriculture 267
  schooling 268
  see also education
cisterns 211–12
citrus
  Nile Delta 250–1
  pests
    insects 75
    mites 76
classification of soils, Nile Delta 231
climate 16–25
  and air mass movements 16–17
  and crops 23–5
    cotton 24–5
    wheat 23–4
  evapotranspiration 22–3
  hot ecosystem provinces 39, 40
  Northern Littoral Region, north-west zone 210
  precipitation 20–2
    arid provinces 280, 281
    north-west coastal zone 210
  temperature 17–20
    and khamsin wind 17–19
    means 17, 18
    soils 19–20
clothing
  exports 479
  gellabiya 232
clover, *see bersim*
coal 286
combine harvesters 402
comparative advantage theory 182
concentrates 384, 385–8
  imports 392
  non-traditional feeds 387
    manufacturing plants 389
  subsidized 256, 258, 259
  unified feed 385–7
    manufacturing plants 388
    subsidy reduction 391

see also feedstuffs
cooking oil, ration 425
co-operatives 128, 150
  agricultural 172, 235–7
  equipment hire 404
  food 119, 159, 426
  see also retailing of food, regulated
Coptic community, fasts 117
cotton
  area 427–8
    annual trend 204, 480
  and climate 24–5
  establishment as main crop, and public debt 194–5
  exports 185, 186, 243, 476
    annual trend 481
    as clothing 479
    as textiles 478–9
  fertilizers 242
    foliar 378
  imports 476–7
    annual trend 482
  input subsidies 155, 156
    impact 162
  introduction, 19th century 171, 189
  irrigation requirements 31
  lint
    production 480
    yield 481
  and maize production 477–8
  marketing system 428–9
  Nile Delta 243
  pests
    control 155, 156, 198, 242, 252
    insect 75
    on other crops 198
  picking 242
  prices 242, 428–9, 477
    adjustment 165
    procurement pricing 153
  production
    decline 206, 271, 475–6
    system development 193
    trend 204
  seed/sowing 241–2
  trade expansion 194
  varieties 241, 242
  vegetable cropping after 477
  wind damage 24–5
  yields
    increases 173, 174, 204
    Nile Delta 242, 243
cottonseed cake
  production trend 392
  in unified feed 386, 388
cottonseed oil 117, 243
councils, local 132
Council of Sugar Products 272

credit, agricultural 155, 156
  for desert reclamation 309
  for drainage, World Bank 134
  for equipment 412–13
  for inputs 236, 237
    fertilizers 372, 373
  reform under Sadat 204, 206
  subsidized 155, 156
crops
  and climate 23–5
    cotton 24–5
    wheat 23–4
  for desert farming 337
  as feedstuffs
    by-products 388–9
    fodder, see fodder crops
    residues 389–91
  import/export 182–6
  intervention in production/procurement 152–4
    impact 160–1
    input programmes 153
    legislation 152–3
    liberalization 165
    output intervention 153
    trade policies 154
  irrigation requirements 31
  major 395
  Nile Delta 241–52
    bersim 245
    Cairo vegetable zone 240
    cotton 241–3
    cropping pattern 233–4
    fruit 250–2
    maize 243–4
    northern rice zone 239
    old-new lands 240
    productivity development 262
    rice 244–5
    southern zone 239
    wheat 245–6
    winter 247–50
  Nile Valley, yields 269–71
  Northern Littoral Region, barley 216–18
  oasis 54
  pricing policy 237
  for reclaimed desert, priority 309
  rotations 193, 196, 197, 231–2
  yield increases, recent 173–6
  see also production systems development
  see also by particular crop
cultivation
  for barley 217
  mechanized, for sugar cane 274
  Nile Delta methods 241

dairy products, from cattle/buffalo 255, 433

consumption 117
Dakhla depression 317
Damietta Branch Nile Delta distributary 6
dams, see Aswan High Dam; Aswan Low Dam; Mohammad Ali Dam
  see also barrages
date palms, Sinai 312
dates
  Nile Delta 251
  Northern Littoral Region 220
  Western Desert, New Valley area 319, 320
Delta, see Nile Delta
desert development 297–310
  cost 307, 308
  crop priority 309
  Eastern Desert
    and canal construction 284
    plan 138–9
    projects 313–14
  economic aspects 308
  failure 204
    analysis 338–9
  history of 301–6
    1900–52 301
    1952–80 302–4
    1980–87 304–5
    1987–92 305–6
  integrated approach 299–300, 308–9
  investment sources 309
  management, labour- versus capital-intensive 309–10
  Nile Basin, vegetation 60
  oases 55
  plans 14, 278
    goals 337
  policy changes required 338
  private versus public sector 307–8
  research 340–1
  scientific/technical knowledge need 310
  settlements 298
    see also South Tahrir Western Desert settlement
  Sinai, Land Master Plan 312–13
  stages in 306
  subsidies 298
  success/failure analysis 338–9
  sustainable development 339–40
  technological status 299
  UN Desertification Conference 297
  and weeds 60
  see also development plans; irrigation; reclamation of land; rural development; South Tahrir Western Desert settlement; Western Desert
Desert Development Center (DDC), integrated systems approach 340–1
desert farming systems 110, 310-37
  potential 336–7

desert farming systems (*cont*)
  problems 336
  Sinai 310–13
  traditional 296
  see also production systems development;
      Western Desert
desertification 297
Desert Research Institute/Centre 341, 462–3
deserts 278–344
  climate 280–1
    climatic provinces 278, 279, 280, 287–8
  development, *see* desert development
  farming, *see* desert farming systems
  fauna 292–6
    birds 294
    invertebrates 295–6
    mammals 293–4
    reptiles 294–5
  flora 288–92
    arid Mediterranean Region 291–2
    Eastern Desert 44–6, 288–90
    Sinai Peninsula 47–9
    Western Desert 50–5, 290–1
  geomorphology 284–6
    Eastern Desert 284–5
    Sinai Peninsula 285–6
    Western Desert 285
  human impact
    recent 297
    traditional land use 296
  mineral resources 286–7
    iron ore, Bahariya depression 316
    natural gas 369
    phosphate 317, 369
    of Sinai 311
  phytogeographical territories 41
  social system, traditional 296–7
  soil microbiology 287
  sub-divisions 278
  water resources 281, 283–4
    canals 283–4
    groundwater 283
    Nile River 281, 283
    rainfall 281
  see also Eastern Desert; Western Desert
development plans 136–40
  aims 136–7
  Plan for the Year 2000 137–9
  see also desert development; economic
      growth, and agriculture; Five Year
      Plans; government intervention;
      rural development, government
      and
diarrhoeal disease 124
diesel engines
  manufacture 406, 408
  pump sets 402, 403
diesel fuel, price 396

diet 114–18
  see also food
DINA Farm project 333–4
diseases
  animal 261
  of dates 251
  human, nutrition-related 124
  of maize 243
donkeys 72
  Nile Delta 260
dragonflies 295
drainage 348
  credit for 134
  and increasing irrigation 198
  water re-use 283
drought
  and Aswan High Dam 332–3
  Nile source 262
ducks 73, 261
  duck-cum-fish farming project 227
Dutch Disease phenomenon 96
dykes, in water conservation 211

earthworms 295–6
Eastern Desert 43–4, 313–15
  area 278
  development
    and canal construction 284
    plan 138–9
    projects 313–14, 353–4
  ethnic groups 83
  geomorphology 284–5
  habitat types 44–6
  physiography/geology 6–7
  plant resources 288–90
  see also deserts; Western Desert
Echinochloa stagnina 385
economic growth, and agriculture 177–9
  economic forces 178
  employment trends 179, 181
  GDP from agriculture 178–9
  population trends 179, 180
  and taxation 177–8
  see also development plans
economy, and oil boom 104–5
ecosystem provinces 39, 40
education 445–54, 467
  administration 448, 453
  development 445, 446
  organization 453–4
  policy 445–6
  schools 268
  the system 446–8
    higher 446, 448, 449, 450–1, 452
    and labour status 447
  see also graduates

eggs 260, 261
   production 435
Egyptian clover, see bersim
Egyptian Milk Company 433
Eid ul-Azha 117
El Bustan Canal 284
El-Diffa Plateau 4
electricity generation
   from Aswan High Dam 133, 134–5, 136
   solar/wind 281
electrification 134–5
   Qattara Depression project 138
   see also infrastructure
elephant grass, for fodder 384
El Faiyum, geology 4
El-Horreya Canal, serving Maryut project 351
El-Mullak Canal 353
El-Mullak reclamation project 353–4
El-Nahda reclamation project 351–3
El Salam Canal 284, 312, 313
   and Sinai reclamation 305–6
emigration, during oil boom 104–5
   see also migrant workers
employment
   trends, and economic growth 179, 181
   unemployment 268
   wage, Nile Valley 275–6
   see also wages
engines, see diesel engines
equipment
   credit availability 412-13
   imports 411–12
   manufacture 405–11
      agricultural machinery 406, 408–9
      diesel engines 406, 408
      government assistance 411
      local constraints 409–11
      tractors 405–6, 407
   repair/maintenance 413–14
   research 415, 416
   testing 415–16
   training of mechanics/operators 414–15
   see also mechanization
erosion, wind, after barley cultivation 217
Erythropygia galactotes 294
European Community, milk exports 433
exports 182–6
   annual trend 472
   citrus 251
   cotton 185, 186, 243, 476
      annual trend 481
      as clothing 479
      as textiles 478–9
   dates, Western Desert New Valley 320
   fertilizers 489–91
   and food subsidies/rationing 438
   oranges 473

potatoes 249, 474
rice 184, 185, 186, 244–5
vegetables/fruits 436–7
   Western Desert 334–5
see also imports
extension 465, 467
   bottlenecks 466
   ineffectiveness 237–8
   mechanization 416
   see also research

factories, milk 258–9
falafel 118
Farafra Oasis 316–17, 4–5
   vegetation 52
farm holding size 395
   see also desert farming systems;
      fragmentation of holdings;
      production system development
Farm Machinery Training Center 414–15
farmyard manure 199
   for maize 243
fateer bread 115, 117
fauna, desert 292–6
   birds 294
   invertebrates 295–6
   mammals 293–4
   reptiles 294–5
   see also livestock
feedmills 387, 388, 389
feedstuffs 383–94
   concentrates, see concentrates
   crop by-products 388–9
   crop residues 389–91
   minerals/supplements 391
   subsidies 155
   trends 391–2
      imports 392
   see also fodder crops; livestock
fellah, myth of 188
fellahin, in Nile Delta 232–3
fertilizers 363–82
   animal dung 199
      for maize 243
   for cotton 242, 378
   credit for 372, 373
   distribution 364, 365, 366, 369–70, 372–3
      control 153
   exports 489–91
   foliar 377–8, 379
      micronutrient 378, 379, 380
   forms distributed 363, 366, 367
   future 378–9
      policy changes 378–9
      requirements/availability 378, 492, 493
   imports 369, 370, 371, 489–91

fertilizers (*cont*)
  imports (*cont*)
    foliar 377, 378, 379
    subsidy 374
  industrial/agricultural coordination 379–80
  intervention policy failure 176
  for maize 243
  Nile water value as 195
  nitrogen
    consumption, 1980s 173
    production expectations 492
    tabulated data 489
  phosphorus
    production expectations 493
    tabulated data 490
  potassium 363–4
    tabulated data 491
  for potatoes 249
  prices
    control 153
    unsubsidized 375, 376
  production 367–9, 489–91
    companies 368
    increase 363–4, 365
  requirements
    laboratory estimation 377
    statutory versus actual use 374, 377
  sebakh koufri 195
  subsidies 154–5, 156, 162, 373–4, 375, 376
  tabulated data 489–93
figs, Northern Littoral Region 220
fish/fishing 221–7
  consumption 117
  farming 226–7
  lakes 221–2
  marketing/handling 225–6
    waste in 440
  production 74–5
  river/canals 222–3
  sea 223–5
  Sinai 312
Fish Marketing Company (FMC) 226
fishmongers 225
Five Year Plans 131, 139–40
  1959/60–1964/65 130
  1982–87 304
  current 305, 308
  priority goals 298
flax 247
  see also linseed
flora, *see* vegetation, natural
flour
  consumption 114–15
  distribution 119, 121, 157
    flow chart 486
  imports, annual trend 485
  mills 429–30
  subsidies 157
  see also bakeries; bread
fluvisols, calcaric 12
fodder crops 384–5
  bersim 198, 201, 245, 384, 385
  maize 243, 384, 385
    see also maize
  medic 218
  Nile Delta, summer 247
  see also feedstuffs
foliar fertilizers 377–8, 379
food
  black market 159
  consumption
    calories 122
    per caput 116
    protein 123
    subsidy expenditure 120
  livestock as 199
  meat, *see* meat
  production, recent 176
  retailing, regulated 121
    butchers 434
    consumer co-operatives 119, 159, 426
    fish 226
    horticultural crops, price share 436
    licensed grocers 159, 425
    government food-stores 119, 157, 426
  self-sufficiency decline 420
  subsidies 118–21, 157–9, 423, 424–7
    expenditure 119, 120, 157
    impact 121, 162–4
    procurement agencies 424–5
    rationed commodities 425–6
    ration system 119, 159
    regulated outlets 119, 121, 157, 159, 425
    semi-rationed commodities 426
    unintended consequences 437–8
  see also diet; marketing
forage, *see* fodder crops
foul madames 118
fragmentation of holdings
  Delta region 235
  Upper Egypt 271–2
  see also farm holding size
Free Officers Coup 172
Friesian cattle, for crossbreeding 66, 68
fruit
  consumption 117
  export 436–7
    Western Desert 334–5
  marketing 252, 435–6
    waste in 439–40
  Nile Delta 250–2
    Cairo vegetable zone 240
  pests 75, 76
  production 435
  retail price sharing 436
fruit fly 76

fruit juice drinks 118
fungi in desert soils 287

gammayas 425, 426
  fish retailing 226
  see also retailing of food, regulated
gardens
  rock gardens, Sinai Massif 48
  vegetable 219
Gebel Katherina 8
Gebel Musa 8
Gebel Shayib el Banat 7
Gebel Uweinat 5
  vegetation 52
geese 73, 261
gellabiya 232
General Authority for Fisheries Resources Development programmes
  five-year 227–8
  rice field fish culture 226–7
General Authority for Investment and the Free Zones (GAIFZ) 132
General Authority for Project Reclamation 302
General Authority for Supply Commodities 424
geology, *see* physiography/geology
geomorphology
  Eastern Desert 284–5
  Sinai Peninsula 285–6
  Western Desert 285
  see also physiography/geology
Ghard Abu Muharik 4
ghee 117, 433
gibna beyda 433
Gilf el Kebir Plateau 2, 5
goats 69–71
  Nile Delta 260
  Northern Littoral Region
    north-west zone 215
    population 214
Gordiodrilus siwaensis 296
government intervention 128, 146–69, 172–7, 192–3, 422–4
  crop production/procurement 152–4, 173
  failure 176–7, 201–6
  food retailing 121
    butchers 434
    consumer co-operatives 119, 159, 426
    fish 226
    horticultural crops, price share 436
    licensed grocers 159, 425
    government food-stores 119, 157, 426
  food subsidies, *see under* food
  historical background 118, 146, 173
    1940s 147–8
    Kitchener Act 1913 192
  impact 160–4
    food subsidies 121, 162–4
  input subsidies programme 161–2
  procurement policy 160–1
  input subsidies 154–7, 238
    impact 161–2
  land tenure/agrarian reform 105–6, 148–52, 166, 172, 192–3, 235
  liberalization 164–5, 166–7
  and livestock boom 105–6
  marketing 427–31
    beans/lentils 430–1
    cereals 429–30
    cotton 243, 428–9
    sugar cane 430
  mechanization policies 417–18
    liberalization 418–19
    machinery rental 404–5
  Nile Delta 234–8
    co-operatives 235–7
    extension/research 237–8
    land tenure 235
    pricing 237
  sub-periods 148
  rural development 140–4
    Agency for the Reconstruction and Development of the Egyptian Village (ARDEV) 140–1
    and US aid 143–4
    village development programmes 141–3
    see also subsidies
government structure 129–32
  see also development plans; infrastructure; policies, education
graduates
  in labour market 465–6
  quality 466
  reclaimed land distribution 104
  see also South Tahrir Western Desert settlement
grain
  pests 76
  storage
    historical 118
    in pits 217–18
grapes
  Nile Delta 251–2
  retail price sharing 436
  waste in marketing/handling 440
gravel desert
  Eastern Desert 44
  plants 292
Great Sand Sea 4
greenhouses, plastic 404
'green revolution' 173
gross domestic product, from agriculture 178–9

groundnut meal 389
groundnuts
  marketing, Western Desert settlement 330
  Nile Delta 246
groundwater
  availability/use 283
  Nile Valley/Delta aquifers 35
  and playa formation 5
  Western Desert fossil 35–7, 286
guava 252
Gulf of Aqaba, coastal plain vegetation 49
Gulf of Suez
  coastal plain vegetation 48–9, 291–2
  fishing 312
gummosis 250
gypcrete 11
gypsisols 11

hamada sufaces, Western Desert 5
harvest 268
Helwan Diesel Engines Company 406
herpetofauna 294–5
High Dam, *see* Aswan High Dam
highlands, *see* mountains
Hippolais pallida 294
hiring of equipment 404–5
history 95–9, 170–2
  19th century 96
  of desert reclamation 301–6
    1900–52 301
    1952–80 302–4
    1980–87 304–5
    1987–92 305–6
  of government intervention 118, 146, 173
    1940s 147–8
  izbah system 96, 98
  land-holding peasantry 98
  land tenure 95–9, 191–3
    19th century 96, 191
    izbah system 96, 98
    Kitchener Act 1913 192
    land-holding peasantry 98
    land ownership 1900 97
    modern periods 96
    post-revolution 192–3
    pre-reform inequities 99
  Nile river development 188–91
    20th century 190
    Muhammad Ali regime 189–90
  pre-reform inequities 99
  production systems development 193–8
    cotton as main crop 194–5
    crop area changes, 1844–1970s 193–4
    crop rotations 193, 196, 197
    intensification problems 195, 198
    recent 96, 201–7

holding cards 153, 236
  see also farm holding size; fragmentation of holdings
horses 72
house-building, in village development 141
household unit 267
  roles in 267–8
housing, Nile Delta 232
hydrology, *see* groundwater; irrigation; Nile river

illiteracy, smallholders 324
import bans, agricultural equipment 411
  diesel engines 408
  tractors 405, 411
imports 182, 183, 184
  agricultural equipment 411–12
    duties 410
  cotton 476–7
    annual trend 482
  fertilizers 369, 370, 371, 489–91
    foliar 377, 378, 379
    subsidy on 374
  flour, annual trend 485
  food, and government policy 423
  and food subsidies/rationing 438
  maize 182, 184
    annual trend 487
  milk 392, 433
    trends 392, 473
  wheat 182, 184, 429–30
    annual trend 485
IMT 35 tractor 406, 407
industry
  and desert development 300
  equipment manufacture 405–11
    agricultural machinery 406, 408–9
    constraints on local production 409–11
    diesel engines 406, 408
    government assistance 411
    tractors 405–6, 407
  fertilizers 367–9
    companies 368
    Upper/Middle Egypt 276
infant mortality 121, 125
infrastructure 132–6
  drainage 133–4
  electrification 134–5
  Qattara Depression project 138
  irrigation, *see* irrigation
  Nile Delta villages 233
  postal services 135–6
  social, Western Desert settlement study 327–9
    institutional shortcomings 329
    institutional structure 328–9

infrastructure (*cont*)
  social, Western Desert settlement study (*cont*)
    social differentiation 327–8
    study aims 327
    telephones 135
    transport 135
insects 75–6
International Desert Development Commission (IDDC), conference 340
intervention policies, *see* government intervention
invertebrates, desert 295–6
iron ore, Bahariya depression 316
irrigation 28–35, 132–3, 347–62
  basin system 28–9
    early government control 146
  control structures 33–5, 345, 347
    see also Aswan High Dam; Aswan Low Dam; barrages; canals
  of cotton 31
  Delta lands 231–2
  development 171–2
    19th century 171, 189–90
    20th century 190
    problems arising 172
    review 133
  drip, for orchards 212–13
  equipment
    pressurized 403–4
    pumps 402–3
  Kharga Oasis 53–4
  Nile fluvisols 12
  Northern Littoral Region, north-west zone 212–13
  old lands 356–60
    efficiency improvements 347, 358–60
    major system 356
    minor system 356–7
    on-farm problems 358
    on-farm study sites 357–8
  of olives 219
  perennial system 29–30
  planning, and evapotranspiration information 22
  of rice 31, 244
  Sinai 312
  subsidies 157
  system, water delivery/removal 348, 349
    management improvements 350
  use efficiency, and land tenure 106–7
  water balance 346
  water resource allocation, Nile River 345
  Western Desert
    groundwater scheme 37
    New Valley area 319
  see also desert development
Isma'iliya Canal 284, 313
Isthmic Desert, vegetation 47
izbah land labour system 96, 98

Jaculus jaculus 293
Jonglei Canal 33, 283

kankurs
  in Gebel Uweinat Mountains 5
  Western Desert Gebel Uweinat, vegetation 52
khamedj 251
khamsin winds 17–19, 280–1
  crop damage
    barley 216
    cotton 24–5
    olives 219
Kharga Oasis 53–4, 317–19
  sand dunes 317–18
Kirloskar engines 408
  pump sets 403
Kitchener Act 1913 192
koshari 118

laboratories, fertilizer requirements 377
ladybirds 76
lakes
  Bardawil, fishing 312
  for fish culture 226
  fishing 221–2
  Nasser 30–1
  saline, Western Desert 5
Land Master Plan, regions for reclamation 305
land ownership distribution 395, 396
land tenure 94–113
  development, and equity 95
  farm size distribution 108–10
    large holdings 110
    and mechanization 111
    medium holdings 108, 110
    small peasant holdings 108
  fundamental features 102
  and government policy 105–6
  history 95–9, 191–3
    19th century 96, 191
    izbah system 96, 98
    Kitchener Act 1913 192
    land-holding peasantry 98
    land ownership 1900 97
    modern periods 96
    post-revolution 192–3
    pre-reform inequities 99
  and land reclamation 104
  reform 95, 99–102, 172–3, 191–3, 235
    consequences 101–2

land tenure (*cont*)
  co-operatives 100–1
  laws 100, 192
  pattern of redistribution 99–100
  political motivation 99
  and mechanization 107
  oil boom period 104–5
  and political system 95
  redistribution of reclaimed land 104
  and resource allocation 94
  tenancy 107
  Upper Egypt 271–2
  and water use efficiency 106–7
laser levelling 274, 403
laws
  Agrarian Land Reform (1952) 148–50
  Agricultural Act, Law No, 53 (1966) 152–3
  fixed price procurement 153
  land reform 100
  local government 131–2
  presidential decree 129
  restrictive, 1940s 147
leaf hoppers 75
leather 259
lentils
  marketing control 430–1
  Nile Delta 250
  Western Desert New Valley 320
  yields, and location 269, 270
leptosols, lithic 11–12
Levantine ethnic group 83
levelling of land
  equipment, tractor-powered 403
  for sugar cane 274
libraries, university 464
Libyan Desert, vegetation 50–1
Libyan Plateau 4
linseed
  cake 389
  oil 247
lint
  production 480
  yield 481
lithic leptosols 11–12
livestock 63–74, 383
  feed subsidies 155
  fodder, *see* fodder crops
  graduate farms 323
  input value 431
  intensity, and farm size 108, 110
  Nile Delta
    Cairo vegetable zone 240
    donkeys/camels 260
    northern rice zone 239
    poultry 260–1
    sheep/goats 260
    southern zone 239
  Northern Littoral Region 214–16
  overgrazing 214
  sheep 214–15
  output value 431
  ownership pattern 383
  in peasant agriculture 199–201
  population 63–4
  production
    efficiency, and holding size 432
    historical trend 199, 200
    recent boom 105
    and government policy 105–6
  Sinai 311
  and smallholder system 64
  trade potential 186
  Western Desert New Valley 321
  see also fauna; feedstuffs; fish production; fodder crops; milk
  see also by particular animal
loans, *see* credit; Structural Adjustment Loan (SAL)
locusts 75
lucerne 384

machinery rental stations 404, 405
  see also equipment; mechanization
madinas 131
maize
  in baking 115
  and cotton production 477–8
  fertilizer 243
  as fodder 243, 384, 385
  imports 182, 184
    annual trend 487
  marketing 430
  Nile Delta 243–4
  production trend 205, 486
  stalks, as feed/fuel 390
  Western Desert New Valley area 320
  yellow, subsidy 155, 156
  yield increases 173, 174
malnutrition 121, 124
Mamelukes 82, 171
mammals, desert 293
  see also livestock
mangal, Red Sea Coast 288
mangoes, Nile Delta 251
Manzala, Lake, islands 43
marketing 420–44
  beans/lentils 430–1
  cereals 429–30
  channels
    regulated 420–1
    unregulated 421
  cotton 243, 428–9
  fish 225–6
  food-subsidy/distribution system 424–7
    problems 437–8

marketing (*cont*)
  procurement agencies in 424
  rationed commodities 425–6
  retail outlets, *see* retailing of food, regulated
  food-subsidy/distribution system (*cont*)
    semi-rationed commodities 426
    unrationed commodities 427
  government policies 422–4
  liberalization programme 421
  meat 434
  milk 433
  poultry 435
  problems/inadequacies 437–40
    competition inadequacy 439
    food subsidy/rationing system 437–8
    losses/waste 439–40
  Sinai produce 313
  sugar cane 430
  vegetables/fruits 252, 435–7
  see also retailing of food, regulated
marshes, *see* reed swamps; salt marshes
Maryut Lake 221–2
Maryut reclamation project 351
meat
  cattle/buffalo, Delta 254, 259
  consumption 117
    and subsidy 121
  imports 392
  marketing/handling 434
    waste in 440
  poultry 260, 261
  prices 434
  ration 426
  veal 65, 433
mechanization 173, 397–405
  Agricultural Mechanization Research Institute 415
  in barley cultivation 217
  combine harvesters 402
  degree of 397, 398
  draft cattle replacement 66
  extension 416
  and farm size 111
  government machinery rental 404–5
  incidence study 396
  labour shortage and 396
  land levelling/shaping 403
  and land tenure 107
  large farms 395
  number of machines 398–401, 399, 400
  policies 417–18
    failure 176–7
    liberalization 418–19
  pressurized irrigation 403–4
  pumps 402–3
  threshers 401, 402
  tractors, *see* tractors

  see also equipment; machinery rental stations
Medicago spp.
  lucerne 384
  medic 218
Mediterranean coast 39
  flora 40–2
  habitat types 42–3
  see also Northern Littoral Region
melons, Nile Delta 251
mice 76
Middle Egypt, *see* Nile Valley
migrant workers 92, 104–5, 265
  fishermen, and fishing decline 224
  remittances from 105, 265
  from Upper/Middle Egypt 275
milk
  from camels 216
  from cattle/buffalo 65, 66, 255
    production 258
    selling 258–9
  city distribution 258
  concentrates as payment 387
  consumption 117, 432
  from goats 71, 215
  importance to Delta farmer 254
  imports 392, 433
  marketing 433
  processing 433
  production 383
    systems 432–3
  from sheep 69, 215
  yields/costs 433
mills
  feed 387, 388, 389
  flour 429–30
  olives 220
  see also windmills in north-west zone
mineral feed supplements 391
mineral resources 286–7
  iron ore, Bahariya depression 316
  natural gas 369
  phosphate 317, 369
  of Sinai 311
mining, and desert development 300
ministries, governmental 130
  Agriculture and Land Reclamation (MALR) 455
  Electricity, research centres 463
  Land Reclamation 463
  Planning 130
  Public Works and Water Resources 348
mites 76
Moghara Oasis, geology 4
molasses 388–9
  production trend 392
  supplements 391
mole crickets 75

INDEX 509

morabiteen 213
mountains
  Eastern Desert 6–7, 284
    plants 45–6, 289
  Isthmic Desert, plants 47
  Sinai Peninsula 8, 285–6
    plants 290
  soils, lithic leptosols 11–12
  Western Desert 5, 285
Mount Sinai 8
mudbricks, termite tunnelling 295
muhafazas 131
Muhammad Ali 171
  agricultural development under 147, 189–90
  irrigation 133
  social change under 191
  land tenure 96
Muhammad Ali Dam 190

Nannodrilus staudei 296
Nasr Canal 283–4
Nasr tractors 405, 407
National Agricultural Research Project 415, 466–7
National Aquaculture Research Centre 463–4
National Research Centre 461–2
National Specialized Councils (NSP) 130
natural gas, as fertilizer raw material 369
nectarines 252
nematodes 76
Nesokia indica 293
Nile Basin vegetation 56–60
  farmlands 56–9
    cultivated land 56–8
    roadsides 58–9
    waste land 58
  irrigation–drainage system 59
  reclaimed desert 60
  riverain flora 56
Nile Delta 55, 229–64
  agricultural areas 238–40
    Cairo vegetable zone 239–40
    northern rice zone 238–9
    old-new lands 240
    southern 239
  animal health 261–2
  animal production 253–61
    cattle/buffalo 253–60
    donkeys/camels 260
    poultry 260–1
    sheep/goats 260
  aquifers 35
  area 229
  climate 229, 231
  cropping pattern 233–4
  crop production 241–52

bersim 245
  cotton 241–3
  fruit 250–2
  maize 243–4
  rice 244–5
  summer crops (other) 246–7
  wheat 245–6
  winter crops (other) 247–50
cultivation methods 241
development
  future 137, 262
  history 188–91
and government policy 234–8
  co-operatives 235–7
  extension/research 237–8
  land tenure 235
  pricing 237
irrigation 231–2
map 230
marketing fruit/vegetables 252
people 232–3
physiography/geology 6
plant species 40
soils 231
  calcaric fluvisols 12
see also Nile Basin vegetation
Nile river 188–91, 348
  discharge 281, 283
  fishing 222–3
  floods 271
  flow 26–8
  history of development 188–90
    20th century 190
    Muhammad Ali regime 189–90
  hydraulic works 33–5, 345, 347, 348
  see also Aswan High Dam; Aswan Low Dam; irrigation
Nile Valley 26, 55, 265–77
  aquifers 35
  crops
    cash crops 273
    historical changes 271
    sugar cane 272–5
    yields 269–71
  development 190
  governorates 265, 266
    cultivated areas in 267
  history, and agricultural production 271
  households, roles in 267–8
  industries 276
    small-scale, for village development 276–7
  land area/quality 269
  land tenure, Upper Egypt 271–2
  migration of workers 275
  physiography/geology 5–6
  population 268
    ethnic identity 267
  soils, calcaric fluvisols 12

510 THE AGRICULTURE OF EGYPT

Nile Valley (cont)
  standard of living 275-7
  wage employment 275-6
  see also Nile Basin vegetation
Nile Water Agreements
  Anglo-Egyptian (1929) 30
  with Sudan 345
Nimos Farm Project, Western Desert 336
nitrogen fertilizers
  consumption, 1980s 173
  production expectations 492
  tabulated data 489
  see also fertilizers
norag 268
Northern Littoral Region 209-28
  north-east zone 209
  north-east zone fishing 221-7
    fish culture 226-7
    lakes 221-2
    marketing 225-6
    river/canals 222-3
    sea 223-5
  north-west zone 209-20
    area 209
    barley production 216-18
    climate 210
    fruit production 219-20
    livestock husbandry 214-16
    medic for forage 218
    population 213-14
    soils 209-10
    vegetable production 218-19
    water conservation 210-12
    water distribution 212-13
  planned development 227-8
Nubariya Canal
  serving El-Nahda project 353
  serving Maryut project 351
Nubian Desert, vegetation 51-2
Nubian ethnic group 83
Nubian goat 70, 71
Nubian sandstone 2
  aquifer 35-7, 283, 286
Nuclear Research Centre 463

oases, Western Desert 52-5, 285, 316-18
  Berber population 83
  vegetation 53-5
Oenanthe spp. 294
oil (mineral)
  pollution
    Lake Maryut 221-2
    sea, and fishing decline 224
  resources 286
oil (vegetable)
  cooking, ration 425
  cottonseed 117, 243

linseed 247
olive 219
soyabean 247
oil boom 104-5
old-new lands, Nile Delta 234, 240
oligochaetes 295-6
olives, Northern Littoral Region 219-20
onions
  Nile Delta 249-50
  Northern Littoral Region 218-19
  Western Desert New Valley 320
  yields, and location 269, 270
oral rehydration salts 124, 125
oranges
  exports 473
  Nile Delta 250-1
orchards 206
  drip irrigation 212-13
Ossimi sheep 67, 68
Ottoman Empire 170-1
overgrazing, Northern Littoral Region 214

palm groves, oasis, weeds 54
paper industry 276
partridge 294
pasha class 95, 172
  izbah land labour system 96, 98
  origins 96
peaches 252
Pearl Farm project, Western Desert 335-6
pears, Nile Delta 252
pellagra 124
Penman evapotranspiration formula 22, 24
pest control
  as constraint in desert farming 326
  cotton 242
    by hand 198, 252
    subsidies 155, 156
pests
  of barley 216
  of citrus 250
  of cotton 242
  insect 75-6
  of maize 243
  non-insect 76
  and production intensification 198
  rodents 76, 293-4
phosphate, rock 317, 369
phosphorus fertilizers
  production expectations 493
  tabulated data 490
  see also fertilizers
physiography/geology
  Eastern Desert 6-7
  Nile Delta 7
  Nile Valley 5-6
  Sinai Peninsula 7-8

physiography/geology (*cont*)
 Western Desert 1–5
 see also geomorphology
pigs, 261
 see also swine
planning, *see* development plans; Five Year Plans
Plan for the Year 2000 137–9
playas, Western Desert 5
policies, education 445–6
 see also government intervention
political system, and land tenure 95
 see also government structure
pollution
 Lake Maryut, oil/industrial 221–2
 sea, oil, and fishing decline 224
population 191, 192
 Ababda tribe 314–15
 bedouin, *see* bedouin
 Berber 83
 distribution 80, 88
 ethnic minority groups 83, 91
 ethnic origins 81–2
 growth 345, 346
  Cairo 276
  and cropped area 31, 33, 91–2
 Middle/Upper Egypt 268
 physical characteristics 82–3
 trends 84–91, 91–2, 179, 180
  age distribution 86, 90
  censuses 84, 85
  dependency ratio 90–1
  and employment 91, 92
  growth rate 84–6, 95
  rural density 89
  sex distribution 90
  urbanization 87, 89–90, 91
postal services 135–6
potassium fertilizers 363–4
 tabulated data 491
 see also fertilizers
potatoes
 exports 249, 474
 fertilizer 249
 Nile Delta 248–9
 retail price sharing 436
 marketing/handling 439
  waste in 440
poultry 72–3
 broiler/layer industry 384
 broiler production 434–5
 consumption 117
 feed subsidy 155
 graduate farms 323
 marketing 435
 Nile Delta 260–1
  diseases 261
 Western Desert New Valley 321

 see also eggs
poverty, rural 107–8
power
 solar 281
 wind 281, 282
 see also electricity generation; electrification; windmills
prawns, decline 223–4
precipitation 20–2
 arid provinces 280, 281
 north west coastal zone 210
prices
 bread 125
 cotton 242, 428–9, 477
  adjustment 165
  procurement pricing 153
 diesel fuel 396
 fertilizers
  control 153
  unsubsidized 375, 376
 fixing 153, 237
  as tax 160–1
 fruit/vegetables 436
 meat 434
 sugar cane 272–3, 274–5
 Western Desert settlement 331
 wheat straw 391
Principal Bank for Development and Agricultural Credit (PBDAC) 153
 credit 155, 156
  for agricultural equipment 412–13
  for fertilizer 372, 373
 in fertilizer distribution 364, 365, 366, 369
 in food procurement 424–5
production systems development 193–8
 cotton as main crop 194–5
 crop rotation introduction 193
 crop rotations 196, 197
 intensification problems 195, 198
 small farmers 206
 see also desert farming systems; Nile Delta; Northern Littoral Region
Psammotermes spp. 295
pumps 402–3
 small, for farm use 357

qameradeen 117
qariyas 131
Qattara Depression 4, 5
 electrification project 138
quotas in crop procurement 153
 impact 160–1

rabbits 73–4, 261
Rahmani sheep 68, 70
railway network 135
 building, 19th century 171

rainfall, *see* precipitation
Ramadan, food consumption in 117
ration cards 159, 425, 426
rationing of food 119, 159, 423
  commodities 425–6
  impact 163
  unintended consequences 437–8
  see also subsidies
rats 76
reclamation of land
  desert, *see* desert development
  inferences from project experience 355
  Nile Delta 262
  projects 350–5
  public programme 152
  redistribution of land 104
  and total cultivated area 104
Red Sea
  coast
    tourism development 138
    vegetation 44–5, 288
  wadis, plants 289–90
Red Sea Mountains 6–7
reed swamps
  oasis 54
  Red Sea coast 45, 288
  see also salt marshes
regosols 13–14
remittances from migrant workers 105, 265
renting of equipment 404–5
rents, and land tax 150, 152
reptiles, desert 294–5
research 237, 455–65, 467
  Academy of Scientific Research and Technology 460–1
  Agricultural Research Centre 455, 456–60
  desert development 340–1
  Desert Research Institute/Centre 462–3
  equipment 415, 416
  extension coordination problems 466
  history 455
  Ministry of Agriculture and Land Reclamation 455
  Ministry of Electricity centres 463
  Ministry of Land Reclamation 463
  National Agricultural Research Project 466–7
  National Aquaculture Research Centre 463–4
  National Research Centre 461–2
  prioritization problems 466
  private units 465
  universities 464–5
  Water Research Centre 462
  see also extension; universities
resorts
  in Nile Delta development plan 137
  in Western Desert development plan 138

see also tourism
retailing of food, regulated 121
  butchers 434
  consumer co-operatives 119, 426
  fish 226
  horticultural crops, price share 436
  licensed grocers 159, 425
  government food-stores 119, 157, 426
  see also bakeries; marketing; supermarkets, retail
revolution, 1952 172–3
  land reform after 192–3
rice
  consumption 117
    trend 203, 487
  import/export trade 184, 185, 186, 244–5
  irrigation 31, 244
  Nile Delta 244–5
  northern Delta zone 238–9
  production
    control 430
    trend 203
  ration 426
  straw 389, 390
    for feed 256–7
  Western Desert New Valley 319–20
  yields
    increases 173, 175
    and location 269, 270
rice fields, for fish culture 226–7
Rice Mechanization Centre 415, 416
road network 135
roadsides, Nile Basin, vegetation 58–9
rock gardens, Sinai Massif 48
rodent pests 76, 293–4
Roman cisterns 212
rufous warbler 294
rural development, government and 140–4
  Agency for the Reconstruction and Development of the Egyptian Village (ARDEV) 140–1
  and US aid 143–4
  village development programmes 141–3
  see also desert development

sabkhas
  Sinai Peninsula coast 8
  vegetation 55
Saidi buffalo 65
Saidi sheep 68
St. Catherine Monastery 48
Salhiya Canal 284, 313
Salhiya project 313–14
salination, from lack of drainage 134
salinity, and soil management 13
salt marshes
  Gulf of Suez, plants 291–2

salt marshes (cont)
  Red Sea coast 44–5, 288–9
  Western Desert, plants 290–1
  see also reed swamps
Samalut reclamation project 354–5
sand dunes
  Sinai Peninsula 286
  Western Desert 285
    barchan, Kharga 317–18
    plants 290–1
saqqiyas 190, 357
sardines, decline, and Aswan High Dam 223
schools 268
  see also education
sea fishing 223–5
sebakh koufri 195
seed policy 155
services, see infrastructure
sesame, Nile Delta 246–7
settlements in desert reclamation 298, 306
  see also South Tahrir Western Desert settlement
shaduf 357
shami bread 115
sharecropping 150
sheep 67, 68–9, 70
  Nile Delta 260
  Northern Littoral Region 214–15
shops, see bakeries; marketing; retailing of food, regulated; supermarkets, retail
Sinai 46–7, 310–13
  area 278
  economy 311
    agriculture 311–12
    fisheries 312
    Land Master Plan 312–13
    markets for produce 313
  geomorphology 285–6
  habitat types 47–9
  physiography/geology 7–8
  plants 40, 41–2
    Sinai Massif 47–8
    Sinai Plains 291
  population 310
    settlement pattern 310–11
  reclamation 305–6
Siwa Depression 5
slaughtering
  for brucellosis/tuberculosis 261
  for meat 434
smuggling, sheep into Libya 215
social fund 125–6
Socialist Public Prosecutor 129–30
social system
  change under Muhammad Ali 96, 191
  Western Desert settlement study 327–9
    institutional shortcomings 329

institutional structure 328–9
social differentiation 327–8
study aims 327
traditional desert 296–7
village development programmes 142
  see also land tenure
soft drinks 118
soils 8–14
  calcaric fluvisols 12
  calcisols 11
  gypsisols 11
  lithic leptosols 11–12
  microbiology, deserts 287
  Nile Delta 231
    Cairo vegetable zone 239–40
    northern rice zone 238
    old-new lands 240
    southern zone 239
  Northern Littoral Region 209–10
  regosols 13–14
  sandy, reclamation 303
  Sinai 312, 313
  soil association distribution 10
  solonchaks 12–13
  temperatures 19–20
solar power 281
solonchaks 12–13
soya bean, as feed 389
sorghum 384, 385
  Western Desert New Valley 320
South Tahrir Western Desert settlement 321–36
  graduate settlers 321–2
  graduate survey 322–3
  marketing studies 329–32
    dairy products 330–1
    groundnuts 330
    horticultural crops 330
    phases 330
    prices/costs 331
    supply/demand study 331–2
  smallholder survey 323–7
    community development 324
    constraints to development 324–6
    families 323–4
  social infrastructure study 327–9
    institutional shortcomings 329
    institutional structure 328–9
    social differentiation 327–8
    study aims 327
  water/energy problems 332–3
soyabeans
  import/export trade 184
  Nile Delta 247
straw, as feed 389–91
  baling 390
  chemical improvement 391
  rice 256–7, 389, 390

strawberries
  Nile Delta 252
  Western Desert, for export 334–5
street foods 118
Streptopelia turtur 294
Structural Adjustment Loan (SAL) 125
subsidies
  agricultural inputs 154–7, 236
    concentrates 256, 258, 259, 386–7
    credit 155
    feed 155
    fertilizers 154–5, 373–4, 375, 376
    impact of programme 161–2
    irrigation 157
    pest control 155
    seeds 155
  desert development 298
  food 118–21, 157–9, 423, 424–7
    expenditure 119, 120, 157
    impact 121, 162–4
    procurement agencies 424–5
    rationed commodities 425–6
    ration system 119, 159
    regulated outlets 119, 121, 157, 159, 425
    semi-rationed commodities 426
    unintended consequences 437–8
  reduction 391
Sudan, Nile water rights 30, 345
Sudd, and Nile flow 27
Suez Canal 171
  see also Gulf of Suez
  Suez Canal University 464
sugar
  consumption in beverages 118
  ration 425
sugar beet 247–8
sugar cane
  marketing control 430
  in Upper/Middle Egypt 272–5
    area 272, 273
    pressure to grow 274
    prices 272–3, 274–5
    problems for small holdings 273–4
    productivity improvements 274
    water usage 275
    and world market fluctuations 272
  yields
    increases 173, 176
    and location 269, 270
  sugar cane stemborer 75
sun breads 115
supermarkets, retail 439
swine 117
  see also pigs

tambour 357
tamiya 118

tamween 425
taxes
  crop quota/delivery programme as 160–1
  and economic growth 177–8
  land, and land rent 150, 152
  local 132
  Mameluke collectors 171
tea
  consumption 118
  ration 425
telecommunications 135–6
telephones 135
temperatures
  climatic 17–20
    and khamsin wind 17–19
    means 17, 18
  desert provinces 280
  north-west coastal zone 210
  soils 19–20
tenancy, and agrarian reform 150, 152
termites 295
threshers 401, 402
  financing 413
threshing, maize 268
tomatoes
  importance 435
  marketing/handling 435–6
    waste in 440
    wholesale 439
  Nile Delta 248
  Northern Littoral Region 219
  processing 437
  retail price sharing 436
  Upper Egypt 273
tomato sauce 437
Toshki Canal 138
tourism
  and desert development 300
  development, Red Sea coast 138
  Sinai 311
  see also resorts
Tractor and Machinery Testing Center 415–16
tractors
  countries of origin 414
  credit for purchase 412–13
  distribution policy failure 176–7
  manufacture 405–6, 407
  Nile Delta 241
  population, increase 398
  repair/maintenance 413–14
  training of mechanics/operators 414–15
trade, agricultural 179, 182–6
  see also exports; imports
training, equipment mechanics/operators 414–15
  see also education
transport infrastructure 135–6
Trifolium alexandrinum, see bersim

tuberculosis, slaughter policy 261
tuberworms 75
turkeys 73, 261
Turkish rule 171
turtle doves 294

unemployment 268
  see also employment
unified feed 385–7
  manufacturing plants 388
  subsidy reduction 391
  see also concentrates
United Nations
  Desertification Conference 297
  Development Programme, for drainage materials aid 134
United States, cotton imports from 476–7
United States Agency for International Development (USAID)
  Agricultural Machinery Project 415
  Agricultural Production and Credit Project 416
  drainage materials aid 134
  in village development 143
universities 446, 448
  courses 454
  in research 464–5
  teaching 453–4
  see also research
Upper Egypt, see Nile Valley
urbanization 87, 89–90, 91

vaccination of livestock 261
veal, buffalo 65, 433
vegetables
  Cairo vegetable zone 239–40
  after cotton 477
  export 335, 436–7
  insect pests 75
  marketing/handling 252, 435–6
    waste in 439–40
  Nile Delta 248–50
  Northern Littoral Region 218–19
  processing 437
  production 435
  retail price sharing 436
  Upper Egypt 273
  Western Desert, for export 335
  see also by particular crop
vegetation, natural 39–62
  Eastern Desert 44-6, 288–90
  Mediterranean coast 40–3
  Nile Basin 56–60
    farmlands 56–9
    irrigation-drainage system 59
    reclaimed desert 60

riverain flora 56
phytogeographical regions/territories 39, 41
Sinai Peninsula 47–9
Western Desert 50–5
  arid Mediterranean Region 292
  Gebel Uweinat 52
  habitat types 42–3
  hyperarid region 290–1
  Libyan Desert 50–1
  Nubian Desert 51–2
  oases 52–5
veterinary science, university faculties 448, 452
veterinary services
  Nile Delta 261–2
  Northern Littoral Region 216
villages
  banks 230
  development programmes 141–3
  see also Agency for the Reconstruction and Development of the Egyptian Village (ARDEV)

Wadi El Moulak 314
Wadi el Natrun, geology 4
wadis
  arid Mediterranean Region, plants 292
  Eastern Desert 7, 284–5
    plants 46, 289–90
  Isthmic Desert, plants 47
wages, agricultural
  1938-74 102, 103
  1970-90 105, 106
  post-revolution 150
  Upper/Middle Egypt 276
  see also employment
warblers 294
wasteland, Nile Basin, vegetation 58
Water Master Plan 284, 304
watermelons
  Nile Delta 251
  Northern Littoral Region 219
  Sinai 312
Water Research Centre 341, 462
weeds
  Nile Basin farmland 56–8
  Nile Delta, control 241
  oasis farmland 54
  reclaimed land 60
wells
  Kharga Oasis 53
  in north-west zone 211, 212
Western Desert 49–50, 315–36
  agribusiness case studies 333–6
    DINA Farm 333–4
    Nimos Farm 336

Western Desert (*cont*)
  agribusiness case studies (*cont*)
    Pearl Farm 335–6
    Project and Investment Company 334–5
  aquifer 35–7, 283, 286
  area 278, 315–16
  Berber population 83
  development 305
    and canal construction 283–4
    El-Nahda project 351–3
    Maryut project 351
    plan 137–8
    Samalut project 354–5
  East Oweinat area 318
  fauna 292
  geomorphology 285
  irrigation, groundwater 37
  New Valley area 316–18
    agriculture in 319–21
    Bahariya depression 316
    Dakhla depression 317
    Farafra depression 316–17
    Kharga depression 317–19
    population 318–19
  physiography/geology 1–5
  settlements, *see* South Tahrir Western Desert settlement
  vegetation 50–5
    arid Mediterranean Region 292
    Gebel Uweinat 52
    habitat types 42–3
    hyperarid region 290–1
    Libyan Desert 50–1
    Nubian Desert 51–2
    oases 52–5
  see also Eastern Desert
wheat 77
  bran as feed 388
  and climate 23–4
  distribution flow chart 486
  imports 182, 184, 429–30
    annual trend 485
  irrigation requirements 31
  mills 429–30
  Nile Delta 245–6
  production/consumption trends 202, 483, 484
  self-sufficiency decline 429
  straw 390
    price trend 391
  subsidies 157
  Western Desert New Valley area 320
  yields 269, 270
    annual trend 484
    increases 173, 175
  see also flour
White Nile 26
  dam 190
  flow 27
wind blast of barley 216
  see also khamsin winds
wind erosion, after barley cultivation 217
windmills, in north-west zone 212
  see also mills
wind power 281, 282
women
  anaemia 124
  bedouin 315
  graduate settlers' wives 322
  smallholders' wives 324
  role 267, 267–8, 268
    livestock care 432
    milk/dairy products 255, 433
wool production 69
workshops, for equipment manufacture/repair 409
World Bank loans, for drainage 134

Youth Canal 284, 313

zabaleen 117
Zaraibi goat 70, 71